International Symposium on Aviation Psychology (ISAP 2023)

Rochester, New York, USA
31 May - 3 June 2023

ISBN: 979-8-3313-0346-4

Printed from e-media with permission by:

Curran Associates, Inc.
57 Morehouse Lane
Red Hook, NY 12571

Some format issues inherent in the e-media version may also appear in this print version.

Wright State University – Conferences & Events
3640 Colonel Glenn Highway
Dayton, Ohio 45435
USA

Phone: (937) 775-1000

http://www.wright.edu

Additional copies of this publication are available from:

Curran Associates, Inc.
57 Morehouse Lane
Red Hook, NY 12571 USA
Phone: 845-758-0400
Fax: 845-758-2633
Email: curran@proceedings.com
Web: www.proceedings.com

TABLE OF CONTENTS

ISAP 2023 Practitioners' Panel

The ISAP Practitioners' Panel is a key feature of the Symposium. These panelists represent broad and important aspects of how Aviation Psychology contributes to flight safety and the advancement of aviation. The Panel moderator is Dr. Sherry Chappell.

Dr. Kathy Abbott

Dr. Kathy Abbott is the US Federal Aviation Administration's (FAA) Chief Scientific and Technical Advisor for Flight Deck Human Factors, with over 40 years of work on human performance and human error. Dr. Abbott has led the integration of human engineering into FAA/international regulatory material and policies for flight guidance systems, avionics, all-weather operations, Required Navigation Performance, crew qualification, data communication, instrument procedure design criteria, electronic flight bags, electronic displays, organizational culture, design-related pilot error, flight crew alerting, manual flight operations, and other areas. She has been involved extensively in accident, incident, and other safety data analysis.

Dr. Abbott came to the FAA from the National Aeronautics and Space Administration (NASA), where she was responsible for leading analytical, simulation, and flight studies with the specific objective of improving aviation safety and operational efficiency. She is a Fellow of the Royal Aeronautical Society, an Associate Fellow of the American Institute of Aeronautics and Astronautics, and a Member of the Livery of the Honourable Company of Air Pilots. She is a certificated private pilot, with familiarization training in several large transport aircraft. Dr. Abbott earned her B.S. in Mathematics and Information Science from Christopher Newport College, an M.S. in Computer Science from George Washington University, and a Ph.D. in Computer Science from Rutgers University.

Dr. Igor Dolgov

Dr. Igor Dolgov currently works as a Staff Human Factors Engineer at Joby Aviation, where he is responsible for flight deck human factors, passenger experience, and human-systems integration. He is an accomplished professional with a diverse background. In his previous role as the Lead Human Factors Engineer at Uber Elevate, he played an active role in shaping the Uber Copter and drone delivery projects.

Prior to joining industry, Dr. Dolgov was a tenured associate professor of Engineering Psychology at New Mexico State University, where he conducted NASA- and DoD-sponsored research in the areas of Unmanned Aircraft Systems (UAS), augmented reality, and human-machine interaction. With a transdisciplinary Ph.D. from Arizona State University and a B.S.E. in Computer Science along with a certificate in Robotics and Intelligent Systems from Princeton University, Dr. Dolgov has extensive knowledge and unique expertise in his field.

Dr. Brian Hilburn

Dr. Brian Hilburn is Principal Consultant at the Center for Human Performance Research, CHPR (Netherlands and USA). He is also a consulting Human Factors Expert at MITRE's Center for Advanced Aviation System Development (CAASD).

Formerly, he was Head of Human Factors for the Netherlands National Aerospace Lab (NLR), and VP of R&D for Engineering and Information Technology, Inc (EIT). Brian has a Ph.D. in Applied Experimental Psychology, and has held adjunct professorships at Lund University, Sweden (School of Aviation), Technical University of Delft, Netherlands (Aerospace Engineering), and Shanghai Jao Tong University, China (Electrical Engineering).

Most of his research career has been in the areas of air traffic management, neuroergonomics, hybrid human-machine cognition, and automation design. He is also an active, instrument-rated pilot, and is almost ready to take his commercial checkride.

Dr. Barbara E. Holder

Dr. Barbara Holder is a Presidential Fellow at Embry-Riddle Aeronautical University and an Associate Professor in the School of Graduate Studies in the College of Aviation. She is responsible for growing the aviation applied human factors research capability of the university. She teaches graduate level courses in aviation human factors and human-centered design. Before joining ERAU, she worked at Honeywell Aerospace as a Technical Fellow, and prior to that, at Boeing as Associate Technical Fellow and Lead Scientist of Boeing's Flight Deck Concept Center.

Dr. Holder has over 20 years of experience researching aviation safety and flight deck human factors. One of her current research programs investigates ways to reduce the risk of loss of control in flight during the go-around maneuver by designing the go-around procedures to direct pilot attention to appropriate flight path parameters. Another of her current research programs is investigating the cognitive consequences of flight deck automation, with the goal of identifying and representing the essential cognitive skills used by airline pilots while performing flight path management tasks.

Dr. Holder has served on numerous industry committees. Currently, she chairs the Subcommittee on Human Factors for the FAA's Research, Engineering, and Development Advisory Committee. She is also a member of the FAA's Air Carrier Training Aviation Rule Making Committee's working group on Flight Path Management. Dr. Holder is a Fellow of the Royal Aeronautical Society. She completed her Ph.D. and M.S. in Cognitive Science at the University of California, San Diego, under the guidance of Professor Edwin Hutchins. For fun, Dr. Holder enjoys flying aerobatics with her husband in their Extra 300. Dr. Holder holds a Private Pilot's License and is currently pursuing an instrument rating. She also enjoys sea kayaking, playing pickleball, and running.

AIRPORT PERSONNEL PERCEPTIONS OF AIRPORT ENVIRONMENTAL PROGRAMS

Navya Nikhita Agasam, M.S.

Deborah S. Carstens, Ph.D., PMP

Florida Institute of Technology

Melbourne, Florida

Abstract

Aviation is a rapidly growing industry that is believed to account for 25% of the carbon footprint by 2050 (Graver et al., 2019, as cited in Alfaro, V.N. & Chankov, S., 2022). Countries are making environmentally friendly changes to save the earth. Humans are the think tanks of any process. Airport personnel should consider implementing these changes. By considering the complex and dynamic nature of human beings, this study uses a survey approach to understand the attitudes and behavior of airport personnel in the transition towards airport environmental programs. The study focuses on factors influencing human behavior and their willingness to save the earth. The study findings will be discussed along with future research recommendations. These results would pave the way for further studies and serve as a guide for improving safety culture and policymaking in the aviation industry.

Introduction

The current research aims to understand consumer perceptions of new technologies and environmental programs. In business environments, operations must be consumer-centric. The FAA (2021) funded 44 airports across the U.S. to reduce the environmental footprint of the airports. This suggests the need for an increase in implementing environmental programs at airports. The U.S. aviation climate action plan to achieve net zero emissions by 2050 needs increased action to meet this mission. This study investigates the current gaps in airport environmental programs by administering a survey. Furthermore, this study seeks to understand airport personnel's perceptions regarding their views on new projects and environmental programs in airports.

Literature review

Past studies have investigated environmental efforts within airports. Cremer, Rice, Gaenicke and Oyman (2016) measured consumer perceptions on reusing water for landscape and drinking water. The study findings suggest that consumers viewed reusing water for landscapes was positive but reusing water as drinking water was not positively viewed. Sitorus and Manik (2021) conducted a study in 2020 in Indonesia of Silangit Airport of Lake Toba to understand stakeholders' perceptions of 23 social implications categorized by human rights, working conditions, cultural heritage, social-economic repercussions, and governance. A Social Life Cycle Assessment (SCLA) was the methodological framework utilized in the survey using a seven-point Likert Scale to measure gaps between expectation and perception of corporate sustainability. The Society of Environmental Toxicology and Chemistry/United Society of

Environmental Program Code of Practice was utilized as a guideline in conducting this research. The survey examined stakeholders' perceptions of which social criteria were important, airport operational social activities that have been experienced and which are expected, and social sustainability hotspots requiring further research and policy. The survey respondents recruited were selected based on their insight and experience with the airport and were comprised of local government, academics, community leaders, and non-government organizations. The study findings suggest that the stakeholders met socioeconomic criteria but that improvements were necessary concerning living conditions and the level of transparency on social and environmental issues.

Reitinger et al. (2011) conducted research to critically review literature concerning the area of protection (AoP) and the impact categories in social life cycle assessment (SLCA) which builds upon life cycle assessment (LCA), common methodology for quantifying sustainability. AoP consists of human health, natural environment, natural resources and man-made environment. Over the past few decades, there is greater awareness of the environmental problems endangering our planet to include the well-being of humans now and in the future. There are many negative side effects caused by human actions in our natural environment, sustainability is becoming more important. There are three different pillars consisting of environment, economy, and society to consider when implementing sustainable development initiatives. Applying SLCA can be helpful in measuring environmental aspects and is useful for strategic analysis in structuring complex decision-making processes for identifying optimization potentials within a organization.

Murphy (2004) conducted a survey in Minnesota consisting of 13 questions to gain knowledge of water issues and environmental literacy of residents aged 18 or older and to examine literacy changes from the previous survey where a baseline of environmental literacy was obtained for residents. There were 1,000 survey respondents that conducted the survey through phone calls. Survey respondents self-reported on specific topics consisting of demographics, attitudes towards their environment, environmental laws and regulations (we can also use airport environmental program), responsible environmental organizations, and residents' daily behavior. The findings suggest an increase in general environmental knowledge from the previous 2001 survey, where 68% of Minnesota adults have average knowledge about the environment. More interesting and aligned with our current study is the portion of the survey regarding knowledge of water issues. Of the survey respondents, 45% of Minnesota adults have at least an average level of knowledge regarding water issues, including laws and regulations preventing water pollution and causes of water pollution, etc. Survey respondents had 61% knowledge of the benefits of wetlands, which help store water before it enters bodies of water such as lakes and streams. Of the survey respondents, 22% answered correctly that the source of mercury in lakes is due to coal-burning power plants. The survey respondents had 53% knowledge of water entering storm sewers going into their wetlands, lakes, and rivers. Of the survey respondents, 45% correctly answered that the phosphorous had a major environmental impact by promoting excessive plant and algae growth within Minnesota lakes and rivers. The findings suggest a need for environmental education on the environmental health of Minnesota wetlands, lakes, and rivers. Furthermore, education could be conducted to assist residents on what part they can do to enhance the water health of the state.

Methodology

The study focuses on airport personnel in the U.S. The research methodology used is convenient sampling through an online survey administered through Qualtrics. The survey began with respondents agreeing to participate by advancing to the survey questions after checking informed consent. The survey consists of nineteen questions with a five-point Likert scale and three demographic questions. The four categories of Likert scale questions included daily life practices, experience with and exposure to environmentally friendly programs, and willingness to contribute to society. The university's Institutional Review Board identified the research as exempt. A pilot study was conducted to assess the face and content validity of the survey. The survey was modified based on the feedback from the pilot study. Survey respondents were recruited utilizing a snowball technique through emails and social media pages. Survey data was collected over five weeks. Descriptive statistics were used in analyzing the survey results.

Results

The survey was responded to by 73 participants, with 49 completed and seven incomplete responses. 82% of those who took part were between 26 and 65 years old. 63% male, 31% female. 78% of participants work in higher-level roles in airports.

Figure 1 shows the results of the daily pro-environment practices followed by the participants. More than 90% of the participants claim to practice safe littering and 75% use reusable materials. Around 88% of participants are concerned about the depleting environment. Only 52% of the participants have participated in environment-friendly programs in the past decade, but 64% are willing to encourage others to practice, and 82% are willing to learn more. This shows their interest in the environment, as seen by the nature of the graph, which is skewed to the right.

Figure 1

Q1 - Pro-environment practices in daily life.

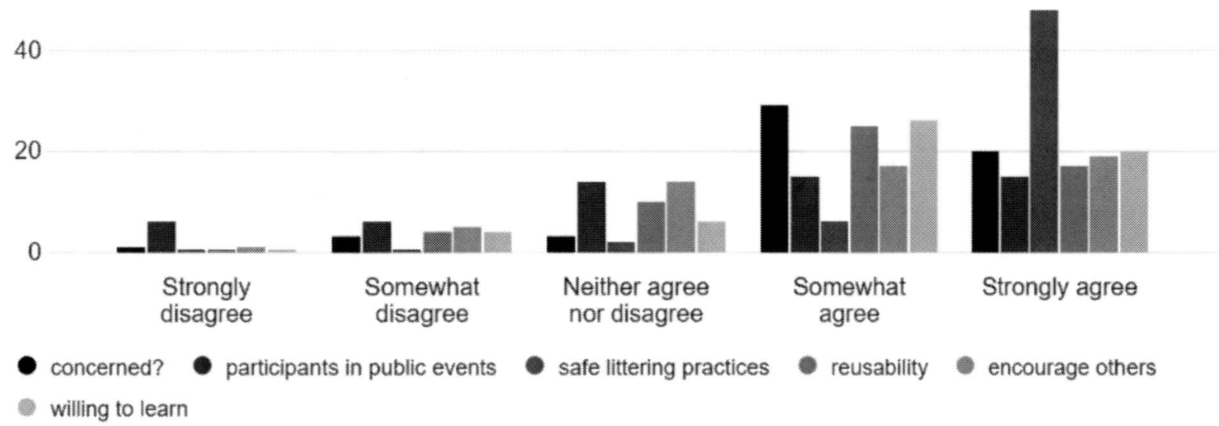

The next category of questions shown in Figure 2 were asked to understand the impact of industry and government competition to generate new ideas, including the environment. 60% of respondents knew about some competitions, and 30% of participants attended them. 70% of participants believe such competitions are beneficial, and half feel they effectively bring ideas to the world. Around 60% of participants felt these were not accessible to all people due to various factors, which can also be supported by the fact that only 18% had generated new ideas through such platforms.

Figure 2

Q2 - Competitions to promote new ideas for airport environment

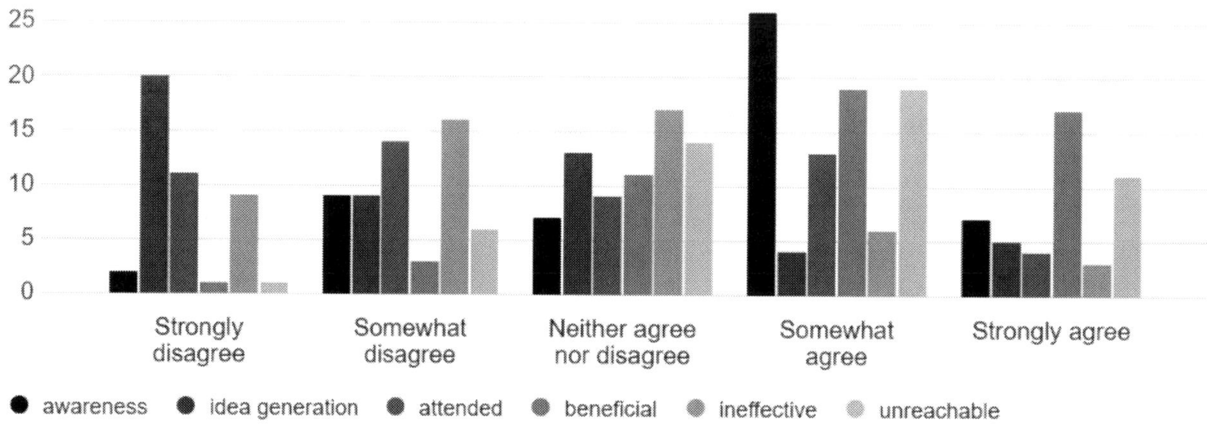

The third category of questions in Figure 3 is to understand the participants' awareness of the airport where they work. The graph being positively skewed in general explains a significant awareness of the participants about the airport they are working at. Around 84% of the participants claim to be aware of the influence of reuse, reduce and recycle of waste. Almost everyone knows the source of electrical energy and the types of trash segregation at their airport. 86% of participants know the source of water they are drinking or using at various airport locations.

Figure 3

Q3 - Awareness of the airports you work

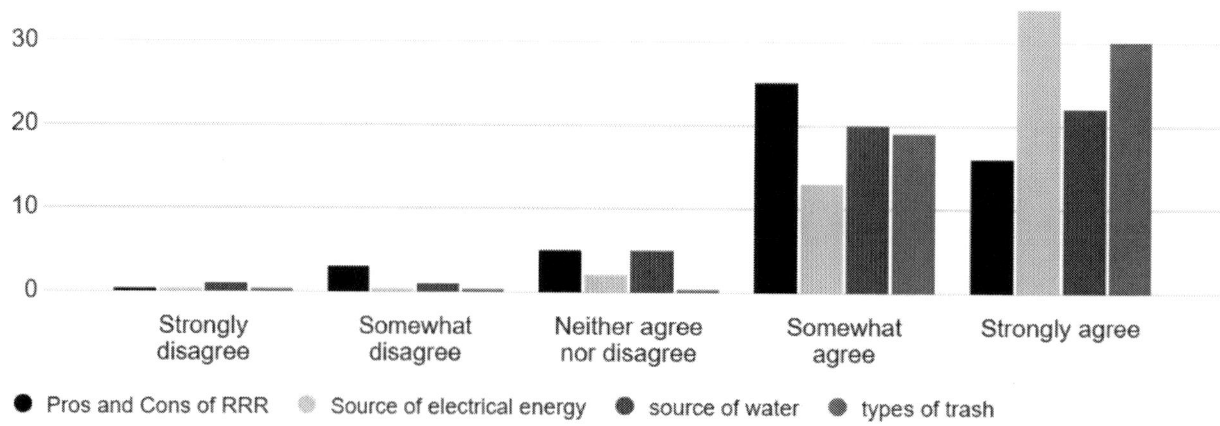

The last category of questions is to understand the extent to which people are willing to support new ideas and their creation regarding environmentally friendly projects. Around 60% of participants have seen new ideas implemented in real life. Around 86% of participants are willing to adapt to change and help generate new ideas. This is also seen in the positively skewed nature of the graph.

Figure 4

Q4 - Ready to support

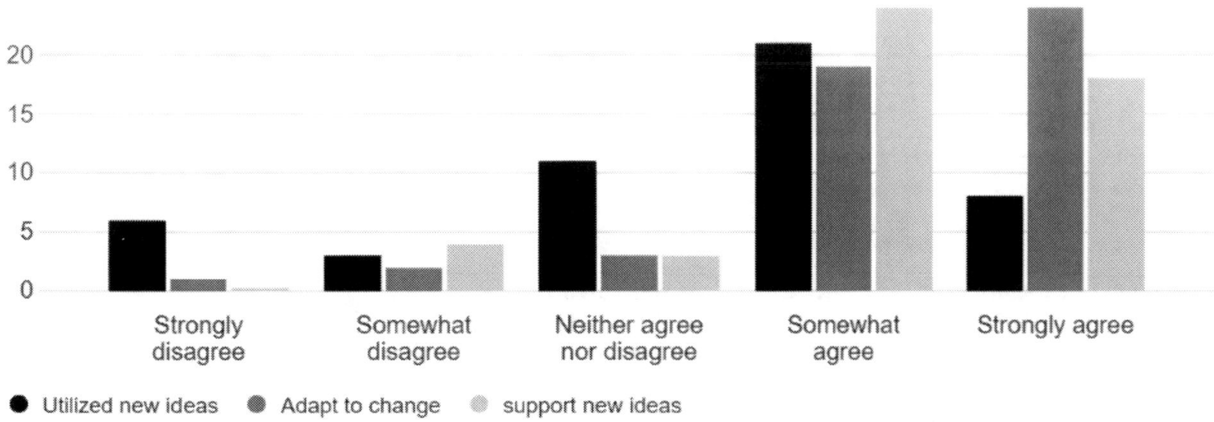

Conclusion

The study focused on understanding the attitude of airport personnel toward airport environmental policies. The study addressed the influence of new ideas being generated and how far they are being implemented. The survey is designed to analyze the results qualitatively, as people's perceptions are always subjective and tend to change. The significant majority of the participants are of working age and have higher-level roles in the airports they work at. This shows a close relationship with the target population for the study. Participants show general understanding and awareness but lack participation. They are also willing to learn more and support new ideas when necessary. They also believe that the reach of programs and competitions that generate new ideas is insufficient. Increasing the reach of talent hunt and programs to create new ideas such as creating more programs, reaching out to various institutions other than universities, etc.

Based on the research findings, several future areas of research were identified. The current study could be expanded through a larger sample size to understand airport employees' perceptions of airport environmental programs. This could include employees that work in retail or food services within airports. Furthermore, airline passengers could be surveyed to understand their perceptions of environmental programs implemented within the airports they use for travel. A future study could aim to understand factors such as airport personnel decision-making with regard to which environmental programs are implemented and maintained. It would also be beneficial to identify which factors constitute successful programs. Researchers can derive results specific to groups or airports using more demographic questions to compare between groups and different questionnaire methods for advanced statistical analysis.

References

Alfaro, V.N. & Chankov, S. (2022). The perceived value of environmental sustainability for consumers in the air travel industry: A choice-based conjoint analysis. *Journal of Cleaner Production, 380 (2).* https://doi.org/10.1016/j.jclepro.2022.134936.

Cremer, I., Rice, S., Gaenicke, S., & Oyman, K. (2016). *Consumer Affect and Type of Water Recycling Projects: Implementation at Airports*. Springer International https://doi.org/10.1007/978-3-319-34181-1_6.

Federal Aviation Administration (2021). https://www.faa.gov/airports/environmental/sustainability#:~:text=These%20documents%20include%20initiatives%20for,provided%20grants%20to%2044%20airports.

Murphy, T.P. (2004). The second Minnesota Report Card on Environmental Literacy. *Minnesota Office of Environmental Assessment.*

Retinger, C., Dumke, M. Barosevcic, M. & Hillerbrand, R. (2011). A Conceptual Framework for Impact Assessment within SCLA. *International Journal of Life Cycle Assessment*, https://DOI 10.1007/s11367-011-0265-y.

Sitorus, G. S., & Manik, Y. (2021). Socio-economic Life Cycle Assessment of Silangit Airport in Lake Toba Area. *Turkish Journal of Computer and Mathematics Education, 12*(8), 117-3124. https://portal.lib.fit.edu/login?url=https://www.proquest.com/scholarly-journals/socio-economic-life-cycle-assessment-silangit/docview/2623461165/se-2

DEVELOPMENT AND VALIDATION OF A VIRTUAL UAM TRANSPORTATION SYSTEM

Stacey M. Ahuja
Thomas Z. Strybel
Kim-Phuong L. Vu
Panadda Marayong
Praveen Shankar
California State University, Long Beach
Long Beach, CA 90840
Vernol Battiste
San Jose State University at NASA Ames Research Center
Moffett Field, CA 94035

Urban Air Mobility (UAM) refers to a system of passenger and cargo air-transportation vehicles within an urban area that is currently being designed to reduce demands for surface transportation. Their success depends on whether many obstacles to UAM operations are overcome. An important challenge to UAM success is the inability of the current air traffic management system to manage urban airspace, and new procedures and operating concepts are needed for coordination of UAM vehicles with existing commercial airspace traffic. Moreover, all systems currently under development initially will require remote or onboard pilots, and these pilots will need significant training to become certified for UAM operations. To evaluate ATM concepts of operation, cockpit interfaces and operator performance, we are developing a UAM vehicle and simulation environment.

Urban Air Mobility (UAM) is a conceptual transportation and infrastructure system that facilitates on-demand air transportation services for passengers and cargo in urban (and surrounding) areas. The UAM concept under development in the United States by the National Aeronautics and Space Administration (NASA), the Federal Aviation Administration (FAA), and other industry stakeholders seeks to revolutionize transportation in urban areas by establishing a connected and increasingly autonomous system for the quick and efficient aerial routing of passengers, cargo, and packages in and around metropolitan areas (FAA, 2020). A fully-mature UAM system will involve infrastructure and operational regulations and conditions that will be almost entirely novel in the air travel domain. For instance, there will be innovative flight operation and control systems with varying degrees of autonomy, new pilot training requirements (including reduced expertise resulting from UAM system automation), new aircraft displays and monitors, and new regulations and procedures for integrating into and navigating the airspace (Lombaerts et al., 2020). As early-stage development of the UAM concept is already underway, empirical research on discrete components of the complex system is also underway and evaluation of every operational component of the proposed UAM system throughout its developmental lifecycle will be critically important. As such, this proposed study seeks to expand existing work on validating a virtual simulation tool for the assessment of UAM vehicle operations within a revolutionary transportation system.

Conceptual vehicles in the UAM system are electric-powered vertical take-off and landing (eVTOL) aircraft; thus, they are quieter and more environmentally conscious than traditional fuel-powered aircraft and do not require an expansive runway infrastructure. In addition, feasibility studies conducted by industry stakeholders indicate that UAM systems have the potential to be profitable and to realize a host of other benefits such as reduced commute times, fewer vehicles on the roadways, and an environmentally clean mode of transport at a reasonable price (Marayong et al., 2020; Preis & Hornung, 2022).

UAM Implementation Challenges

However, for all its anticipated benefits and enormous potential for revolutionizing urban transportation, the UAM concept faces substantial challenges. For instance, the implementation of a UAM transportation system will occure in densely packed urban environments and infrastructure planning and design are likely to be the most challenging aspects of realizing a well-performing UAM system. Along these lines, it is currently unclear how operators in the UAM system would navigate government regulations surrounding what amounts to a web of low-flying aircraft above and across urban population centers, highlighting the need for an effective strategy to integrate UAM operations into the existing National Airspace System (NAS) and to garner passenger and community acceptance (Strybel et al., 2022).

Early on, the market for UAM will serve as airport shuttles that fly to and from airports along fixed routes. This concept is similar to earlier helicopter flight routs in large metropolitan areas like New York's, Pan Am flights to and from the city's major airports. These early UAM, operations will consist of low-tempo, low-density flights along a small set of routes between a few takeoff and landing areas. As such it is expected to be heavily dependent on existing air-traffic- management (ATM) rules and procedures. In the near-term, it is expected that communications will be based on analog voice and existing data-link systems for safety critical information. Navigation will be based on GPS, INS, Loran and very high frequency omnidirectional range (VOR)/distance measuring equipment. Air traffic services and management is proposed to be similar to VFR flight services provided by air traffic control (ATC). In sum, near-term UAM operations will require human operators, licensed human pilots operating under VMC conditions and under VFR/IFR rules with the supervision of air traffic controllers (ATCos).

Although it is assumed that UAM operations will be fully autonomous, development of autonomous UAM operations will most likely evolve over stages of increased automation. Early-stage operations are expected to be integrated into the NAS with certified pilots flying within the current operational environment. As new automation is developed, less skilled operators will be able to fly UAM vehicles, with automation assistance. Further development will result in UAM-operator – autonomous-system teams and finally fully autonomous vehicles supervised by on ground UAM flight managers. These developmental design stages will be accompanied by new operatizing concepts and regulations (FAA, 2020). As such, the need for UAM aircraft will initially be operated by expert, highly-qualified pilots and these pilots and UAM operators will be an important element in the design of new automated systems. As new systems are introduced, pilots will initially server a fail-safe function, in case of automation failures. Pilots will also be required for testing automated systems for their suitability in different scenarios, environmental conditions and airspaces. One advantage of skilled on board skilled operators is the ability to communicate with designers and flight managers in real time. In summary, although UAM vehicles will be increasingly automated, pilots and UAM operators will be essential to the design of these systems.(Strybel et al., 2022).

Simplified Vehicle Operations

One solution to the anticipated shortage of certified pilots and costly training requirements for UAM, is the concept of simplified vehicle operations (SVO). SVO is a key concept in a well-functioning UAM system whereby pilot skill and training requirements begin at expert level in the early stages, but gradually transition to the level of trained operators of semi-autonomous aircraft, and then eventually phased out as the aircraft become fully autonomous in later stages (Lombaerts et al., 2020). Fully evolved SVO will reduce incidents related to pilot error by replacing the pilot with end-to-end automation;

however, successful reduction in pilot knowledge, skill, and training requirements will depend on well-developed and validated SVO concepts of novel flight command and control systems (Lombaerts et al., 2020).

SVO will assist human pilots/operators by reducing the complexity of flight-system interfaces, operations and training which should reduce workload and increase safety. Originally, SVO was focused solely on aircraft handling but more recently SVO has been expanded to include the use of advanced automation for mission management, flightpath management and tactical operations, thus changing the pilot/operator's role from that of human-in-the-loop to human-on-the-loop (Wing et al., 2020).

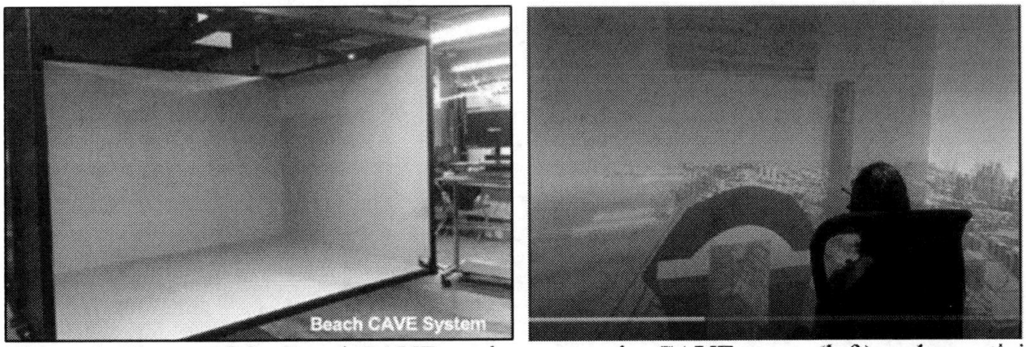

Figure 1. Photos of the BeachCAVE environment: the CAVE space (left) and a participant seated in the UAM vehicle pilot's chair.

UAM Simulation Test Bed

Given the need for tests of new ATM procedures and SVO, it is essential that simulation tests of concepts of operation be performed. Because human pilots or operators initially will operate eVTOL aircraft from onboard the vehicle during the transition to full autonomy, it is necessary to identify which operator functions can be safely simplified and to develop an approach for evaluating such simplifications. In response to this need, we have initiated a virtual UAM test bed for examining both SVO and new ATM procedures. Development of this system is being achieved by researchers at California State University Long Beach, San Jose State University Research Foundation, and NASA Ames Research Center (Marayong et al., 2020; Shankar et al., 2022; Strybel et al., 2022).

The UAM testbed is being developed in the BeachCAVE laboratory at California State University, Long Beach (see Figure 1). This facility consists of a VisCube™ M4 CAVE Immersive 3D Display that has approximate dimensions of 8'h x 8'd x 12'w (Visbox, Inc.). The BeachCAVE includes a four-wall projection system, an eight-camera advanced real time full body motion capture system, surround sound, and a graphics workstation using a 12-core Intel Xeon E5-2650 v4 processor with Nvidia Quadro P5000 graphics card (Visbox, Inc.). The CAVE system (as opposed to a head mounted display) is appropriate for applications where a wide field of view facilitates a greater sense of immersion in the virtual environment while still allowing participants to interact with physical controls and experience the space as if they were sitting in a real cockpit.

The virtual environment utilizes the Unity 3D game engine (Unity Technologies, Inc.) in combination with the MiddleVR plugin to render 3D content across multiple screens. The virtual UAM vehicle was adapted using Blender, an open-source graphic software (The Blender Foundation), from a quadcopter base model purchased through the Unity Asset Store (Unity Technologies, Inc.) and has been customized via code to enable easier participant control of the aircraft and out-the-window views. UAM operators wear special glasses to facilitate viewing of the 3D simulation with head tracking to automatically adjust the operators view. The aircraft can be flown in autonomous or manual mode and dimensions of the cockpit display were set to conform to the point of view of a seated operator.

Figure 2. Simplified flight controls (A) and display (B) in the virtual cockpit: the joystick controls vehicle heading, speed, altitude, flight mode, zoom, and messaging; the cockpit interface displays a map, heading, speed, altitude, and flight mode.

The vehicle is controlled with an integrated Attack 3 joystick (Figure 2A), which controls all flight parameters (heading, speed, altitude, etc.) and enables sending predefined messages to and from ATC. The flight stick also supported adjusting the map display and switching between operational modes, either Flight Mode or Ground Mode. In Flight Mode, participants used the joystick to move forward, change speed, climb/descend and control the heading of the quadcopter. Ground Mode was used for the final approach to the landing pad; the participants could move the vehicle forward and backward, laterally, and rotate the vehicle. Moreover, turning, rotating and accelerating could be achieved at slower speeds. The joystick flight control arrangement approximated SVO concepts described in Wing et al. (2020) for a single joystick based on NASA's EZ-Fly Concept for simplifying V/TOL flight handling with some exceptions. The joystick configuration approximates some of the EZ-Fly concepts for vehicle control, although differences in joystick hardware and the addition of a "ground Mode" created some differences. The cockpit display (Figure 2B) also contained an integrated display of current and assigned flight parameters and a map showing vehicle position over the city of San Francisco. In one condition, the map display also showed a route overlay of the pilot's planned flight path.

The airspace environment is currently located in the city of San Francisco. It was created with the World3D application programming interface, which provides a real time interactive 3D mapping of various cities. The World3D API keeps the map data accurate and current through various location services. The simulation was updated with streets and buildings currently located in San Francisco. For additional details on the development and design of the simulated UAM vehicle and test environment, see Marayong et al. (2020), Shankar et al. (2022).

Initial Validation Test

An initial validation test of the UAM vehicle was performed with certified-pilot and student/non-pilot participants. Operator performance and workload was evaluated for pilots and students when a route map overlay was present versus no route map overlay. The scenario consisted of flights from downtown to San Francisco International Airport (SFO) and return. It was shown that the map overlay significantly reduced flightpath deviations (relative to automated flights). Moreover, non-pilot participants reported significantly higher ratings of workload when no overlay was present, suggesting that the route map overlay was a helpful tool, especially for novices. Both pilots and non-pilot participants rated the vehicle as easy to fly but pilots rated scenario realism significantly lower than non-pilots and suggested adding weather, more interactions with air traffic controllers, pre-flight planning tasks, and additional traffic in order to increase the realism of the scenarios.

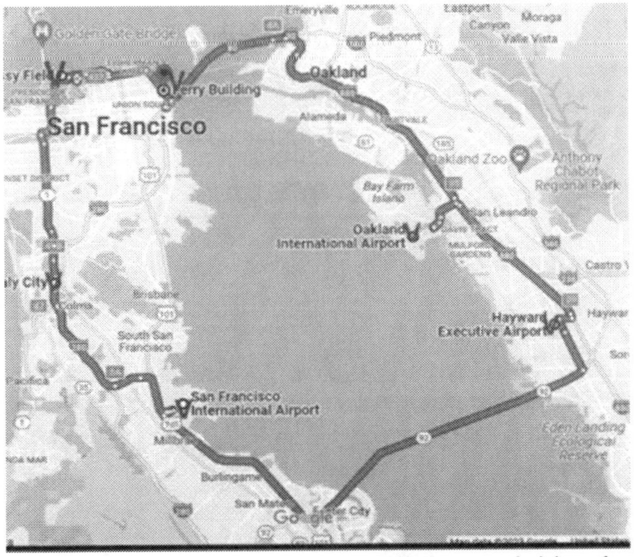

Figure 3. Route for test of a UAM transportation system. Flights are initiated at the Ferry Building in downtown San Francisco and continue around the San Francisco/Oakland area (either clockwise or counterclockwise) with six vertiport stops (marked by the letter "V").

Validation Test of UAM Transportation System

The initial test of our UAM testbed suggests that it is a promising simulation method for evaluating UAM vehicles and airspace operation concepts. However, the initial test was asimple flight, with few communication requirements that were mostly automated. Therefore, we are developing an expanded test of the system that would more closely approximate a UAM vehicle as one component of a UAM transportation system in the near term. The purpose of this test is to further expand UAM vehicle operations validity testing by using a longer and more realistic flight route around the San Francisco area with multiple vertiport stops as shown in Figure 4. Six vertiport locations along the route and participants will be required to land at each vertiport, then depart and follow the route to the next vertiport until the Ferry Building vertiport is reach again. Because it is anticipated that early-stage UAM vehicles will operate within existing NAS regulations, participants will be required to communicate with ATC and vertiport managers using either text or voice communication modes. Note that each vertiport will have its own manger, and the UAM pilot must be aware of the next vertiport manager and adjust their communication protocol accordingly. In addition, flight plan changes will be issued by ATCos at various times during the flight, for example, speed changes may be issued as the UAM vehicle crosses the bay.

A prototype text-based communication system that will be used by the pilots. In addition, flight plan changes will be issued by ATCos at various times during the flight, for example, speed changes may be issued as the UAM vehicle crosses the bay. Messages between the pilots and ATC will appear on the simulation screen and on an pad provided to the participants that will be strapped to the participants' leg in the same manner as a traditional flight kneeboard.

Status and Future Development

The simulation described above will be completed this year, as data collection will commence in the upcoming months. We expect to expand our test bed even further, to create a realistic simulation facility for testing UAM operations and pilot performance based on near term developments as well as evaluate new automous systems expected to improve the efficiency of a UAM transportation system. Some of our developments will include:

- Enhance scenario realism by adding additional traffic, greater interactions with air traffic control, and environmental conditions.
- Improve the aircraft model to enhance the fideltiy of current-day UAM vehicle controls as well as future control concepts based on SVO.
- Integrate the UAM vehicle in NASA's Multiaircraft System (Prevot & Mercer, 2007) to enable tests of both UAM operators and ATC personnel, both current day and anticipated new roles.

References

Federal Aviation Administration (FAA). (2020). *Urban Air Mobility, Concept of Operations v1.0.* Washington, DC

Lombaerts, T., Kaneshige, J., & Feary, M. (2020, June 15-19). Control concepts for simplified vehicle operations of a quadrotor eVTOL vehicle. In *AIAA AVIATION 2020 FORUM,* virtual event. https://doi.org/10.2514/6.2020-3189

Marayong, P., Shankar, P., Wei, J., Nguyen, H., Strybel, T. Z., & Battiste, V. (2020). Urban air mobility system testbed using CAVE virtual reality environment. *2020 IEEE Aerospace Conference, 1.* doi: https://doi.org/10.1109/AERO47225.2020.9172534

Preis, L., & Hornung, M. (2022). Vertiport operations modeling, agent-based simulation and parameter value specification. *Electronics*, *11*(7), 1071. https://doi.org/10.3390/electronics11071071

Strybel, Z., Battiste, V., Marayong, P., Shankar, P., Viramontes, J., Nguyen, H., Cheung, J., (2022, October 10-14). *Preliminary validation of a virtual UAM vehicle and simplified cockpit interface.* Human Factors & Ergonomics Society 66th International Annual Meeting, Atlanta, GA, United States.

Shankar, P., Marayong, P., Strybel, T., Battiste, V., Nguyen, H., Cheung, J., Viramontes, J. (2022). Urban Air Mobility: Design of a Virtual Reality Testbed and Experiments for Human Factors Evaluation. ASME International Mechanical Engineering Congress and Exposition, Columbus, OH.

Wing, D., Chancey, E., Politowicz, M., & Ballin, M. (2020, June 15-19). Achieving resilient in-flight performance for advanced air mobility through simplified vehicle operations. In *AIAA AVIATION 2020 FORUM*, virtual event. https://doi.org/10.2514/6.2020-2915

APPLYING HUMAN-CENTERED DESIGN TO AI-ENABLED PILOT SCHEDULING[1]

Amy L. Alexander, Audrey Haque, Michael Snyder
MIT Lincoln Laboratory
Lexington, MA
Rachael Kusiak, Brice Okubo, Kathie Chung
Revacomm
Honolulu, HI
Eric Robinson
Department of the Air Force Artificial Intelligence Accelerator
Cambridge, MA

Air Force mission and training scheduling is an immensely complex, time-consuming, and significantly manual process. A scheduling tool known as Puckboard has been developed to help C-17 squadrons transition from moving pucks across large whiteboards to utilizing technology to dynamically plan and deconflict resources in the presence of complex constraints. The overarching goal of incorporating artificial intelligence (AI) into this tool is to empower schedulers to quickly produce more efficient schedules that promote unit readiness, with more pilots completing their training syllabi faster, and with fewer disruptions to missions, training, and aircrew personal life. Our AI efforts focused on refining a neural network approach combining reinforcement learning with linear programming to generate optimal schedules across varying timeframes. The development of this AI-enabled pilot scheduling tool involved applying human-centered design best practices, namely actively involving end-users to inform persona generation, tool functionality, existing and AI-enabled workflows, and wireframe development and iteration.

Flight and crew scheduling presents a complex and time-consuming problem, especially in the United States Air Force (USAF) context with competing mission and training priorities, where bureaucracy and churn add further complexity. Unlike traditional scheduling - which typically involves known constraints, decisions, and values - military scheduling involves unpredictable and dynamic scenarios that require the capability to adjust to new constraints or decisions as they arise. For instance, on any given day there could be a natural disaster, an invasion, or other unexpected events that require immediate response. In addition, aircrew members are not just employed to operate aircraft; they also hold leadership roles, resulting in conflicting schedules between different decision authorities managing people in different ways. The scheduling process therefore requires back-and-forth negotiation to achieve desirable results without burning relationships, and any scheduling technology must consider heavily the "human" element involved. The complexity of this problem requires an iterative design process that considers the evolving needs of the end-users in the Air Force.

---

[1] DISTRIBUTION STATEMENT A. Approved for public release. Distribution is unlimited. Research was sponsored by the United States Air Force Research Laboratory and the Department of the Air Force Artificial Intelligence Accelerator and was accomplished under Cooperative Agreement Number FA8750-19-2-1000. The views and conclusions contained in this document are those of the authors and should not be interpreted as representing the official policies, either expressed or implied, of the Department of the Air Force or the U.S. Government. The U.S. Government is authorized to reproduce and distribute reprints for Government purposes notwithstanding any copyright notation herein.

Puckboard - the Mobility Air Force scheduling tool - has been developed to help squadrons transition from moving pucks across large whiteboards to utilizing technology to dynamically plan and deconflict resources in the presence of complex constraints. Our approach to Air Force mission and training scheduling is unique in its emphasis on human-centered design. We included USAF subject matter experts on our team to ensure that our technology aligns with the domains and use cases of our end-users. We worked in short batches, adjusting the algorithm, code, and design with feedback from users approximately every two weeks. Our soft rollout experiments on real data in production helped us get feedback under representative service (i.e., usage) load.

We focused on building a system that maximizes user acceptance and incorporates explainable reasoning. To achieve this, we leveraged the psychology behind how scheduling is historically accomplished, suggesting various options with explanations for why specific aircrew are recommended for specific seats, and aligning the recommendations that are shown to both aircrew and scheduler. This approach addressed the complex and dynamic nature of Air Force mission and training scheduling, making it easier for users to understand recommendations and make decisions with confidence. The overarching goal of incorporating artificial intelligence (AI) into this tool was to empower schedulers to quickly produce more efficient schedules that promote unit readiness, with more pilots completing their training syllabi faster, and with fewer disruptions to missions, training, and aircrew personal life due to last-minute changes.

The goal of this paper is to provide a summary of how we applied human-centered design best practices to developing an intelligent scheduling recommendation capability, known as Puckboard.AI.

Method

The development of Puckboard.AI involved applying key human-centered design best practices, namely actively including end-users in project decision-making and conducting user research regularly throughout the course of the project. In this way, end-users informed tool functionality by providing input to AI-enabled workflows and wireframe iteration. Our research methods included interviews and collaborative working sessions, resulting in personas, workflow diagrams, and wireframes.

User Research Activities

One of the first steps in a human-centered design effort is to understand the users that will ultimately incorporate a tool or system change into their daily tasks. At our project kickoff meeting, we met with stakeholders and end-users to discuss the entire problem space, including scheduler pain points, priorities, and potential new features. We brainstormed solutions to pain points with a "How Might We" statement, generating trans-disciplinary ideas from a wide range of participants (see Figure 1). Ideas spanned a broad range of topics, for example, providing individual event suggestions to aircrew to nudge them to volunteer for existing events. Other ideas generated during that session included enabling currency notifications, logging schedule changes and justifications within particular events, and providing multiple schedules associated with different priorities (e.g., mission readiness, burden distribution).

Figure 1. "How Might We…" ideation during project kickoff

Other follow-up user research activities involved mapping out users' current scheduling processes while using Puckboard and other tools to understand existing goals and pain points, and how schedulers might utilize algorithmic recommendations to optimize their ability to create optimal and more efficient schedules. As we drafted workflows showing how schedulers might utilize such intelligence, we hosted feedback sessions to get feedback about the proposed process, soliciting comments and change requests prior to building any wireframes. Throughout the course of the project, we continually checked in with end-users via MatterMost and Zoom meetings to course-correct process flows and wireframes when necessary (see Figure 2).

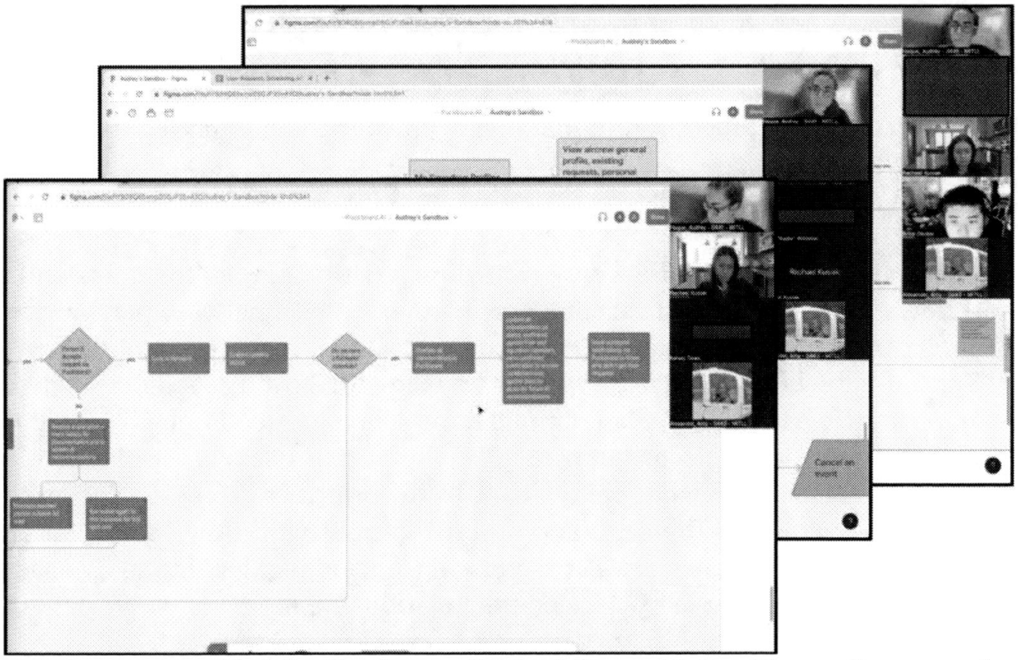

Figure 2. End-users provided feedback on a "Scheduling with Intelligence" proposed workflow

These user research activities resulted in the creation of the artifacts (i.e., personas, workflow maps, wireframes) presented below.

User Research Artifacts

Generate Personas

Personas are important in user interface design because they remind project teams to create products that are tailored to the needs of their end-users (Cooper et al., 2014). A persona is a fictional character that represents a particular user group or segment, based on data-driven research and analysis. By creating personas, designers and developers can gain a deeper understanding of the users they are building for, including their goals, motivations, and challenges or pain points. Personas can also help project teams make informed decisions about features, functionality, and user flows by providing a common reference point for the team. Figure 3 presents a series of personas developed in collaboration with subject matter experts as part of the Puckboard.AI project, focusing on key end-users, namely a C-17 pilot, Loadmaster, and Scheduler.

Figure 3. Personas developed from initial kickoff meeting and interviews with end-users

Capture Workflows

In order to effectively propose a new workflow, we first needed to understand users' current workflows. We worked with schedulers and other subject matter experts to create and verify a general overview of C-17 squadron scheduling as it currently exists. We then iterated on proposed "Scheduling with Intelligence" workflows, showing how a scheduler might utilize AI-assistance to more efficiently produce an optimal crew schedule (see Figure 4). A key aspect of this workflow included allowing schedulers to set the importance levels of solver parameters for a given time period or specific event, relating back to the priorities discussed during the kickoff meeting (e.g., mission readiness, burden distribution). Various workflow considerations related to locking aircrew into crew seats were also captured, for example, locking in by order of aircrew requests received versus locking in by AI recommendations first.

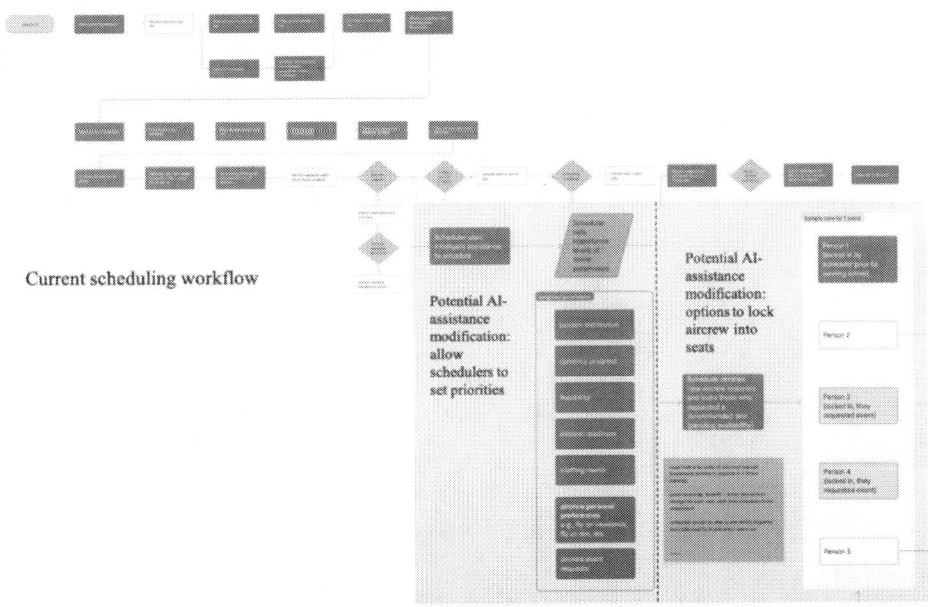

Figure 4. "Scheduling with Intelligence" proposed workflow modifications showing how a scheduler might incorporate AI-assistance into their workflow

Create Wireframes

The next step in our design process was to generate wireframes for incorporating optimized crew member recommendations into the Puckboard application. Wireframes provide a visual representation of the layout, structure, and components of the UI design while focusing on functionality and user workflow (Garrett, 2011). Figure 5 presents a series of wireframes created at different stages of the Puckboard.AI effort, specifically focused on assisting schedulers in assigning the optimal crew to a specific event.

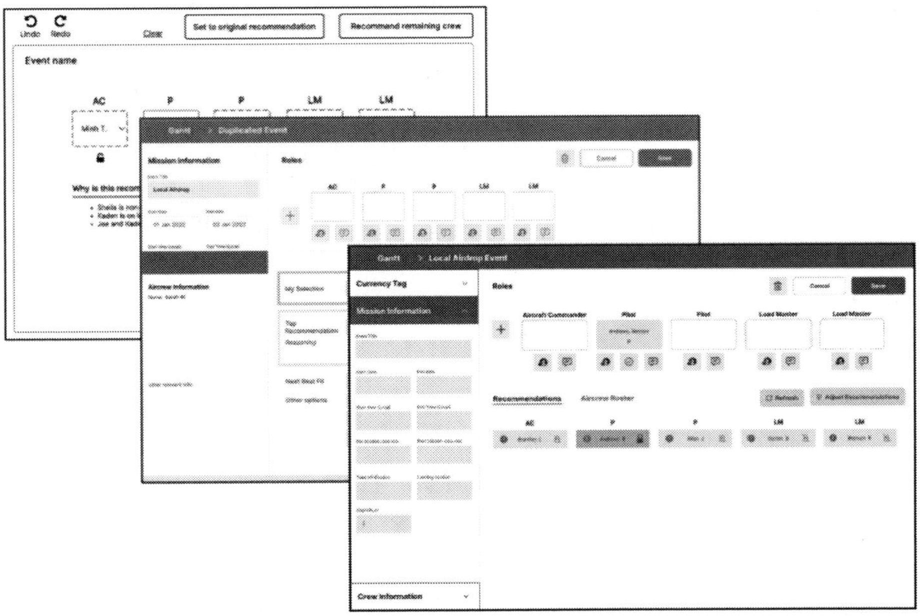

Figure 5. Progression of wireframe development to display the optimal crew for a specific event

AI Optimization

The research performed by our team explored a novel hybrid of reinforcement learning and linear programming whereby the weights for personnel assignments learned by a reinforcement learning model were supplied to a linear program to increase the robustness of solutions. This work, published in Association for the Advancement of Artificial Intelligence 2022 (Kenworthy et al., 2022), is being implemented and integrated into Puckboard first with the linear programming capabilities, and over time as data is collected to train the reinforcement learning model, the machine learning component will be introduced. With the initial linear programming capabilities, schedulers are able to designate various objective functions (e.g. burden distribution, minimal qualification, training progress) that either maximize or minimize progress towards those objectives, where the justification and explainability of the solutions – critical for trust-building – is communicated via the human-centered design work previously discussed.

Conclusions/Future Work

Within the context of Air Force training and mission planning, our approach to incorporate algorithm-aided recommendations into C-17 squadron scheduling allows schedulers to plan their assignments in seconds while considering complex objectives like maintaining training requirements and distributing flight hours. This work required careful integration with existing planning workflows and user interfaces to ensure that not only did the solutions meet the users' needs, but also that they were explainable and configurable. While this paper focused on C-17 crews, scheduling in the presence of constraints applies to many domains and AI-assistance can be applied to many complex workflows.

In summary, this paper discusses the unique challenges posed by Air Force mission and training scheduling, and presents Puckboard and Puckboard.AI, a scheduling tool that incorporates intelligent recommendations to empower military aircrew to produce more efficient schedules. The paper highlights the importance of incorporating human-centered design in the development of such tools, including involvement from domain experts, and the need for iterative design processes to ensure that the tool evolves to meet the evolving needs of the end-users in the Air Force.

References

Cooper, A., Reimann, R., Cronin, D., & Noessel, C. (2014). About face: The essentials of interaction design (Fourth Edition). Indianapolis, IN: John Wiley & Sons, Inc.

Garrett, J.J. (2011). The elements of user experience: User-centered design for the web and beyond (Second Edition). Berkeley, CA: New Riders.

Kenworthy, L., Nayak, S., Chin, C., & Balakrishnan, H. (2022, June). NICE: Robust scheduling through reinforcement learning-guided integer programming. In *Proceedings of the AAAI Conference on Artificial Intelligence, 36*(9), 9821-9829.

Stereoscopic Depth Cues for Enhancing Pilot Interpretation of the Artificial Horizon

Dale-Allen Arrundell[1], Annemarie Landman[1,2], Olaf Stroosma[1], René van Paassen[1], Eric Groen[2], Max Mulder[1]

[1] Delft University Of Technology, Delft, The Netherlands, [2] Netherlands Organisation for Applied Scientific Research, Soesterberg, The Netherlands

Background. Previous studies and accident analyses have shown that pilots can make roll reversal errors when responding to bank angles shown by the artificial horizon in the Primary Flight Display (PFD). In the current study, we tested whether adding stereoscopic depth cues to the artificial horizon may lead to better bank angle representation due to an improved figure-ground separation between the symbols. **Method.** Stereoscopic depth cues were created by using a half-silvered mirror multi-layer PFD, which presented the horizon symbol on a lower layer and the aircraft symbol on a higher layer. A group of 23 non-pilots and 18 general aviation pilots were shown left or right bank angles on this multi-layer PFD as well as on a normal single-layer PFD, with the task to roll the wings level using a joystick. **Results.** In the pilot group, a similar amount of roll-reversal errors was made with both displays (median = 3.3%) with no significant difference, $p = 0.635$. In the non-pilot group, fewer roll-reversal errors were observed with the multi-layer display, but this difference did not reach significance either (median = 3.3% vs. 5.0%, $p = 0.182$). In both pilots and non-pilots, the reaction time was longer in the multi-layer display, which reached significance in the non-pilots ($p = 0.016$) but not in pilots ($p = 0.215$). Participants noticed that the depth was only visible during the start of the session. **Conclusions.** The results suggest that using stereoscopic depth cues are not a viable manner to enhance the figure-ground relation in the artificial horizon.

Spatial disorientation is still one of the major causal factors in cases of loss of control in flight (LOC-I). It was determined to have contributed to 17% of LOC-I accidents in transport and commuter aircraft between 1981–2016, with no signs of a decreasing trend (Newman & Rupert, 2020). The Primary Flight Display (PFD) is the main instrument with which pilots can prevent or counteract spatial disorientation. Several studies have indicated that the bank indication of the artificial horizon is suboptimal, as it can lead to misinterpretations of the bank angle direction as well as incorrect roll inputs known as roll-reversal errors (RREs). In simulator studies, airline pilots were shown to make RREs in 6.9-8.7% of the cases when being shown a PFD with an unforeseen bank angle and attempting to correct to wings-level flight (Müller, Sadovitch, & Manzey, 2018; Van den Hoed et al., 2022). When spatially disorienting roll cues preceded PFD presentation, this percentage increased to

20% of total cases and 39% of the first encounter (Van den Hoed et al., 2022). Examples of accident cases that have been associated with spatial disorientation-induced RREs include Kenya Airways Flight KQA507 (Cameroon Civil Aviation Authority, 2010), Flash airlines flight 602 (Bureau d'Enquêtes et d'Analyses pour la Sécurité de l'Aviation Civile, 2009), and Crossair flight 498 (Aircraft Accident Investigation Bureau, 2002).

These misinterpretations of the artificial horizon are thought to be caused by suboptimal display design. According to the so-called "figure-ground" principle, we normally perceive objects moving in the foreground against the fixed horizon in the background. In standard artificial horizon displays, the horizon symbol is the part that moves, while the aircraft symbol is fixed, which hampers quick interpretation (Grether, 1947; Johnson & Roscoe, 1972; Roscoe, Corl, & Jensen, 1981). The format of current displays does allow for integration with head-up displays. Hence, it is relevant to investigate display adaptations that may enhance the PFD in its current format.

In the current study, stereoscopic depth cues are investigated as a means to improve the figure-ground relation in the artificial horizon, in order to prevent RREs. These cues are presented using a multi-layer display (MLD), presenting the aircraft symbol and horizon symbol on two different layers. As this produces a different visual image in each eye, the cues are stereoscopic. MLDs have previously been used to separate categories of information to improve users' information uptake and to prevent clutter (Dünser, Billinghurst, & Mancero, 2008; Hayes, Moore, & Wong, 2006). They also have been applied to make information more salient to improve search performance (Wong, Joyekurun, Nees, Amaldi, & Villanueva, 2005). However, to our knowledge, MLDs have not yet been used to improve figure-ground separation in cockpit displays.

Method

Design

The experimental tasks were first performed by a group of non-pilots to obtain more information on optimal depth between two layers of the MLD. These non-pilots performed the tasks with a single layer display (SLD) PFD (baseline condition) and with a MLD with either low or high depth between the two layers (two groups, randomly assigned). The effect of MLD and that of MLD depth were tested using a mixed-model design.

Using these outcomes, one layer of depth was chosen to further test with a group of private pilots. This group performed the same tasks in the baseline condition and with the MLD with the depth that had the most effect in the non-pilots. The effect of the MLD was tested using a within-subject comparison.

Participants

Of the non-pilot participants (n = 23, 18/5 male/female, mean age = 24.8, *SD* = 6.9), 5 had never seen nor used an artificial horizon, 4 knew what it is but had never used it, 10 had some experience with flight simulators at different fidelity levels, and 8 had experience with glider flying. None had experience controlling powered aircraft of any category.

The participating pilots (n = 18) held a private pilot license (PPL, all male, mean age = 43.5, *SD* = 16.1). Flight experience was on average 324 flight hours, *SD* = 245 hours, and 9.8 years of being active as a private pilot, *SD* = 7.4.

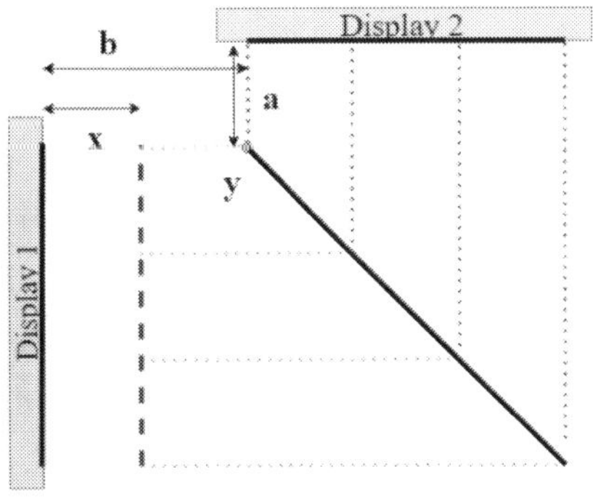

Figure 1. Left: the experimental setup with the display and joystick. Right: The configuration of the MLD with two displays and mirror (the black diagonal line).

All participants were right-handed and had normal or corrected-to-normal eyesight. This experiment was approved by the ethical committee of the Delft University of Technology, and all participants provided informed consent.

Apparatus

A desktop setup was used with no outside visuals (see Figure 1, left). A half-silvered mirror-type display was used for the MLD. A schematic of the inner workings of this display can be seen in Figure 1 (right). For each display layer, a standard LCD-type display was used with a resolution of 1152 x 864. These displays were placed perpendicular to each other, and a half-silvered mirror was placed at a 45°between the displays. This creates a virtual display (dashed blue line in Figure 2) at some distance x in front of display 1. Distance x could be set by moving display 2 to increase distance a. For the low depth condition, X was 1.6 cm, and for the high depth condition 2.1 cm. For the single-layer display (SDL; baseline) condition, display 2 was turned off and all information was displayed on display 1. A standard Logitech Extreme 3D pro joystick was used as input device. The maximum angular deflection on the roll axis of this joystick was measured to be 20°.

A Boeing 747-based PFD was used (see Figure 2). In the MLD, the aircraft symbol, sky pointer and bank angle scale are presented on the upper layer, as well as the speed and altitude tapes. The horizon and pitch ladder were presented on the lower layer. A simplified aerodynamic model was used. The model had a fixed speed of 120 knots, altitude always indicated 10,000 ft., and the attitude was controllable in the pitch and roll axis with sensitivity and dampening resembling that of a small single-engine piston aircraft.

Procedure and tasks

After a briefing and an intake questionnaire, participants were first showed the SLD and MLD version of the PFD. Non-pilots were explained the symbols and how the display is used to control an airplane. Participants then familiarized themselves with the flight dynamics for 5 minutes by flying several turns and level changes. The familiarization was followed by 10 practice runs of the experimental task with the SLD and 10 with the MLD. Each run started with a black screen displayed for 5 seconds. Then, the display showed a PFD

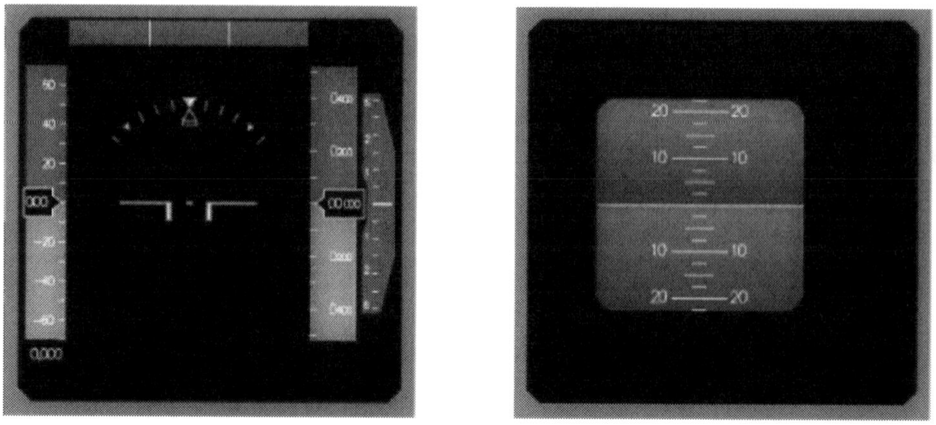

Figure 2. Left: The image presented on the upper layer of the MLD. Right: the image presented on the lower layer of the MLD.

indicating a bank angle at 30°to the left or to the right. When the display appeared, the participant was to respond immediately by rolling the wings level using the joystick. Ten seconds after the appearance of the PFD, the run ended. This was immediately followed by the black screen at the start of the next run. The participant's hand was always placed on the joystick. Following practice, this task was performed in two blocks of 30 runs with the SLD and with the MLD, with a short break in between. The order of conditions was counterbalanced between participants.

Dependent measures

Data on the roll angle and roll control input were logged and analyzed with MATLAB. From this, the following dependent measures were obtained:

- Error rate. A roll reversal error is recorded when the first roll control input was towards the same side of the displayed bank angle. The threshold of input detection was set at 1.5°stick deflection. The error rate is the percentage of runs with an error over the total number of runs.
- Reaction time. This is the time between PFD presentation and the start of the first roll control input.

Lower error rates and faster reaction times were expected with the MLD compared to the baseline condition, as the MLD was expected to facilitate quicker and more accurate recognition of the figure-ground relationship.

Statistics

In both groups, the effect of MLD was tested using a paired-samples t-test between MLD and baseline. If not normally distributed, the Wilcoxon signed rank test was used.

In non-pilots, the effect of the MLD depth between layers was additionally tested using a mixed-model ANOVA with display type (baseline, MLD) as the within-subject factor, and MLD depth (low depth group, high depth group) as the between-subject factor.

Results

Data collection

A number of runs (1% of total) were excluded from analysis due to incorrect detection of an input immediately following PFD presentation due to the stick position not being centered within the limits of 1.5°stick deflection.

Non-pilots

The Wilcoxon signed rank test showed that there was no significant difference in error rates between the baseline (median = 5% errors) and MLD condition (median = 3.3%, $Z(22) = -1.33$, $p = 0.182$). The mixed-model ANOVA showed no significant difference in improvement between the two MLD depths used ($F(1{,}21) = 1.059$, $p = 0.315$). Nevertheless, as the low depth showed the largest improvement, we decided to use this depth for the pilot group. The paired-samples t-test indicated a significant difference in reaction time between the baseline condition (mean = 591 ms) and the MLD condition (mean = 605 ms, $t(22) = -2.608$, $p = 0.016$), a difference which was opposite to the expected direction.

Pilots

A Wilcoxon Signed Rank test showed that there was no significant difference in error rates between the baseline (median = 3.3% errors) and MLD condition (median = 3.3%, $Z(18) = -0.475$, $p = 0.635$). A Wilcoxon Signed Rank test t-test indicated no significant difference in reaction time between the baseline condition (median = 625 ms) and the MLD condition (median = 644 ms, $Z(18) = -1.24$, $p = 0.215$).

Discussion

The results did not indicate a significant improvement in performance when using the MLD compared with the SLD. In contrast, reaction times in non-pilots were longer when using the MLD than the SLD, suggesting that it was more difficult to read the bank angle quickly with the MLD. Participants reported that they indeed perceived depth in the MLD, although this depth perception was mostly present at the start of the experiment or any time they moved their head. The requirement of head motions indicates that the MLD was unsuccessful in presenting stereoscopic cues, as no head motions should be required for such cues. This also makes it impractical for use in the cockpit.

The required distance between layers in the MLD to obtain a stereoscopic effect would cause the pitch and roll indications to become inaccurate, as the position of the symbols relative to each other would then shift greatly depending on head position. Several participants mentioned that they thought the aircraft model in the MLD condition was slower to react than that in the SLD condition, which was not the case. We do not know what may have caused this perception. The results of the experiment lead us to conclude that stereoscopic depth cues achieved thought MLDs are not suitable for enhancing the figure-ground representation in the attitude indicator.

Subsequent research into optimizing the PFD for attitude representation could focus instead on monoscopic cueing. There are several types of monoscopic cues that could be implemented without making serious changes to the PFD designs currently in use in commercial aviation. Examples of this are the use of a horizontal color gradient to simulate "aerial perspective" (Gibson, 1950), linear perspective lines, ground texture, extending the horizon behind the speed and altitude tapes, or adding a dark line under the aircraft symbol to simulate shadow. With each of these additions, it is important to ensure that the salience of the horizon, aircraft, and pitch and roll ladders remains intact. Additions to the sky and ground should not contrast too much with the colors of these surfaces, and thickness of added lines should be minimal. Empirical studies are needed to evaluate the effectiveness of these design changes, and possibly to fine-tune the optimal use of added symbols.

References

Aircraft Accident Investigation Bureau. (2002). *Final Report of the Aircraft Accident Investigation Bureau on the Accident to the Saab 340B Aircraft, Registration HB-AKK of Crossair Flight CRX 498 on 10 January 2000 Near Nassenwil/ZH.* https://www.sust.admin.ch/inhalte/AV-berichte/1781_e.pdf.

Bureau d'Enquêtes et d'Analyses pour la Sécurité de l'Aviation Civile. (2009). *Final report on the accident on 1st June 2009 to the Airbus A330-203, registered F-GZCP, operated by Air France, Flight AF 447 Rio de Janeiro–Paris.*

Cameroon Civil Aviation Authority. (2010). *Technical Investigation into the Accident of the B737-800 Registration 5y-Kya Operated by Kenya Airways that Occurred on the 5th of May 2007 in Douala.* https://reports.aviation-safety.net/2007/ 20070505-0_B738_5Y-KYA.pdf.

Dünser, A., Billinghurst, M., & Mancero, G. (2008). Evaluating visual search performance with a multi layer display. In *Proceedings of the 20th Australasian Conference on Computer-Human Interaction: Designing for Habitus and Habitat* (pp. 307–310).

Gibson, J. J. (1950). *The Perception of the Visual World.* Boston (MA): Houghton Mifflin.

Grether, W. F. (1947). Discussion of pictorial versus symbolic aircraft instrument displays. *USAAF AMC, Engineering Division, Aero Med Lab., TS EAA.*

Hayes, J., Moore, A., & Wong, B. L. (2006). Information layering to de-clutter displays for emergency ambulance dispatch. In *Proc. 13th European conference on Cognitive ergonomics: trust and control in complex socio-technical systems* (pp. 10–16).

Johnson, S. L., & Roscoe, S. N. (1972). What Moves, the Airplane or the World? *Human Factors: The Journal of the Human Factors and Ergonomics Society*, *14*(2), 107-129.

Müller, S., Sadovitch, V., & Manzey, D. (2018). Attitude indicator design in primary flight display: Revisiting an old issue with current technology. *The International Journal of Aerospace Psychology*, *28*(1-2), 46–61.

Newman, R. L., & Rupert, A. H. (2020). The magnitude of the spatial disorientation problem in transport airplanes. *Aerosp. medicine and hum. perf.*, *91*(2), 65–70.

Roscoe, S. N., Corl, L., & Jensen, R. S. (1981). Flight Display Dynamics Revisited. *Human Factors*, *23*(3), 341–353.

Van den Hoed, A., Landman, A., Van Baelen, D., Stroosma, O., van Paassen, M. M., Groen, E. L., & Mulder, M. (2022). Leans illusion in hexapod simulator facilitates erroneous responses to artificial horizon in airline pilots. *Human Factors*, *64*(6), 962–972.

Wong, B. W., Joyekurun, R., Nees, A., Amaldi, P., & Villanueva, R. (2005). Information layering, depth and transparency effects on multi-layered displays for command and control. In *Proc. Human Factors and Ergonomics Society* (Vol. 49, pp. 352–356).

EXPLORING INFORMAL LEARNING STRUCTURES IN AIRLINE OPERATIONS

Barth Baron, Jr.
San José State University Foundation
San Jose, CA

Dorrit Billman
NASA Ames Research Center
Moffett Field, CA

Randall J. Mumaw
San José State University Foundation
San Jose, CA

Airline pilot training is extensive, highly structured, and driven by aircraft and airspace system operating requirements, yet pilots describe a tradition of between-pilot knowledge transfer and self-directed learning. While industry and regulators focus on "formal learning" systems, pilots report relying on this "informal learning" to build operational expertise, suggesting gaps in how successfully formal learning prepares pilots to handle operational complexities. The community that researches learning has extensively studied informal learning, including in a workplace setting, and its characteristics align with how pilots report increasing their skills and knowledge informally. However, no research into informal learning practices among airline pilots seems to exist. In this paper we provide some examples of informal learning in commercial aviation, show how they fit into two existing frameworks for workplace learning, and propose that researching informal learning might help identify opportunities to improve formal aviation learning systems.

Airline pilot training is typically driven by Advanced Qualification Program (AQP) job task analyses. Each task is linked to instructional objectives and learned through standardized computer-based training modules, classroom lectures, and simulator briefings. These tasks and performance objectives are designed to give pilots the skills and knowledge to safely operate within their aircraft's operating limitations and the airline's operations specifications. Despite this structured formal learning environment, another tradition of learning plays an outsized role in the way pilots attain operational expertise. We refer to this learning, which occurs outside the formal structures, as "informal learning." Various definitions of Informal learning (IL) exist; however, one relevant definition is that IL can include talking with others, self-directed learning, observing others, and reflecting on actions (Lohman, 2005), and pilots report a long tradition of these practices in aviation. Commonly used terms for IL illustrate how prevalent it is in pilot culture: "tribal knowledge" refers to knowledge obtained outside of training that is needed to operate effectively in line operations, and "hangar talk" refers to the informal process of sharing information between pilots. Awareness of IL's value in promoting safety is reflected in an FAA report that stated a commitment to facilitating experience transfer among pilots (Federal Aviation Administration, 2010). Such learning between pilots can be found throughout the industry, as in the following example of talking with others:

> A captain completes transition training to a new airplane with a lower wing loading and higher residual thrust. During line flying in gusty winds, the captain experiences unexpected deviations of speed and path when using the standard procedures he learned during training. A colleague with more experience on that airplane suggests using a reduced flap setting and disconnecting the auto throttles during approach.

IL in aviation takes various forms and evolves along with a pilot's experience, perhaps starting from reading publications geared towards novices, including accounts of errors and mishaps. Later the setting changes to hotel van rides and layover dinners, but the foundation remains the same: Pilots relating their lived experiences to colleagues and learning by observing work in practice. In this paper we review some relevant aspects of IL and how it can improve knowledge and skill in the commercial aviation context. We know of no research on IL in an aviation context; thus, in this paper we rely on informally gathered anecdotes. Because of this lack of research literature exploring IL in commercial aviation, we borrow two existing research frameworks characterizing workplace learning.

Throughout, we consider the roles and contributions of IL in commercial aviation around this question: Does the persistent use of IL by airline pilots indicate gaps in airline training? We also identify areas for future research, including whether a better understanding of successful IL practices at airlines may suggest ways to design and deliver better formal learning systems.

Examples of Informal Learning at Airlines

Pilots typically claim that much of their operational expertise was gained through informal learning activities. This is unsurprising because research indicates that IL strategies integrate learning directly from relevant contexts, optimizing transfer to a degree not guaranteed by formal training (Moore & Klein, 2019). In contrast to the now standard use of passive, computer-based training (CBT), by its nature IL is an active and learner-centric sociocultural activity with learning constructed by building on prior knowledge through social interaction. Such interactions, as discussed by Vygotsky, Bruner, and others, help build meaningful learning (see also Mayer, 2009) and play a role in creating highly contextualized learning, where learners interact with their personal, sociocultural, and physical environments (Falk & Storksdieck, 2005; Riedinger & Storksdieck, 2023). However, airline training predominantly uses passive learning strategies (previously classroom slide presentations and now online CBT) removing the opportunities for such contextualized learning anchored in sociocultural interactions. In some cases, pilots may feel inadequately prepared by a formal learning program and seek a peer for additional help, as in this example of talking with others:

> A captain has been flying short-haul domestic operation for 20 years and will now simultaneously learn a new widebody airplane built by a different manufacturer (increasing the difference in system and flight control design to master) and learn oceanic and extended-range operations for the first time. A friend recently went through the same curriculum and for the first time failed a validation event. Concerned about the stress of failing in training, this pilot contacts a pilot with experience in the airplane and long-range flying. Weeks before training starts, they meet occasionally to review the aspects of the airplane that some pilots struggle to master. The systems are described in the context of how they will be used, various fault modes they experience, and how these impact operations. Hearing it from a friend in a relaxed setting before starting the curriculum, the pilot feels more confident and performs well during training.

This sort of peer-to-peer directed learning is deeply rooted in the operational context and provides an excellent match of learner needs to instructional delivery. However, one consequence of this example is that the learner is unlikely to provide feedback to the training department that would improve its delivery methods to help similarly concerned pilots. After this pilot's simulator portion of training and preparing for international and extended-range operations training, the pilots meet again and use another form of IL, reflecting on actions:

> The same pilot reads a deidentified safety report of a captain new to oceanic flying who made a critical decision-making error on an oceanic flight, putting the passengers and airplane at risk. Our example pilot feels concerned about the amount of learning needed for flying in a non-radar environment, using long-range navigation, under international regulations. To organize the differences and how to study for it, the friend tells personal stories about times that oceanic trips have been challenging, the actions taken, and they discuss what the pilot could have been done better. They use these stories to connect to the various locations in the manuals where the relevant information resides, providing the transitioning pilot with a series of narratives that can anchor the new knowledge.

Frameworks for Understanding Workplace Learning

The field of *workplace learning* can be applied to analyze IL. Workplace learning views the workplace as an environment where participants engage in a socially-situated and highly contextual community of practice (Rainbird et al., 2004). Such strategies integrate learning directly from relevant contexts, optimizing transfer to a degree not guaranteed by formal training (Moore & Klein, 2019). So impactful are these experiences that it appears as if the preponderance of learning in the workplace takes place informally, despite the frequency of formal workplace learning occurrences (Marsick & Watkins, 2001). Research into the role of IL in the workplace is accelerating (Smet et al., 2022), providing opportunities to improve our understanding of how pilots use IL at the airlines.

We can consider an airline as a workplace and apply two existing frameworks to explore how pilots use IL to develop expertise. These frameworks allow us to highlight various forms of IL to supplement knowledge or skills that did not adequately transfer from their airlines' training. One example of a conceptual framework to understand the interplay between formal and informal learning in the workplace considers three variables: Where the learning occurs, whether the learning is structured, and the role of the instructor/facilitator, as shown in Table 1 (Jacobs & Park, 2009).

Table 1.

Jacobs and Park Workplace Learning Framework.

Location of learning --	At work	Away from work
Degree of planning --	Structured	Unstructured
Role of Instructor --	Passive	Active

These three factors combine to describe eight broad learning contexts, several of them informal. Consider the following as an example of at-work/unstructured/passive instructor role, which is also an "observing others" activity in our IL definition:

> A new hire-pilot is assigned a "familiarization flight" to sit in the cockpit jump seat for a day, observing others prior to operating passenger flights under the supervision of an instructor. The new pilot observes the line pilots perform their duties and handle operating issues that occur on that flight. The trainee pilot also hears the line pilots describe their strategies for addressing issues that may not be fully addressed in training, such as ATC challenges at some airports.

Here a learner is socialized into a community of practice, highlighting the sociocultural aspect of learning. In a survey asking pilots about the effectiveness of their airlines' training program in developing FMS skills (Holder, 2013), a majority (62%) reported that they did not feel comfortable with the FMS until gaining at least three months of line experience, and 21% required more than six months of experience to feel comfortable with the FMS. Pilots may compensate for this perceived gap in their training through various forms of IL. Among these IL strategies is self-directed learning, where learners are personally responsible for constructing learning outcomes (Garrison, 1997). Again, using this framework, the following example describes the away-from-work/unstructured/passive instructor role:

> An airline receives a new aircraft type into its fleet. During descent with autopilot connected and while following standard procedures, the crew observes that the airplane is diverging above the desired path with the autopilot and FMC correctly configured. Without changing auto flight modes, they will violate their ATC clearance. The crew changes to a lower level of automation to comply with the clearance. During the layover the crew reviews their manuals together. On the Company's training website, they discover a supplemental reference document to the FMS describing the anomalous auto flight behavior that had not been covered in the training program.

As an example of an at-work/unstructured/active instructor, a simulator instructor makes use of extra time to help new pilots develop a skill they must demonstrate during line operations, but which is not covered by the simulator curriculum:

> Two first officers are in their first jet aircraft simulator training program. After the qualification check rides are complete, the instructor lets them use extra simulator time to practice energy management strategies during arrival. This is not in the airline's approved training program, and the instructor lets them experiment with the scenarios until they have developed some confidence in this skill, explaining that instructors observe new pilots struggle with this during line operations.

Here, the instructor identified a gap between the performance standards in the training curriculum and the skills needed to operate in busy airspace and adds this to the simulator experience.

Another formulation of workplace learning provides a different strategy to categorize workplace learning activities and the role of IL within them. This framework entails six categories of learning by first differentiating between learning that is intentional/planned and unintentional/unplanned. The information learned might be things that are already known to others, the development of an existing capability, or learning that which is new or treated as new, see Table 2 (Hodkinson & Hodkinson, 2004):

Table 2.
Hodkinson & Hodkinson Workplace Learning Framework.

	Intentional/planned	Unintentional/unplanned
Learning that which is already known to others	(1) Planned learning of that which others know	(2) Socialization into an existing community of practice
Development of existing capability	(4) Planned/intended learning to refine existing capability	(3) Unplanned improvement of ongoing practice
Learning that which is new in the workplace (or treated as such)	(5) Planned/intended learning to do that which has not been done before	(6) Unplanned learning of something not previously done

Note. These categories can help identify classes of IL to analyze common strategies learners use to attain the desired outcome.

The following example of Type 2 learning, an unplanned learning of information known to others, illustrates this type of informal learning in an airline context:

> A crew pushes back from the gate and experiences an ignition fault during engine start. The first officer is new to the airline and believes they must return to the gate for maintenance attention, but the captain knows to apply the rarely practiced abnormal start and alternate maintenance deferral procedures. Helping the first officer find the correct references allows the crew to continue the flight without incurring a delay.

Here an informal learning experience fills a gap in one pilot's formal training. This unplanned development of an existing capability, where the first officer's application of maintenance procedures at a new airline is developing, creates a highly contextual and sociocultural experience where the learner directly interacts with the environment and engaged colleague. Type 6 learning is unplanned and can refer to something completely new or treated as new, like a pilot who transitioned to oceanic flying after the introduction of satellite automated position reporting and must contend with an unexpected loss of satcom functionality while enroute. While the pilot may have learned the alternate procedure during initial training, the high reliability of satellite communication may have caused that pilot to disregard the need to remain proficient on high frequency radio position reporting procedures.

Discussion and Directions for Future Research

Our examples illustrate that pilots use various forms of IL to fill certain gaps in airline training. We believe there is value in better understanding these IL activities to help identify areas where current training practices fall short of providing pilots with the knowledge and skills they need to perform competently. From an airline training policy perspective, two questions emerge: First, can airlines use this understanding to fill identified learning gaps in their training? The recent adoption of pilot mentoring training recognizes the value of peer learning, but the knowledge and skills transferred there are not tracked by the training system. Second, can airline training take better advantage of the learning strategies that that characterize IL and incorporate them in training? Sociocultural and constructivist strategies underlying IL also succeed in formal learning, though airline training currently tends to be passive and CBT-based and not the contextual, sociocultural experience that characterizes successful learning designs which are attainable by applying modern pedagogy.

Such designs can result in developing supportive attitudes and habits and build foundations for future learning (National Research Council, 2009), and this affective domain learning could include supporting perpetual learning and safety attitudes. Were airlines to develop IL-inspired formal learning systems they could control the learning delivery and assessments within that training, assuring continuity of content and measurement of outcomes. We can see that much of IL is self-directed, derived from a pilot's desire to understand the system better and perform at a high level. This attitude is a strong component of the beliefs behind resilience, in which pilots develop methods for managing the variability and complexity in the operational environment. Developing measures of IL outcomes may be valuable, independently or in coordination with training. This can build understanding of how pilots contribute to the development of expertise.

Acknowledgments

The work reported here was funded by NASA's System-Wide Safety Project, part of the Aeronautics Research Mission Directorate's Aviation Operations and Safety Program.

References

Falk, J., & Storksdieck, M. (2005). Using the contextual model of learning to understand visitor learning from a science center exhibition. *Science Education*, *89*(5), https://doi.org/10.1002/sce.20078

Federal Aviation Administration. (2010). *Answering the Call to Action on Airline Safety and Pilot Training*. https://downloads.regulations.gov/FAA-2008-0677-0338/attachment_1.pdf

Garrison, D.R. (1997). Self-directed learning: Toward a comprehensive model. *Adult Education Quarterly*, 07417136, Fall97, Vol. 48, Issue 1

Hodkinson, & Hodkinson. (2004). The complexities of workplace learning: Problems and dangers in trying to measure attainment. In *Workplace Learning in Context*. Routledge.

Holder, Barbara (2013). *Pilot Perceptions of Training Effectiveness,* Boeing Commercial Airplanes.

Jacobs, R. L., & Park, Y. (2009, May 15). A Proposed Conceptual Framework of Workplace Learning: Implications for Theory Development and Research in Human Resource Development. *Human Resource Development Review*, *8*(2), 133–150. https://doi.org/10.1177/1534484309334269

Lohman, M. C. (2005). A survey of factors influencing the engagement of two professional groups in informal workplace learning activities. Human Resource Development Quarterly, 16(4), 501–527.

Marsick, V. J., & Watkins, K. E. (2001). Informal and Incidental Learning. *New Directions for Adult and Continuing Education*, *2001*(89), 25. https://doi.org/10.1002/ace.5

Mayer, R. E., (2009). *Multimedia Learning*. (2nd ed). Cambridge University Press.

Moore, A. L., & Klein, J. D. (2019, December 6). Facilitating Informal Learning at Work. *TechTrends*, *64*(2), 219–228. https://doi.org/10.1007/s11528-019-00458-3

National Research Council. (2009). *Learning Science in Informal Environments*. Committee on Learning Science in Informal Environments. Philip Bell, Bruce Lewenstein, Andrew W. Shouse, and Michael A. Feder, Editors. Washington, DC: The National Academies Press.

Riedinger, K., Storksdieck, M. (2023). Application of the Contextual Model of Learning and Situated Identity Model in Informal STEM Learning Research. In: Patrick, P.G. (eds) How People Learn in Informal Science Environments. Springer, Cham. https://doi.org/10.1007/978-3-031-13291-9_10

Rainbird, H., Fuller, A., & Munro, A. (Eds.). (2004, January 1). *Workplace Learning in Context*. Routledge. https://doi.org/10.1604/9780415316309

Smet, K., Grosemans, I., De Cuyper, N., & Kyndt, E. (2022). Outcomes of Informal Work- Related Learning Behaviours: A Systematic Literature Review. *Scandinavian Journal of Work and Organizational Psychology,* 7(1): 2, 1–18. DOI: *https://doi. org/10.16993/sjwop.151*

ANALYSES OF THE BOEING 737MAX ACCIDENTS: FORMAL MODELS AND PSYCHOLOGICAL PERSPECTIVES

Immanuel Barshi
Human Systems Integration Division
NASA Ames Research Center, California
Asaf Degani
GM Advance Technology Center
Hertzeliya, Israel
Robert Mauro
Decision Research and University of Oregon
Eugene, Oregon
Randall J. Mumaw
San Jose State University Research Foundation
NASA Ames Research Center, California

Two fatal accidents involving the B737MAX resulted from the flight crews' inability to overcome the effects of the Maneuvering Characteristics Augmentation System (MCAS). MCAS was designed to mimic the control column feel pressure and pitching behavior of the B737NG, which was the certification basis for the B737MAX. We briefly describe the potential role of formally modeling different perspectives during system design, and how such modeling can reveal gaps and conflicts between perspectives. We also discuss some of the relevant human factors issues involved in these accidents and how the aircraft's behavior may have affected the pilots' psychological states. Implications for automation design are considered.

The Boeing 737 has been the most successful airliner model in the history of aviation. At any given moment, there are more B737s flying in the world than any other aircraft. In spite of its enviable safety record, the two fatal accidents of the B737MAX-8, one in Indonesia in October of 2018 (KNKT, 2019) and the second in Ethiopia in March of 2019 (EAAIB, 2022) shook the world and led to the unprecedented world-wide grounding of the MAX fleet.

In both accidents, the pilots were unable to understand what was happening to their aircraft. Although MCAS, the Maneuvering Characteristics Augmentation System, has been a major focus of numerous discussions of these accidents, the confusion that rendered the pilots unable to successfully diagnose and remedy the problems began before MCAS was activated. To understand why these accidents happened, one must consider the situations the pilots encountered from the pilots' perspectives.

The B737 was originally designed, in the 1960s, as a "federated" system. Separate and redundant aircraft avionics, on the right and left sides of the aircraft, supply data to a corresponding set of flight displays; left side for the Captain and right side for the First Officer. Although some comparators were added as the aircraft evolved, the fundamental federated design concept remained. Similarly, each side has separate flight control computers. Should a problem occur on one side, control can be transferred to the other side's computers and safely continue the flight. However, in this design, the burden is on the flightcrew to communicate about what is happening on each side, to determine which set of equipment is functioning properly. Accidents can occur when the flightcrew fails to understand which side has failed.

Formal Modeling

Every human-machine system can be viewed from different perspectives. These different perspectives can be characterized as "models," including the human's mental model of how an aircraft and its systems work (Degani et al., 2022). The design begins with a "conceptual model" that exists in the mind of the designer(s). This model—not necessarily fully detailed, accurate, or complete—portrays the thinking behind the system and is the vital first step. Next is the "machine model" which concretizes how the design team understands the conceptual model. The machine model is not necessarily complete, but the "system dynamics model" incorporates how the system works in its operational environment and is verified using system engineering tools and flight simulators. Other models such as requirement/specification models and software implementation models may also be produced.

An "interface model" represents the information the user is expected to need to operate the system. Thus, it necessarily abstracts the detailed behavior behind indications seen on displays; it is augmented with aircraft manual and training informaiton. The next model is called the "user model." This model, an abstracted version of the interface model, characterizes an individual user's "mental" model of the system and its workings. The user model is based on the information obtained and the user's understanding of it. It can decay with time and lack of recurrent experience. User models are also subject to degradation and loss due to fatigue and stress. Common examples of such degradation are cognitive tunneling and inattentional blindness (Levin & Baker, 2015), in which the user's attention is focused on one thing and ignores other potentially relevant data.

Formally modeling these different perspectives allows for the identification of gaps and potential conflicts between models. Such gaps and conflicts could lead users to confusion and to mistakes. Early research on pilot interactions with cockpit automation showed that the inability to understand what the automation was doing constituted the most critical concern (Billings, 1997; Parasuraman, et al., 2000). Cockpit observations by Earl Wiener when automated flight control systems were first introduced into commercial aviation showed that pilots wanted answers to four key questions: "what's it doing now, why is it doing it, how did I get here, and what will it do next" (Wiener & Curry, 1989). Albeit somewhat colloquial in nature, these types of questions are still being asked in modern-day cockpits.

For the sake of brevity, we focus here on one model and one accident (for a detailed analysis, see Barshi et al., 2023). We focus on the way in which a description of the machine model can expose a conflict with the user model. Exposing these conflicts while the aircraft is being designed could lead to solutions that could mitigate the risks associated with such conflicts. For a discussion of the pilots' experience, we focus on the first accident, Lion Air flight 610, because that crew did not know about MCAS (KNKT, 2019). The crew of the second accident, Ethiopian Airlines flight 302 supposedly knew about MCAS and was refreshed in its training of the proper procedure to disable it (EAAIB, 2022).

Figure 1 below presents a simplified version of a small portion of the machine model of the electric pitch trim system of the B737MAX (for a detailed analysis, see Barshi, et al. 2023). This system controls the movement of the horizontal stabilizer to trim the pitch attitude of the aircraft and includes a manual and an electric activation. The manual activation is performed using a hand-operated wheel in the cockpit that is physically connected with cables to the stabilizer and allows the flight crew to directly control the movements of the stabilizer. Electric activation is performed using an electric motor that can receive commands from the flight crew by use of thumb switches mounted on each yoke. The electric motor can also receive commands from the flight control computer, which houses three components that can activate the trim: the autopilot, the speed trim system, and MCAS (NTSB, 2019). The stabilizer can be moved to trim the aircraft nose up (ANU) or aircraft nose down (AND). The trim is used to relieve pressures from the control column for any given pitch attitude, power setting, and speed.

The stabilizer can be moved by automated systems (the autopilot and the speed trim system) that can fail and cause a runaway trim situation where the aircraft is forced into a dangerous pitch attitude (either too high leading to a stall, or too low leading to a dive). To stop a runaway trim, a mechanism is installed under the cockpit floor (known as the *floor switch*), at the base of the control column, that disengages the electric trim motor in case the column is moved in a direction opposite to the movement of the trim. For instance, if the autopilot fails and causes an excessive nose-up trim, pushing the control column forward stops the motor from moving the stabilizer. However, because MCAS is designed to produce forward pressure on the control column when the pilot is pulling the control column back, the floor switch is disabled when MCAS is active (MCAS_input = true in Fig. 1), leaving MCAS free to continue moving the stabilizer in an AND direction (NTSB, 2019).

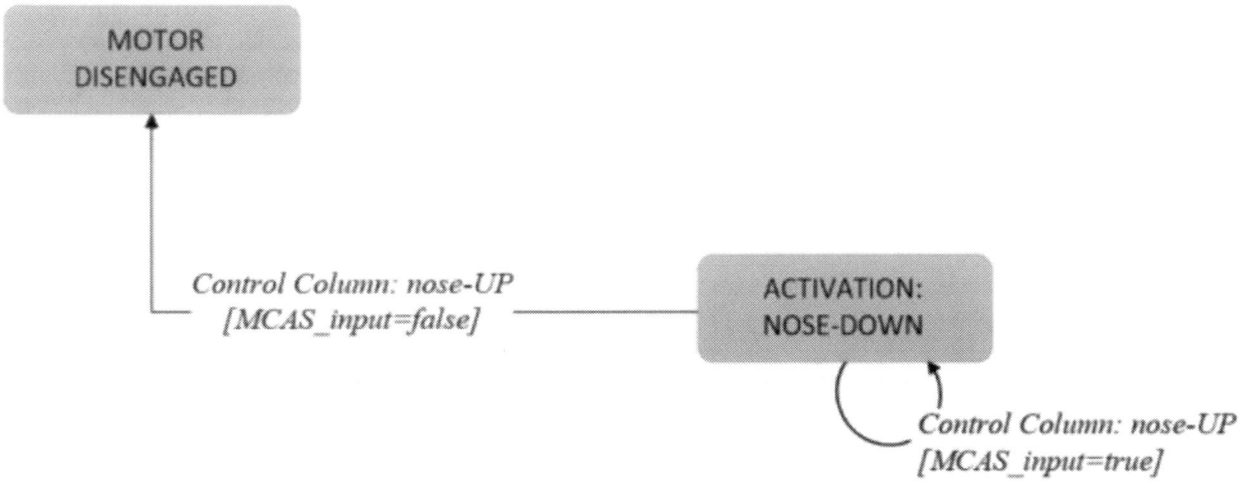

Figure 1. Modeling the behavior of the column-activated floor switch.

Presenting the machine model as seen in Figure 1 shows that the design creates a situation that to the pilot would appear *non-deterministic*; the pilot cannot predict the behavior of the system, even if it is completely predictable to the designer. If the autopilot is trimming AND, pulling back on the yoke stops the movement. If MCAS is trimming AND, pulling back on the yoke does nothing. Since the pilot may not know which system is causing the AND runaway trim, the pilot cannot predict whether pulling back on the controls is going to help or not. From the pilot's perspective the behavior of the system is unpredictable, and the pilot cannot answer the question of "what is it doing now?" nor the question of "what will it do next?" Thus, presenting the machine model, as in Figure 1, reveals this apparent non-determinism and provides the designers with an opportunity to develop a mitigation to resolve it.

This apparent non-determinism falls under the category of "mode error" (Woods et al., 1997; Sarter & Woods, 1995); the user is unable to determine what mode the system is in, and to predict the implications for ongoing control of the aircraft. Mode errors resulted in a long series of automation-related accidents starting in the 1980s (Mumaw, 2021). The method described here of detailing the different models involved in the human-machine system can expose specific gaps between the machine model and the user model and thus lead to specific design solutions.

Psychological Perspective

In the case of Lion Air flight 610 (KNKT, 2019), the problems on the flightdeck started as the aircraft began to rotate for its takeoff. At this point, the Captain's "stick shaker" activated. The stick shaker is designed to warn pilots that the aircraft is about to stall and cease flying. At any altitude, it demands immediate attention. When the aircraft is only a few feet above the ground, it warns of an impending disaster. The immediate emotional response is fear. The concomitant psychological and physiological changes would reduce the pilots' functional working memory, making it difficult to reason through complex problems (Davies & Parasuraman, 1982; Moran, 2016).

At this point in the takeoff, there is likely insufficient runway left to put the aircraft safely back on the ground. An attempt to abort the takeoff would likely cause the aircraft to overrun the runway, seriously damaging the aircraft and probably causing injuries to the passengers. Pilots are trained to continue the takeoff at this point. But if the aircraft stalls when near the ground, there is insufficient altitude to recover, and the aircraft will crash. The immediate actions that pilots must take when the stall warning is activated are well-rehearsed. Lower the nose and add power to break the stall. In the simulator, during training, a successful stall recovery is defined in part by a minimum loss of altitude. At takeoff, the aircraft is near full thrust, so there is little power to add. The aircraft is also so low that there is little altitude to lose. Thoughts of the Northwest 255 (NTSB, 1988) and Spanair 5022 (CIAIAC, 2011) accidents may jump to mind. In these accidents, the aircraft was improperly configured for takeoff and crashed immediately thereafter. One might expect that the pilots of Lion Air 610 immediately checked that their aircraft was properly configured. A glance at the cockpit indications would confirm that it was and furthermore the aircraft was flying and gaining altitude. Meanwhile, the stick shaker continued to shake the Captain's yoke and arms while making a loud racket. The only option at this point is to try to gain additional altitude and diagnose the problem or at least determine that the aircraft would be able to return to the airport and land safely.

There is no explicit indication from the Cockpit Voice Recorder (CVR) that either pilot noticed that only the Captain's stick shaker was activated, but it is hard to ignore. There could be a malfunction with the equipment on one side of the aircraft, but which side and what is the nature of the malfunction? Other alerts appeared seconds later: altitude disagree, airspeed disagree, and feel pressure differential. All indicate that some of the information calculated by the right-hand computers disagreed with the information calculated by the left-hand computers, but again, the burden in this federated architecture is on the flightcrew to determine which side is correct.

The immediate inference that one could make from these alerts is that something is seriously amiss with the aircraft systems. Airspeed is directly relevant to the potential stall problem, so determining which airspeed display (left or right) is correct would be a high priority after maintaining control of the aircraft. So the Captain called for the First Officer (FO) to carry out the memory items for the "Unreliable Airspeed" non-normal checklist. The FO failed to respond. A short while later, the Captain called for the checklist itself. The First Officer had trouble locating the checklist. These problems are likely symptoms of substantial stress and anxiety (Maloney et al., 2014; Moran, 2016). It is very unusual for the stick shaker to operate continuously. In normal operations, it rarely activates; when it does, it is only active momentarily, ceasing when the triggering condition is corrected. In addition to the noise, the constant reminder that the aircraft could cease flying at any moment could take a toll on the crew.

After the Unreliable Airspeed checklist was located it could have been used to effectively troubleshoot, locate the reliable airspeed indicator, and determine that the aircraft was not in danger of stalling. But at this point, the Captain asked the FO to request an Air Traffic Control (ATC) clearance to a holding point; he was likely looking for a safe space to troubleshoot the problem. This may indicate that the Captain had concluded that the aircraft was not in imminent danger of stalling and that he wanted to confirm that the aircraft could be safely operated before attempting to land, but this can't be confirmed from the CVR transcript available in the accident investigation report.

The FO complied and also suggested raising the flaps from 5 to 1. This action would be in line with normal procedures after takeoff but would have been a possible problem due to the loss of lift if the aircraft were on the verge of a stall. The Captain's agreement might further indicate that he thought the aircraft was not about to stall, despite the stick shaker.

The Captain then requested that the FO take over the controls, perhaps to allow him to be free to troubleshoot. The FO replied for him to standby and suggested raising the flaps the rest of the way.

Avoiding taking control of the aircraft and sticking to the normal procedures in a non-normal situation might be additional signs of narrowed attention and reduced cognitive functioning (Maloney et al., 2014; Moran, 2016). The Captain agreed to retract the flaps. Unbeknownst to the crew, this action armed MCAS. The MCAS program, like other programs running on the left-side aircraft computers were receiving and relying upon erroneous angle of attack information from the angle of attack sensor on the left side of the aircraft. The stick shaker and the various alerts were all symptoms of this malfunction.

Shortly thereafter, MCAS began to exert downward pressure on the controls through inputs to the stabilizer pitch angle. As Figure 1 shows, just pulling back on the controls was not going to stop MCAS. The Captain ordered a return to flap 1 and retrimmed, countering the effects of the previous MCAS command. Had the flaps remained deployed, MCAS would not have reactivated, and the flight could have landed safely.

But the flaps were raised again. The CVR transcript provided in the accident report (KNKT, 2019) does not include any discussion or commands to raise the flaps. There is no evidence that the airspeed unreliable checklist was ever completed. Yet, the Captain maintained appropriate pitch with trim, using the thumb switch, stopping MCAS and compensating for MCAS initiated nose down trim. Yet, he never verbalized what he was doing, and the FO may have had no awareness of these actions and the Captain's struggles. Perhaps, he was not completely conscious of it. With his hands shaking throughout the flight from the stick shaker, he might not have been fully aware of the forward pressure on the control column. In any case, the Captain managed to return the horizontal stabilizer to its climbing trimmed angle following each MCAS activation, and thus kept the aircraft flying safely.

While maneuvering for a return to a landing, and after 21 successive MCAS activations, but without making any reference to the extensive use of the trim, the Captain asked the FO again to take control of the aircraft. He might have wanted to take a break to prepare for the landing or do the trouble-shooting that the FO had been unable to conduct. He might have been saturated. When control was transferred, the aircraft was properly trimmed and flying. But the flaps were up and MCAS activated. The floor switch was disabled, the FO failed to compensate sufficiently with the use of the yoke-mounted thumb switch and eventually the MCAS' AND inputs overwhelmed the FO's attempts to manage pitch, leading to an unrecovered dive into the water.

Conclusion

Evaluating formal models during the design of a system can help identify gaps between models, such as the gap of apparent non-determinism between the machine model and the user model described above. Such a gap could be made visible to the crew, for instance, through a salient mode annunciation, alerting, or through education. Furthermore, understanding the operational context of use and some of the psychological aspects of the user can help elaborate the user model and possibly expose additional gaps, particularly between the user model and the interface model. Although the analysis presented here was done post-hoc, after the aircraft was already produced and after the accidents had occurred, the methodology can be applied during the design, testing and verification of systems and thus help prevent such accidents from happening again.

References

Barshi, I., Degani, A., Mauro, R., & Mumaw, R.J. (2023). *Analyses of the Boeing 737MAX accidents: Formal models and psychological perspectives*. Manuscript in preparation.

Billings, C. E. (1997). *Aviation Automation: The Search for a Human-Centered Approach. Mahwah*, NJ: Erlbaum.

CIAAIC (2011). Accident involving a McDonnell Douglas DC-9-82 (MD-82) aircraft registration EC-HFP, operaed by Spanair, at Madrid-Barajas Airport, on August 2008. Comision de Investigacion de Accidentes e Incidentes de Aviacion Civil, Report A-032/2008. Madrid, Spain.

Davies, D. R., & Parasuraman, R. (1982). *The Psychology of Vigilance*. London, Academic Press.

Degani, A., Shmueli, Y. & Bnaya, Z. (2022). Equilibrium of Control in Automated Vehicles: Driver Engagement Level and Automation Capability Levels. *Proceedings of the 4th IFAC Workshop on Cyber-Physical & Human-Systems*. Houston, TX: IFAC.

EAAIB (2022). Investigation Report on Accident to the B737- MAX 8, ET-AVJ, 10 March 2019. Ethiopian Aircraft Accident Investigation Bureau, Report No. AI-01/19. Addis Ababa, Ethiopia.

KNKT (2019). Aircraft Accident Investigation Report. PT. Lion Airlines Boeing 737 (MAX); PK-LQP Tanjung Karawang, West Java, Republic of Indonesia 29 October 2018. Komite Nasional Keselamatan Transportasi (National Transportation Safety Committee), Report No. 18.10.35.04. Jakarta, Indonesia.

Levin, D. & Baker, L. (2015). Change blindness and inattentional blindness. In. J. Fawcett et. Al. The Handbook of Attention, MIT Press.

Maloney, E. A., Sattizahn, J. R., and Beilock, S. L. (2014). Anxiety and cognition. *Wiley Interdisciplinary Reviews Cognitive Science, 5*, 403–411. doi: 10.1002/wcs.1299

Moran, T. P. (2016). Anxiety and working memory capacity: a meta-analysis and narrative review. *Psychological Bulletin, 142*, 831–864. doi: 10.1037/bul0000051

Mumaw, R.J. (2021). Plan B for eliminating mode confusion: An interpreter display. *International Journal of Human-Computer Interaction, 37* (6).

NTSB (1988). Aircraft Accident Report: Northwest Airlines, Inc. McDonnell Douglas DC-9-82, N312RC, Detroit Metropolitan Wayne County Airport, Romulus Michigan, August 16, 1987. National Transportation Safety Board. Washington, DC.

NTSB (2019). System Safety and Certification Specialist's Report. National Transportation Safety Board, NTSB ID No.: DCA19RA017. Washington, DC.

Parasuraman, R., Sheridan, T.B., & Wickens, C.D. (2000). A model for types and levels of human interaction with automation. *IEEE Transactions on Systems, Man, and Cybernetics, 30* (3), 286-297.

Sarter, N.B. & Woods, D.D. (1995). "How in the World Did We Ever Get into That Mode?" Mode Error and Awareness in Supervisory Control. *Human Factors, 37*, pp. 5-19.

Wiener, E. L., & Curry, R. E. (1989). Cockpit Automation and Crew Coordination. NASA Contractor Report 196099. NASA Ames Research Center, Moffett Field, CA

Woods, D., Sarter, N., & Billings, C. (1997). Automation surprises. In G. Salvendy (Ed.), *Handbook of Human Factors and Ergonomics*, pp. 1926-1943. New York: John Wiley.

EXTRACTING LESSONS OF RESILIENCE USING MACHINE MINING OF THE ASRS DATABASE

Immanuel Barshi
Human Systems Integration Division
NASA Ames Research Center, California
Bryan Matthews
KBR Inc.
NASA Ames Research Center, California
Jolene Feldman
Human Systems Integration Division
NASA Ames Research Center, California

NASA's Aviation Safety Reporting System (ASRS) database is the world's largest repository of voluntary, confidential safety information provided by aviation's frontline personnel. The database contains close to 2 million narratives, many of which describe everyday situations in which people saved the day. In these situations, people's resilient behavior solved a problem, dealt with a malfunction, and maintained a safe operation despite a serious perturbation. To be able to extract lessons of such resilience from this large database, the use of machine learning algorithms is being explored. In this report, we describe a comparison between two such algorithms: BERT and Word2Vec. An identical search using both programs was done on a database containing approximately 270,000 ASRS reports. The comparison reveals some of the strength and weaknesses of each algorithm as well as the challenges inherent in using such algorithms to extract lessons of resilience from the ASRS database.

Aviation safety is often examined in terms of errors leading to incidents and accidents. There is much to be learned from such events, but these events represent an extremely small portion of flight operations. In the vast majority of commercial aviation operations, all goes well in spite of various perturbations. Moreover, in the vast majority of inflight malfunctions of any sort, the crew is able to solve the problem and complete the flight safely. Such resilience as demonstrated in everyday operations can also be a source of much learning.

Learning how to be resilient from what goes well has not been part of common flight training programs. As a result, there are no established methodologies to collect relevant data and to extract relevant lessons. Yet, the aviation industry collects vast amounts of data, especially with the advent of Safety Management Systems (SMS). Thus, it behooves us to make the most of existing sources of data for this purpose of learning resilience from what goes well. One such existing source of data is NASA's Aviation Safety Reporting System (ASRS).

The ASRS database is the world's largest repository of voluntary, confidential safety information provided by aviation's frontline personnel, including pilots, air traffic controllers, mechanics, flight attendants, dispatchers, and other members of the aviation community and the public. The database contains close to 2 million narratives, many of which describe everyday situations in which people saved the day. In these situations, people's resilient behavior solved a problem, dealt with a malfunction, and maintained a safe operation despite a serious perturbation. Hence, these narratives provide a rich source of potential lessons about being resilient in the face of adversity. But the database is very large and extracting such lessons can be challenging.

The online ASRS database has search tools built in. Airline-based Aviation Safety Action Partnership (ASAP) programs which are modeled after the ASRS also have such tools. These databases can be searched in a variety of ways, but depending on the size of the database and the search terms used, searches may yield a voluminous number of reports, well beyond the ability of a human analyst. Feldman et al. (2021) note that "key word searches can be informative, but can fail to detect resilient behaviors that are not specifically named." Moreover, what would count as 'resilient behavior' is context-dependent; the very same action can be resilient in one situation and catastrophic in another. For instance, an old aviation adage says that the first thing to do in an emergency is to "wind the clock" (it goes back to the times when a mechanical clock requiring winding was installed in the cockpit) to allow the pilot a moment to pull back from the situation and think slowly to properly identify the problem. In many in-flight emergencies, taking a moment to think rather than react immediately can be life-saving. However, some situations such as a rejected takeoff in case of a power failure prior to V1 do require an immediate response and winding the clock at that moment can be life-ending.

Furthermore, resilience is often implicit in narrative reports and cannot be easily identified by keywords or key phrases. To go beyond keywords or phrases, the ASRS search tools allow the use of codes (e.g., ASRS coding taxonomy), and various filters. Thus, it is possible to limit the search to particular situations or events of interest (e.g., Chandra et al., 2020). Beyond such searches, advanced software tools are needed (Paradis et al., 2021).

The first such software tool, specifically designed to support searches of the ASRS database, was Perilog (McGreevy, 2005). Developed by Michael McGreevy at the NASA Ames Research Center, home of ASRS, Perilog is still one of the best text mining tools for studying the ASRS narratives. Functions such as "key word search," "phrase search," and "search by example" make Perilog an excellent search tool, whereas functions such as "review vocabulary," "review phrases," and "phrase generation" make Perilog an exciting discovery tool. One of the unique features of Perilog is the "search by example" function, in which an ASRS report, or any text of any size, can be used as the "search term."

Analysts and researchers may want to search the ASRS database to find information about a particular type of event (e.g., automation surprises or unstable approaches), a particular phase of flight (e.g., descent or approach), or a particular type of operations (e.g., general aviation of scheduled airlines). Operators may have different needs. An airline's safety officer may come across an ASAP report and want to know if there are similar reports in the database. The similarity might be in terms of the particular event at hand, a particular piece of equipment, or something about the circumstances leading to the reported event. Likewise, identifying a particular resilient strategy in a report can serve as the basis for finding the use of that strategy under different circumstances, or for finding different strategies that can be used under similar circumstances. Thus, being able to search the database using a report as an example can be very useful.

Below, we describe a comparison between two new algorithms: Word2Vec and Bidirectional Encoder Representations from Transformers (BERT). An identical search-by-example using both programs was done on a database containing approximately 270,000 ASRS reports submitted between 1988 and 2022. The comparison reveals some of the strength and weaknesses of each algorithm as well as the challenges inherent in using such algorithms to extract lessons of resilience from the ASRS database.

Method

Software Mining Tools

The Word2Vec algorithm (Mikolov et al., 2013) is a natural language processing (NLP) algorithm used to model term similarity between two words in a multi-dimensional embedding space. The algorithm accomplishes this by training a neural network to learn word associations. The algorithm is unsupervised, meaning that no labels are provided by a subject matter expert to train the model. There are two approaches to learning the word associations: 1) using a continuous bag of words (CBOW), and 2) skip-gram. CBOW uses the surrounding words to predict the probability of a target word in the middle of a window. Windows are typically 3 or 5 words in length. The skip-gram method is the inverse task, namely, learning to predict the probability of surrounding words from the target word. Neither method considers word ordering other than the target word being in the center of the window. Both methods use the same neural network architecture with a fully connected neural network layer of input size equal to the entire term corpus mapped to a 300-dimension hidden layer. The final layer maps the hidden layer to the predicted output word space with a softmax activation function (Bridle, 1990) to convert the output to a classifier. A classifier model is used to learn word similarity becasue if the classifier can accurately predict the target word(s) with the contextual word(s), then the embedding space is presumed to be well organized by term similarity. Common NLP techniques such as removing stopwords and stemming are applied to all the ASRS reports before training. Stopwords are typically pronouns, articles, prepositions, and other words that do not add significant value to the text's meaning but often indicate grammatical relations. Stemming involves the use of stem-words for the different forms words can take such as using "friend" for friends, friendly, and friendship. For our work, we have been using Python's NLTK package's 'english' stopword list (Loper & Bird, 2002). Once the model is learned, the hidden layer can be leveraged to extract word associations. Each word is mapped into the embedding space and word similarities can be computed using the cosine similarity function between any two word vectors. Term Frequency-Inverse Document Frequency (TF-IDF) weighting, a commonly applied technique, is also applied to the embedding vector for each of the words. This weighting approach attempts to de-emphasize words that appear across a majority of the documents (and therefore their presence is less informative than infrequent terms) while boosting terms that occur frequently within a document. For example, if a term appears multiple times in a report and is very uncommon across the rest of the reports then its weighting is high. The inverse is true for common words that appear in both the report and the rest of the dataset. An entire report embedding can be represented by computing the average word embedding across the report with TD-IDF weighing. This vector representation allows comparisons among reports.

Bidirectional Encoder Represenations from Transformers (BERT) algorithm (Devlin et al., 2018) can also perform this task. Using the same concept as Word2Vec, the BERT algorithm maps a report into an embedding space. Similar to Word2Vec, the algorithm's architecture is based on a neural network; however, the BERT network is much deeper than Word2Vec with 12 fully connected multi-headed self-attention layers (Vaswani et al., 2017) with a hidden layer of 768 dimensions. The self-attention layers capture the bi-directional context of a word, using the words prior to as well as following a target word to predict the output sequence of words. Another difference from Word2Vec is that the embedding dimensions are applied to an entire sentence and not at the word level. The entire report embedding vector is obtained by calculating the average sentence embedding. Cosine similarity is also used as the similarity metric for ranking reports against the query. The pretrained Microsoft *mpnet* (Song et al., 2020) model with fine tuning on an additional 1.17B data tuples was used to perform the sentence embedding. This open-sourced model is publicly available (Espejel, 2021).

Runway Safety Narrative

Runway safety has been a high-priority safety concern in flight operations at an international level (ICAO, 2017). The Flight Safety Foundation launched a Global Action Plan for the Prevention of Runway Excursions (GAPPRE; FSF, 2021), and is currently engaged in launching a similar Global Action Plan for the Prevention of Runway Incursions (GAPPRI; FSF, personal communication). Given these efforts, our search of the ASRS database focused on issues related to runway incursions.

Rather than search the database for a report of a runway incursion, a narrative was drafted to be used in a search-by-example. Writing up such a narrative can be a very productive approach to mining the database. An airline's Safety Officer, or a safety researcher can imagine a situation of interest and write up a narrative as if experiencing the situation and writing an ASRS or an airline-internal ASAP report about it. Writing up such a narrative presents an opportunity to fashion the report along the specific aspects of interest. Moreover, different reporters often use different words and phrases to describe similar and even identical situations. Writing up a narrative allows the researcher to use multiple phrases and styles within a single report to increase the likelihood of finding relevant reports in the database.

Search-By-Example Process

For both Word2Vec and BERT any text/report can be used to query for similar reports. The Word2Vec process involves stemming and dropping stop words. Then each remaining word in the report is mapped to the model's 300-dimension embedding vector space and the TF-IDF weight for that word is applied to that vector. This process is repeated for all words in the report and the average embedding vector is calculated. Prior to the query, this process was applied across all reports in the sample ASRS database to calculate each report's average embedding vector. The cosine similarity function was used to compute the angle of similarity between the query report's average embedding vector and each of the ASRS report's average embedding vectors. This approach allows queries to be agnostic of report length and therefore a query can be a single term or an extensive report of any length. The process is similar for BERT, however stemming and stop word filters are not applied and instead of computing the average embedding vector across words, the average embedding vector is computed across sentences. The embedding vector for the large BERT model is 768-dimensions and the cosine similarity function is used to find the closest matches. With both algorithms, the top 10 most similar reports were analyzed.

Two search runs were employed. In the first run, the sample writeup was used to search for similar reports in the sample ASRS dataset. In the second run, the top most similar report from each search was used in a second round of searches. Thus, 38 reports in all were analyzed for similarity with the initial sample writeup.

Results

The top 10 most similar ASRS reports to the written-up runway incursion narrative produced by the Word2Vec algorithm were all related in some way to runway safety issues. Not all reports deemed similar were of the same type of operation; the written-up narrative described an airline operation and some of the reports found involved a general aviation operation. Furthermore, not all reports involved a runway incursion; some involved landing at the wrong airport, a takeoff without a clearance, or being stuck on a taxiway. However, they all did have sufficient similarity to be of potential interest.

Only 2 of the top 10 most similar ASRS reports to the written-up runway incursion narrative produced by the BERT algorithm were related in some way to runway safety issues. Most of the reports involved an in-flight anomaly. What's more, none of the reports produced by BERT were also in the top 10 most similar reports produced by Word2Vec.

Because the top most similar report produced in the first search was used to drive the second search, and because this report was an ASRS report, as expected, both algorithms returned the same report as the most similar to the example used for the search. Of the additional 9 reports returned by Word2Vec, only 2 were also among the 10 reports produced in the first search. Most reports involved surface operations though not necessarily a runway incursion. Similarly, only 2 of the 9 most similar reports returned by the second search in BERT were among the reports produced in the first search. Most of the reports returned by BERT involved in-flight anomalies.

The most striking similarity across all 38 reports was their length. The average length of the narratives in the sample ASRS database used in this study was 230 words (with a median of 191 words). The written-up runway incursion narrative used as the example in the first search had 763 words. The average number of words in the 10 most similar reports returned by Word2Vec was 668 words, and 813 words as the average for the narratives returned by BERT. The second search with Word2Vec returned an average length of 946 words, whereas BERT returned an average narrative length of 847 words. All these reports are significantly longer than the vast majority of reports in the database, and certainly longer than the average report length in the database (see Fig. 1 below).

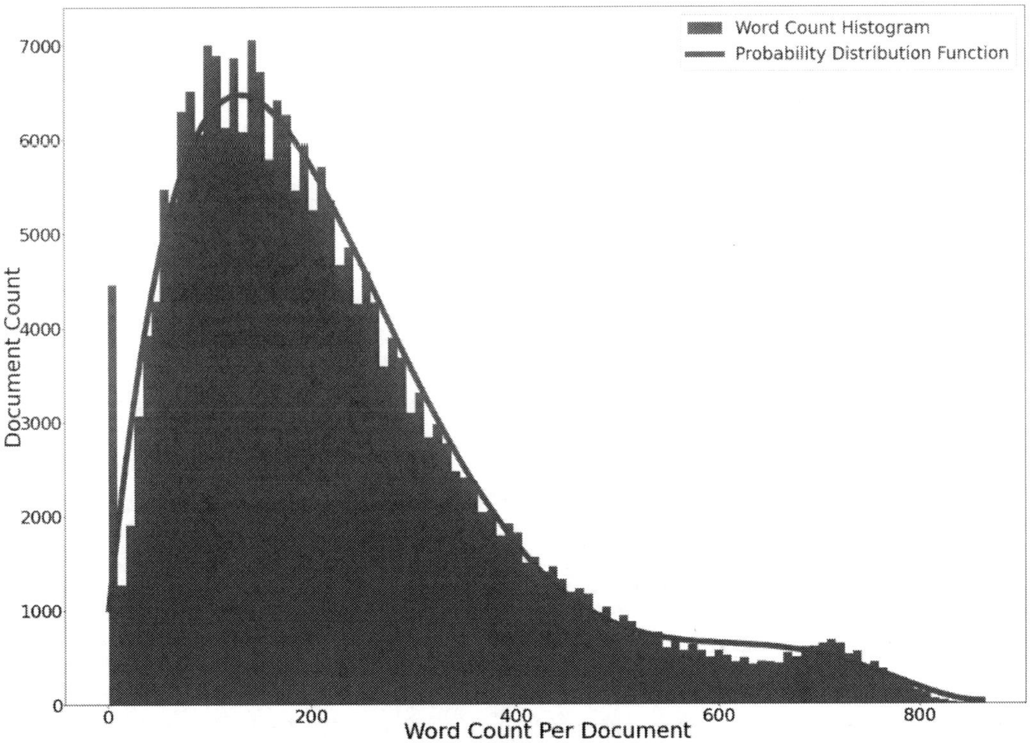

Figure 1. Distribution of narrative (document) length, in terms of number of words, in the ASRS database used in this study.

Discussion

The longer the narrative, the richer it is. Rich ASRS reports typically include much detail about the circumstances involved, including various pressures the crew might have been under such as fatigue or schedule constraints. Thus, the richer the report, the more elements in it could be used by an algorithm to determine similarity independent of the specific event or malfunction at the core of the event. These similarities are of potential interest as they could involve similar resilient strategies. However, that determination is left to the human analyst. Similarly, because of the high context-dependency of determining any behavior as "resilient," as discussed above, current search algorithms may not be sensitive enough to support the extraction of lessons of resilience from a narrative database such as ASRS. These algorithms can help narrow the search to some extent and thus allow the human analyst to focus on the most relevant reports to one's interest. But that relevancy too must be examined as a report could be deemed "similar" based on parameters outside the analyst's interest.

The difference in relevancy to the initial runway incursion narrative writeup between the two algorithms might be explained in part by the different texts used in their initial training. An algorithm trained on a database of newspaper articles or scientific articles might return very different results from those returned by an algorithm trained on social media posts. Thus, when choosing to use an algorithm in the analysis of narrative texts, one must be mindful of the database used in the training of the algorithm, and ensure that the vocabulary, grammatical structures, and language style are appropriate to the texts to be analyzed.

Acknowledgements

The work reported here was funded by NASA's Human Contribution to Safety part of the System-Wide Safety Project, of the Aeronautics Research Mission Directorate's Aviation Operations and Safety Program.

References

Aviation Safety Reporting System: ASRS Database Online. (2023). https://asrs.arc.nasa.gov/search/database.html. Accessed April 16, 2023.

Bridle, J. S. (1990). Probabilistic Interpretation of Feedforward Classification Network Outputs, with Relationships to Statistical Pattern Recognition. In: Soulié, F.F., Hérault, J. (eds) *Neurocomputing*. NATO ASI Series, vol 68. Springer, Berlin, Heidelberg. Retrieved from https://doi.org/10.1007/978-3-642-76153-9_28

Chandra, D., Sparko, A., Kendra, A., & Kochan, J. (2020). *Operational complexity in performance-based Navigation (PBN) arrival and approach instrument flight procedures (IFPs).* Retrieved from https://rosap.ntl.bts.gov/view/dot/43835

Devlin, J., Chang, M.-W., Lee, K., & Toutanova, K. (2018). *BERT: Pre-training of Deep Bidirectional Transformers for Language Understanding.* Retrieved from https://arxiv.org/abs/1810.04805.

Espejel, O. (2022). *Train and fine-tune sentence transformers models* [Data model]. Retrieved from https://huggingface.co/sentence-transformers/all-mpnet-base-v2/blob/main/README.md

Feldman, J., Barshi, I., Smith, B., & Matthews, B. (2021). Reports of resilient performance: Investigating operators' descriptions of safety-producing behaviors in the ASRS. In *Proceedings of the 2021 International Symposium on Aviation Psychology* (pp. 122-127).

Flight Safety Foundation. (2021). *Global action plan for the prevention of runway excursions.* Retrieved from https://flightsafety.org/wp-content/uploads/2021/05/GAPPRE-Parts-1-2-2021-FINAL.pdf

International Civil Aviation Organization. (2017). *Runway safety program – Global runway safety action plan.* Retrieved from https://www.icao.int/safety/RunwaySafety/Documents%20and%20Toolkits/GRSAP_Final_Edition01_2017-11-27.pdf

Loper, E., & Bird, S. (2002). *Nltk: The natural language toolkit.* Retrieved from *https://arxiv.org/abs/cs/0205028*

McGreevy, M. W. (2005). *Perilog text mining methods and software.* Moffett Field, CA: NASA Ames Research Center.

Mikolov, T., Chen, K., Corrado, G., & Dean, J. (2013). *Efficient Estimation of Word Representations in Vector Space.* Retrieved from https://doi.org/10.48550/ARXIV.1301.3781

Paradis, C., Kazman, R., Davies, M. D., & Hooey, B. L. (2021). *Augmenting topic finding in the NASA Aviation Safety Reporting System using topic modeling.* AIAA SciTech Forum. Retrieved from https://arc.aiaa.org/doi/abs/10.2514/6.2021-1981

Song, K., Tan, X., Qin, T., Lu, J., & Liu, T.-Y. (2020). *MPNet: Masked and Permuted Pre-training for Language Understanding.* Retrieved from https://arxiv.org/abs/2004.09297.

Vaswani, A., Shazeer, N., Parmar, N., Uszkoreit, J., Jones, L., Gomez, A. N., Kaiser, L., & Polosukhin, I. (2017). *Attention Is All You Need.* Retrieved from https://arxiv.org/abs/1706.03762.

A COMPETENCY FRAMEWORK FOR AVIATION PSYCHOLOGISTS AND HUMAN FACTORS SPECIALISTS IN AVIATION

Sonja Biede
Airbus Operations
Toulouse, France
Frederic Detaille
Belgian Air Force, Aviation Safety Directorate
Beauvechain, Belgium
OptimHum
Beloeil, Belgium
Tina Narotra (Balachandran)
Flourishing Minds Clinic
Lusail, Qatar
Katarina Petrovic
Aviology Inc.
Newark, United States
Hermann Rathje
formerly Eurocontrol
Brussels, Belgium
Alessandra Rea
IT-APA
Rome, Italy
Michaela Schwarz
formerly European Association for Aviation Psychology
Maarssen, The Netherlands
Anna Vereker
Civil Aviation Authority, Aviation House
Gatwick, West Sussex, United Kingdom

The new Commission Regulation (EU) 2018/1042 on technical requirements and administrative procedures includes psychological assessment of flight crew which refers to accredited Aviation Psychologists (AVPSY). However, the associated acceptable meansof compliance and guidance material does not provide a specific definition or requirements of such psychologists. Likewise, the term Aviation Human Factors Specialist (AVHFS) has not yet been legally recognised in Europe. AVHFS lacks a European definition and no pan-European competency-based endorsement exists. This paper reports the development of a competency framework by members of the European Association for Aviation Psychology (EAAP) and intendeduse in accreditation for AVPSY and AVHFS. It was established based on a two-fold approach combining a review of relevant input documentation and structured interviews. A total of 47 stakeholders and subject matter experts were interviewed. The framework specifies essential key competencies for both AVPSY and AVHFS and differentiates related roles.

Following the Germanwings tragedy in 2015, the professions of Aviation Psychology and Human Factors have received more scrutiny from both governmental and aviation industry organisations. Commission Regulation (EU) 2018/1042 (EU, 2018), enacted in February 2021, now requires commercial operators to perform a 'Psychological Assessment' of flight crew *'conducted by a suitably qualified and accredited clinical psychologist with expertise and experience in aviation psychology'* during their selection (AMC2 MED.B.055 Mental health).

There is currently no homogenous definition or set of competencies which define Aviation Psychology or Aviation Human Factors across countries or aviation organisations. Therefore, the European Association for Aviation Psychology (EAAP) has developed a competency framework for application by European aviation organisations. These competencies will be used as part of a European accreditation as an 'Aviation Psychologist' (AVPSY) or 'Aviation Human Factors Specialist' (AVHFS) and will provide further weight and utility to the endorsement for practitioners.

Method

Overall, sixteen EAAP members from a variety of European and Non-European member countries and organisations volunteered to complete this development work. The method included a thorough review of five information sources relevant for AVPSY and AVHFS work and derivation of conclusions as input for a draft competency-based framework. The information sources consulted were international regulations, registration standards, university/ training curricula/ courses/programs, industry stakeholder views and job description and advertisements provided by accredited AVPSY and AVHFS. A summary of the main findings is included below. Detailed results including raw data as Microsoft Excel spreadsheets are available upon request from the authors. Conclusions drawn from the review are reported in the results section dedicated to describing the draft competency-based framework.

International Regulations and Standards

A survey of existing regulations, standards, and recommended worldwide practice referencing aviation psychology or human factors yielded a total of 90 aviation-related documents subsequently classified as the following:

- *Regulations* (N=26) from EASA, European Commission/ Union related to HF in aircraft design and maintenance,training, licensing, selection, and rostering of aviation personnel (e.g., flight crew, ATCOs, maintenance technicians), as well as occurrence reporting requirements related to HF (e.g., fatigue)
- *Recommendations/ Best practices* (N=50) from ICAO, EUROCONTROL, EASA, FAA, CAA UK, IFATCA andCANSO describing HF and Human Machine Interaction (HMI) principles, HF tools and psychometric methods or
  ′HF safety considerations′, (operational) performance, training, selection.
- *Standards* (N=14) concerning civil / military system or aircraft design (in engineering, safety management and /or system validation)

The review of international regulations indicated the broad spectrum of applications of work and organisation psychology and human factors 'principles', 'considerations' and 'concepts' in aviation. However, detailed specifications and definitions of the competence requirements or qualifications could not be found.

International Registration Standards

The review included registration standards and guidelines offered by professional associations (e.g., the British Psychological Society, EFPA, the Health and Care Professionals Council, the Chartered Institute of Ergonomics and Human Factors) and some national professional councils. Evidence demonstrates that certificates for psychologists and human factors in general do exist but not specifically in aviation-related areas of practice.

University and Training Curricula

A total of 35 university and training curricula in aviation psychology and human factors disciplines worldwidewere collected. Twenty-seven courses were classified as academic and completedby a bachelor's or master's degree. The reviewconfirmed that most degrees are labelled as Human Factors degrees, aiming at providing human factors knowledge and skills (no attitudes) including psychological concepts, but not solely dedicated to aviation. Worldwide only three professional associations provide a dedicated postgraduate aviation psychology curriculum (AAPA, 2020; AEPA, 2019; INMAE, 2021).

Job Descriptions and Job Advertisements

EAAP members and accredited aviation psychologists and human factors specialists were asked to submit their own job titles and associated job descriptions, tasks, and job requirements as input for the formulation of competencies. In total 19 job roles and descriptions were submitted, most of them having the term 'human factors (HF) or human performance (HP)' in the job title, rather than 'aviation psychologist'.

The submitted job roles and descriptions provided a definition of tasks and responsibilities but did not furtherdefine or specify associated competencies. It was also apparent that most jobs are labelled using 'HF or HP' rather than 'psychology'.

Industry and Stakeholder Views

A total of 94 stakeholders were identified via professional networks and contacted via email, with 47 ultimately participating. Interviews took place mostly by telephone, with some submissions made by email. This is a representative sample as it captures a global view from both European and non-European members. Participantsincluded 17 air/ground operators (with consulting and staff associations), nine research and academic institutions, eight from Defence, Authorities, or international groups, seven from industry (including consulting companies), and six from other safety critical industries such as rail and maritime.

Three questions were asked during the stakeholder interviews related to either occupational psychology or human factors roles:

- Does your organisation employ HFS/OP? If yes, how many? If no, do you see a need in the future? Why?
  - o If Yes: What is the role of HFS in your organisation?
- What key competencies (knowledge, skills, attitudes, experience) are expected from an OP/ HFS in the day-to-daywork in your organisation?
- If you were providing advice to a new OP/ HFS: What tricky things would you want them to know about inadvance so they can be of rapid help in your organisation?

After transcription of the interview data a qualitative content analysis was applied, and data was allocated within an initial draft framework.

Results

Competency Framework

The knowledge, skills, and attitudes (KSA) identified by this work were structured into five groups. These included aviation domain knowledge (e.g., global aviation systems, operations, regulations etc.), knowledge about humans, methodological approaches and tools (e.g., design/validation/diagnostics), application practice areas (e.g., selection, design, workforce management etc.), and common supporting skills and attitudes (e.g., critical thinking, reflective practice, communication, curiosity). Figure 1 contains the Venn-diagram showing the overarching framework.

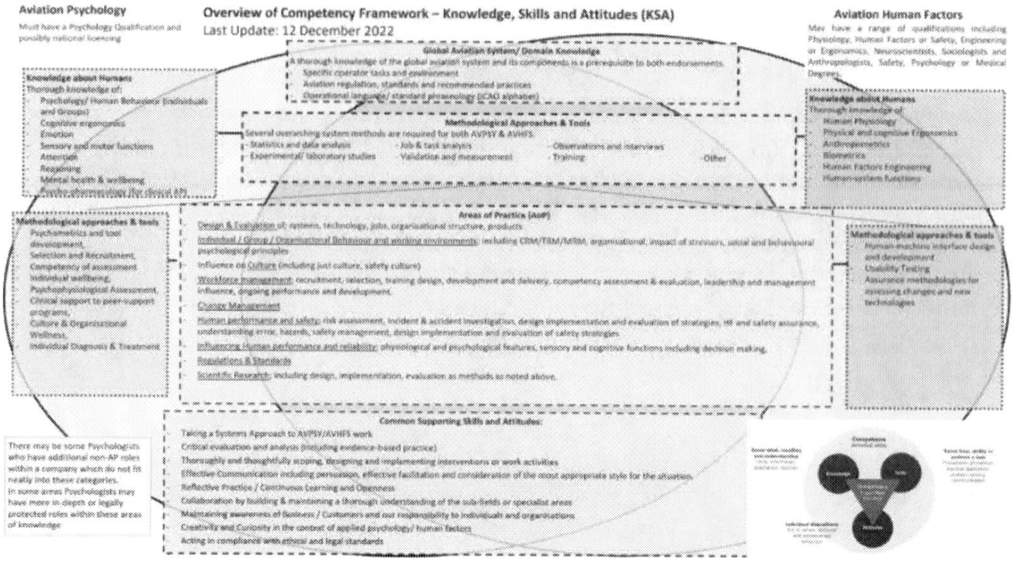

Figure 1. Competency Framework (Venn Diagram)

Global aviation system / domain knowledge. The selection of a Venn-diagram (overlapping circles) stems from the observation that both AVPSY and AVHFS may work on the design of functional systems, tools, technology, jobs, organisational structure, procedures, and products in aviation.

Knowledge about humans. Each professional group has a 'core area' of specific requirements either based on their academic training or European requirements for psychologists. For psychologists, education and knowledge about humans specifically cover human behaviour (individuals and groups), cognitive ergonomics, emotion, sensory and motor function, attention, reasoning, mental health; and psychopharmacology (for clinical AVPSY). AVHFS may have a range of qualifications including physiology, human factors and safety, engineering or ergonomics, neuroscience, sociology and anthropology, safety, psychology, or medical degrees.

Methodological approaches and tools. AVPSY have methods and approaches specific to psychologists; for the design of jobs / roles, an AVPSY may perform a job / task analysis by gathering and analysing data and information using specific usability testing methodsor employing psychometric tools such as personality or aptitude tests, which an AVHFS without psychological background would not be eligible to perform. AVPSY may also engage in a specialisation in the individual, group, and organisational behaviour and working environments, practice area relating to psychometric tool development anduse (especially in personnel selection and recruitment and workforce management by an occupational AVPSY or diagnosis of mental health issues by clinical AVPSY).

AVHFS have core methods and approaches including risk management from design to implementation, evaluation of risk and safety strategies to risk assessment (for example hazard analysis of routine operations or changesto a system), and ongoing aviation incident and accident investigation.

Areas of practice (AOP). The framework identifies nine areas of practice (AOP) in which both AVPSY and AVHFS may engage. These included: design and evaluation; individual / group / organisational behaviour and working environments; culture; workforce management; change management; human performance and safety; influencing human performance and reliability; regulation and standards; and scientific research. There were some methodological approaches and tools identified as specific to either AVPSY (for example the development and use of psychometric tools), or AVHFS (for example human-machine interface design).

Common supporting skills and attitudes. AVPSY and AVHFS share nine common supporting core skills and attitudes including, taking a systems approach towork; evidence based approach and critical evaluation and analysis; robust processes, structure and logic in scope, design, and implementation of work activities; effective communication; reflective practice; building and maintaininga comprehensive understanding of specialist areas; maintaining an awareness of business / customers and acting responsibly with individuals and organisations; creativity and curiosity; and acting in compliance with ethical and legal / judicial principles and laws.

Competency Levels

The framework is linked to two principal competence levels of AVPSY and AVHFS; namely: 1) 'fully endorsed/ accredited AVPSY/ AVHFS' and 2) 'AVPSY/ AVHFS under development'. The latter recognises new graduates at the beginning of their professional careers. This replaces the previous 'junior aviationpsychologist' category. To differentiate between the previous accreditation options and the revised framework, we have changed the designations from AP and HF to AVPSY and AVHFS.

Continuous professional development (CPD)

The competency-based framework was developed to be integrated into a Continuous Professional Development (CPD) approach based on the example of the British Psychological Society (BPS, 2012) and EFPA.

It is envisaged that the competency-based framework is embedded in an electronically supported database that enables individual applicants and evaluators to maintain record keeping for continuous professional development.

Conclusion & Outlook

Findings in the initial development phase indicate the need for and use of psychological and human factors knowledge, expertise, and practice in aviation. The current regulatory and standard situation is unsatisfactory as it leaves a 'qualification and competence gap' attracting a wide audience (outside AVPSY and AVHFS). The competence framework for AVPSY and AVHFS in aviation outlines these links and aims to close the identified gap with clear competence requirements.

Analysis of results and development of the Venn-diagram has produced a robust competency framework which EAAP may now use for initial submission and continuing competency development of AVPSY and AVHFS endorsements. The Venn-diagram particularly highlights the overlap in working areas and methods between AVPSYand AVHFS, and it is anticipated that this will be useful for both practitioners and employers to understand the scopeand professional basis of both endorsements.

The framework provides a detailed description of the knowledge, skills, and attitudes expected from both AVPSY and AVHFS in each area of practice. The four main areas include: aviation domain knowledge, knowledge about humans, methodological approaches & tools, and application practice areas and common supporting skills and attitudes.

The framework is a 'living' document that will continue to be refined as more initial endorsement and CPD submissions are made. It is likely that further work to develop behavioural indicators will be completed in the future,and these would allow more specific development for those seeking an initial endorsement or continued professionaldevelopment. This may also help to create specific developmental tools such as coaching or academic and training courses for professionals in AVPSY and AVHFS.

Acknowledgements

This handbook was developed over a period of two years with significant inputs from subject matter experts withinand outside the European Association for Aviation Psychology (EAAP). The authors would like to specifically thankall developers (core working group) and reviewers within and across the aviation industry for their critical remarks and constructive feedback. Special thanks go to the Italian Aviation Psychology Association (IT-APA) for sharingtheir competency-based framework for aviation psychologists and foundation for this work. The complete study is reported in journal Aviation Psychology and Applied Human Factors, 2023. The authors thank Hogrefe Publishing for reusing part of the content.

References

AAPA (2020). A postgraduate curriculum in Aviation Psychology for Interested Psychologists. Presented at the EAAP 2020 Online Meeting and Virtual Workshop Week 28 September – 02 October 2020. Austrian Aviation Psychology Association.

AEPA (2019). Accredited Psychologist Expert in Aeronautical Psychology (EAP) in Spain. Presented at the 2020 Online Meeting and Virtual Workshop Week 28 September – 02 October 2020. Asociacion Epsanola de Psicologia de la Aviacion.

British Psychological Society (2012). Continuous Professional Development Policy. Retrieved from https://www.bps.org.uk/professional-development

EFPA (2021), EuroPsy – the European Certificate in Psychology. EFPA – European Federation of Psychologists Associations. Retrieved from Brussels/BE website: https://www.europsy.eu/_webdata/europsy_regulations_december_2021_virtual_ga.pdf / https://www.skillup-air.eu/

EU (2018). COMMISSION REGULATION (EU) 2018/1042 of 23 July 2018 amending Regulation (EU) No 965/2012, as regards technical requirements and administrative procedures related to introducing support programmes, psychological assessment of flight crew, as well as systematic and random testing of psychoactive substances to ensure medical fitness of flight and cabin crew members. Retrieved from https://eur- lex.europa.eu/legal-content/EN/TXT/HTML/?uri=CELEX%3A32018R1042&from=EN

INMAE (2021). Aviation Psychology degree at the University of Buenos Aires in cooperation with the National Institute of Aeronautical and Space Medicine (INMAE) - Argentine Air Force. Retrieved from https://www.modestoalonso.com.ar/

Kearns, Suzanne K. ; Timothy J. Mavin & Steven Hodge (2016),Competency-Based Education in Aviation Exploring Alternate Training Pathways. Taylor & Francis Ltd; New edition (13 April 2016)

IATA (2021). Competency Assessment and Evaluation for Pilots, Instructors and Evaluators, Guidance Material. First edition. IATA, Quebec, Montreal, Canada.

ICAO (2013). Manual of Evidence Based Training Doc 9995 AN/497. First Edition. ICAO, Montreal, Canada. www.icao.int

SESAR Consortium (2014). Proactive Pilot's Competency Analysis Tool for the support of designing concept and training anticipation - Guidelines. Project 16.06.05: Impact of Future Systems and Procedures on Selection, Training, Competence and Staffing (including Advanced Automation). Brussels: SESAR Joint Undertaking.

McClelland, D. C. (1973). Testing for competence rather than for "intelligence." American Psychologist, 28(1), 1-14. Retrieved from https://doi.org/10.1037/h0034092

SURVEY ASSESSMENT AND INITIAL DATA: FLIGHT CONTEXT AND PILOT TECHNIQUES IN EVERYDAY FLIGHTS

Dorrit Billman
NASA Ames Research Center
Moffett Field, CA
Alan Hobbs, Lucas Cusano, & Nóra Szládovics
San Jose State University Research Center
Moffett Field, CA

The aviation industry is recognizing that flight crews routinely contribute to system safety in ways that go beyond adherence to standard operating procedures (SOPs). Our research goals were to explore a) whether a survey could shed light on pilots' contributions to adaptation and resilience in everyday flights and b) relevant assessment methods. The survey focused on challenges faced by pilots in normal operations, and on the ways that pilots anticipate and monitor those challenges. We collected responses concerning revenue flights from two pilot groups; one group also provided responses concerning a simulated scenario. The results indicated that relatively few flights proceeded exactly as in the original flight plan. Pilots routinely anticipated and adapted to changing circumstances. We discuss some design and assessment challenges encountered for a survey on this topic, we provide 5 approaches to assessment, and we present example findings as illustrations. We expect assessment methods such as these will lead to useful surveys of resilience in flight.

Much of our knowledge about human performance in flight safety has come from the analysis of undesired events, whether accidents, incidents, or crew behaviors identified via flight exceedance monitoring or observational techniques. Recent years have seen an acknowledgement that operational personnel are not merely sources of "human error," but also make a unique human contribution to safe outcomes. In a few celebrated cases, this takes the form of "heroic saves," but on many more occasions, operational personnel contribute to safety through everyday, barely-noticed, actions that turn potentially hazardous situations into non-events.

An emerging approach to safety, frequently referred to as "Safety II," proposes that the positive human contribution is an important, largely untapped source of safety information. Some airlines have successfully trained observers to identify and record the positive behaviors exhibited by the flight crew (American Airlines, 2020). In other cases, flight crew are interviewed about good practices. However, each of these methods are relatively limited in scale and resource intensive. A survey could provide a relatively low-cost approach to systematically gather this information on a larger scale.

Our research focus is methodological, investigating prospects and challenges for surveying "Safety II" activities. This paper does not review the survey responses, but reports on our assessment of the interest and accuracy of the survey. We include example responses to survey items to illustrate our assessment methods and the potential benefits of a survey on the positive human contributions. Our goal is that a survey of this type will be useful to researchers and the aviation industry.

Survey Development and Response Collection

This paper describes the iterative development and, particularly, assessment of a survey to examine the human contribution to resilience in routine airline operations. Each survey version was critiqued by airline pilot advisors and completed by a sample of airline pilots. Several design considerations shaped the scope and prioritized the coverage of the survey:

- Our survey was directed at adaptive behaviors that are not specified in SOP or standard practices.
- Resilient behavior has been described as monitoring, anticipation, responding, and learning (Hollnagel, 2015). Our survey focused on the more proactive over reactive aspects, in part because this is less studied than reactions to triggering events.
- We limited the initial scope of the survey to the descent phase of flight as we anticipated that this would provide us with numerous opportunities for resilient pilot behavior. For example, Standard Terminal Arrivals (STARs) can require complex interactions with the autoflight system, well-timed actions, and an understanding of automation, ATC, and the airspace.
- To understand the intent of pilot behavior, it is necessary to understand the operational context in which the behavior occurred. Therefore, we included some situational questions, primarily about ATC actions and weather.

An important methodological topic, briefly summarized, is the principles guiding the organization and design of questions, to make them as clear and easy to answer as feasible:

- We aimed to avoid abstract terminology or jargon that might be used in the research community but not necessarily familiar to pilots. For example, a major airline (American Airlines, 2020) uses specially trained personnel who observe flights from the jump seat and record instances of resilient performance using a standard set of terms. Pilots lacking specialized training might vary widely in how they interpreted such terms.
- Our focus was on adaptive activities in ordinary circumstances that were unlikely to be particularly striking or memorable. Therefore, to minimize interference, we focused on the most recent flight.
- Unless phrased carefully, questions about resilient behavior can imply a "correct" or desirable answer. For example, a survey question asking if a potential threat was included in a briefing could imply that the threat should have been included. We framed the majority of questions to be "matter of fact" descriptions about the flight and what the crew did.

We used a variety of question formats, including checkbox items, rating scales, and free text responses. A checkbox *item* consists of a question and *response* choices, allowing multiple choices. Throughout the survey development process, airline pilots with research backgrounds helped us to ensure that questions were relevant and phrased appropriately. Survey development was guided by considerations of content, question design, and question format. We iterated through four major cycles of development and testing. Versions 3 and 4 were very similar and are reported here. All respondents are airline pilots, sampled by convenience not randomly. "SOTERIA" pilots participated in a flight simulation study conducted at NASA Langley Research Center as part of NASA's SOTERIA[1] study (Stephens et al., 2021). Table 1 gives group and Table 2 gives item characteristics.

Table 1. Characteristics of survey response sets.

	Group Name (n)	Flight Type	Version	Pilot Source
1	SOTERIA-rev (25)	revenue	V3-long	simulator study
2	LCP-rev (65)	revenue	V4	line-check pilots
3	SOTERIA-sim (22)	simulator	V3-short	simulator study

Table 2. Survey question content and format for the 4th iteration (used by LCP-rev pilot

Format	What happened Op. Context	What did you do						Evaluate	
		Proactive/Anticipatory				(Re?/active)	Explc. "Learn"		
		briefing	info gathering	assessment	"monitoring"				
Checkbox	6	3	2	3	2	5			21
Rating		1:eval	2:eval		1 + 6:eval			2 +9	12
Text	1	2			1		2	4	10
	7	5 (+1)	2 (+2)	3	4 (+6)	5	2	14	43

[1] System Wide Safety Operations and Technologies for Enabling Resilient In-Time Assurance (SOTERIA)

Results and Assessment

The survey gathered findings on an important, underinvestigated topic. However, their value depends on the credibility of our survey. Do our questions ask about things that are both interesting and that pilots can report? How clear are our questions so respondents' understanding matches our intent? We describe five *approaches* to assessing the value of this survey, considering validity and reliability. We use selected results as illustrations.

Approach 1: questions reviewed individually for interesting but reasonable findings. For an individual question there were few or no cases where responses seemed inconsistent with how the world is, though many provided novel information; this is reassuring. Text responses were coded into categories and sub-categories based both on our expectations of what might be reported and what was observed. Example 1A: SOTERIA-rev pilots described what was most challenging and in the next question how they managed it. Responses from the 25 participants were coded into 1 or more sub-categories, grouped into more general categories. Figure 1 shows the dominant challenge was Operations, specifically, Scheduling/Delays/Timing Out, with Fatigue a close second. It is striking that CRM category was identified as a management method in almost 3/4's of the reports, with the proactive strategy of using an Extended Briefing in almost 1/4 of reports.

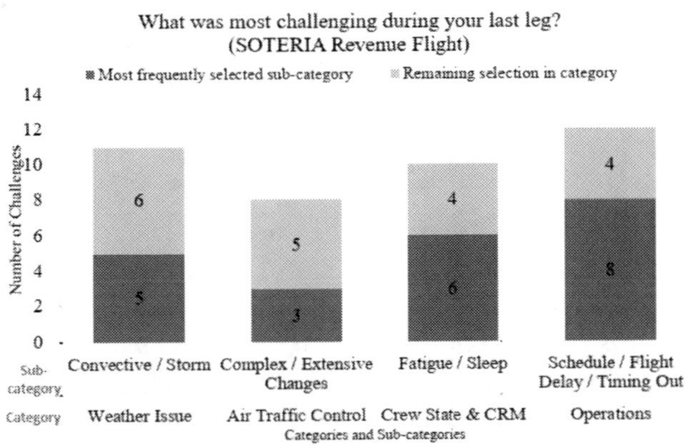

Figure 1A: Most challenging aspects.

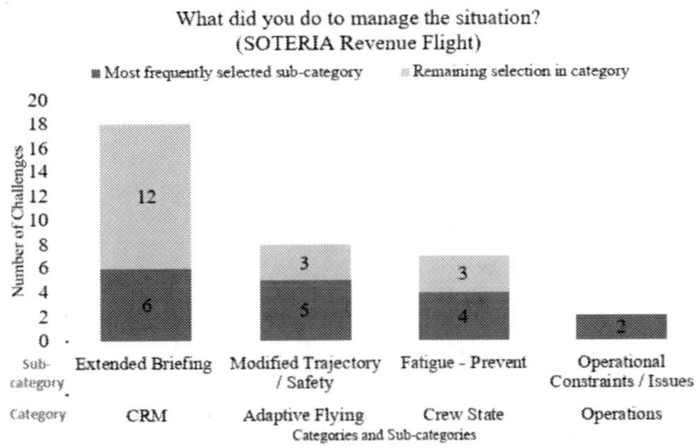

1B: Method for management.

Example 1B: If a pilot said they had learned something that might help on a future flight (32 of 65 did), they described what that was. Their responses were classified into one of 9 categories (see Figure 2). Choices were diverse, but the most common (1/5 of the group) addressed communication in the cockpit, again highlighting the prominence of CRM in pilot experience.

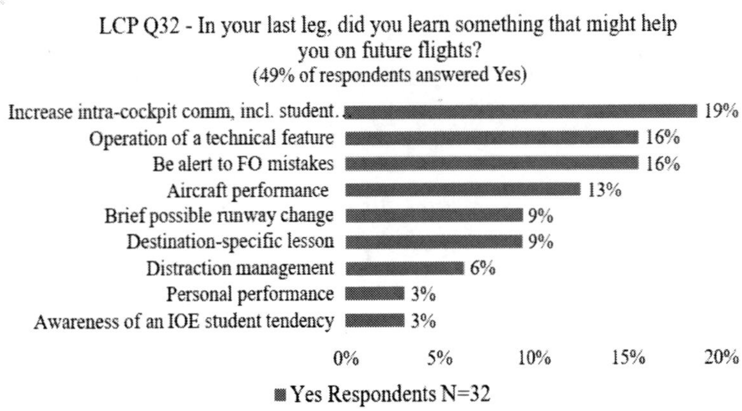

Figure 2. Categories of what pilots learned (LCP data).

Example 1C: Several checkbox questions asked about what ATC did, the weather, and other aspects of the operational environment. As Figure 3A shows, Q16 asks about ways ATC might modify an arrival, plus a "none" and other option (as on all LCP checkboxes). Strikingly, only 13.9% of arrivals were not modified by ATC. Thus, it is a small minority of arrivals where STARs are flown as published (and the large majority where pilot response is required). The rather high proportion of runway changes is also interesting.

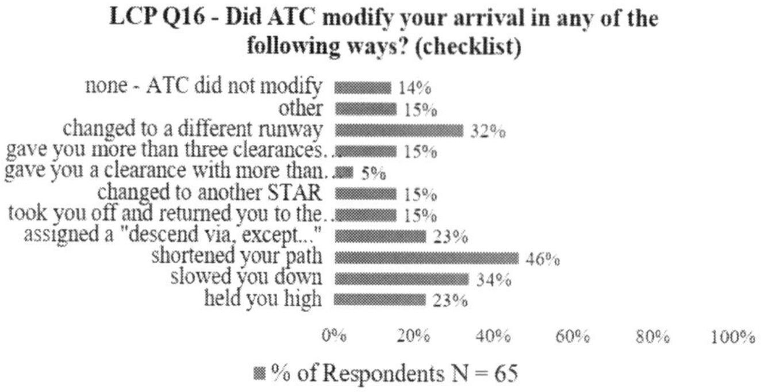

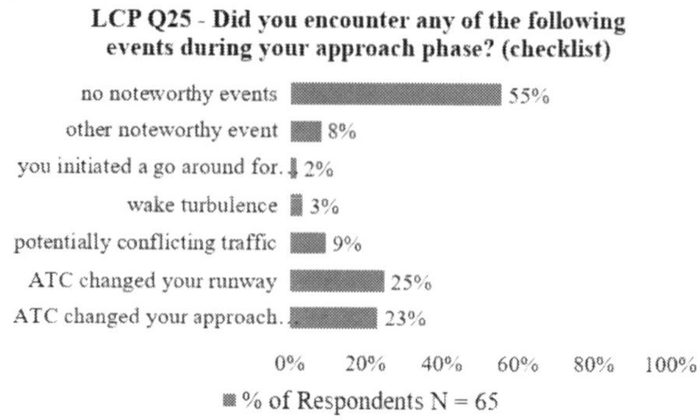

Figure 3A: Arrival Phase - ATC modifications (LCP). 3B: Approach Phase - events encountered (LCP).

Approach 2: consistency across related questions. The relations between responses to different questions may be associated in expected or in surprising ways; a surprise may challenge assumptions about the world or about the basis for answering the question. We give example findings of an expected pattern, of surprising patterns, and of absence of clear relations where we thought they might occur. Example 2A: we thought flights judged more challenging than normal might be more likely to provide something to learn. Of the LCP pilots who judged the flight more challenging than normal, 60% said they learned something (and 40% did not), while of the pilots who said it was a normal or less level of challenge, 40% said they learned something (and 60% did not). The correlation between the challenge rating (5- pt scale) and pilot learning was r(65)=.69. Thus, the pattern of responses to these two items was consistent with the expected relation.

It may be hard to tell whether a surprising finding is accurate or an artifact of the question design. Consider the reports of runway changes shown in Figure 3A&B. If the percentage of runway changes on Arrival (Fig. 3A) and on Approach (Fig. 3B) are summed, the total is over 50% (33% +25%). Looking at individual responses shows everyone who checked the 'runway on approach' response also checked 'runway on arrival' response. Possibly there were two runway changes. Alternatively, the respondents counted the same change twice. Using *Approach 5* on the SOTERIA simulation data provides additional hints, described below.

Example 2B: we asked pilots about the percentage of time spent on different activities, as shown in Table 2. Items a and b in Table 3 are not explicitly reverse coded, but we expected these two would sum to about 100%, which they do. However, it is highly likely that when working on systems (44%>> 16%) one is not also specifically attending to the progress of the flight (44 + 84>> 100). This apparent inconsistency may suggest difficulty of reporting about interleaved tasks or alternatively, a strong belief in the ability to truly multitask.

Table 3. Percents of flight time pilots judged as allocated to different activities. (LCP data)

During the descent phases, estimate the % of time in which... *[respondent clicked on a 1-100 timeline]*	Mean %
a) I "mentally flew" the aircraft, even when the autopilot, or the other pilot, was controlling it.	84
b) I was NOT specifically attending to the progress of the flight.	15
c) I was working on systems management (e.g., entering values in FMS) or communications (e.g., radio settings, talk with ATC).	44

Example 2C: We had hypothesized we might see clear associations between events (e.g., ATC clearances) and pilot actions (e.g., input to the autopilot). However, the complexity of possible relations was not easy to trace out in relations among these responses. This may suggest that the combined operational complexity of how pilots adapt will benefit from a more focused, contingent inquiry: asking whether an event occurred, and if so, asking about possible actions for managing or anticipating it.

Approach 3: compare response patterns across different groups. Comparing frequency of responses across different groups provides some indicators of stability. For example, in the LCP group the proportion of flights where ATC did not modify descent was low (14%); turning to the SOTERIA revenue flights, 20% did not have an ATC modification, a similar though somewhat higher percent. Of course, differences may reflect actual differences between groups as well as less meaningful variability. Turning to the pilot monitoring (PM) versus pilot flying (PF) within the LCP group, we set a heuristic criterion of 20% difference between the two roles to consider noteworthy. None of the responses to any of the 6 items about what happened and only 4 responses in the more than 75 responses across the 21 items about pilot action differed by this criterion. These broad patterns are not particularly diagnostic but suggest that findings do not differ majorly when a flight is reported by PM or PF.

Approach 4 & 5 compare ratings of the same situations. These are feasible for SOTERIA crews in simulator events, for responses to checkbox items. In *Approach 4*, ratings of same-crew PM and PF can be compared using standard reliability measures; we explored several and settled on percent agreement. We looked at the agreement between PM and PF on whether they selected a particular response on checkbox questions. We scored whether a given crew agreed on a given response and averaged these to get a percent agreement a) across crews for a response and b) across responses for a crew. Agreement scores for individual crews ranged from 72% to 86%. Agreement scores for individual responses ranged from 36% to 100%. The overall agreement level was 77%.

Factors that seem to contribute to high reliability of a response include being highly standard actions or SOPs (Table 3 #1) and being highly salient, observable events (Table 3 #2). Factors contributing to low reliability include reference to standards SOP; it may be unclear what is the standard level of automation, or SOP (Table 3#3, #4), and actions which may fall close to such a boundary (Table 3 #4); a response may have low reliability both because it is hard to decide what category the question refers to, and to decide if the actual events fit in that category. Table 3 shows examples.

Table 3. Responses With High and Low Agreement (SOTERIA -sim data).

<table>
<tr><td rowspan="2">Highest Agreement</td><td>#1 What did you do to assess how your autoflight system would handle your STAR?
--checked that the values in the flight management computer matched values on the chart--</td><td>91%</td></tr>
<tr><td>#2 Did you encounter any of the following events during your arrival?
--ATC changed your runway--</td><td>100%</td></tr>
<tr><td rowspan="2">Lowest Agreement</td><td>#3 Did you fly any part of the approach manually, or at lower levels of automation than standard for your airline?
--No/Not Applicable--</td><td>36%</td></tr>
<tr><td>#4 During descent, the PM:
--provided positive confirmation of expected actions or states, beyond SOP--</td><td>36%</td></tr>
</table>

Approach 5: comparison to an observer. The most effective way for observers to review and rate a crew's flight is by reviewing video recordings of the sim session. We plan to conduct *Approach 5* assessment in the future. Nevertheless, we can gain some clues about validity without an extensive review of simulator events. ATC clearances were scripted elements of the scenarios, delivered by a member of the research team. Two of the event scenarios, seen by 6 total crews, included a single, scripted runway change. Although only one runway change occurred, 11 of the 12 pilots reported two, one during arrival, one during approach. This suggests that in these scenarios, pilots were not distinguishing when a runway

change occurred, and that the question might be better framed by asking about whether any runway change(s) occured, and then asking in what phase of flight.

Discussion & Conclusions

The primary purpose of the research was to develop and assess surveys, as a little-used method for caputring crews' activities in normal flights and the operational perturbations routinely encountered. The assessment provided both information about the flights and information about what questions might merit revision. For example, despite much iteration on this topic it was hard to ask pilots questions involving behavior that went beyond standard performance, one of the ways we tried to communicate resilience. How much difficulty comes from how the respondent understands the question intent or from how the respondent categories the situation or their behavior (e.g., 'just doing my job') is hard to determine. The multiple assessment approaches tried to tease apart where variations were due to difference in external circumstances of flight (e.g. between groups), in perspective (e.g., between PM vs PF participants), or in how the question was understood or information retrieved. We were extremely fortunate to have data from two groups of pilots, and from simulated as well as revenue flights, including pilot pairs crewing the same simulated flight. This gave us the opportunity to use the data to assess the survey using several *approaches*. *Approaches 1-3* depend on making sense of how responses fit in with, yet extend, what we know, broadly, its validity. This can be done by looking at individual items and responses, by looking for patterns of coherence between items, and by looking for consistency or meaningful differences between groups replying to the survey. *Approaches 4* and *5* measure agreement between pilots in the same crew or compare crew responses to observers equipped to make a best estimate of 'ground truth.' This agreement measure would be a further measure of validity. We are not aware of this style or degree of assessment of surveys in the aviation domain. We find our current results both encouraging and a guide for further survey evolution.

As the presented examples suggest, responses also provided sensible and interesting information about prevalence of situations or behaviors, for example, the pervasiveness of ATC changes during descent and the association between how challenging the flight was and learning something new. Future reports will provide more comprehensive coverage of findings. We also plan to summarize suggestions about survey design relevant to understanding resilience, pilot activity, and its context. Our goal is that survey assessment will result in trust-worth, useful surveys for measuring pilot contributions to resilience.

Acknowledgements

Our thanks to the airline pilots who gave their time to complete the surveys. Thanks also to several pilots who provided valuable input to survey development, particularly Capt. Barth Baron, Capt. Dan Kiggins, and Cap.t Rob Kotesky. Thanks to Immanuel Barshi for critical networking, to Jon Krosnick for discussion, and to Jon Holbrook, Lawrence Prinzel & Chad Stephens for support of the SOTERIA project. This research was funded by NASA's System Wide Safety Project.

References

American Airlines' Department of Flight Safety (AA DFS). (2020). Trailblazers into Safety-II: American Airlines' learning and improvement team, a white paper outlining AA's beginnings of a Safety-II journey. Retrieved from https://www.skybrary.aero/sites/default/files/bookshelf/5964.pdf

Hollnagel, E. (2015). RAG—Resilience Analysis Grid. Retrieved from http://erikhollnagel.com/onewebmedia/RAG%20Outline%20V2.pdf

Stephens, C., Prinzel, L., Kiggins, D. Ballard, K., & Holbrook, J. (2021). Evaluating the Use of High-Fidelity Simulator Research Methods to Study Airline Flight Crew Resilience. 21st International Symposium on Aviation Psychology, 140-145.

EFFICACY OF USING VIRTUAL REALITY TO SUPPORT THE TRAINING OF THE EMERGING PILOT WORKFORCE

James G. Birdsong
Kurt L. Reesman
JoEllen M. Sefton
Matthew W. Miller
Auburn University
Auburn, AL, United States

Classroom instruction, computer-based training, flight simulation, and aircraft are used to train pilots. New immersive technologies and associated learning methods are used to train military pilots and may have value in civilian pilot training. This paper describes a study to explore the efficacy of using Virtual Reality (VR) Head-mounted Displays (HMDs) to support the training of the emerging pilot workforce. Participants were *ab-initio* civilian pilot students enrolled in a collegiate aviation program. Participants learned commercial aircraft preflight tasks using one of three methods. The control group used a combination of traditional classroom lectures and simulator sessions; one experimental group used VR to augment classroom instruction and simulator training; the other received classroom instruction and used VR to replace simulator training. Participants' performance was evaluated at a partner air carrier. Researchers completed a preliminary data analysis to understand the effectiveness of the training provided with results presented in this paper.

The current 14 CFR Part 121 air carrier pilot workforce comprises four generations: Baby Boomers (born 1946-1964), Generation X (born 1965-1980), Generation Y (born 1981-1996), and Generation Z (born 1997-2012) (FAA, (n.d.) *U.S. Civil Airmen Statistics* & Pew Research Center, n.d.). Baby Boomers will exit the workforce within the next ten years, and Generation X will retire between 2030-2045, leaving an emerging workforce of Generation Y and Z pilots (FAA, (n.d.) *U.S. Civil Airmen Statistics)*.

Current pilot training methods include classroom instruction, computer-based training (CBT), flight simulation, and aircraft training. New immersive virtual reality (VR) technologies have demonstrated value in military pilot training (Lewis & Livingston, 2018; Pennington et al., 2019; McFarland, 2020; & Mishler et al., 2022) and may have value in civilian pilot training, specifically in the development of procedural knowledge, filling a niche between CBT and traditional flight simulation (Bauer & Klingauf, 2008) and supplementing classroom instruction and simulator procedure training activities (Cross et al., 2022).

This research effort evaluates immersive technology's potential human factors, benefits, and limitations for training pre-flight tasks in a transport airplane. This is part of a larger FAA research effort to understand the characteristics of the emerging pilot workforce and various training and checking methods that might be effective for the emerging pilot workforce.

Method

Experimental Design

This effort evaluated interactive, immersive asynchronous eLearning using a VR headset and software to train general Airbus A320 flight deck orientation and operation of systems preflight tasks described in FAA Airline Transport Pilot Airman Certification Standards and FAA Advanced

Qualification Program (AQP) Job Task Listings Terminal Proficiency Objectives (TPOs) and Supporting Proficiency Objectives (SPOs) examples found in FAA Advisory Circular 120-54A Advanced Qualification Program.

The control group used a combination of traditional classroom lectures and simulator sessions in a Flight Deck Solutions Airbus A320 Procedures Trainer to learn system preflight tasks. In contrast, the experimental groups used VR Pilot A320 flight deck procedural trainer software installed on Oculus Meta Quest 2 virtual reality headsets to learn and rehearse tasks in a guided virtual environment. One experimental group used VR to augment classroom and simulator training; the other used VR to augment classroom training and used VR as a replacement for simulator training. The three groups are summarized below.

- Control group: access to traditional classroom training and simulator training.
- Experiment group #1: access to classroom training, simulator training, and VR on a head-worn device to augment baseline classroom and simulator training.
- Experiment group #2: access to classroom training, no simulator training, and VR on a head-worn device as a replacement for simulator training.

Tasks instructions were presented in two modules developed for this project, based on two aircraft preflight checklist procedures, Preliminary Flight Deck Prep (Module 1) and Flight Deck Prep (Module 2), outlined below.

Module 1 TPO is Preliminary Flight Deck Prep and includes 7 SPOs (with 29 subtasks):

1. Aircraft setup (5 subtasks)
2. Batteries, external power (2 subtasks)
3. APU fire test, start (5 subtasks)
4. Cockpit lights (1 subtask)
5. EFB initialization (3 subtasks)
6. Aircraft acceptance (4 subtasks)
7. Before walkaround (9 subtasks)

Module 2 TPO is Flight Deck Prep and includes 5 SPOs (with 46 subtasks):

1. Overhead panels (15 subtasks)
2. Center instrument panel (4 subtasks)
3. Pedestal (12 subtasks)
4. Glare shield (6 subtasks)
5. Lateral console, instrument panel (9 subtasks)

Sample

The population for this study was *ab-initio* civilian pilot students enrolled in the Auburn University School of Aviation, a 14 CFR Part 141, FAA R-ATP approved institution, who have completed their instrument rating as a minimum, are working towards or completing the commercial certificate and multi-engine rating and enrolled in the AVMF 4320 AIRLINE TRANSPORT CATEGORY SYSTEMS AND PROCEDURES capstone course. This course is a 4-credit (3 classroom hours + 1 lab hour) course focusing on Airbus A320 systems and operational procedures and 14 CFR Part 121 air carrier flight and crew management. Procedural content presented in Modules 1 and 2 were not previously taught in this course and were added for this study. AVMF 4320 students are members of the emerging pilot workforce. They will join the 14 CFR Part 121 air carrier pilot workforce upon reaching as little as 1,000 flight hours, typically within 12-18 months of graduating from Auburn University.

To increase the number of research participants, the study spanned two semesters of AVMF 4320. The baseline control group included Fall semester 2022 AVMF 4320 students. The experimental groups included students from the Spring semester 2023 AVMF 4320 course. Spring AVMF 4320 sim blocks were predesignated (before student signup) for Module 1 and Module 2 instruction methods, either using head-worn VR devices to augment classroom and simulator training (experimental group #1) or classroom with no simulator training and VR head-worn device as a replacement for simulator training (experimental group #3). Participants self-selected their AVMF 4320 simulator schedule based on their academic schedule, not knowing which experimental group they would be assigned to. All students had approximately two weeks to learn Modules 1 and 2 before being evaluated. 26 students opted into the study, beginning with 7 in the control group, 10 in experimental group #1, and 9 in experimental group #2. Four students in experimental group #2 opted not to complete the field test, leaving a total of 22 participants who completed the field test.

Tools

The VR device was the Oculus Meta Quest 2 virtual reality headset, a fully mobile, wireless/untethered, all-in-one VR system (headset and two controllers) that allows the user to train anywhere, anytime. A personal computer (PC) or console was not required. This system is inexpensive to purchase compared to higher-end VR devices and is widely available for approximately $399. The US Navy used an earlier version of this system in their Project Avenger VR study (Mishler et al., 2022). The Meta Quest 2 VR system weighs about four pounds, uses a Fast-Switch LCD Display, has 1832 x 1920 resolution per eye, and supports 60 Hz, 72 Hz, or 90 Hz refresh rates. This system is eyeglass compatible and uses six degrees of freedom tracking. The headset tracks the movement of both head and body, then translates them into VR with realistic precision.

The experimental groups used VRflow software developed by VRpilot. The VRflow A320 provides a 3D virtual flight deck that responds and behaves like a real aircraft (including sounds, lights, screens, etc.) to provide the user with a realistic training environment. VRflow provides "learning" and "exam" modes to guide procedural learning and check if the procedure has been learned correctly. VRflow provides student feedback on procedure performance and shows the time taken to complete the procedure. Training content is stored locally on the VR device and can be transmitted to a learning management system (LMS) with Wi-Fi connectivity or collected manually. VR headsets were managed using the ManageXR platform and "locked down" to prevent other software use on the VR devices. Devices were distributed at the beginning of the AVMF 4320 course, along with a quick start VR user's guide developed by a graduate student.

The Flight Deck Solutions Airbus A320 Procedures Trainer simulator is a spatially correct, functionally accurate, precise tactile feel flight deck with an instrument panel, glare shield, aisle stand, primary and aft overhead, flight controls shell/interior, and crew/observer seating. The flight deck is integrated with a Q4 Services SupraVue Collimated 10' dome visual system (Level D compliant), ProsimA320 Aviation Research Professional Suite flight data package, and Lockheed Martin's Prepar3D Pro Plus imaging system. The trainer uses Brunner control loaded rudders and a sound system with dual 800-watt subwoofers and custom surround sound to simulate accurate and directional wind noises, vibrations, engine sounds, and aerodynamic drag. This A320 procedures trainer is FAA Level 4 compliant but not FAA-certified since Auburn University does not type-certify students in the A320. The simulator is used primarily for human factors training focusing on crew resource management (CRM) using Line Oriented Flight Training (LOFT) scenarios.

Data Collected

Participants took a pretest to measure A320 systems and procedures knowledge and a pretest survey to collect the following information:

1. Demographics
2. Flight experience
3. Career pathway program participation
4. VR experience
5. Gaming experience
6. Inventory of Learning Styles
7. Eyeglasses or contact lenses use.

Control and experimental group participants participated in a modified Line Oriented Simulation (LOS) in a Level 7 Flight Simulation Training Device (FSTD) at a partner 14 CFR Part 121 air carrier using Aircrew Program Designees (APDs) as evaluators. The APDs used grading sheets for each module, developed by the researchers, based on modified AQP grading criteria (2 – not proficient (error not managed), 3 – competent (managed error), 4 – proficient (no errors)) to evaluate individual task proficiency and the overall performance of experimental and control group members. APDs were blind to the training intervention and did not know whether the participants they evaluated were from an experimental or control group. Airbus Original Equipment Manufacturer (OEM) A320 procedures were used to standardize all devices' training and evaluations.

The experimental groups had an early morning LOS and spent the night near the air carrier's training center to minimize fatigue. The control group used one FSTD, and the experimental group used two FSTDs simultaneously for evaluations. APDs initialized the flight deck the same for each participant's LOS, with either a graduate assistant or faculty member familiar with A320 aircraft procedures in the left seat and the research participant sitting in the right seat. Participants performed both checklists alone initially at their own pace. When complete with the checklist, the individual in the left seat read the checklist with challenge and response. As researchers observed, air carrier APDs scored task performance during each participant's LOS. Researchers collected LOS scores from the APDs and recorded the time to task completion.

Following the LOS, participants completed a post-test survey to capture user experience data:

1. Self-confidence in execution (Likert, 1-7)
2. Time (hours) spent studying to feel confident in the procedural ability
3. Perceived difficulty (Likert, 1-7)
4. Intrinsic motivation inventory (Isen & Reeve, 2005)
5. Simulator Sickness Questionnaire (SSQ) (Kennedy, Lane, Berbaum, & Lilienthal, 1993)
6. VR experience questionnaire

Results

Data Analysis

A series of Kruskal-Wallis tests were used to determine if there was a difference in Preliminary Flight Deck Prep and Flight Deck Prep task performance and combined task completion time based on training method (Classroom and Simulator Training, Classroom and Simulator Training with VR to augment, and Classroom with No Simulator Training and VR as a replacement for Simulator Training). Preliminary Flight Deck Prep task performance by training method is shown in Table 1, Flight Deck Prep task performance by training method is shown in Table 2, and combined task completion time by training method is shown in Table 3.

Table 1

Preliminary Flight Deck Prep Task Performance by Training Method.

Preliminary Flight Deck Prep Task Performance Score	*M*	*SD*	*n*	*df*	*H*	*p*
Classroom and Simulator Training	3.80	.11	7	2	2.81	.246
Classroom, Simulator, and VR Training	3.69	.12	10			
Classroom, No Simulator, and VR Training	3.62	.24	5			

Note: N = 22

Table 2

Flight Deck Prep Task Performance by Training Method.

Flight Deck Prep Task Performance Score	*M*	*SD*	*n*	*df*	*H*	*p*
Classroom and Simulator Training	3.77	.14	7	2	1.31	.519
Classroom, Simulator, and VR Training	3.75	.09	10			
Classroom, No Simulator, and VR Training	3.62	.21	5			

Note: N = 22

Table 3

Combined Task Completion Time by Training Method.

Combined Task Completion Time (Minutes)	*M*	*SD*	*n*	*df*	*H*	*p*
Classroom and Simulator Training	13.27	2.11	7	2	5.54	.063
Classroom, Simulator, and VR Training	15.37	1.65	10			
Classroom, No Simulator, and VR Training	18.39	4.52	5			

Note: N = 22

Discussion

All groups learned the required tasks and had mean scores between the competent and proficient levels. No significant difference was found in Preliminary Flight Deck Preparation task performance based on the training method ($H(2) = 2.81, p = .246$), Flight Deck Preparation task performance based on the training method ($H(2) = 1.31, p = .591$), or the task completion time ($H(2) = 5.54, p = .063$).

Conclusion

Preliminary task performance results indicate that all three types of training were equally effective, suggesting Virtual Reality (VR) Head-mounted Displays (HMDs) may have value in learning commercial aircraft preflight task procedures. Further data analysis, including user experience, is forthcoming.

Acknowledgments

This research is derived from a more extensive research effort sponsored by the Federal Aviation Administration (FAA) through an aviation research grant with the Center of Excellence for Technical Training and Human Performance. The views expressed herein are those of the authors and do not reflect the views of the United States (U.S.) Department of Transportation (DOT), the FAA, or Auburn

University. The authors thank the FAA and our various industry and airline partners for their generous support of this research effort.

References

Bauer, M., & Klingauf, U. (2008, August 18). Virtual-Reality as a Future Training Medium for Civilian Flight Procedure Training. *AIAA Modeling and Simulation Technologies Conference and Exhibit*. AIAA Modeling and Simulation Technologies Conference and Exhibit, Honolulu, Hawaii. https://doi.org/10.2514/6.2008-7030

Cross, J. I., Boag-Hodgson, C., Ryley, T., Mavin, T., & Potter, L. E. (2022). Using Extended Reality in Flight Simulators: A Literature Review. *IEEE Transactions on Visualization and Computer Graphics*, 1–1. https://doi.org/10.1109/TVCG.2022.3173921

Federal Aviation Administration [FAA]. (n.d.). *U.S. Civil Airmen Statistics*. Retrieved September 11, 2022, from https://www.faa.gov/data_research/aviation_data_statistics/civil_airmen_statistics

Isen, A. M., & Reeve, J. (2005). The Influence of Positive Affect on Intrinsic and Extrinsic Motivation: Facilitating Enjoyment of Play, Responsible Work Behavior, and Self-Control. *Motivation and Emotion*, *29*(4), 295–323. https://doi.org/10.1007/s11031-006-9019-8

Kennedy, R. S., Lane, N. E., Berbaum, K. S., & Lilienthal, M. G. (1993). Simulator Sickness Questionnaire: An Enhanced Method for Quantifying Simulator Sickness. *International Journal of Aviation Psychology*, *3*(3), 203. https://doi.org/10.1207/s15327108ijap0303_3

Lewis, J., & Livingston, J. (2018, November 26-29). *Pilot training next: Breaking institutional paradigms using student-centered multimodal learning* [Paper presentation]. Interservice/Industry Training, Simulation, and Education Conference (I/ITSEC), Orlando, FL, United States.

McFarland, C. (2020, July 31). Aviator training next – an experiment in virtual reality. *Army Aviation, pp.* 44–45. https://5abe1488a536b7d66554-40ebbf4e472cfd77f5021bc42c60f8a3.ssl.cf1.rackcdn.com/ug0zdvvmvo6406hmqgkajx62nrvumt-optimized-pub.pdf

Mishler, A., Severe-Valsaint, G., Natali, M., Seech, T., McCoy-Fisher, C., Cooper, T., & Astwood, R. (2022). *Project Avenger Training Effectiveness Evaluation*. Retrieved October 16, 2022, from https://apps.dtic.mil/sti/citations/AD1162306

Pennington, E., Hafer, R., Nistler, E., Seech, T. & Tossell, C. (2019, April 26). *Integration of advanced technology in initial flight training* [Paper presentation]. 2019 Systems and Information Engineering Design Symposium (SIEDS). Charlottesville, VA, USA. DOI: 10.1109/SIEDS.2019.8735628.

Pew Research Center (n.d.). The generations defined. Retrieved from https://www.pewresearch.org/wp-content/uploads/2018/03/ST_18.02.27_generations_defined.png

USABILITY SHORTCOMINGS IN DEPARTMENT OF DEFENSE FUNDED SYSTEMS

Jerry Burpee
Intuitive Research and Technology Corporation
Huntsville, Alabama

Government developed systems, in particular Department of Defense (DoD) systems experience usability shortcomings that are not seen as often with systems developed for public or commercial use. The factors discussed include different funding models, limited competition, usability requirements and specifications, usability metrics, utilization of specialized users, lack of usability specialists, and non-revenue generation. By understanding and addressing these shortcomings, the usability in DoD developed systems can be improved to save the tax-payer money, minimize project and system risks, and improve user acceptance and satisfaction.

Usability is Lacking in Department of Defense Funded Systems

Usability is the "extent to which a system, product or service can be used by specified users to achieve specified goals with effectiveness, efficiency, and satisfaction in a specified context of use." (ISO 9242-11, 2018)

In this paper, government developed DoD systems, are systems that are developed for government use and funded by the government. An example would be crew-stations used by government or DoD personnel for monitoring situational awareness and performing tasks associated to a military operation. Implementation tends to follow innovation. This paper concentrates on the implementation or developmental phases of a system, but many of the statements can be applied to the innovation phases as well.

Funding Models are Different

Note: When discussing the funding model in this paper, the concepts presented are generalized.

Generally, the funding model determines whether a product or system is "Governmental Developed" or "Public Developed." With government developed systems, governmental representatives first identify a need for the system and then define how the system needs to be developed. This is done by generating a scope of work with requirements and specifications that requires a minimum of three qualified private sector contractors to bid on the project or it is developed internally by government sponsored employees. In the bid situation, the contractor awarded the project is selected based on price and a commitment that they can meet the scope of work within the timeline stipulated. The contractor may be hesitant in providing any additional features, even if they can improve or innovate the product or system. If they include these features, it could affect the contractor's resource availability, time commitments, and eventually their ability to win the bid.

Basically, government projects are funded first and THEN developed. This means the funding arrives and then the project is developed (*Figure 1*). This is opposite from the private sector where a product is developed and THEN purchased (funded). There is more risk in the private sector because investment in the development of the product may not be returned if the product or system is not purchased. How does this affect usability? Since the products are already developed for the private sector, a key purchase point is determinate upon the product or service's user friendliness. With a government funded product or service, if it meets the defined requirements and specifications (assuming usability metrics were not clearly defined in the requirements) a non-user-friendly product can be delivered. It can even be argued that there is incentive in not having a user-friendly product since additional funding may be awarded to make the product more user-friendly. The developer is not intending for the end-user to have a horrible user experience utilizing their product, but usability is not a focal point in the development of government systems.

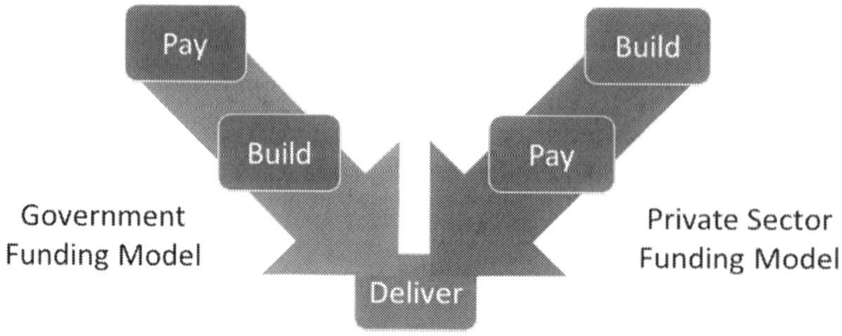

Figure 1: Funding Model of Government vs Private Sector

Limited Competition

Competition for developing a government system is concentrated during the bidding process before the project is awarded. Once the project is awarded, external competition generally comes to an end. Usability problems become evident when the project undergoes tests with end-users and the user struggles to use the product. This often occurs in the later stages of development. Unfortunately, if usability requirements and specifications were not specifically defined in the bid process, usability issues discovered later are not disqualifiers.

Competition begins prior to and during bidding phases (*Figure 2*). Before any funding is awarded, contractors may not assist the government with developing requirements and specifications. If a contractor assists the government in developing the requirements and specifications, the contractor may be at a competitive disadvantage during bidding because they would need to recoup some of their services and expenses spent in assisting the government. Additionally, if the requirements and specifications were perceived to be written to favor the assisting contractor in any way, their competition could take exception by offering lower priced options or challenging the bid as being non-competitive.

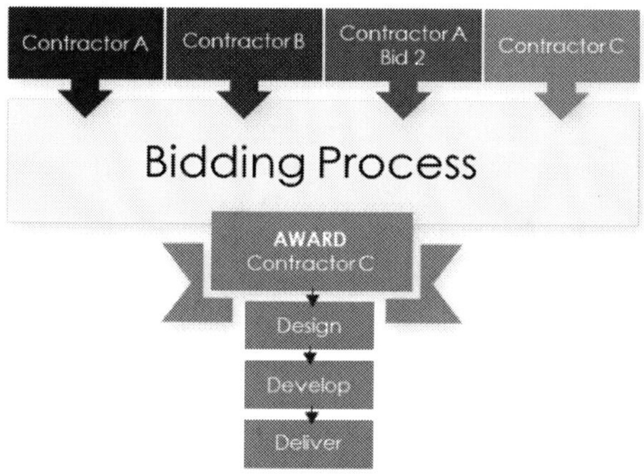

1. Request for bids based on requirements and specifications is solicited.
2. Several entities submit bids.
3. A contract is awarded to a winning bid prior to system design.

Figure 2: Government Bidding Process Example

A competition of concepts based on end-user usability amongst qualified contractors may provide a potential solution in determining who should be awarded the design and development contracts. Contractors would be funded to provide their best design and development product or system for government review, approval, and acceptance. The system or design that best meet the government's requirement would be approved and awarded the contract.

Usability Requirements and Specifications Hard to Define

Properly written usability requirements and specifications are key factors to reaping the benefits of a user-friendly product. Requirements dictate the needs of a system, but specifications provide the directions needed for the system. A requirement example would be: "The system shall display a map on the screen." A specification example would be: "The map shall have the ability to display terrain". Most requirement and specifications are objectively written with defined metrics to be met, proven, or confirmed during verification and validation testing. These can be simple binary responses such as pass/fail, or defined to fall within a set threshold, such as ±1 degree. Using the above example, the requirement and specifications would be met if the system displayed a map on the screen and a method was implemented to toggle the terrain on.

Usability requirements and specifications are frequently too vague or too restrictive. In the previous example, "the map shall have the ability to display terrain," the specification language is too vague. In the implemented design of the system, this specification may require the user to perform unnecessary steps to turn the terrain on or off, such as having to access separate menus or navigate to a different area of the interface. The design would meet the specification, but it would not be very effective or efficient for the user to complete the task. An example of a too restrictive specification is: "The map shall have the ability to display terrain by using a slider bar in the map filter menu." This may appear to be better written for usability, however, there may be another way to accomplish the task that is more conducive to the workflow, i.e., using a check box. Restrictive specifications restrict the developer's flexibility in designing a system to meet a user's needs.

Basic Usability Scope of Work (SOW) Requirements

An effective method to address usability requirements early is to influence the SOW prior to bidding the project. Examples of usability methods written into the SOW requirements include:

- Project shall have a dedicated User Experience (UX) Professional(s) involved "*from initial user requirements through the program life cycle to system disposal*" (DoDI 5000.95, 2022).
- The UX Professional(s) shall define user needs prior to design and development.
- The UX Professional(s) shall be involved in the product research and discovery phase, and provide recommendations to design, development, and training.
- The UX Professional(s) shall conduct "*usability and other user testing to support and inform human and machine interface analysis under operational conditions*" (DoDI 5000.95, 2022) prior to or during specified milestones.
- The UX Professional(s) shall provide the design and development team heuristic analysis of user interactions and interfaces.
- The UX Professional(s) shall conduct a benchmark evaluation at stage *XX* (or *XX*% product completion) and another one at completion of the product development.

These recommended requirements do not provide a measurable metric or an acceptable level of usability, however, including these SOW guidelines can greatly improve the usability of the product.

As of April 1, 2022, the Office of the Under Secretary of Defense for Research and Engineering (USD(R&E)) issued DoD Instruction 5000.95 – Human System Integration (HSI) in Defense Acquisition (DoDI 5000.95). DoDI 5000.95 states "*The DoD will utilize HSI in defense acquisition to provide a disciplined, unified, and interactive approach to integrate human considerations across system design to optimize total system performance and minimize life-cycle costs.*" This directive provides usability requirements that DoD is directed to implement along with a DoD HSI Guidebook that was published in May 2022 that provides additional usability implementation guidance.

Usability Metrics

Usability is often interpreted to pertaining to the aesthetics of the interface display. However, usability involves the entire interaction between the system and the user. Usability metrics use Key Performance Indicators (KPIs) such as:

- Task success or completion rate
- Time on task
- User error rate
- Product usability
- Product usefulness
- User workload
- User situational awareness
- Learnability

To transpose usability KPIs into requirements and specifications, acceptable KPI thresholds must be defined. Stakeholders may be able to provide some direction with the thresholds. Examples include: "The user must complete the task of turning on the map terrain 90% of the time, with an acceptable 10% error rate, within 1 minute. The user must rate the usability of this task with a minimum System Usability Scale (SUS) score of 70."

While defined parameters help in writing the requirements and specifications, how these metrics are measured must be defined so the method and process can be repeated for verification. Without defining the data capture and analysis procedures, the evaluation methods can be manipulated to get desired results. The following considerations include:

- What is being measured (general or specific task workflow interaction with a user interface)?
- How is it being measured (user study, surveys, observations)?
- Other factors, mainly pertaining to environment or biases that also need to be defined.
- Who are the evaluation participants (developers, subject matter experts, end-users)?

Usability can be difficult to define and measure which is why it is not typically included in requirements and specifications. Although more work needs to be done in the field of quantifying usability, having a dedicated UX professional or team of UX professionals will greatly improve the usability of the product.

Utilize Specialized Users

Unlike products or systems designed for the general public, DoD systems are generally designed for trained, specialized users, such as the warfighter. With these systems, it is part of the user's role to interact with the product and use this tool to complete their responsibilities. Too often, when errors are encountered with the system due to poor usability, the response is to train the user on how to avoid the error. This burdens the operator with extra cognitive recall instead of allowing the user to simply rely on recognition with intuitive interface features. If the system is intuitive, and designed to avoid errors from occurring, training can focus less on how to use the interface and more on enhancing the system's capabilities. Savings in training resources, such as instructor and student time commitments and cost, can be realized through an intuitive system or product.

Another factor overlooked is that while designers and developers interact with the system up to 40 hours a week, the warfighter or specialized user has many other duties that they are responsible for and their interaction with the system may be only a few hours a week or there is a large inactivity gaps between uses.

Lack of Usability Specialist

The demand for Usability Specialist Professionals is growing. It is becoming increasingly difficult to meet the labor and expertise demand, with professionals working on government contracts. Particularly with DoD programs, individuals may be required to be a US citizen, obtain a security clearance, and be willing to actively work with offensive and defensive systems.

When discussing a system or project, people often feel like they are human factors or UX experts and will have strong opinions on how the user will react. This unfortunate mentality is often the reason used to justify not having a separate usability expert included in the design and development phases. An often-made comment is "We've been doing it this way for years without a so called UX expert, and our products came out just fine. Plus, everyone on the team already understands what the user wants and needs."

There are three main components of a system: the software, the hardware, and the user. Within a project, there are usually specific disciplines involved, such as software developers, hardware engineers, even contract specialists. All these disciplines care about the end-product and want it to be accepted and embraced by the end-user. However, each of these disciplines' mindset have constraints based on their specialty. The software developer's perceived user needs may be bounded by limitations of the coding software, thereby limiting potential capabilities. A machinist who does not have a die that can produce a feature for a specific prototype may design the feature based on the dies they currently possess and not based on the user's needs. With project roles dedicated to software and hardware, it makes sense to have someone dedicated to the user and the usability of the system. Ideally this would be someone who is not constrained by the limitations of the tools within the software and hardware development processes.

Have a dedicated trained Usability Specialist

When managers realize they need a dedicated usability specialist, they usually convert someone from another discipline or use a former end-user. Having one or several individuals dedicated to only looking at product or system usability is a big step in the right direction. However, without human factors, UX, or usability formal training, their efforts may produce misleading or unintentional incorrect results.

Non-Revenue Generation

Most government and DoD systems are not developed for the purpose of generating revenue. They are designed to perform or accomplish a specific and complex task. In the private sector, products that provide a function are often designed to sell volumes of that product (e.g., Microsoft or Apple products, video games). Other private sector products may be developed as a method to sell items (e.g., retail websites) or their sustainment may rely on selling advertisements (e.g., Facebook, news outlets). When revenue is involved, additional KPIs can be utilized including knowing total number of clicks, number of pages viewed, order amounts, revenue, and more.

Conclusion

There are several factors contributing to usability shortcomings with DoD government funded systems development programs. Upfront funding model and lack of competition during the development phases are practices that are difficult to change. Having a UX Professional involved early, and if appropriate usability SOW requirements and specifications are generated, system usability should greatly improve. Even if the usability requirements and specifications do not stipulate any usability metrics, outlining good usability practices, procedures, and accountability will greatly advance the system. Requiring usability evaluations with end-users throughout the design and development phases will identify usability shortcomings early and greatly improve the overall system and the system's acceptance by the user.

About The Author

Jerry Burpee is a Usability Research and Evaluation Senior Principal Engineer for *INTUITIVE* Research and Technology Corporation in Huntsville, Alabama. He is a U.S. Navy veteran and is the Human-Machine Interaction (HMI) subject matter expert for many U.S. Army's Air and Missile Defense (AMD) and unmanned air vehicle (UAV) programs. For the past 15 years he and his team have conducted numerous military system usability evaluations with both Army and Air Force Warfighters. He has designed several usability evaluation laboratories, including the AMD HMI Mobile laboratory, a trailer built to go directly to the Warfighter to perform usability evaluations.

References

DODI 5000.95: Human Systems Integration in Defense Acquisition. (2022), Office of the Under Secretary of Defense for Research and Engineering, Department of Defense Instruction

ISO 9241-11:2018 Ergonomics of human-system interaction - Part 11: Usability: Definitions and concepts. (2018), International Organization of Standards

CREATING USABLE RESEARCH FOR THE DESIGN AND EVALUATION OF FLIGHT DECK SYSTEMS AND HUMAN INTERFACES

Divya C. Chandra
United States Department of Transportation (USDOT) Volpe Center
Cambridge, MA

This paper offers advice to researchers who want their research to be used by regulatory and industry practitioners to design and evaluate flight deck systems and their human interfaces. First, I present a few examples of success and review existing guidance. Next, I explain a design-thinking paradigm that views the design of a research study as a product. In this context, the users of the research (i.e., the product) are the practitioners. Researchers can smooth the path from research to practice by using this paradigm.

Many researchers in aviation human factors are motivated to make an impact in the real world. We do our best work with good intentions, thoughtful studies, and thorough analyses, managing hurdles along the way. Then we are surprised and disappointed to find that people who are in a position to apply our data to real systems do not do that, as though the results are not useful. How can we anticipate and prevent this scenario and, instead, smooth the transfer of research to practice? The benefits would be significant because we could improve safety, justify the investment in research, and perhaps even amplify the scope and longevity of its impact.

Here I share insights on how to create research for the flight deck that is usable and useful, not just to a pilot, but to potential users of the research results who are practitioners in industry (e.g., avionics manufacturers) and regulatory organizations (e.g., the Federal Aviation Administration, FAA, and other international aviation authorities). The principles are generalizable, but here I focus on the design of research aimed at improving flight deck systems, including their human-system interfaces. This is just one of several research needs for aviation practitioners. The same reasoning can be applied to other research needs (e.g., how to make flight operations more efficient while maintaining safety, or how to ensure that pilots have the right training and procedures to use existing flight deck systems effectively).

Human factors researchers working on flight deck systems typically see the pilot as the end user, but that is true only at one level. At another level, researchers can treat the *design* of a research study as a usability problem itself, where practitioners are the users of that product. When the design of the research study is seen as a *product*, it clarifies what steps researchers can take to help ensure a path to practice. I begin by presenting examples of research that have successfully impacted real flight deck systems. I also review advice on this subject from the FAA. Then I show how to apply a design-thinking paradigm to this problem.

Examples of Success and Existing FAA Research Guidance

The FAA Aviation Safety Office (AVS) is one main sponsor and user of flight deck human factors research. This office posts regulations and policy related to flight deck human factors issues within the FAA AVS website. Some research studies that have transferred to flight deck systems are listed on the FAA NextGen Human Factors Division website. For example, multiple studies done by the Civil Aerospace Medical Institute supported the development of

FAA regulations for low-visibility operations and use of Enhanced Flight Vision Systems (EFVS). Be aware, however, that regulatory and industry documents do not necessarily cite research reports; citations are not a requirement for successful transfer of research to practice.

Yeh et al. (2016) elaborates on what the FAA needs from research products (Appendix B, pp. 283-292). The report describes the roles, responsibilities, and background experience of FAA personnel who may use research products (Table B-1, p. 283). It also describes various uses of research, such as direct support of the design/evaluation of systems. An important point is that, to be legally enforceable, FAA *minimum* requirements for equipment must be tied to a specific regulation or to a Technical Standard Order (TSO) for specific components (e.g., avionics equipment). This may be unintuitive to researchers who try to improve systems, not just meet minimum standards.

Yeh et al. (2016) lists four categories of research needed by the FAA, with examples of exemplary products. Three of the four categories cover the synthesis and consolidation of existing knowledge: developing checklists for system evaluations, developing recommended requirements and guidelines (which may form a basis for the checklists), and industry reviews that help the FAA to understand how the market might drive approval needs. The fourth category focuses on experiments related to flight deck systems. Yeh et al. references Zuschlag, Chandra, & Grayhem (2013) as a positive example of such research. This study compared different symbol-fill options for flight deck displays of traffic information to examine the effect of symbol design on rating of threat by pilots. The report has clear research objectives and makes a direct connection to FAA guidance documents. The work was documented in both a full government report (Zuschlag et al., 2013) and a short paper presented at an engineering conference (Zuschlag, Chandra, & Grayhem, 2011). Both versions are publicly accessible online.

Design-Thinking Paradigm

Norman (2013) points out that design thinking is central to all innovation regardless of the domain. There are four main activities in human-centered design: observation, idea generation, prototyping, and testing. Design thinking also emphasizes iterative progress; the activities reoccur in a cycle. Table 1 below translates each design activity into activities for the design of human factors research to support practitioners. While the analogy does not transfer perfectly, the paradigm does remind researchers of helpful steps. I add one step to this list: communication of the results. Each section below takes a closer look at these activities.

Observation: Understanding the Practitioner

Understanding the needs of the user is a standard human-centered design activity. Here, the user is a practitioner who can apply human factors research. Regulatory practitioners regulate or evaluate flight deck systems. Industry practitioners might represent an operator or be involved in developing software or hardware to support flight deck tasks. Industry systems may be subject to review by regulators. All practitioners care about safety and minimum standards.

Table 1.
Applying a design-thinking paradigm to the creation of research to support practitioners.

Design Activity	Application to Design of Research
Observation	Understand and identify what practitioners know, what they do, and what they need.
Idea Generation	Generate multiple possible technical approaches and research methods.
Prototyping	Evaluate different possible technical approaches and methods through thought experiments. Imagine what the results might be and discuss the potential patterns of results with practitioners. Assess whether the different patterns would affect decisions or not.
Testing	After running the study, explain the results to a variety of stakeholders. Listen to the feedback and questions. Refine the takeaway messages by addressing criticisms and limitations of the work.
Communication	Present results clearly and succinctly, focusing on the practitioners' needs.

Many practitioners are not trained researchers and instead have a different knowledge base. Practitioners might be trained as engineers or pilots. Most are not trained formally on human factors subjects. Some field evaluators might have trained in the military or may not be college graduates. Despite the knowledge differences, both regulators and industry practitioners are familiar with FAA policies and guidance documents, and industry guidance. They understand the capabilities of the technology, how it may evolve, and how it would be used. Yeh et al. (2016) provides more detail on the different types of FAA users of human factors research.

Regulators may use human factors research to update regulatory documents including policies and guidance to help evaluate systems. Different types of evaluations happen at different levels of system maturity (e.g., bench tests in a laboratory setting, flight-simulator tests, in-flight tests, and operational implementation). FAA and industry practitioners may use human factors research to develop standards and recommendations (e.g., voluntary industry standards, or standards that are cited by as a possible means of compliance by FAA).

Flight deck systems are typically evaluated by different parts of the FAA's Office of Aviation Safety. In brief, the Aircraft Certification Service approves that the equipment works "as intended" and does not interfere with other equipment. (See Yeh et al., 2016, for an introduction to the concept of "intended function.") The Flight Standards Service reviews the equipment from the perspective of the human operator; it approves the use of the equipment by the flightcrew. The operational approval establishes the crew training necessary to operate the equipment under various scenarios, including full or partial failures.

The goals of practitioners become clearer in the context of their job responsibilities and demands. For example, they may have to ensure that the system meets minimum standards for the intended function in (sometimes messy) real-world situations. They consider all conditions under which that system may be used in the context of the full flight task. In contrast, researchers often prefer to study a limited set of conditions to isolate the issues of interest. For example, pilots constantly switch between different flight deck systems. How are human interfaces that support different functions integrated within the limited physical (and screen) real estate of the flight deck? In turbulence, or with smoke in the flight deck? These are real conditions that practitioners consider, which researchers might not.

Practitioners are less able to take the time to understand details that may be important to researchers (e.g., about the study design or statistical analysis). They appreciate short documents (or just Executive Summaries) that highlight key points, especially if these writeups make direct connections to regulatory and guidance documents that they use. Human factors researchers can help practitioners absorb key messages from research studies quickly and effectively so that they know how, when, and whether to apply the results of the research.

Idea Generation: Designing a Technical Approach

Researchers have different goals for applied research. They might test a system to determine if it (a) could work *at all*, (b) could work *better than* an existing design, either in terms of efficiency or functionality, or (c) meets guidelines or minimum standards. Option (c) is important to regulators but unrelated to optimization or efficiency. Industry researchers may also be interested in conducting the types of studies suggested by options (a) and (b).

The study needs to address a problem statement with a clear scope, clear motivation, and clear purpose for the results. Researchers benefit from the engagement of practitioners in developing the problem statement. Together they can refine and craft an initial proposal. For example, how will the data inform specific government policies or industry products? Sometimes problem statements need to be broken down into reasonable steps towards addressing a bigger issue. Sometimes initial problem statements are overly specific and could promote studies that are not generalizable. As with other design exercises, developing a clear problem statement is an iterative process.

When designing a study for practitioners, researchers should be creative and flexible, considering a range of options. Often, researchers have specialized knowledge of some methods and less expertise in others, but it is important to customize the research method for the problem the practitioner needs to address. The researcher should be willing to apply unfamiliar or even novel approaches. They should consider the full range of data that could be gathered, including subjective data, objective data, quantitative data, and qualitative data. They should even consider whether an existing data set could produce useful insights with further analysis. Collecting data on human performance is important (because we know that preference does not equal performance), but we need to collect data that are most relevant to the practitioners' goals.

For research to transfer to practice, data should have face validity and operational significance. One means of increasing face validity is for study participants to be realistically representative of the real users (e.g., airline pilots) as well as possible within cost constraints. Operational significance can be affected by the level of task and equipment fidelity; researchers should consider various levels carefully to maintain operational relevance while staying within budget. Sometimes less expensive, lower fidelity approaches are better because they focus on the problem statement at hand. For example, using a flight simulator may introduce distractions that draw attention away from the task of primary interest (e.g., using aeronautical charts).

Prototyping: Assessing Potential Research Methods

Once researchers have sketched out multiple study approaches, they should "test drive" these approaches with thought experiments before finalizing the specific methodology. In a thought experiment, the researcher hypothesizes all potential patterns of results and determines, in consultation with the practitioners, how each of the patterns might affect practitioner decisions. If none of the possible result patterns would change what the practitioner does, then that approach is not useful. Sometimes, practitioners unknowingly advocate for research approaches that, in the end, would not affect their policy/evaluation decisions (e.g., requesting use of a flight simulator when it may not be necessary). Researchers may need to sketch out all the possible results and their lack of impact on decisions to demonstrate that those studies are not worth doing. One good strategy is to ensure that there are lessons to be learned regardless of the pattern of results. Perhaps different result patterns would support different recommendations and considerations, if not specific policies and guidance.

Sometimes thought experiments reveal weaknesses in the problem statement or the research method. If it is not easy to hypothesize potential results and interpretations, then the problem statement needs to be clarified. Keep revising it until you can state what practitioner need(s) will be addressed and how the data collected will be used. It is also useful to document and critique methods that were considered but discarded. For example, why did that method not meet the needs at the time? What would have to change for that method to be useful?

Testing: Refining the Research Takeaways

After the research results are in, but before takeaways are finalized, socialize the study and its interpretation. Present the study even as the analysis is in progress. Present to a variety of stakeholder audiences. Gather feedback on preliminary takeaways, i.e., conclusions, highlights, and recommendations. This is especially important if the results connect to guidance documents because guidance needs to be clear and acceptable to many audiences. The draft takeaways, and even the analyses if needed, should be enhanced based on stakeholder questions and feedback.

Recommendations are a particularly important type of takeaway. Well-constructed recommendations are easy to read, actionable, they make sense, and are believable (i.e., have face validity). They are clear about their scope and limitations. Maximize the detail in the recommendation without going beyond what the data support. Flight deck human-interface recommendations should be traceable to their impact on pilot tasks. Ideally, recommendations should converge with findings from other sources (e.g., data from other studies or industry working group discussions).

However, keep in mind that recommendations from research do not always give specific *solutions* to flight deck problems because they do not take into account all the constraints for the problem. Let the appropriate stakeholders (not researchers) determine who implements a solution and how. Industry practitioners are partners; they want to understand what the problems are and the principles and rationale that move towards a solution, without being overly constrained. Industry is happy to use insightful findings that they can tailor to their situation.

Communication: Documenting the Study

Once key takeaway points are settled, researchers must communicate them effectively to practitioners. The communications should be brief, to the point, clear, and public. Researchers typically document studies for other researchers who may want to replicate a study, but practitioners just want to know why the results matter. The most useful elements of a report for practitioners are the Executive Summary and recommendations. An Executive Summary summarizes the study and key takeaways in one page ideally, or a few pages at most. It succinctly explains the purpose of the research and the main results relevant to regulatory policies or industry products. These points also should stand on their own because, realistically, they might be the only aspects of the research that circulate widely among practitioners.

Researchers also need to critically assess and acknowledge the limits of their study. An honest assessment will greatly help practitioners determine when and whether to apply the results in practice, improving trust between researchers and practitioners.

Summary

The generalized human-centered design process is iterative and combines observation, idea generation, prototyping, and user testing. I add one final step to this list, to communicate the results and fine tune the takeaway messages. To create usable research, treat the research as a product that will be used by a practitioner.

Although flight deck human factors research problems will change, this design paradigm for creating usable research will apply in general. Cooperative and open dialogue between practitioners and researchers is necessary. Researchers need to do their homework to learn about practitioners' needs, and they need to be creative and willing to learn new ways to think about the problem. The overall goal is to realize the safe and effective use of flight deck equipment in operations. Design thinking will lead the way.

Acknowledgements

The views expressed herein are those of the author and do not necessarily reflect the views of the Volpe National Transportation Systems Center or the United States Department of Transportation (USDOT). I would like to thank Bill Kaliardos, Tracy Lennertz, Sherry Chappell, Janeen Kochan, Don Fisher, Maura Lohrenz, and Stephen Popkin for their support and review of early drafts of this paper. This work was funded by the USDOT Volpe Center.

References

Norman, D. (2013) The Design of Everyday Things. Basic Books.

Yeh, M. Swider, C., Jo Y.J., and Donovan, C. (2016) Human factors considerations in the design and evaluation of flight deck displays and controls. Version 2.0 DOT/FAA/TC-16/56. https://rosap.ntl.bts.gov/view/dot/12411

Zuschlag, M, Chandra, D.C., and Grayhem, R. (2013). The Usefulness of the Proximate Status Indication as Represented by Symbol Fill on Cockpit Displays of Traffic Information, DOT-VNTSC-FAA-13-03; DOT/FAA/TC-13/24. https://rosap.ntl.bts.gov/view/dot/9982

Zuschlag, M., Chandra, D.C., & Grayhem, R. (2011). The Use and Understanding of the Proximate Status Indication in Traffic Displays. *Proceedings of the 30th Digital Avionics Systems Conference*, 16-20 October 2011, Seattle, Washington. https://rosap.ntl.bts.gov/view/dot/9513

INTERNATIONAL STUDENTS SENSE OF BELONGINGNESS AND MOTIVATION ON ACADEMIC AND FLIGHT PERFORMANCE

Sophie M. A. Chanoux
Embry-Riddle Aeronautical University
Daytona Beach, Florida
Andrew R. Dattel
Embry-Riddle Aeronautical University
Daytona Beach, Florida

Motivation, confidence, and internal achievement factors such as locus of control (LOC) and self-efficacy are important in successful learning. A feeling of belongingness might affect students' confidence, therefore affecting flight training performance. This study explored the relationship between self-reports of social activities and confidence with academic performance and flight performance. Nineteen international students (13m, 6f) with a mean age of 21.42 (SD = 2.29), currently enrolled in a flight training program at a university answered a survey. Significant correlations were found between LOC and confidence in the English language; self-efficacy and number of failures at the end of the Private Pilot course; confidence in the English language and social involvement; and flight training confidence and social involvement. Males reported significantly higher levels of flight training confidence than females. A regression model showed that flight training confidence can be significantly predicted by students' self-assessed sense of belonging, academic confidence, and LOC.

There are many challenges international students face that domestic students may not face, which can include language barriers, new environments and cultures, and homesickness (Madden-Dent et al., 2019). The purpose of this study was to analyze the independent variables that can influence international students' level of achievement in flight training programs operationalized as the number of checkride failures they have had. The level of achievement in flight training was the dependent variable. The independent variables were the students' sense of belonging, the students' attendance to social events on campus, the students' attendance to social events with friends, the students' confidence in academic performance, the students' confidence in flight training performance, the students' confidence in using the English language, the students' involvement with the community, the students' academic performance, the students' self-efficacy, and the students' locus of control.

Previous research (Madden-Dent et al., 2019) has focused on international students' academic achievement, but not on international flight students' flight training achievement. The purpose of this study was to find out if factors such as the students' sense of belonging, attendance to social events on campus or with friends, confidence in academic performance, flight training performance, and the English language, involvement with the community, academic performance, self-efficacy, and locus of control are related to the students' performance in flight training.

Research Questions and Hypotheses

The following research questions were developed.

R1: Do international students' sense of belonging, level of involvement with the community, attendance of social events on campus, attendance of social events with friends on and off campus, academic confidence, flight training confidence, English level confidence, locus of control, and self-efficacy affect the students' number of checkride failures?

R2: Does the gender of international students affect their flight training confidence and academic confidence?

R3: Do locus of control, sense of belonging, and academic confidence predict flight training confidence?

The following null hypotheses were tested.

$H_0 1$: There is no significant relationship between sense of belonging in international students and number of checkride failures.

$H_0 2$: There is no significant relationship between level of involvement with the community in international students and number of checkride failures.

$H_0 3$: There is no significant relationship between international students' attendance of social events on campus and number of checkride failures.

$H_0 4$: There is no significant relationship between international students' attendance of social events with friends on and off campus and number of checkride failures.

$H_0 5$: There is no significant relationship between academic confidence in international students and number of checkride failures.

$H_0 6$: There is no significant relationship between flight training confidence in international students and number of checkride failures.

$H_0 7$: There is no significant relationship between English level confidence in international students and number of checkride failures.

$H_0 8$: There is no significant relationship between locus of control in international students and number of checkride failures.

$H_0 9$: There is no significant relationship between self-efficacy in international students and number of checkride failures.

$H_0 10$: The means for international males and females in terms of flight training confidence are equal.

$H_0 11$: The means for international males and females in terms of academic confidence are equal.

$H_0 12$: There is no significant relationship between international students' locus of control, sense of belonging, academic confidence, and flight training confidence.

Review of the Relevant Literature

Prior research has demonstrated that four factors determine international students' confidence of academic success. Those factors are community acceptance, language ability, academic ability, and financial stability (Telbis et al., 2014). International students entering a large collegiate flight program with higher English scores are more successful in obtaining their private pilot flight certification, perform better academically, and perform better in their flight courses (Dusenbury & Bjerke, 2013). Belonging, self-efficacy, behavioral, and emotional engagement play an important part in STEM courses (Wilson et al., 2015). In a study conducted in the aviation industry, it was found that self-efficacy, work engagement, and human error were significantly correlated with each other (Li et al., 2021). Locus of control plays an important part in aviation because it has been linked to hazardous events. A positive correlation was found between having an external locus of control and being involved in hazardous events (Joseph et al., 2013). The main obstacles female pilots face are lack of acceptance, self-efficacy, lack of social support, and stereotyping (Germain et al., 2012). Campus resources, such as writing and student success centers, or counseling centers, can help international students face challenges such as English proficiency or homesickness (Banjong, 2015).

Methodology

A survey was distributed by email and consisted of 68 questions. There were 14 demographic questions, 10 questions on academic, flight, and language experience, seven questions asking to self-assess seven variables on a Likert scale, a 29-item Rotter's Locus of Control Scale, and an eight-item self-efficacy scale. Two-tailed Pearson's correlation coefficients, independent samples *t*-tests, and a regression model were computed to test the null hypotheses. The survey was sent to every international flight student enrolled in the Aeronautical Science Bachelor's Degree at Embry-Riddle Aeronautical University in Daytona Beach, Florida. The participation requirements were that the participant must be enrolled as an international flight student and must have completed at least their Private Pilot course, but no more than their Instrument course, at ERAU. The data collection devices were a computer and the program Statistical Package for the Social Sciences (SPSS). The software package SPSS was used to analyze the results and see if there were any significant relations between the independent variables and the dependent variable. The variables that needed to be coded into groups were gender, continent based on the countries, first language, class standing, participation in organizations, and language test requirement. The two scales needed to be scored. Hypothesis testing was conducted through two-tailed Pearson's correlations for $H_0 1$ through $H_0 9$, independent samples *t*-tests for $H_0 10$ and $H_0 11$, and a regression model for $H_0 12$. 15.8% were from South America, 42.1% were from Asia, 5.3% were from Africa, 15.8% were from Europe, and 21.1% were from North America.

Results

Descriptive Statistics

Demographics results. There were 19 participants (male: n = 13, female: n = 6) who completed the survey. 15.8% of participants were from South America, 42.1% were from Asia, 5.3% were from Africa, 15.8% were from Europe, and 21.1% were from North America. 21.1% had English as a first language, and 78.9% did not. Table 1 depicts the mean, standard deviation, minimum and maximum for age, years spent in the United States, GPA, number of clubs, and private failures.

Table 1.

Descriptive Statistics for the Demographics of the Participants

Variable	*M*	*SD*	*Min.*	*Max.*
Age	21.42	2.29	19	28
Years in US	3.62	3.34	1	12
GPA	3.72	0.35	2.50	4.00
Clubs	1.68	1.70	0	6
Private Failures	0.26	0.56	0	2

Self-assessments results. The seven self-assessments variables in the survey were sense of belonging in the university, involvement in the university social life and community, attendance to social events on campus, attendance to social events with friends on and off campus, confidence in academic performance, confidence in flight training performance, and confidence in the ability to speak, read, write and understand the English language. Table 2 depicts the mean, standard deviation, minimum, and maximum for each of the variables.

Table 2.

Descriptive Statistics for the Self-Assessments

Variable	*M*	*SD*	*Min.*	*Max.*
Sense of Belonging	5.79	1.47	2	7
Involvement	4.84	2.24	1	7
Campus Events	3.68	2.00	1	7
Friends Events	5.68	1.70	1	7
Academic Confidence	6.26	1.24	3	7
Flight Confidence	5.52	1.58	2	7
English Confidence	6.21	0.92	5	7

Hypothesis Testing

Private failures. A two-tailed Pearson's correlation was computed to test the null hypothesis $H_0 9$, stating that there is no relationship between number of private failures ($M = 0.26$, $SD = 0.56$) and self-efficacy ($M = 4.39$, $SD = 0.63$). At a .05 alpha level, there was a significant positive correlation, $r(17) = -.479$, $p = .038$. Therefore, the null hypothesis was rejected. Table 3 describes the correlations between the number of private failures and the seven self-assessed variables and two scales.

Table 3.

Pearson Correlation Matrix for Private Failures

Private Failures	Pearson Correlation	Sig. (2-tailed)
Sense of Belonging	-.479	.038
Involvement	.079	.748
Campus Events	-.070	.775
Friends Events	-.431	.065
Academic Confidence	-.344	.149
Flight Confidence	-.290	.228
English Confidence	-.221	.363

Gender, flight training confidence, and academic confidence. The null hypothesis $H_0 10$, which stated that there will be no significant difference in flight training confidence between international females and males, was statistically analyzed through an independent samples *t*-test. The assumption of equality of variance was tested. Levene's test of equality of variance for flight training confidence was not significant ($p > .05$). Equal variances were assumed. The mean of self-assessed flight training confidence for females ($M = 4.33$, $SD = 1.75$) was lower than the mean of self-assessed flight training confidence for males ($M = 6.08$, $SD = 1.19$). An independent samples *t*-test was significant at the alpha level of .05, $t(17) = 2.565$, $p = .020$. Therefore, the null hypothesis was rejected.

Locus of control, sense of belonging, academic confidence, and flight training confidence. The null hypothesis $H_0 12$ stated that there is no significant relationship between international students' locus of control, sense of belonging, academic confidence, and flight training confidence. Multiple linear regression was used to test if locus of control, sense of belonging, and academic confidence significantly predicted flight training confidence. The fitted regression model was Y = .419(Sense of Belonging) + .56(Academic Confidence) – 1.38 (Locus of Control) + 1.34. The overall regression was statistically significant ($R^2 = .789$, $F(3, 15) = 23.441$, $p < .001$). It was found that sense of belonging significantly predicted flight training confidence ($\beta = .392$, $p = .004$). It was found that academic confidence significantly predicted flight training confidence ($\beta = .440$, $p = .003$). It was found that locus of control significantly predicted flight training confidence ($\beta = -.381$, $p = .006$). Therefore, the null hypothesis was rejected.

Discussion, Conclusions, and Recommendations

Significant correlations were found between LOC and confidence in the English language; self-efficacy and number of failures at the end of the Private Pilot course; confidence in the English language and social involvement; and flight training confidence and social involvement. The correlation between locus of control and confidence in the English language shows that students who display an internal locus of control also tend to display higher confidence in their ability to speak, read, and write the English language. The correlation between self-efficacy and the number of failures at the end of the Private Course means that students who had a low self-efficacy score also tend to have 1 or 2 checkride failures, as opposed to 0. The correlation between flight training confidence and involvement shows that as international students rate their confidence in flight training higher, they also tend to feel more involved in the university life. The independent-samples *t*-test showed that males have higher confidence than females when it comes to flight training confidence. The regression model showed that flight training confidence can be predicted with students' self-assessed sense of belonging, academic confidence, and locus of control scores. It is important to know what affects flight confidence, even though it is not the same as performance.

The analyses that were run and were significant, such as the regression model on what affects flight training confidence and the difference in levels of confidence for flight training performance based on gender, are both interesting to consider and refer to for further research on these topics. Additionally, there were results that were significant when looking at the independent factors and analyzing them with each other. The significant results from the correlations strongly reinforce what has been seen in the available literature on self- efficacy and locus of control. This is useful in determining that the available literature can be applied to a non-traditional collegiate international flight student population. Adapting the study to include different factors that can quantify success, such as number of days to complete the course, or total cost of training, may help design a study that will yield more insight on the topic. Recommendations for colleges interested in enhancing the experience and success of their international students include the creation of more community-oriented clubs for international students, providing international students with resources and knowledge of the country they are moving to, and setting up mentorship programs that pair new international students with domestic students to help them become more familiar with their new environment.

Acknowledgements

Throughout the development of my research study, I have received support from many different people and organizations. Thank you, Dr. Thropp, for your expertise, knowledge, support and advice which have been vital to the completion of this project. I would also like to thank all faculty and staff of the College of Aviation. I would like to thank my advisor BeeBee Leong, Dr. Dattel, my classmates, and all the graduate assistants for their constant support, patience, knowledge, and kindness.

I would also like to thank the flight department at Embry-Riddle Aeronautical University for allowing me to send out my survey to the college's flight students and funding my attendance at the International Symposium on Aviation Psychology. Thank you as well to the Office of Undergraduate Research for awarding me with a Spark Grant for this project. It should be noted that the views of the research reported do not reflect the views of the granting organization.

Thank you to all my friends, and to my family, especially my Mom and Dad. You've believed in me and helped me follow my dreams wherever they lead me since the day I was born, and I am incredibly grateful for that.

References

Banjong, D. N. (2015). International Students' Enhanced Academic Performance: Effects of Campus Resources. *Journal of International Students*, *5*(2), 132–142. https://doi.org/10.32674/jis.v5i2.430

Dusenbury, M., & Bjerke, E. (2013). Predictive power of English testing: Training international flight students. *Journal of Aviation/Aerospace Education & Research*, *23*(1), 13–22. https://doi.org/10.15394/jaaer.2013.1601

Germain, M.-L., Herzog, M. J. R., & Hamilton, P. R. (2012). Women employed in male-dominated industries: lessons learned from female aircraft pilots, pilots-in- training and mixed-gender flight instructors. *Human Resource Development International*, *15*(4), 435–453. https://doi.org/10.1080/13678868.2012.707528

Joseph, C., Reddy, S., & Kashore Sharma, K. (2013). Locus of Control, Safety Attitudes and Involvement in Hazardous Events in Indian Army Aviators. *Aviation Psychology and Applied Human Factors*, *3*(1), 9–18. https://doi.org/10.1027/2192-0923/a000036

Li, Y., Liu, Z., Lan, J., Ji, M., Li, Y., Yang, S., & You, X. (2021). The influence of self- efficacy on human error in airline pilots: The mediating effect of work engagement and the moderating effect of flight experience. *Current Psychology (New Brunswick, N.J.)*, *40*(1), 81–92. https://doi.org/10.1007/s12144-018-9996-2

Madden-Dent, T., Wood, D., & Roskina, K. (2019). An Inventory of International Student Services at 200 U.S. Universities and Colleges: Descriptive Data of Pre- Departure and Post-Arrival Supports. *Journal of International Students*, *9*(4), 993–1008. https://doi.org/10.32674/jis.v9i4.346

Telbis, N. M., Helgeson, L., & Kingsbury, C. (2014). International students' confidence and academic success. *Journal of International Students*, *4*(4), 330–341. https://files.eric.ed.gov/fulltext/EJ1054787.pdf

Wilson, D., Jones, D., Bocell, F., Crawford, J., Kim, M. J., Veilleux, N., Floyd-Smith, T., Bates, R., & Plett, M. (2015). Belonging and Academic Engagement Among Undergraduate STEM Students: A Multi-institutional Study. *Research in Higher Education*, *56*(7), 750–776. https://doi.org/10.1007/s11162-015-9367-x

HOW AVIATION STUDENTS USED AN ESCAPE ROOM TO ASSESS SA FROM A COLLABORATIVE AND INDIVIDUAL APPROACH

Andrew R. Dattel
Hui Wang
Corey Spitz
Miles Moyer
Kyhan Gulsen
Embry-Riddle Aeronautical University
Daytona Beach, FL

A class on situation awareness (SA), taught at an aeronautical university, has been a favorite graduate course for the past 7 years. One requirement for the course is to conduct a non-aviation related SA class project. The most recent class assessed how collaborative efforts compared to individual efforts affect SA at an escape room. The team of five participants instructed to solve puzzles individually showed significantly better SA than the team of five participants instructed to solve the puzzles collaboratively. The collaborative group spent more time conversing with each other about solving the puzzles than those instructed to work individually. Students commented that the experience helped them understand not only how to design a SA research project, but to have a better understanding of SA. As with previous classes, these research projects developed by the students continue to be engaging and an effective pedagogical approach to students understanding of SA.

In complex and dynamic safety critical industries, the importance of situation awareness (SA) can be frequently found in the industry's respective selective and training materials. SA conceptualizes how one can comprehend the relevant elements in conditions where information and status can rapidly change (Durso et al., 2007). Endlsey's (2015) three interchangeable levels include a person's perception (the ability to perceive stimuli relevant to the condition), understanding (what that stimulus means or foretells), and projection (how the situation may or can change in the future).

SA has its roots in aviation and where SA research seems to be most prolific (Endsley & Jones, 2016). It would be exceedingly rare that a veteran pilot, or even a student in flight training would not have a definition of what SA is and even a willingness to give an example. In fact, aviation handbooks frequently mention SA (Federal Aviation Administration, 1991). Classroom instruction in collegiate aviation courses (e.g., Private Pilot training, Air traffic Control) typically devote substantial time that may include lively discussion about SA. However, it is uncommon to find a college course solely dedicated to SA.

In 2016, I began teaching a class titled, *Situation Awareness and Performance in the Aerospace/Aviation Industry*. This graduate course, taught at an aeronautical university, attracts students from different majors and of different interests (e.g., flight instructors, maintenance technicians, human factors specialist). This popular course is taught annually, and demand now requires it to be taught for two consecutive semesters (at least for this year). One important criterion about the format of the course is that it is presented in a seminar format. Class settings include a conference room, rather than a traditional classroom. Although there are no exams in the course, the material and assignments can be rigorous and challenging.

Because learning is an active process, the course is designed to encourage critical thinking and inquiry-based learning. This process requires forging, exploration, and discovering ideas; and then the application of those ideas (Levy & Petrulis, 2012). Inquiry-based learning, when coupled with individual and classroom projects, helps students understand, apply what is learned, analyze one's actions and decisions, then evaluate for successful and improved performance (Bloom, 1956), Finally, students had to explore some non-aviation material and apply SA concepts to those different areas. These goals were achieved by designing the format of the course based on the following:

- Readings and group discussions of about 15 to 20 peer-reviewed articles and book chapters
  - These readings start with seminal and theoretical articles and move to more applied articles (including non-aviation topics)

- Individual presentatiods of National Transportation Safety Board (NTSB) Report where SA was listed as a primary or secondary factor
  - All NTSB reports had to be non-aviation

- Ask friends or family members to provide examples of good and bad SA that they experienced
  - Student would report to the class what their friend or family member said, then evaluate if that example reflected what SA is

- Individual research proposal of an experiment using a published SA measurement (e.g., SAGAT, SPAM)

- Class project, where the class collects SA data

Feedback from students throughout the years has been positive. Many flight instructors who have taken the course commented that at the beginning of the course they felt confident they knew everything about SA (because they are flight instructors); however, by the end of the course they realized how much more there was to know about SA. Some flight instructors even commented that the course helped them to become better instructors. Some selected student course evaluation comments from the most recent class are:

- "The class "research project" was the assignment that most helped me learn the content in this course. Rather than summarizing or analyzing situations where situation awareness was high/low/lacking/etc. (i.e., with the SA interviews/NTSB accident presentations), the project required everyone to synthesize all course learning outcomes to plan, coordinate, and execute the project and identify ways to measure our classmates' SA during the project experience"

- "The group discussion, instead of lecture style"

- "To my surprise not only did the in class discussions greatly improve my understanding of Situation Awareness, but also the reading materials. The professor provided an important moderator role in the class that kept the class not only engaged but also did not ruin the primacy of others. Another unique element was the escape room. As far as I could tell this gives students who are new to the program a great opportunity to see how research is done."

- "It was a great experience overall, and this is the way that I wish academia was taught"

As can be seen from the student comments, the class project (i.e., Escape Room) was highlighted as a favorite and effective activity. During the course the students must develop and execute a project where data is collected, and SA is measured. These class projects have varied throughout the year and include collecting data from patrons and employees at EPSCOT (Dattel et al., 2017); attending a minor league baseball game where students interviewed players and "worked" in concession stands; observed SA of players on teams at a Paintball game; interviewing the captain, collecting data of customers, and interviewing dealers on a gambling boat casino; and observing vehicles and pedestrians remotely at crosswalks throughout the world (Dattel et al., 2021).

Method

The most recent class conducted an experiment where the students in the class were participants. The students decided to use an Escape Room as a setting to collect data. Escape Room are typically commercial businesses where customers have up to 1 hour to figure out 13 puzzles or clues to escape the room. The theme used for this Escape Room was to investigate Dr. Brown, a research scientist employed at Area 51 (US Air Force Nevada Test and Training Range) to determine if he was hiding some suspicious research. Customers had access to Dr. Brown's office while he was at lunch. The customers had to search for clues in his office to find any covert, suspicious research he may have been conducting.

Participants

Twelve students participated in the research (one student was unable to attend the venue). Five students were randomly assigned to one of two teams (the Axolotl Team and the Capybara Team). The other two students conducted the experiment by deciding on the manipulation and collecting the data.

Materials and Procedure

The manipulation was to instruct one team that the best method to figure out the puzzles and clues was to "divide and conquer" (the Capybaras). That is, they were instructed to determine the answers to the clues and puzzles individually, but to work as a team when needed. The other team (the Axolotls) was instructed that the best method to complete the puzzles and clues was to cooperate and work as a team. The participants were asked 10 SA questions during the time they were in the escape room (e.g., How many clues do you have left? How do you reset the directional lock before entering the code? How much time is left for you to escape?). SA questions were asked over a loudspeaker set up in the room. Participants were instructed to answer questions as fast and as accurately as possible. Aligned with the Situation Present Assessment Measure (SPAM), accuracy and response time were collected as measurements for SA (Durso & Dattel, 2014). Participants on both teams were instructed to answer the SA questions individually, and to consult each other for the answer only when necessary.

Results

There were no significant differences between the teams for the total time to escape the room, with the Divide and Conquer team taking 57 minutes and 35 seconds and the Cooperation team taking 59 minutes and 24 seconds. The Divide and Conquer team (M = 1.033 seconds, SD = 0.583) answered SA questions faster (See Figure 1) than the Cooperation team (M = 2.729 seconds, SD = 1.551). Thus, the Divide and Conquer team has better SA than the Cooperation team $t(8.945) = 2.894$, p =.018 (after adjusting for unequal variances). The total time teams spent talking to each other was measured in 5-minute intervals. The Cooperation team spent more time talking to each other (M = 136.45 seconds, SD = 74.70) than the Divide and Conquer team (M = 89.91 seconds, SD= 31.08). $t(20) = 1.908$, p = .07 (See Figure 2).

Figure 1.
Response Time to SA Questions for Divide and Conquer Team and Cooperation Team

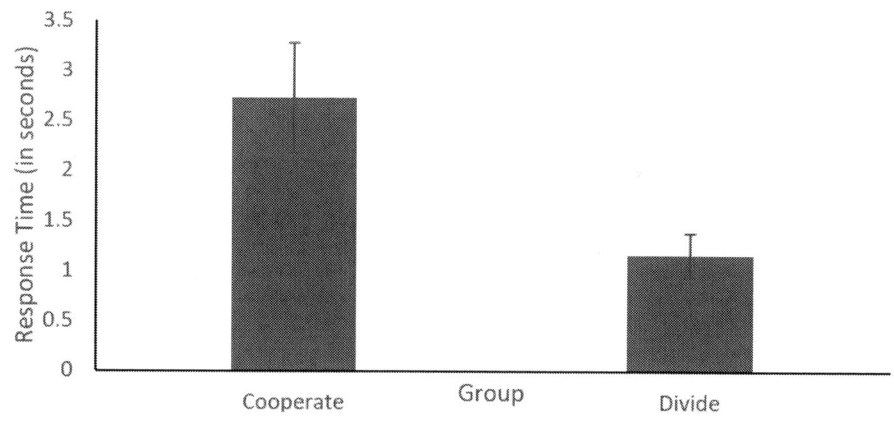

Figure 2.
Average Time Team Members Spent Talking to Each Other in 5-minute Intervals

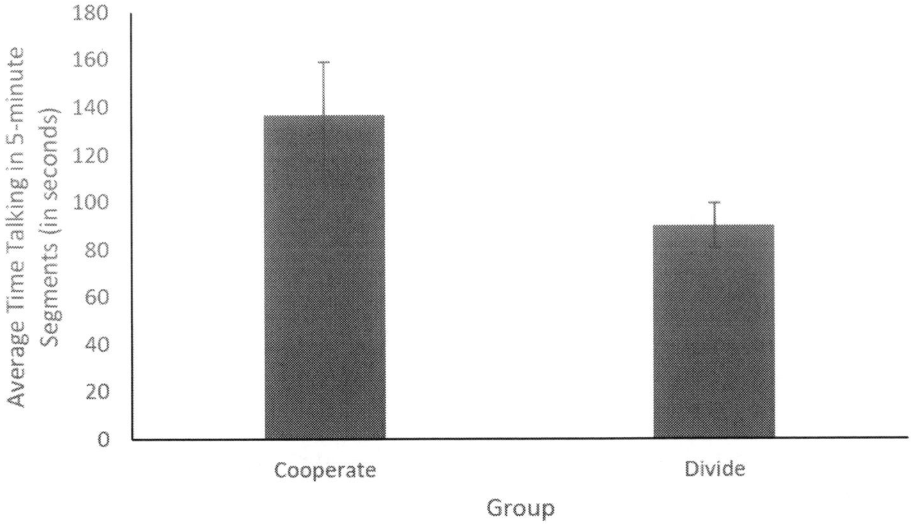

Discussion

The students in this SA and Performance class successfully achieved the course objectives. Although the class project is only part of the course requirements, the students in this class designed an experiment, created stimuli, participated in the study, and analyzed data that showed interesting and significant results between team manipulation. It should be noted that each class designs their own class project, and the project must be different from any previous year.

The tools and elements of this course resonate with the students where they report that their understanding of SA has greatly improved. In this class, students had the opportunity to experience all aspects of conducting an SA study, in addition to the other class requirements. These students should be commended for designing a study that illustrated how to successfully conduct SA studies. Not only do the students gain a better understanding of SA in this course, but I, as the instructor, also appreciate the effort of each student's commitment and accomplishments.

Acknowledgement

The authors would like to thank Kayla Taylor for her contribution to the design and execution of this study.

References

Bloom, B. S. (1956). *Taxonomy of educational objectives* Vol. 1: *Cognitive Domain*. McKay.

Dattel, A. R., Babin, A., Li, T., Dong, Z, Fussell, S. G., & Yang, Q. (2017). A pedagogical approach to teach aviation students how to conduct situation awareness research. Proceedings paper published in the Proceeding of the *2017 International Symposium on Aviation Psychology*, 443-448, Dayton, OH.

Dattel, A. R., Wang, H., Booker, N., Matveev, A., Haris, S. R. M., & Xie, H. (2021). How to teach college aviation students about situation awareness in a virtual setting. Published in the Proceedings of the *2021 International Symposium on Aviation Psychology*, 98-103.

Durso, F. T., Rawson, K.A., & Girotto, S. (2007). Comprehension and situation awareness. In F.T. Durso, R. S. Nicerson, S. T. Dumais, S. Lewandowsy, & T. J. Perfect (Eds.). *Handbook of applied cognition* (2nd ed.) (pp. 163-193). Wiley & Sons.

Endsley, M. R. (2015). Situation awareness misconceptions and misunderstandings. *Journal of Cognitive Engineering and Decision Making* *9*(1), 4-32.

Endsley, M. R. & Jones, D. G. (2016). *Designing for situation awareness: An approach to user-centered design* (2nd ed.). CRC Press.

Federal Aviation Administration. (1991). *Aeronautical decision making* (AC No. 60-22). U.S. Department of Transportation. https://www.faa.gov/documentLibrary/media/Advisory_Circular/AC_60-22.pdf

Levy, P., & Petrulis, R. A. (2012). How do first-year university students experience inquiry and research, and what are the implications for practice of inquiry-based learning? *Studies in Higher Education, 37*(1), 85-101. https://doi:10.1080/03075079.2010.499166

PREVENTING SCENARIO RECOGNITION IN HUMAN-IN-THE-LOOP AIR TRAFFIC CONTROL RESEARCH

Gijs de Rooij, Clark Borst, M. M. (René) van Paassen and Max Mulder
Aerospace Engineering - Delft University of Technology, Delft, The Netherlands

In academic air traffic control research, traffic scenarios are often repeated to increase the sample size and enable paired-sample comparisons, e.g., between different display variants. This comes with the risk that participants recognize scenarios and consequently recall the desired response. In this paper we provide an overview of mitigation techniques found in literature and conclude that rotating scenario geometries is most frequently used. The potential impact of these transformations on participant behavior, as described in this paper, is however not sufficiently addressed in most studies. As an example we, therefore, analyze previously collected eye tracking data from ten professional air traffic controllers, each presented with three repetitions in various rotations of several distinct scenarios. Results imply that researchers wishing to repeat scenarios should more carefully consider whether mitigation techniques might have an impact on their results.

Introduction

In air traffic controller (ATCO) training and airspace redesign trials, simulation scenarios are designed to be as realistic as possible with many different flights over a prolonged period of time. High face validity enables the ATCOs to execute their tasks as they would in an operational setting. Academic research, however, often benefits from simplified, more constrained scenarios that are presented to novices or experts while tracking their behavior e.g., when using different display variants. Constructing alike scenarios, where the scenario itself is not an independent variable, is a major task, requiring considerable effort and input from subject matter experts. As an alternative, identical traffic scenarios are, therefore, often repeated to obtain paired-samples at the risk of scenario recognition. Depending on the aim of the study, this can be undesirable as participants may recall their earlier responses rather than coming up with an independent solution, aggravating learning effects. This applies especially to studies that measure ATCO consistency, such as in the personalization of conflict resolution advisories (Westin et al., 2016). Finding a balance between using alike scenarios and preventing recognition is not trivial.

In this paper we, for the first time, provide an overview of techniques used to mitigate scenario recognition in existing air traffic control (ATC) studies. A straightforward and frequently employed method is to rotate and/or mirror scenarios. While these transformations result in identical scenarios in terms of conflict angles, traffic densities and patterns etc., the change in orientation may unconsciously impact participant behavior. This may not reveal itself in the final outcome, e.g., solving a conflict, but it can elicit different visual scan patterns to arrive at this outcome. Visual search is an essential process that ATCOs use to continuously update their mental picture (Fraga et al., 2021). Changes in this process may lead to faster or slower conflict detection in otherwise identical scenarios, affecting related objective measures. Furthermore, perceived workload may be affected (e.g., due to unusual traffic directions, especially for experts) and action sequences or conflict resolutions might change due to different fixation orders.

These effects are, to the best of our knowledge, not sufficiently identified and recognized in literature. Authors often merely mention that scenarios are transformed to ‘prevent recognition’ without further detailing their considerations or the transformation’s implications. In addition to our literature survey on mitigation techniques, we therefore analyze eye tracking data from a previously executed experiment that featured scenario transformations. The data consists of ten professional ATCOs who each performed conflict detection and resolution in 15 distinct scenarios, of which five were selected for this analysis. Each scenario was presented three times to them with different transformations. By comparing the order in and speed at which flights were fixated, we empirically describe the participants’ behavioral consistency when presented with transformed repetitive scenarios. To conclude we argue on the implications that researchers should consider when repeating scenarios, based on these initial findings.

Mitigation Techniques

A literature survey resulted in the identification of three categories of techniques to prevent scenario recognition, explicitly described in 20 ATC studies and summarized in Table 1: geometric, textual and temporal. Most studies used a combination of techniques, with rotating scenarios as the most popular technique, employed in 15 studies.

Table 1: *Scenario recognition mitigation techniques explicitly mentioned in existing research.*

	Geometric		Textual		Temporal	
Study	Rotation	Mirroring	Renaming callsigns	Renaming waypoints	Time shifting	Reordering
Abdul Rahman, 2014	✓	-	-	-	-	-
Albuquerque et al., 2008	-	?	-	-	✓	✓
Borst et al., 2017	✓	-	-	-	-	✓
Borst et al., 2019	✓	✓	-	-	-	-
Cummings et al., 2005	✓	-	-	-	-	-
Harrison et al., 2014	-	-	✓	-	-	-
Hilburn et al., 2014	✓	-	-	✓	-	-
IJtsma et al., 2022	✓	-	-	-	-	-
Jans et al., 2019	✓	-	-	-	-	-
Jasek et al., 1995	-	-	-	-	-	✓
Jha et al., 2011	✓	-	✓	✓	-	-
Kim et al., 2022	✓	-	✓	✓	-	-
Klomp et al., 2016	✓	-	✓	-	-	-
Major and Hansman, 2004	✓	✓	-	-	-	-
Metzger and Parasuraman, 2006	✓	-	-	-	-	-
Rovira and Parasuraman, 2010	✓	-	✓	✓	-	-
Sollenberger and Hale, 2011	-	-	✓	-	-	-
ten Brink et al., 2019	✓	-	-	-	-	-
Trapsilawati et al., 2021	✓	-	✓	✓	✓	-
Wilson and Fleming, 2002	-	-	✓	-	-	-

Geometric When a scenario is rotated or mirrored, its (objective) taskload formed by the traffic density, conflict geometries etc. remains the same, but its (subjective) workload might change. Especially with experts, accustomed to traffic streams from certain directions, changing the principal axis can have an impact on their perceived workload, as it requires a change in scan pattern.

Geometric transformations can only be done when the sectors are relatively symmetric, which is generally not the case in operational environments. Furthermore, on a widescreen monitor, rotations other than 180° may result in a reduced look-ahead range for flights coming towards the sector. Square-shaped monitors (or simulated windows), as found in many ATC centers, eliminate this problem. Only rotation multiples of 90° were found in the studies, presumably because this generates sufficient transformations and is easy to execute. Albuquerque et al. (2008) mention that they 'invert the route structure', without further detailing what is meant by that.

Textual Changing callsigns and waypoint names is a simple technique that can be widely applied, does not change the taskload and has proven to be sufficient on its own in some cases, such as the study by Wilson and Fleming (2002). When realistic callsigns and aircraft performance data are used, the callsign should match the flight's characteristics (e.g., no big airliner for small airlines or non-standard destinations). Similarly, when using operational airspaces, waypoints may need to be left unaltered to match operational routes. Neither are a problem when using airspace-naive novices.

Temporal Shifting occurrences of, for example, conflicts in time is a feasible technique for relatively long scenarios, where chunks of traffic entering the sector can be shuffled (Albuquerque et al., 2008; Trapsilawati et al., 2021). Such temporal transformations do, however, risk ignoring cognitive built-up and its associated impact on (perceived) workload. This technique is, therefore, mostly used to construct realistic scenarios from recorded flight data, by shifting flights to create a plausible scenario that is denser or has more conflicts than the recording.

When an experiment consists of multiple scenarios per test condition, their order can be changed. If, for example, display variants are tested that are sufficiently distinct from each other, participants may be predominantly occupied by the changed visuals and/or tasks, making it even less likely for them to recognize repeated scenarios at all (Jasek et al., 1995).

An extreme case of re-ordering chunks of traffic is to add dummy scenarios in between measurement scenarios, as done by Borst et al. (2017). If planning allows, measurements for each participant can even be split over multiple days. This requires good planning (difficult when using experts) and is more prone to introducing confounds due to a lack of control over variables such as participant energy levels or between-session (professional) experiences. It is therefore not often used, except in longitudinal studies such as by Hilburn et al. (2014).

A technique not explicitly found in literature is the shifting of all flights up or down in altitude. The individual contribution might be marginal, as people predominantly recognize plan-view patterns, but in combination with other techniques it can require participants to not completely rely on their memory. Care must be taken not to alter the altitudes too much, as changes in flight level have an effect on ground speeds and thus closing rates, impacting the time a loss of separation occurs and/or conflict warnings will be issued.

Data Description

As an example of the potential impact of scenario transformations, we revisit and analyze eye tracking data from a previously executed experiment designed for task analyses. To prevent scenario recognition it involved static scenarios featuring several geometric and textual transformations, dummy scenarios and a varying scenario order.

Participants and Apparatus

Ten professional en-route ATCOs (age: $\mu = 42.7$, $\sigma = 6.8$, years of experience: $\mu = 18.8$, $\sigma = 6.2$), from Eurocontrol's Maastricht Upper Area Control Centre (MUAC) voluntarily participated in a simulator experiment, as approved by the Human Research Ethics Committee of TU Delft under number 2754. All participants provided written informed consent. A TU Delft-built medium-fidelity simulator was designed to mimic the MUAC interface on a 1920 x 1920 pixels 27" display with a computer mouse for control inputs, shown in Fig. 1. Although the scenarios were static, participants could measure predicted minimum separation between flights and display extended flight labels.

Figure 1: Experiment set-up with participant (left) and observer (right) positions.

Gaze data was recorded using a head-worn Pupil Labs Core eye tracker (Kassner et al., 2014) with Pupil Capture v3.5.1. The forward facing scene camera recorded at 30 Hz and the pupils were recorded at 120 Hz. Eight AprilTag markers were placed along the edges of the screen to relate gaze to screen pixels. Clusters of gaze points that were close in location and time were classified as fixations through the Python version of I2MC by Hessels et al. (2017), with a minimal duration threshold of 60 ms as used by Fraga et al. (2021). The fixations were correlated to flights by drawing voronoi-like areas of interest around each flight's symbol, speed vector and label.

Scenarios

Participants assessed five distinct static scenarios, intermingled with ten dummy scenarios that were not included in the current analysis. Each scenario was shown three times with different transformations and featured an artificial, octagonal 80 x 80 NM sector, with four waypoints in the cardinal directions. This made sure that the professionals would not fully rely on their trained scan patterns and that scenarios could not be recognized based on the sector shape. Four flights were present on direct routes to their exit points. In the dummy scenarios there were only two or three flights, for which measures like fixation orders would be less robust. Variants were created by applying any (combination) of the following transformations:

- Rotation: 90, 180 or 270 degrees,
- Mirroring: flipping along the x- or y-axis,
- Altitude shift: all flights up or down by 1,000 or 2,000 ft.

Callsigns were randomized for all variants and flight labels were always placed at a 90 degree offset to the direction of travel. Figure 2 shows an example of a scenario with corresponding transformations. Note that flights in the center of the sector were invariant to all geometric transformations and always appeared at the same location on the screen.

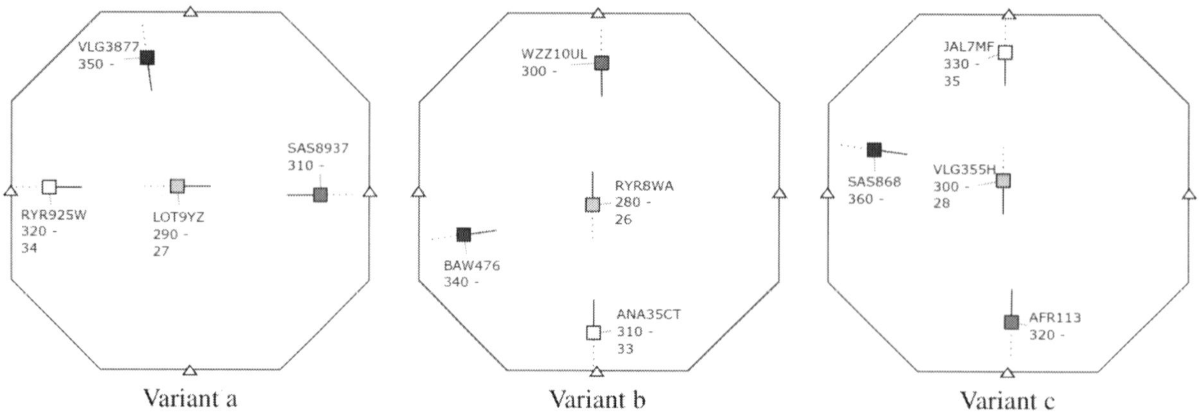

Figure 2: Three transformations of Scenario 5. Colors relate to the same flights in each transformation.

All participants got to see the same order of transformations, but the scenario ordering was counterbalanced between them to account for learning effects. This ordering of scenarios was defined in the previously executed experiment and might, in hindsight, have been suboptimal for the current study. The experiment started with six training scenarios.

Participant Task

Participants were asked to first indicate for each scenario whether there were any conflicts and to consequently solve these through altitude clearances only. Some flights had to leave the sector at a different flight level, requiring a clearance that would generally also solve any conflict(s). An intermediate level was needed in some cases to not create a conflict. If there were no (remaining) conflicts and all flights were at or cleared to the correct flight level, the participant could advance to the next scenario by clicking a button in the lower right corner of the screen. This button was carefully placed to ensure a common first fixation point, not related to any flights, when a scenario loaded.

Results and Discussion

After the experiment, some participants mentioned that they did recognize the repetition of certain conflict geometries, but none of them recalled that it were identical scenarios apart from the applied transformation(s). Our present analysis stays away from concluding whether the recognition mitigation has worked and instead focuses on the consistency of fixation behavior. Since participants showed vastly individualized behavior, no between-participant comparisons are performed and all observations discussed here relate to the three scenario repetitions per individual.

Fixation Order

Conflict detection time is directly driven by the order in which flights receive attention, especially when scenarios include many flights. After all, if an ATCO fixates flights in a different order, he/she might observe a conflicting pair earlier or later. To this end, Fig. 3 shows for each scenario's three repetitions the flight that was first fixated by each ATCO. The level of consistency, in terms of identical first fixations for all three transformations (visible as a row of three similarly colored squares), varied per ATCO from zero (Participants 5 and 10) to three scenarios (Participants 7 and 8). A similar variance can be seen between the scenarios, with consistent first fixations for one (Scenarios 1 and 2) to five (Scenario 5) ATCOs. This suggests that the rotations may have had an impact on the fixation order, and that this can differ per individual and traffic layout. On closer inspection, in 80% of the runs, the first fixated flight in Scenario 5 was located in the center of the sector (and therefore in the exact same location for all repetitions). Conversely, Scenario 1, the only one with no flight near the center, shows the lowest level of consistency.

To illustrate individual differences, complete orders of fixation for two participants on either extremes of the aforementioned consistency scale are shown in Fig. 4. Note how Participant 8's complete fixation sequence is consistent for all variants of Scenario 3. This, in combination with the inconsistent fixation orders seen in other scenarios or with other participants, further hints at a non-negligible influence of scenario rotation on the processing of traffic scenarios. For more insight in the relevant mechanisms, an analysis of scan patterns at different transformations would be useful, but this requires scenarios with more flights. The static, low density scenarios used in this study imply that the results are not necessarily applicable to dynamic and/or denser scenarios.

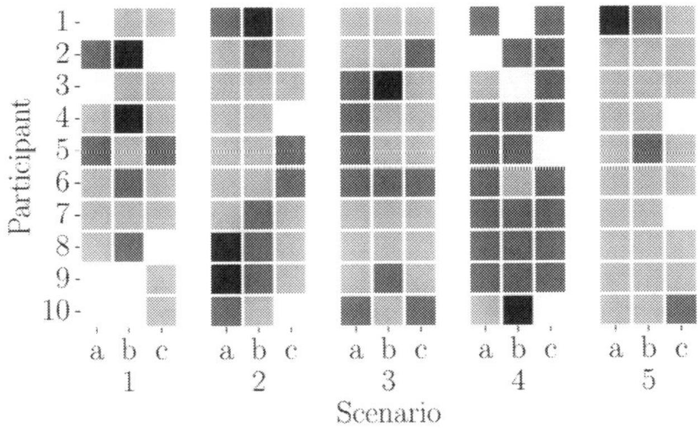

Figure 3: First fixated flight per participant. Colors represent specific flights in a scenario (see Fig. 2 for Scenario 5).

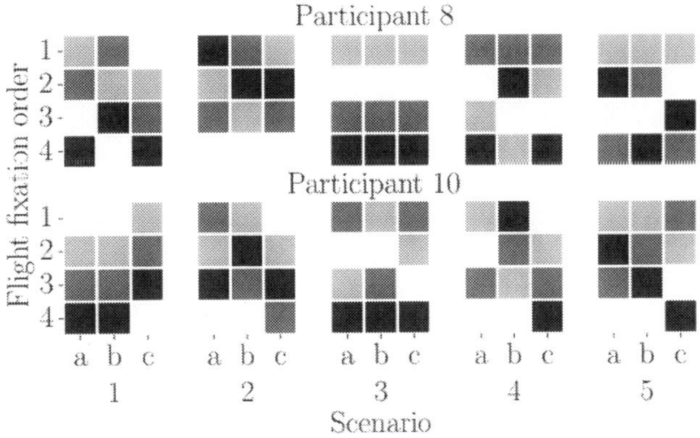

Figure 4: Complete flight fixation orders of two participants. Colors represent specific flights in a scenario (see Fig. 2 for Scenario 5).

Fixation Speed

To further illustrate the potential influence of rotations on fixation sequences and duration, Fig. 5 shows the standardized time till specific flights in Scenarios 3 and 5 had been first fixated. Results imply that the rotational-influence on this measure is dependent on the researcher's flight of interest. This is most visible in Scenario 5b, where Flight 1 shows significantly different means compared to the other two rotations. Akin to the fixation order, differences between individuals are again considerable, reflected in the wide spread of most data.

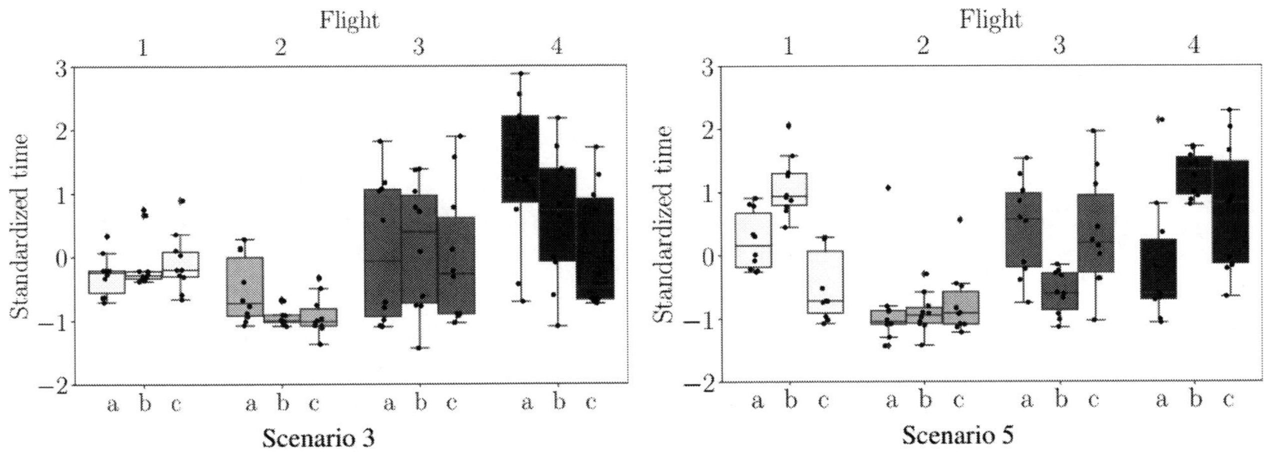

Figure 5: Standardized (per participant) time till flights have been first fixated in two scenarios, split per transformation. Colors represent specific flights in each scenario (see Fig. 2 for Scenario 5).

While the order of scenarios was counterbalanced between participants, the order of their repetitions was not (i.e., all participants first saw a, followed by b and then c). While this resulted in a clearly visible reduction in total fixation time over the three repetitions, this reduction is not (always) reflected in the results presented here. We therefore conclude that this speed-up was mostly caused by the participants getting more acquainted with the task at hand and advancing to the next scenario, rather recognizing the specific scenarios. To further isolate the effect of purely the transformation, future studies should include duplicate scenarios where no transformation has been applied.

Conclusion

Scenario transformations such as rotation and mirroring are proven techniques to create paired-samples in human-in-the-loop ATC research, but the potential impact on results is not always sufficiently recognized. We showed that the most popular technique, rotating scenarios, does risk eliciting different eye fixation behavior from participants, potentially confounding objective measures such as conflict detection time. Whether this is a problem strongly depends on the research at hand and requires careful consideration. No definitive conclusions regarding the size of these effects can be made on the basis of the limited analysis presented here. The first indications do warrant further research with more elaborate, potentially dynamic, traffic scenarios and a tailored experiment design.

Acknowledgments

The authors express their gratitude to all involved ATCOs and MUAC for facilitating the experiment.

References

Abdul Rahman, S. (2014). *Solution Space-based Approach to Assess Sector Complexity in Air Traffic Control* (Doctoral dissertation). Delft University of Technology.

Albuquerque, E. A. F., Trabasso, L. G., Sandes, A., de Araújo, M., Li, L., & Hansman, R. J. (2008). Experimental Setup for Air Traffic Control Cognitive Complexity Analysis. *Symposium of Operational Applications in Areas of Defense*, 342–347.

Borst, C., Bijsterbosch, V. A., van Paassen, M. M., & Mulder, M. (2017). Ecological interface design: supporting fault diagnosis of automated advice in a supervisory air traffic control task. *Cognition, Technology and Work, 19*(4), 545–560.

Borst, C., Visser, R. M., van Paassen, M. M., & Mulder, M. (2019). Exploring Short-Term Training Effects of Ecological Interfaces: A Case Study in Air Traffic Control. *IEEE Transactions on Human-Machine Systems, 49*(6), 623–632.

Cummings, M. L., Tsonis, C. G., & Cunha, D. C. (2005). Complexity Mitigation Through Airspace Structure. *International Symposium on Aviation Psychology*, 159–163.

Fraga, R. P., Kang, Z., Crutchfield, J. M., & Mandal, S. (2021). Visual Search and Conflict Mitigation Strategies Used by Expert en Route Air Traffic Controllers. *Aerospace, 8*(7).

Harrison, J., Izzetoğlu, K., Ayaz, H., Willems, B., Hah, S., Ahlstrom, U., Woo, H., Shewokis, P. A., Bunce, S. C., & Onaral, B. (2014). Cognitive workload and learning assessment during the implementation of a next-generation air traffic control technology using functional near-infrared spectroscopy. *IEEE Transactions on Human-Machine Systems, 44*(4), 429–440.

Hessels, R. S., Niehorster, D. C., Kemner, C., & Hooge, I. T. (2017). Noise-robust fixation detection in eye movement data: Identification by two-means clustering (I2MC). *Behavior Research Methods, 49*(5), 1802–1823.

Hilburn, B., Westin, C., & Borst, C. (2014). Will Controllers Accept a Machine That Thinks like They Think? The Role of Strategic Conformance in Decision Aiding Automation. *Air Traffic Control Quarterly, 22*(2), 115–136.

IJtsma, M., Borst, C., van Paassen, M. M., & Mulder, M. (2022). Evaluation of a Decision-Based Invocation Strategy for Adaptive Support for Air Traffic Control. *IEEE Transactions on Human-Machine Systems, 52*(6), 1135–1146.

Jans, M., Borst, C., van Paassen, M. M., & Mulder, M. (2019). Effect of ATC Automation Transparency on Acceptance of Resolution Advisories. *IFAC PapersOnLine, 52*(19), 353–358.

Jasek, M., Pioch, N., & Zeltzer, D. (1995). Enhanced Visual Displays for Air Traffic Control Collision Prediction. *IFAC Proceedings Volumes, 28*(15), 553–558.

Jha, P. D., Bisantz, A. M., Parasuraman, R., & Drury, C. G. (2011). Air traffic controllers' performance in advance air traffic management system: Part I-performance results. *International Journal of Aviation Psychology, 21*(3), 283–305.

Kassner, M., Patera, W., & Bulling, A. (2014). Pupil: An Open Source Platform for Pervasive Eye Tracking and Mobile Gaze-based Interaction. *Adjunct Proceedings of the 2014 ACM International Joint Conference on Pervasive and Ubiquitous Computing*, 1151–1160.

Kim, M., Borst, C., & Mulder, M. (2022). Situation Awareness Prompts: Bridging the Gap between Supervisory and Manual Air Traffic Control. *IFAC-PapersOnLine, 55*(29), 13–18.

Klomp, R. E., Borst, C., van Paassen, M. M., & Mulder, M. (2016). Expertise Level, Control Strategies, and Robustness in Future Air Traffic Control Decision Aiding. *IEEE Transactions on Human-Machine Systems, 46*(2), 255–266.

Major, L. M., & Hansman, R. J. (2004). *Human-Centered Systems Analysis of Mixed Equipage in Oceanic Air Traffic Control* (tech. rep.). MIT International Center for Air Transportation. Cambridge, MA, USA.

Metzger, U., & Parasuraman, R. (2006). Effects of automated conflict cuing and traffic density on air traffic controller performance and visual attention in a datalink environment. *International Journal of Aviation Psychology, 16*(4), 343–362.

Rovira, E., & Parasuraman, R. (2010). Transitioning to future air traffic management: Effects of imperfect automation on controller attention and performance. *Human Factors, 52*(3), 411–425.

Sollenberger, R. L., & Hale, M. (2011). Human-in-the-Loop Investigation of Variable Separation Standards in the En Route Air Traffic Control Environment. *Proceedings of the Human Factors and Ergonomics Society 55th Annual Meeting*, 66–70.

ten Brink, D. S., Klomp, R. E., Borst, C., van Paassen, M. M., & Mulder, M. (2019). Flow-based air traffic control: Human-machine interface for steering a path-planning algorithm. *2019 IEEE International Conference on Systems, Man and Cybernetics, SMC 2019*, 3186–3191.

Trapsilawati, F., Chen, C. H., Wickens, C. D., & Qu, X. (2021). Integration of conflict resolution automation and vertical situation display for on-ground air traffic control operations. *Journal of Navigation, 74*(3), 619–632.

Westin, C., Borst, C., & Hilburn, B. (2016). Strategic Conformance: Overcoming Acceptance Issues of Decision Aiding Automation? *IEEE Transactions on Human-Machine Systems, 46*(1), 41–52.

Wilson, I. A. B., & Fleming, K. (2002). Controller reactions to free flight in a complex transition sector re-visited using ADS-B+. *Proceedings. The 21st Digital Avionics Systems Conference, 1*, 5–1.

NASA HUMAN FACTORS RED LIGHT/GREEN LIGHT ANALYSIS: OVERVIEW AND SPACSHIPTWO RL/GL CASE STUDY

Tracy Dillinger, Psy. D.
National Aeronautics and Space Administration
Washington, DC
Isabel Hernandez, M.P.A.
National Aeronautics and Space Administration
Glendale, AZ

Historically, learning from accidents and incidents involves identifying errors of commission or omission and "things that go wrong" and how to reduce their recurrence. While well-intentioned, infrequent failures receive more attention than everyday successes. Human Factors analysis need not limit itself to learning from failures. Learning from "what goes right" involves identifying resilience factors associated with behaviors and processes that strengthen the system and reduce risk. This article provides an overview of how the National Aeronautics and Space Administration (NASA) Human Factors Task Force (HFTF) constructed the NASAHFACS model and the Red Light/Green Light approach. Using SpaceShipTwo as a case study we applied the updated taxonomy and approach to illustrate its use.

Investigators frequently apply accident analysis models to recreate accident scenarios, identify causative factors and assist in developing interventions to avoid similar incidences. However, the foundation of such methods focuses on infrequent failures and errors, offering an incomplete view. This presents an opportunity to understand extraordinary behaviors and highlight what humans do right. These resilient behaviors are often part of everyday operations. Understanding and learning from what succeeds and fails allows for identifying factors through the whole system. The model chosen to analyze an accident determines an investigators perspective. This crucial choice guides the investigation's conclusion, and determines preventative measures from such findings (Hollnagel, 2014). To offer a more holistic view, in coordination with Shappell and Weigmann the Human Factors Task Force (HFTF) created a Red Light/Green Light approach to create balanced results for prevention.

The National Aeronautics and Space Administration (NASA) created a Human Factors Task Force (HFTF) in 2016 to identify, analyze, track, and trend human factors for mishap prevention. The HFTF adopted and modified the Human Factors Analysis and Classification System (HFACS) created by Shappell and Weigmann (2000). NASAHFACS integrated specific concerns (e.g., microgravity), enabling the Agency to capture lessons learned from mishaps and close-call events occurring in the aerospace environment.

Inception fo NASAHFACS Taxonomy

Initially, the structure of NASAHFACS consisted of 4 tiers and 21 categories. Several versions, 1.0, 1.1, 1.2, and 1.3, were modified throughout the years. NASAHFACS retained the original four tiers of Acts, Preconditions, Supervision, and Organization. The HFTF added a unique "Space" environmental category under the Preconditions Tier to capture the distinct impact of space on human performance. Dillinger & Kiriokos (2019), developed a HF handbook in association with the taxonomy to define the terms and application of NASAHFACS.

Figure 1.
Draft of NASA Office of Safety and Mission Assurance Human Factors Handbook V1.4 Procedural Guidance and Tools.

NASA Office of Safety and Mission Assurance
Human Factors Handbook V1.4
Procedural Guidance and Tools

Tracy Dillinger PsyD., FASMA, FAsHFA
NASA Headquarters / OSMA

Isabel Hernandez MPA
NASA Headquarters / OSMA

Andre Karpowich MS
NASA Kennedy Space Center

Samuel Serafini
ARES Corporation

January 2023

Originally, the HFTF applied NASAHAFCS for a three-year "beta test" at one operational Center from 2014-2016. The beta-test results proved helpful in reducing reportable events by targeting prevention efforts to the most significant areas of identified Human Factors (HF) categories. In 2017, NASA Center Points of Contact (POCs) began meeting annually to code center operational events, with each Center contributing to an Annual Report of NASA. Annual results were briefed to Agency senior leaders. All NASA Human Factors Annual reports from 2017-2022 reside in the NASA Scientific, Technical, and Research Information discoVEry System (STRIVES) for internal Agency use.

Evolution of NASAHFACS Taxonomy & Red Light/Green Light

From 2022 - 2023 there was a substantial shift in philosophy. The HFTF consolidated and "neutralized" the taxonomy for version 1.4. While it retained the four tiers of Acts, Preconditions, Supervision, and Organization instead of 21 categories, there were now 19 newly labeled categories, nanocodes, and definitions. The HFTF modified associated training and application materials, including the HF handbook and investigators checklist. The neutralized taxonomy avoids negative labels laden with negative connotations such as "violations." The traditional (red light), can still be used with the neutral term "compliance" to site the violation or can also be used to identify policy insistence to call out a greenlight preventative behavior.

Figure 2.
NASAHFACS Taxonomy V1.4.

The HFTF updated the Human Factors Investigator's Checklist, to succinctly apply the taxonomy. The checklist allows investigators notation of complete details of an event. The investigator summarizes the event followed by the checklist to identify each tier, category and nanocodes if desired. The middle of the checklist lists the neutralized terms. When an error occurs, known as the red light, investigators place it on the left side. When resilient factors exist, known as the green light, investigators note it on the right side.

Figure 3.
Neutralized HF Investigators Checklist.

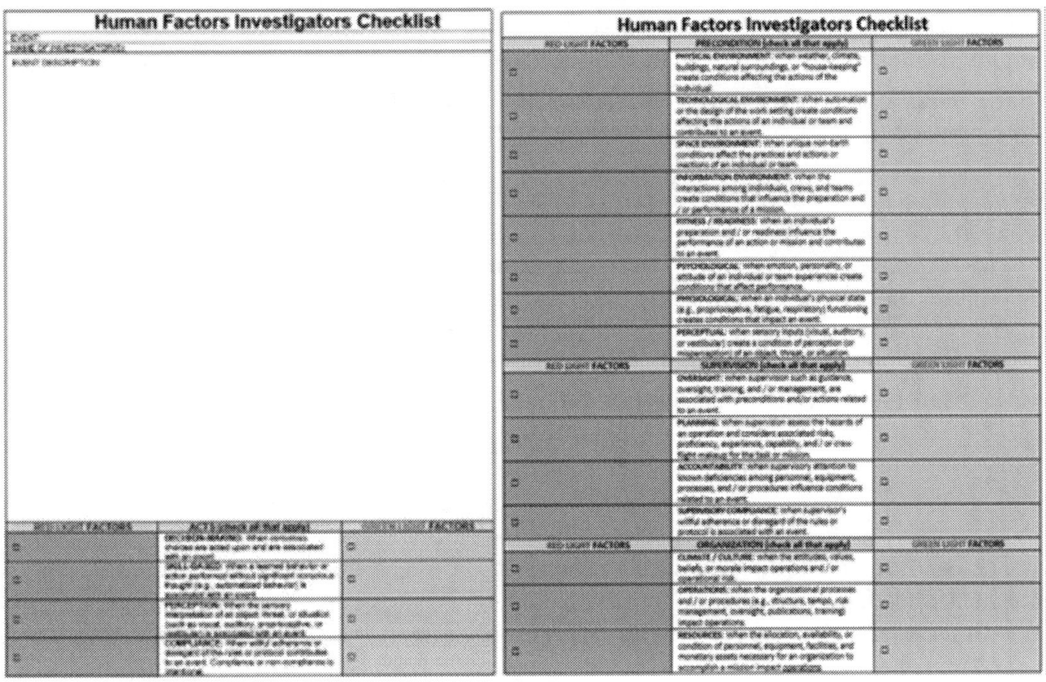

SpaceShipTwo

On October 2014, SpaceShipTwo (SS2), a reusable suborbital rocket, broke into several pieces during a rocket-powered test flight. It impacted terrain across a five-mile range near Koehn Dry Lake, California. No one on the ground suffered injuries, the pilot received serious injuries, and the copilot received fatal injuries.

Analysis

Red Lights. Starting at the Acts tier, the copilot unlocked the feather just after reaching 0.8 Mach instead of waiting per the flight test to achieve the required speed of 1.4 Mach. According to the National Transportation Safety Board (NTSB) report, time pressure can lead to cognitive overload as it increases the rate at which an individual must process information. Due to time pressure a tradeoff of speed versus accuracy often occurs leading individuals to make decisions quickly at the expense of evaluating all the information. The copilot's actions while correct occurred prematurely, which falls under *Activity* and *Skill-Based.*

Moving on to the Preconditions tier, while the pilots' flight experience included several preflight simulations, the copilots' experience included only two glide flights and one powered flight. The NTSB reported that neither of the crew members performed any SS2 flights for the previous nine months due to engine upgrades. These proficiency factors fall under *Individual Factors* and *Fitness/Readiness.* Additionally, the copilots inexperience regarding the extreme time pressure, vibration and loads associated with test flights lead to the premature uncommanded feather extension and aerodynamic overload falling under *Environmental Factors* and *Physical Environment.*

Supervision represents the third tier. The test pilots, engineers, and managers interviewed indicated awareness that unlocking the feather mechanism prematurely presented a catastrophic risk, yet the pilot operating handbook indicated no warning, caution, or limitation. The awareness factor described falls under *Accountability*.

Organization constitutes the fourth tier. Scaled Composites LLC primarily designs and builds vehicles used by highly trained and experienced test pilots. Subsequently, the focus existed more on the reliability of the SS2 feather system than on designing a system that minimized the likelihood of human error. However, the SS2 used pilots who lacked test pilot experience. This factor falls under *Resources.*

Additional red lights involve the experimental permit and regulatory requirements. In order to perform a flight test, the Federal Aviation Administration (FAA) Office of Commercial Space Transportation requires an experimental permit. Part of the regulation process consists of a hazard analysis requirement that requires the Scaled Composites to identify and describe the hazards that could result from human errors. Scaled Composites LLC analysis failed to meet regulatory requirements to identify and document the potential hazard of unlocking the feather prematurely. The FAA Office of Commercial Space Transportation failed to ensure that Scaled Composites complied with the mitigations cited in the waiver from regulatory requirements or determine whether those mitigations would adequately address human errors with catastrophic consequences. The factors fall under *Operations* and *Resources.*

Lastly, the FAA Office of Commercial Space Transportation technical staff and Scaled Composites technical staff needed more direct communication. Time pressure existed to approve the permit application within the 120-day review period. The lack of defined lines between public safety and mission safety, interfered with FAA's ability to evaluate the SS2 permit application thoroughly. The factors fall under *Operations*.

Figure 4.
NASAHFACS SpacShipTwo RL Analysis

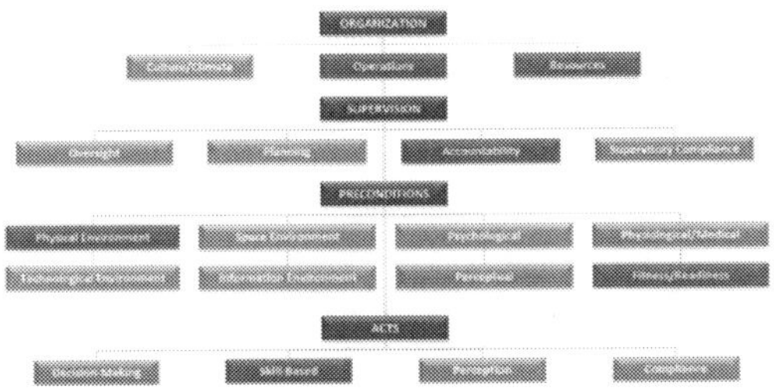

Green Lights. Two Green Lights occurred, one at the Preconditions tier and the second at the Organization tier. The pilot reported not pulling the parachute system's ripcord handle in a post-accident interview. Evidence indicated that once the pilot separated from his seat, the parachute's automatic activation device deployed the parachute. The activation of the device proved instrumental in saving the pilot's life and represents a resilient factor under *Environmental Factors*, followed by the *Technological Environment.* The second Green Light consists of the area of containment. Title 14 Code of Federal Regulations 437.57, states that during each permitted flight the instantaneous impact point (IIP) must be contained within an operating area. Based on the NTSB report it was concluded that on the day of the accident SS2's IIP was consistent with the requirements representative of *Operations.*

Figure 5.
NASAHFACS SpacShipTwo GL Analysis

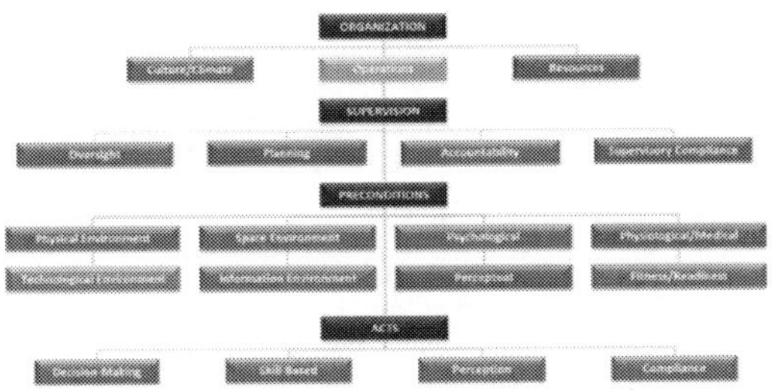

Outcome

The SS2 case study outlines a series of human factors that led to the mishap. Applying the NASAHFACS tool to the SS2 case study resulted in the identification of six RLs in different categories (eleven nanocodes) and two GLs in different categories (two nanocodes). By using NASAHFACS in the traditional way of Human Factors and recognizing factors that add or strengthen resilience to a system or process through the identification of green lights, organizations can encourage and reinforce desired behaviors that sustain and prevent error chains leading to unwanted events.

Figure 6.
NASAHFACS SpacShipTwo RL/GL Analysis

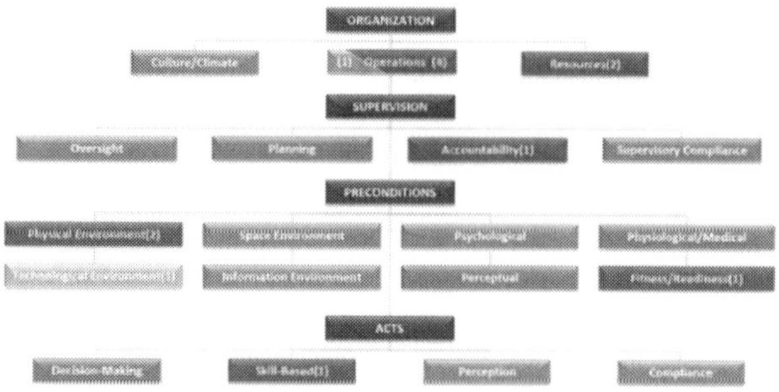

Conclusion

NASA HFTF made great strides, from creating and applying the tool, conducting regular analysis, and out-briefs, identifying the NASA "Dirty Dozen", and reinvigorating the NASAHFACS tool with an eye toward resilience. This new resilience tool retains the goodness of traditional mishap investigation, sometimes referred to as "Safety I, to a resilience-promoting aspect in line with Safety II and the concepts identified by Hollnagel (2014) and others such as Stroeve et al., (2023). Identification and advocacy for resilience human factors offer a way to provide decision-makers with balanced feedback to make risk-informed decisions for prevention.

Acknowledgements

We want to thank Andre Karpowich, Nick Kiriokos, and Samuel Serafini for their presentations, support, and dedication to the Human Factors Program. Disclaimer: The National Transportation Safety Board conducted an official investigation report of the SS2 mishap. This report became the source for the SS2 analysis. The intention is to demonstrate the application of NASAHFACS by highlighting the traditional (Red Light) and resilient (Green Light) factors, not to reinvestigate the mishap. The views expressed in this paper are those of the authors and do not reflect the views of the affiliated institution.

References

Dillinger, T., & Kiriokos, N. (2019). NASA Office of Safety and Mission Assurance Human Factors Handbook: Procedural Guidance and Tools; NASA/SP-2019-220204; National Aeronautics and Space Administration (NASA): Washington, DC, USA.

Hollnagel, E. (2014). *Safety-I and Safety-II: The Past and Future of Safety Management* (1st ed.). CRC Press. https://doi.org/10.1201/9781315607511

National Transportation Safety Board. (2015). In-Flight Breakup During Test Flight Scaled Composites SpaceShipTwo, N339SS Near Koehn Dry Lake, California October 31, 2014 (Aerospace Accident Report NTSB/AAR-15/02 PB2015- 105454). Washington, DC, USA.

Shappell, S. & Wiegmann, D. (2000). The Human Factors Analysis and Classification System (HFACS). Report Number DOT/FAA/AM-00/7. Federal Aviation Administration. Washington, DC.

Stroeve, S., Kirwan, B., Turan, O., Kurt, R. E., van Doorn, B., Save, L., Jonk, P., Navas de Maya, B., Kilner, A., Verhoeven, R., Farag, Y. B. A., Demiral, A., Bettignies-Thiebaux, B., de Wolff, L., deVries, V., Ahn, S. I., & Pozzi, S. (2023). SHIELD Human Factors Taxonomy and Database for Learning from Aviation and Maritime Safety Occurrences. Safety, 9(1), 14. https://doi.org/10.3390/safety9010014

COMPARISON OF DELEGATION METHODS FOR TASK-BASED UAV GUIDANCE

Marius Dudek & Axel Schulte
Universität der Bundeswehr München
Neubiberg, Germany

In this contribution we compare UAV delegation methods from a fighter-jet cockpit. Recent research approaches to UAV mission management have mainly been using touchscreen interactions and little research has systematically analyzed different input methods to delegate tasks. In this article, we present three UAV delegation methods that use touchscreen interactions, voice control, and a combination of eye-tracking and HOTAS buttons. The presented methods were integrated in a fighter-jet simulator and evaluated with ten participants. The performance of participants varied for different combinations of delegation method and task load. Touchscreen interaction was fastest on average, followed by voice interaction. The number of errors for each combination was only slightly different. When participants could select a method to use, they predominantly used voice interaction, followed by eye-tracking. The subjects showed the behavior of using eye-tracking method and switching to another method when it did not work as desired.

With increasing capabilities of automation, the widespread use of unmanned aerial vehicles (UAVs) is getting closer. Manned-Unmanned Teaming is nowadays considered a concept for the future operational use of UAVs in which a pilot guides several UAVs from the fighter-jet cockpit. There is much research dedicated to the question of how to formalize the guidance process in such an operational concept (C. A. Miller & Parasuraman, 2007; Uhrmann & Schulte, 2011) but little research focusses on the question of which input methods are most suitable to guide UAVs in this means (Calhoun, Ruff, Behymer, & Rothwell, 2017; Dudek & Schulte, 2022a; Levulis, DeLucia, & Kim, 2018). Therefore, in this article, we investigate three delegation methods that have been developed for tasking UAVs from inside a fighter-jet cockpit, that are based on touchscreen gestures, voice input, and a combination of eye-tracking and buttons located on the flight control devices (HOTAS). All methods put their emphasis on the fast generation of tasks, rather than the detailed specification of tasks. The presented methods are evaluated with an experimental study to address the following questions:

1. Which delegation method is most suitable to guide UAVs?
2. Is it beneficial to offer multiple delegation methods?

Approach

Multi-UAV Task-Based Guidance

The guidance of unmanned systems requires a common understanding by humans and automation of what is to be done by the automation (C. Miller et al., 2005). To achieve this, we leverage the approach of task-based guidance. In task-based guidance, the pilot delegates high-level tasks to so-called cognitive agents aboard the UAVs which, in turn, decompose these tasks and control their aircraft systems accordingly.

UAV task definition. We define military UAV tasks by a taxonomy first presented in (Dudek & Schulte, 2022b). The tasks in this taxonomy consist of the following components:

- *Type*: Describes the purpose of the task, e.g. reconnaissance.
- *Target-Object*: The object to which the task is connected.
- *Success Criteria*: Define the conditions under which a task is considered successful.
- *Constraints*: Constraints define the "how" of task execution.

Delegation agent. To facilitate the delegation of tasks to multiple platforms, we introduce a central instance, the so-called *Delegation Agent*, that simplifies the definition and delegation of tasks. This delegation agent has the ability to directly deduce the type of tasks, the success criteria, and the constraints based on the target object and its state (e.g. whether it has been classified hostile). In addition, the agent has a scheduler with which it can determine the optimal platform and time for executing a specific task. The pilot can choose between two *Scheduling modes*:

- Team-Mode: The agent is responsible for UAV and timing selection
- UAV-Mode: The user is responsible for UAV selection, the agent selects the timing

The delegation agent carries out the scheduling steps on a copy of the mission plan, the so-called *Modify-Plan*. In this plan, changes can be made without affecting the actual UAV behavior. After scheduling, the modify-plan can either be switched active by the pilot, or the pilot can undo the changes.

Delegation Methods

The delegation agent is guided by the pilot using one of three *Delegation methods*. The methods differ in the granularity of control and the modalities used.

Touch-Gesture. This delegation method is based on the usage of touchscreen gestures. The touchscreen interaction is different for the two scheduling modes. To delegate tasks using the Team scheduling mode, pilots have to longpress on the target object. The delegation agent contributes the other information to the task, namely task-type, success criteria and default constraints. The delegation agent also handles the scheduling according to the Team-Mode. When the pilot wants to delegate tasks using the UAV-Mode, he has to select a team member prior to the longpress on the target object. The pilot can activate the modify-plan or revert the changes with two touch buttons next to the displayed plan in SHDD

Voice. This delegation method uses voice commands whose syntax is based on NATO-brevity codes. The syntax for task delegation is as following:

$$(< UAV\ name >)\ < Task - Type > < Target >$$

The specification of the UAV name is optional, if the pilot specifies a UAV name, the delegation agent will use the UAV-Mode for scheduling, the Team-Mode is used otherwise. In contrast to the other delegation methods, the pilot has to specify a task-type, only the success criteria and the constraints are contributed by the delegation agent. The reason for this is that voice commands are more naturalistic in this way, because commands would miss out the verbs elsewise. After the delegation of tasks, the pilot can activate the modify-plan (Command "*Accept*") or revert the changes (Command "*Decline*").

Gate-HOTAS. The Gaze-HOTAS delegation method uses an eye-tracking system and buttons on stick and thrust lever to delegate tasks. For this, the pilot selects either only the target object (Team-Mode) or a UAV and the target object (UAV-Mode). The selection process of objects is controlled by a button on the center stick. While this button is pressed, the pilot can select objects by looking at their symbols on the tactical map. The selection is locked when the selection button is released. After the selection of one or two objects, the pilot can delegate tasks by pressing a button on the thrust lever. As with Touch-Gesture, the delegation agent contributes task-type, success criteria and default constraints. The already mentioned button is also used to activate the modify-plan after delegating tasks and another thrust lever button is used to reset the selection process and revert changes made.

Experimental Study

Experimental Conditions

Mission task. In the study, the participants had to perform multiple missions, in which the primary task was the delegation of tasks to three UAVs. In each mission, the test subjects had to follow a predefined route using the autopilot. While following the flight route, new objects appeared gradually (either one or two/three close to each other) and the participants had to delegate tasks to their UAVs to investigate or engage the newly appeared objects (depending on the type). The subjects were not required to task with a specific scheduling mode. Experimental conditions varied by mission and phase, with delegation method changing between missions and task load changing with each mission phase. One experimental mission lasted 45 minutes.

Delegation method. In the first three missions, the participants were obliged to use a specific delegation method. In the fourth mission, all delegation methods were allowed for usage and participants were free to decide which one they wanted to use and when.

Task load. Task load varied across different phases with an auditory-verbal secondary task in one phase, a visual-manual secondary task in another, and a third phase without a secondary task. In the auditory verbal task, the subjects had to identify a certain sequence of sounds from an audio signal. Twelve sound sequences were played, three of which were the target sequence. The recognition of a searched tone sequence had to be acknowledged with a "Check" command. In the visual-manual task, the subjects had to monitor an airspace and press a button on the thrust lever if one of two aircraft crossed the border (inbound and outbound). In total, twelve crossings could be detected.

Data Analysis

Mission performance. The mission performance is a dependent variable, which is operationalized by several mission performance measures:

- Delegation time: Duration to insert a task into the active plan after object appearance.
- Delegation error count: The number of incorrect delegations.
- Side task reaction time: Duration until the participant reacts to the side task stimulus.
- Side task error count: The number of wrong reactions to side task stimuli.

Method preferences. Method preferences as another dependent variable were operationalized with the usage percentage for each method. For this purpose, we used the mission in which participants were free to select between delegation methods and counted how often participants use the respective methods.

Experimental Setting

Training. Participants completed two training missions prior to the experimental missions. The first mission familiarized them with the delegation methods, whereas the second mission was held under similar conditions than the experimental mission and included tasking UAVs, performing secondary tasks, and guiding the own aircraft with the autopilot. In total, the training lasted 55 minutes.

Participants. From 18 invited participants, we selected ten participants with the most stable gaze measurement (no loss of track, good detection). All participants were male and between 20 and 26 years old. Nine participants had flight experience and four participants practice more than one hour a week in a flight simulator. Participation was voluntary and uncompensated.

Setup. The study takes place in our MUM-T fighter jet simulator, which consists of three head-down displays and a 210 projection dome as an outside view. The central display is the primary operating and display element, showing the tactical situation and enabling user interaction with world objects. The left display shows the mission plan, with a color-coded background indicating whether the active plan or modify-plan is displayed. The right display is not relevant within the scope of the study.

Results

Mission Performance

Primary task. In the missions, in which the delegation methods were fixed, the participants were delegating fastest using Touch-Gesture method ($t_{MD} = 8.8$), followed by Voice method ($t_{MD} = 12.1$) and Gaze-HOTAS method ($t_{MD} = 16.4$). In the mission, in which participants could choose between delegation methods, the average delegation time was in between ($t_{MD} = 11.5$). Further division by task load shows, that both delegation method and task load have an influence on the delegation time, with the influence of the delegation method being greater (Figure 1). The influence of task load does not show the same trend for each delegation method. The data described and presented refer to the appearance of a single object; the delegation times when multiple objects appeared are not shown because there is no clear trend evident in the data.

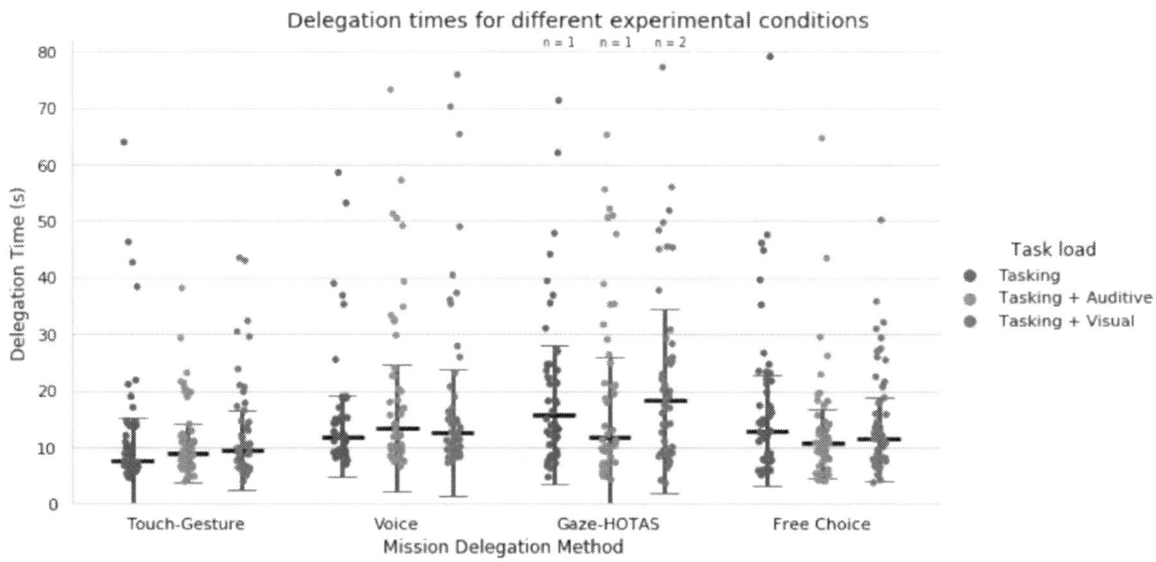

Figure 1. Single task delegation times for different exp. conditions. The median and the median absolute deviation is shown for each condition. In Gaze-HOTAS, the number of outlying samples is shown.

The least delegation errors were made in the Free control condition (four) followed by Touch-Gesture and Voice (five each). The Gaze-Hotas control condition had six errors. A detailed evaluation of delegation errors is omitted because the small percentage of errors makes it difficult to draw conclusions.

Secondary task. The reaction time showed only minor differences between delegation methods. However, the number of errors varied for the different combinations of methods and secondary tasks (Figure 2). The number of errors is higher for the visual task because all samples had to be reported in the visual task and only certain samples in the auditory task. Thus, there were more samples that could slip through in the visual task. The different trend in Voice condition shows, that more errors were made in the secondary tasks when the same modality was used for the primary task (indicating resource conflicts).

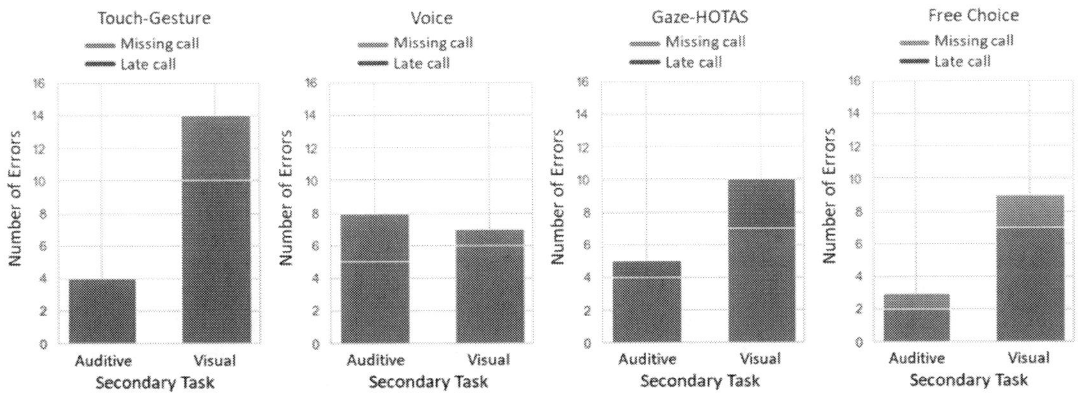

Figure 2. Number of secondary task errors for delegation methods

Method Preferences

Participants predominantly used Voice method (61.9 %), followed by Gaze-HOTAS (29.3 %) when they could choose delegation methods (Touch-Gesture: 8.8%). The method preferences correlated with the selected scheduling mode:

- Team-Mode (34.0 %): Touch-Gesture 4.0 % (of Team); Voice 34.0 %; Gaze-HOTAS 62.0 %
- UAV-Mode (66.0 %): Touch-Gesture 11.3 %; (of UAV); Voice 76.3 %; Gaze-HOTAS 12.4 %

The reason why usage preferences are different between modes could be that the additional effort for specifying the UAV is lower with Touch-Gesture and Voice as compared to Gaze-Hotas. An interesting aspect is, that the test persons often choose the UAV-Mode even though there was hardly any need to do so from mission perspective. This indicates that the participants rather have control for themselves than handing it over to the system.

The pattern frequently emerged that participants initially wanted to delegate tasks using the Gaze-HOTAS method and switched to another method when the delegation did not work as desired. This pattern is also reflected in the data on delegation times. The delegation times for the Voice method and the Touch-Gesture method are increased when the subjects were able to choose a method, while the delegation times for the Gaze-HOTAS method are greatly reduced when the subjects were able to choose freely (Table *1*). The reason for this is that the delegation times of the two methods are adversely affected by the fact that the Gaze-HOTAS method was tried out first.

Table 1. *Effect of Freedom of Choice on Delegation Time with one Delegation Method*

Delegation method	Method fixed	Method selectable
Touch-Gesture	8.8 s	12.8 s
Voice	12.1 s	12.7 s
Gaze-HOTAS	16.4 s	7.6 s

Limitations

A display error occurred during the study, leading to improper display of generated tasks in the mission plan. As these tasks were also shown on the tactical map and there was little feedback from participants, the error is expected to have limited effects. The error may have caused increased delegation times or missing delegations, but the delegation method preference is expected to be less affected. As the error could not be deterministically reproduced, it is unclear which data could be affected.

Conclusion

One question that arises from the results is whether a future fighter cockpit should provide multiple tasking methods. Two conclusions may be drawn from the data. On the one hand, it could be shown on that subjects did not delegate faster when they had several methods at their disposal than when they used the fastest method (Touch-Gesture). On the other hand, the behaviour that test persons use one method and fall back on other methods in case of problems speaks for an implementation of several methods. The identified resource conflicts in the secondary task also argue for offering multiple methods so that pilots can switch methods if the modality of a method used is already under load from another task.

Further research is needed to conclusively clarify this point. For this purpose, the tests should be carried out again with trained fighter pilots, using a mission performance criterion rather than a software ergonomic criterion.

References

Calhoun, G. L., Ruff, H. A., Behymer, K. J., & Rothwell, C. D. (2017). Evaluation of Interface Modality for Control of Multiple Unmanned Vehicles. In D. Harris (Ed.), *Engineering Psychology and Cognitive Ergonomics: Cognition and Design* (pp. 15–34). Cham: Springer International Publishing. https://doi.org/10.1007/978-3-319-58475-1_2

Dudek, M., & Schulte, A. (2022a). Effects of Tasking Modalities in Manned-Unmanned Teaming Missions. In *AIAA SCITECH 2022 Forum*. Reston, Virginia: American Institute of Aeronautics and Astronautics. https://doi.org/10.2514/6.2022-2478

Dudek, M., & Schulte, A. (2022b). *Meaningful guidance of unmanned aerial vehicles in dynamic environments*. Toulouse. Retrieved from https://events.isae-supaero.fr/event/14/contributions/362/attachments/27/64/meaningful_guidance_of_unmanned_aerial_vehicles_in_dynamic_environments_id115.pdf

Levulis, S. J., DeLucia, P. R., & Kim, S. Y. (2018). Effects of Touch, Voice, and Multimodal Input, and Task Load on Multiple-UAV Monitoring Performance During Simulated Manned-Unmanned Teaming in a Military Helicopter. *Proceedings of the Human Factors and Ergonomics Society Annual Meeting*, *60*(8), 1117–1129. https://doi.org/10.1177/0018720818788995

Miller, C., Funk, H., Wu, P., Goldman, R., Meisner, J., & Chapman, M. (2005). The Playbook™ Approach to Adaptive Automation. *Proceedings of the Human Factors and Ergonomics Society Annual Meeting*, *49*(1), 15–19. https://doi.org/10.1177/154193120504900105

Miller, C. A., & Parasuraman, R. (2007). Designing for flexible interaction between humans and automation: Delegation interfaces for supervisory control. *Human Factors*, *49*(1), 57–75. https://doi.org/10.1518/001872007779598037

Uhrmann, J., & Schulte, A. (2011). Task-based Guidance of Multiple UAV Using Cognitive Automation. In T. Bossomaier (Ed.), *Cognitive 2011: The Third International Conference on Advanced Cognitive Technologies and Applications*. Red Hook, NY: Curran. Retrieved from https://www.semanticscholar.org/paper/Task-based-Guidance-of-Multiple-UAV-Using-Cognitive-Uhrmann-Schulte/0921b403c350f4ecbb160957a35b10f3132266f8

DEVELOPMENT OF A CONCEPT FOR ANALYZING MORAL DECISION-MAKING IN HIGH PRESENCE VIRTUAL ENVIRONMENTS

Sissy Friedrich & Axel Schulte
University of the Bundeswehr Munich, Institute of Flight Systems
Munich, Germany

In future military aviation, Artificial Intelligence will play a key role in combat battlefield tactics by reducing workload and taking over decisions exploiting advantages of speed and precision of computers. However, the question of the so-called trigger authority remains the core issue in this field as ethical tensions arise when a machine decides over the use of lethal force. To enable the operator to make the most morally justifiable decision, the most suitable human-automation workshare has to be determined so that he is supported in just the right way and not overloaded nor exposed to automation bias or loss of situation awareness. For this reason, a special kind of mission simulator is developed, which has to be as close to reality as possible to produce the most transferable results. Eye tracking and other behavioral measurements are used to analyze moral decision-making in complex dynamic, uncertain, and non-binary situations.

Challenges concerning the responsible use of weapons have been raised to an accelerating degree with the debate of using unmanned systems, such as unmanned aerial vehicles (UAVs) in Manned-Unmanned-Teaming (MUM-T) missions. While technological advancements offer many benefits, including faster computation, higher precision, and lower costs, using Artificial Intelligence (AI) and thus reduce workload by supporting or even taking over decisions in morally challenging situations raises ethical issues, including questions of responsibility in decision-making. To address these concerns, the European Commission published a list of attributes that AI-based systems should possess to be trustworthy (HLEG, 2020). However, human decision-making remains necessary in morally critical situations to ensure the responsible use of weapons, so humans should have the ultimate trigger authority for lethal force, not automation. The research on this topic includes investigating what human-machine teams should look like in the context of MUM-T missions to address ethical issues. The higher the level of environmental uncertainty, the less suitable is the use of automation, since it cannot necessarily relate the unknown situation to previously learned patterns, whereas a human can achieve expertise by referring to his or her experience (Cummings, 2018). Thus, automation should be used as much as necessary to support the human, but as little as possible to avoid automation bias, complacency, or loss of Situation Awareness (SA).

Therefore, the objective of this research is to conduct analyses of dynamic moral decision-making and differentiate between conscious and non-conscious decisions. As a tool, we develop a mission and cockpit simulator. This way, the research shall then determine the amount and type of information necessary to be provided by the AI for the pilots' decision-making process and use iterative development to analyze the appropriate and helpful human-AI workshare in each phase of the targeting cycle F2T2EA (Jackson, 2006), which is commonly used in the military. Using Rasmussen's Skill-Rule-Knowledge (SRK) model (Rasmussen, 1983), it can be argued that moral decisions should ideally be made at the knowledge-based level, while flying and operating the aircraft's systems should happen on the skill-based and rule-based level. However, in reality also moral decision-making will typically be operationalized on the rule-based level for efficiency reasons, where we certainly have to learn to deal with the pitfalls. By utilizing an iterative development approach, the goal is to identify the most appropriate human-AI workshare to improve the decision-making process and ensure responsible use weapons of unmanned systems.

Decision-Making Analysis

Cognitive decision-making describes a mental process consisting of different phases (Wang & Ruhe, 2007). In aviation, the most common strategy is the FOR-DEC method (Hörmann, 1995), containing the phases facts, options, risks and benefits, decision, execution, and check. But in general, the single cognitive processing steps during decision-making can be subdivided differently.

Decision Types

Decision-making can be divided into different subtypes. Particularly noteworthy at this point is the distinction between tactical decision-making on an organizational level, which is often referred to as the OODA loop (Observe – Orient – Decide – Act) of John Boyd (Osinga, 2007), and individual cognitive decision-making, which is relevant for this contribution. The latter can be further be categorized into analytical / rational or experiential / moral decision (Epstein, Pacini, Denes-Raj, & Heier, 1996). Both categories can additionally be subdivided into decisions in a static or dynamic environment.

Static decision-making, on the one hand, describes a simple, one-dimensional situation, where a, in its simplest form binary, decision has to be made. Several thought experiments show static situations, e.g. a self-driving car about to run over pedestrians due to brake failure, as a variation of the widely known trolley problem (Thomson, 1985). The user must now decide whether the car should swerve to avoid colliding with the pedestrians or continue straight ahead to protect its occupants. Either way, there will be casualties on one or the other side (Awad et al., 2018). This static situation does not change during the decision process and is independent of the decision maker's interaction prior to the decision-making itself. Thus, the decision can be classified as static since physical parameters are not continuously changing. In contrast, dynamic decision-making takes place in highly complex situations, with rapidly changing environmental parameters. Furthermore, the user's interaction influences and changes those parameters as well. The decision usually is time-critical and risky, with several possible outcomes and once a decision is made, the entire situation may change. Even hesitation, as a factor of change, contributes to the complexity of the situation, the effects of which sometimes become visible only far in the future. In a military context, the majority of decisions is of said dynamic, time-critical, and risky type. The overall goal hereby is to stop the enemy, whether by its elimination or limitation of its capabilities.

The above-mentioned examples can further be broken down into analytical and experiential decisions. Analytical or rational decisions rely on the systematic analysis of information to choose the best course of action and thus can be part of the rule-based level. The chosen decision can be classified as right or wrong, unlike moral decisions, which are based on intuitive, experiential thinking. The latter relies on experience and personal judgment and varies according to the individual person.

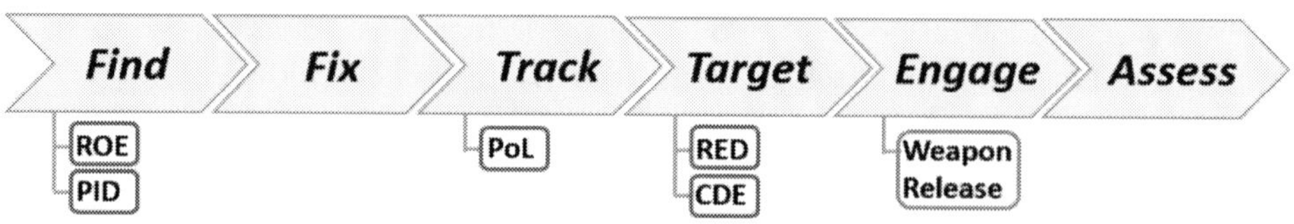

Figure 1. F2T2EA targeting chain (Jackson, 2006) with combined weapon release SOP.

The F2T2EA targeting cycle, as depicted in Figure 1, combines these approaches as the first steps are handled based on analytical pre-decisions, marked in blue, leading up to a moral decision for weapon deployment in the engage phase of the cycle, marked in orange. The mentioned targeting cycle is composed of the Find phase, in which the Rules of Engagement (ROE) are applied for further decisions. After verification, a list of features is processed for Positive Identification (PID) of the hostile target. Once the coordinates of the target have been checked in the Fix phase, the target is monitored in the Track phase and the Pattern of Life (PoL) is analyzed to detect anomalous behavior and to further confirm the classification as a hostile. Weapon selection is done in the following Target phase based on the Collateral Damage Estimation (CDE) and determination of Risk-Estimate Distances (RED). All these steps are operationalized as defined rules with hard thresholds and thus require analytical decisions delimiting the moral decision space for the weapon release in the Engage phase but can also influence the moral decision. The cycle is completed by the last phase of damage assessment.

Method of Decision-Making Analysis

To be able to analyze and draw conclusions from moral decisions, the analytical pre-decisions are evaluated first. As an experiment, automation could be used to identify potential targets using image recognition and a list of characteristics enabling a differentiation between friend and enemy. However, the pilot has to confirm the classification into those categories and ultimately be the one to make the decision for weapon deployment. In order to perform experiments, the automation could intentionally misclassify potential targets putting the pilot into the position of having to recognize the error. As a simple example, an image of an ambulance could be shown with the classification as a hostile military vehicle such as a tank. If the pilot confirms this clearly wrong identification performed by the automated system, the eye tracking path could provide information to further understand the thoughts behind the decision by questioning and confronting the pilot with it. In this way, unconscious decisions that may occur due to attentional tunneling (Wickens, 2005) can be figured out, as possible in the mentioned example. Another reason for this error could be automation bias since the pilot verifies the automation's suggestion without scrutinizing it. This can either happen due to too much trust in the automation, combined with loss of situational awareness, or due to work overload and thus the deliberate handing over of tasks to the automation in order to avoid further increasing work overload. Combining this methodology with interaction analysis, lack of situational awareness or work overload can further be detected or confirmed.

To identify the cause of suchlike incorrect outcomes in analytical decisions, the errors shall be classified according to the error taxonomy (Reason, 1990) based on the SRK-model (Rasmussen, 1983). This way, errors can be distinguished based on intended and unintended actions. Errors based on intended actions and thus causing the problem themselves, can be divided into attention errors, called slips, such as mistiming or attentional tunneling, and memory errors, called lapses, caused by forgetting an original intent, e.g., due to automation bias. This category of errors appears at the skill-based level. Mistakes, that are intentional but unsuccessful attempts to solve problems, can be divided into rule-based and knowledge-based failures. The first can be, for example, the application of a good rule at the wrong time due to lack of knowledge or understanding resulting in a wrong mental model of the situation, whereas the latter can be caused by lack of experience. Another class of error is the so-called violation, in which conscious decision has been made to take a certain action, which violates a rule, for example a ROE. Classifying the analytical errors using this taxonomy, it can be decided which human-AI workshare is helpful in which phase of the targeting chain in order to avoid work overload and still provide enough information necessary for decision-making. Consequently, it can also be determined which human-AI workshare is even influencing the decisions negatively. Furthermore, the way the information is presented should be considered as a factor and be assessed as well as the quantity and type of information necessary to come up with a decision. Based on the preliminary analysis of the rational decisions and thus of the initial situation, a statement about conscious decision-making in moral problems can be made.

However, it remains an absolute necessity for these experiments that the pilot mentally empathizes with the situation and shows a behavior as close as possible to a real-world situation, as otherwise the results of the experiments would not be transferrable to it. Therefore, it is crucial to provide a suitable experimental environment, as described in the following.

High Presence Virtual Environment as a Research Tool

Since experiments cannot be conducted during real military operations, a flight and mission simulator is used. The problem concerning simulators is the risk of the pilot thinking "it's just a simulator" and thus de-emphasizing the situation. To overcome this issue, a special type of flight and mission simulator has to be built, which deeply involves the pilot in the mission, physically as well as mentally.

The Relation between Moral Buffer, Immersion, and Presence

In computer games, simulator flights, or even drone missions, the operator is spatially separated from the situation. This physical and sometimes resulting emotional distance, amplified by the use of a computer interface and not being in the situation physically, creates a moral buffer between the person and the scenario (Cummings, 2004), which is larger or smaller depending on the threshold for resistance to killing, as shown in Figure 2. The smaller the physical and therefore the emotional distance between the person and the target is, the smaller is the moral buffer as the person experiences the results of its decisions in a much more direct and personal way. In contrast to this, if the physical distance is significantly higher, the person emotionally uncouples from the consequences of the decisions as they have close to none or no effect for said person at all, therefore resulting in a higher moral buffering and lower killing-threshold. As a basic requirement for valid decision outcomes, it is necessary to overcome this problem so that the person mentally empathizes with the situation, makes conscious decisions, and does not proceed the decision with indifference. Thus, it does not matter whether it is an ethical distance, i.e. distance from the aspect of a "fake situation", or spatial distance, as in the case of a drone pilot operating a drone from a remotely located control station.

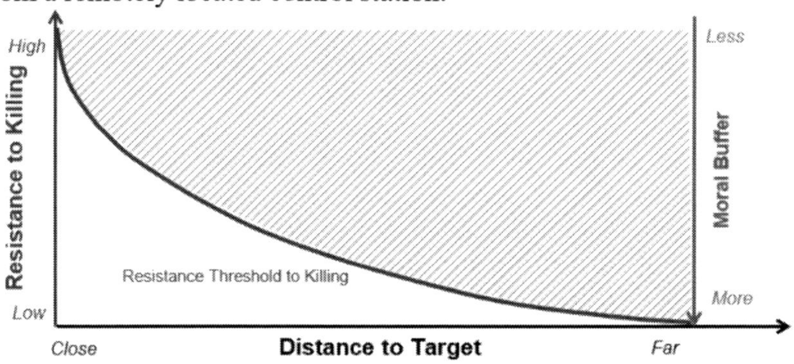

Figure 2. Relationship between the distance of the situation to the human, physically as well as emotionally, and the resulting moral buffer and the resistance threshold to killing (extended from (Cummings, 2004) and applied to simulated situations).

To minimize the moral buffer, the two principles of immersion and presence are used. Immersion is the ability to be physically immersed in a virtual world, whereas presence is the mental or cognitive immersion in the simulated world (Slater, Usoh, & Steed, 1995). A high immersion is reached by the simulator itself by means of the vision-system, the haptics of the cockpit, and the whole setup itself. A high mental presence of the pilot is reached by the right choice of tasks and cueing, which is described in the following.

Concept of a High Presence Virtual Environment

In order to create a high degree of presence in the research mission-simulator, a customized concept is proposed, as depicted in Figure 3. It consists overall of three stages with chronologically ordered subphases.

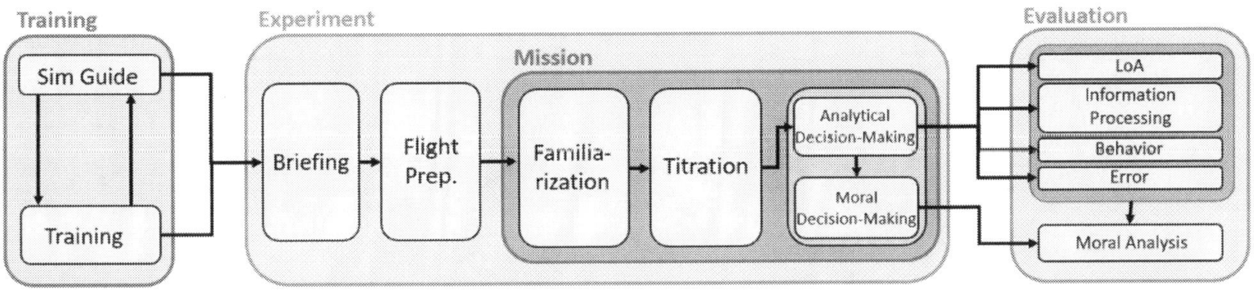

Figure 3. Concept of an experimental procedure in the mission simulator for moral decision analysis.

In order to familiarize the pilot with the simulator and its implemented systems, a training stage is allocated before the experiments take place. During the training, the pilot shall learn how to operate the (generic) aircraft under normal conditions with its designed workflow using a simulator guide and a set of Standard Operating Procedures (SOPs) as resources. The more the training advances, handling of minor non-normals shall be trained. At the end of the training phase, the pilot should have acquired the skill-based and rule-based competences to operate the aircraft safely under normal and non-normal conditions.

The second stage is the actual experiment itself. It consists of several phases derived from a real mission: First, the pilot is briefed on the tactical situation, the weather, and important background information. This briefing should take place in another room to create a spatial separation from the aircraft. Afterwards, the pilot has to prepare the simulator according to the normal workflow and by following the mentioned SOPs, including the usage of checklists and radio-communications. When the aircraft and all systems are ready for mission commencement, the pilot begins the flight into the mission area and starts to pursue the briefed objectives. By his individual interaction with the system, each pilot will influence the scenario in a different way and thus directly affect the mission and shape its outcome in a previously unknown way. Accordingly, it is important that the pilot receives feedback showing the effects of his interaction. In this phase, the pilot should already be mentally involved into the mission and "titration" may be used to amplify his cognitive immersion: unexpected events will now occur in the mission with rising frequency and severity, such as weather changes, tactical problems, or technical malfunctions. In this way, the pilot is challenged with new and quickly changing conditions and needs to constantly re-evaluate the parameters of the mission. This increased workload should intensify his mental presence in the simulation. In the final phase of the experiment, the pilot is confronted with time-critical, risky, moral situations that require the pilot's intervention, which should be handled on the knowledge-based level. These decisions under pressure, as well as the decision-making process leading to it during experiments, is one of the biggest differences of the research simulator in contrast to flight training simulators, used by airlines. The latter has the primary goal of training the crews in the handling of non-normal situations by deepening their procedural knowledge (EHEST European Helicopter Safety Team, 2012). Thus, the training objective is to develop a resilience to surprise moments and to remain calm and proceed according to a defined workflow.

The third phase is comprised of the evaluation of the collected data, where different post-evaluating methods come to play. Those potentially being classic questionnaires and replay logs or even real-time evaluation like behavioral analysis based on gaze tracking allowing to draw conclusions for decision-making, as already mentioned in the previous section.

Conclusion and Outlook

This contribution considers the increasing debate about the use of autonomous weapon systems and the associated trigger authority, which should remain with humans. Since moral decisions are based on analytical pre-decisions, these can be investigated with respect to error and automation bias using gaze tracking and behavioral analysis. However, in order to produce meaningful results in the experiments, the concepts of immersion and presence must be applied. While immersion is created by the setup of the simulator, for presence a procedure plan has to be developed. This starts with training sessions to get familiar with the simulator, followed by the mission building up to acute moral situations that require interventions. Based on the observations, an analysis of moral decisions can be made. This presented concept will be tested in experiments in near future.

References

Awad, E., Dsouza, S., Kim, R., Schulz, J., Henrich, J., Shariff, A., . . . Rahwan, I. (2018). The Moral Machine experiment. *Nature, 563*(7729), 59–64. https://doi.org/10.1038/s41586-018-0637-6

Cummings, M. L. (2004). Creating Moral Buffers in Weapon Control Interface Design. *IEEE Technology and Society Magazine, 23*(3), 28–41. https://doi.org/10.1109/MTAS.2004.1337888

Cummings, M. L. (2018). Informing Autonomous System Design Through the Lens of Skill-, Rule-, and Knowledge-Based Behaviors. *Journal of Cognitive Engineering and Decision Making, 12*(1), 58–61. https://doi.org/10.1177/1555343417736461

EHEST European Helicopter Safety Team (2012). EHEST Leaflet HE 4: Single Pilot Decision Making. Retrieved from https://www.easa.europa.eu/en/document-library/general-publications/ehest-leaflet-he-4-single-pilot-decision-making

Epstein, S., Pacini, R., Denes-Raj, V., & Heier, H. (1996). Individual differences in intuitive-experiential and analytical-rational thinking styles. *Journal of Personality and Social Psychology, 71*(2), 390–405. https://doi.org/10.1037//0022-3514.71.2.390

HLEG, A. I. (2020). *The Assessment List for Trustworthy Artificial Intelligence (ALTAI)*. European Commission.

Hörmann, H. J. (1995). FOR-DEC - A Prescriptive Model for Aeronautical Decision Making. *Human Factors in Aviation Operations. Proceedings of the 21st Conference of the European Association of Aviation Psychology*. (3), 17–23.

Jackson, J. F. (2006). *Targeting - Air Force Doctrine Document 2-1.9*. Maxwell Air Force Base, AL, USA: Air Force Doctrine Center. Retrieved from https://apps.dtic.mil/sti/citations/ADA454614

Osinga, F. P. B. (2007). *Science, Strategy and War: The Strategic Theory of John Boyd. Strategy and history*. London, New York: Routledge.

Rasmussen, J. (1983). Skills, rules, and knowledge: signals, signs, and symbols, and other distinctions in human performance models. *IEEE Transactions on Systems, Man, and Cybernetics, SMC-13*(3), 257–266. https://doi.org/10.1109/TSMC.1983.6313160

Reason, J. T. (1990). *Human Error*. Cambridge: Cambridge University Press. https://doi.org/10.1017/CBO9781139062367

Slater, M., Usoh, M., & Steed, A. (1995). Taking steps. *ACM Transactions on Computer-Human Interaction, 2*(3), 201–219. https://doi.org/10.1145/210079.210084

Thomson, J. J. (1985). The Trolley Problem. *The Yale Law Journal, 94*(6), 1395–1415. https://doi.org/10.2307/796133

Wang, Y., & Ruhe, G. (2007). The Cognitive Process of Decision Making. *International Journal of Cognitive Informatics and Natural Intelligence (IJCINI), 1*(2), 73–85. https://doi.org/10.4018/jcini.2007040105

Wickens, C. D. (2005). Attentional tunneling and task management. *2005 International Symposium on Aviation Psychology*, 812–817. Retrieved from https://corescholar.libraries.wright.edu/isap_2005/121/

PILOTS' PERSPECTIVES ON URBAN AIR MOBILITY SAFETY CHALLENGES AND POTENTIAL SOLUTIONS

Rania Wageh Ghatas
NASA Langley Research Center
Hampton, Virginia
Saeideh E. Samani
NASA Langley Research Center
Hampton, Virginia
Victoria L. Dulchinos
NASA Ames Research Center
Moffett Field, California
Dr. Richard H. Mogford
NASA Ames Research Center
Moffett Field, California

As electric vertical takeoff and landing air taxis make their way to urban airspace operations within the United States National Airspace System, many research efforts are underway to identify and understand pertinent issues needed to support the influx of new, passenger-carrying, air vehicles over highly dense, urban communities. The primary focus of this research effort was to gather subjective data from subject matter experts concerning current-day airspace operations to identify potential gaps and improvements needed to support and sustain near-term UAM operations. These potential gaps and improvements will form the foundation for the development of initial information exchange requirements between the on-board pilot in command and other key entities. This paper focuses on the safety challenges and potential solutions for the in-flight incapacitated pilot scenario from Phase I of the study, which gathered data from helicopter and general aviation fixed-wing pilots.

As the concept, and soon-to-be reality, of air taxis make their way to urban airspace operations within the United States' National Airspace System (NAS), many research efforts are underway to identify and understand pertinent issues needed to support the influx of new, passenger-carrying, air vehicles over highly dense, urban communities.

With the onset of quickly advancing technology, and with a new generation that is looking for ways to travel quickly and efficiently, the concept of on-demand air mobility is now driving the future of flight (FAA, 2021; MITRE, 2020; NASA, 2022). In response to this quickly developing concept, and with safety being the primary concern, the National Aeronautics and Space Administration (NASA) has established an Urban Air Mobility (UAM) subproject that spans multiple NASA centers under the Air Traffic Management – eXploration Project. Many supporting research efforts will be necessary to answer difficult questions regarding airspace management, detect-and-avoid capabilities, public acceptability, and much more for the safe and efficient integration of UAM operations into the NAS (e.g., Arneson & Thiphavong, 2020; Craven et al., 2021; Price et al., 2020; Thipphavong et al., 2018).

There are several research questions that need to be addressed for successful implementation of UAM that include:

a) What initial requirements are needed for information sharing/exchange between the on-board UAM pilot-in-command and other key UAM entities, such as the Provider of Services for UAM (PSU), the aircraft dispatcher (or flight follower), the Operator (company), the vertiport manager, and Air Traffic Control (ATC) to sustain UAM operations?

b) How may current contingency management strategies should(?) must(?) be used to handle future UAM operations during off-nominal and emergency situations?

The purpose of the PSU is to provide a seamless cooperative data exchange between the different service suppliers and the users of the UAM airspace. The goal for these service providers and participating aircraft would be to share data to support operational planning, such as aircraft de-confliction, conformance monitoring, and emergency information dissemination, among others (Federal Aviation Administration, 2020).

The primary focus of the UAM Pilot/PSU Information Exchange and Contingency Airspace Management Procedures (UAM PIE CAMP) study was to gather subjective data from aviation subject matter experts concerning current-day airspace operations. The purpose of the research was to identify potential gaps and improvements needed to support and sustain near-term UAM operations. The identified potential gaps and improvements are intended to help inform the development of initial information exchange requirements between the on-board UAM pilot and other key UAM entities. Additionally, UAM PIE CAMP collected data to support verification of assumptions concerning initial recommendations for airspace procedures during nominal and off-nominal or emergency situations. Data gathered from Phases I and II of the UAM PIE CAMP study serve as the foundation for follow-on pilot- and dispatch-focused studies that will take a deeper look into these assumptions and inform research needs for initial, near-term, UAM operations.

A number of operational scenarios were included in the UAM PIE CAMP study. The present paper focuses on a specific in-flight incapacitated pilot scenario that was presented to twelve pilots during Phase I testing. Pilot participants included nine helicopter pilots (three in helicopter air tours, three in medical evacuation or air ambulance operations, and three in private business air charter services) and three general aviation fixed-wing pilots. Although participants were separated into these four areas, many pilots had experience across multiple categories that increased the diagnostic sensitivity of the data sets collected.

Literature Review

The topic of potentially having an in-flight incapacitated pilot is one that is seldomly discussed during the initial stages of planning for the integration of new aircraft operations in the NAS. However, there is urgent need for this important topic to be explored and researched, especially given the likelihood of a single pilot on-board flying the air taxi vehicle (particularly during early phases of UAM operations in practice). Although single-pilot operations are common among general aviation fixed-wing aircraft and helicopters, the high-tempo and repeated short-distance UAM flights pose many new challenges. One significant one challenge relates to in-flight pilot incapacitation during single pilot UAM air taxi operations within densely populated urban ecosystems.

Pilot incapacitation is the term used to describe the inability of a pilot "to carry out their normal duties because of the onset, during flight, of the effects of physiological factors" (Flight Safety Foundation, n.d.). Although the majority of cases involving an incapacitated pilot are related to cardiovascular disease or gastro-intestinal problems, there are several other causes that could lead to a pilot's inability to perform normal duties. Examples of other causes of pilot incapacitation include hypoxia (insufficient oxygen); a bird strike; smoke or fumes that enter the cabin due to a vehicle malfunction or other issue; or a malicious or hostile act, such as an assault by an unruly passenger or high-powered lasers by persons on the ground. The safety of a flight becomes severely compromised, and loss of control may occur during single pilot operations in the event the pilot becomes incapacitated.

Use-Case Scenarios Activity

The airspace surrounding the Dallas/Fort Worth (DFW) metropolitan area was chosen for this research study due to the complexity and volume of air traffic that is typically experienced in this region. The airspace near the DFW International Airport is designated as a Class Bravo airspace and has two nearby operating airports, including Dallas Love Field (DAL) (Class Bravo airspace) and the Addison Airport (ADS) (Class Delta airspace). Figure 1 provides a visual representation of the DFW airspace area. The corridor system is depicted in purple, vertiports are shown in green, and an extended corridor is depicted in magenta.

On Day 1 of data collection, each pilot participant was presented with 11 use-case scenarios with each scenario comprised of approximately 20 questions. Participants were instructed to play the role of the on-board UAM pilot flying a three-to-five passenger electric vertical takeoff and landing (eVTOL) vehicle, within the DFW metropolitan area airspace with approximately 50 other eVTOL aircraft. The objectives of the scenario narratives were to leverage current-day airspace operational procedures while identifying needed improvements and gaps to address: (a) the future needs of UAM aircraft operating in the NAS, (b) identify information exchange requirements, and (c) inform pilot roles and responsibilities.

In-flight Incapacitated Pilot Scenario

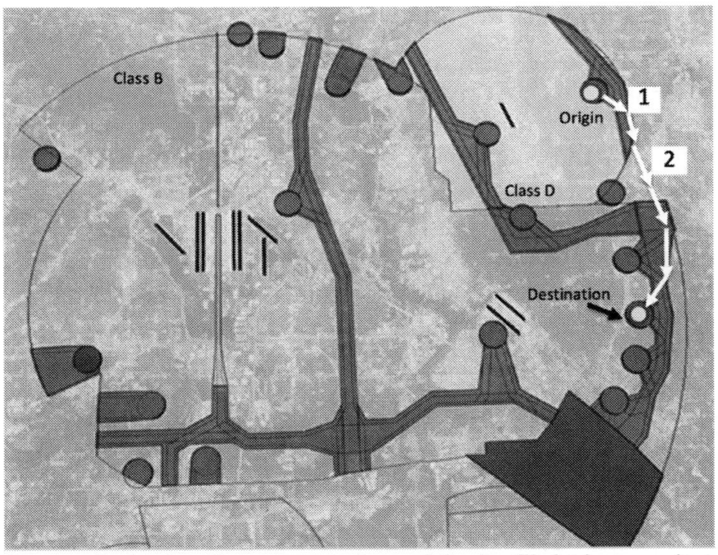

Figure 1. A visual representation of the in-flight incapacitated pilot scenario within the DFW airspace area (Google, 2021).

The in-flight incapacitated pilot scenario, which was the sixth one shown to participants on Day 1 of data collection, represented a UAM flight with a single pilot on board who experiences physically incapacitating symptoms. Flight origin, destination, and planned route are described to the participant using the graphic depicted in Figure 1 with waypoint 1 representing the flight origin. Midway through the flight (waypoint 2 in Figure 1), the scenario revealed to participants that the eVTOL pilot has begun to feel lightheaded, experiencing dizziness, and unable to communicate properly while the flight is transitioning out of one corridor and entering into another corridor system. It was explained to the participant that, in this scenario, the aircraft is quickly losing speed and altitude. The participants are further made aware that there are currently no defined UAM contingency management or safeguards available to address this type of emergency. The pilot, however, could initiate emergency pilot procedures, if able, such as declaring an emergency (e.g., Mayday, "request assistance immediately" procedure, etc.). The aircraft is not fully automated, does not have autoland capability, and cannot be remotely controlled from the ground or by the PSU.

After the researcher read the scenario description out loud, participants were asked a range of questions including what airspace and operational requirements might be necessary to manage this type of medical emergency scenario, and, if able, with what means should the pilot share their incapacitated state and with whom.

Results and Discussion

A significant amount of data was collected from the 12 pilot participants throughout the course of this research effort. Approximately 20 questions were asked, not including follow-on questions, for each use case scenario. The results presented in this paper focus on three of those questions specific to the pilot incapacitation scenario:

1. First question pertained to asking about current-day airspace and operational procedures for situations involving an incapacitated pilot in single pilot operations, and if there are any gaps or areas of improvement that should be considered with the introduction of UAM aircraft into the NAS;
2. Second question pertained to communications; and
3. The third question asked about potential solutions.

Gaps and Challenges

Limited Pilot Incapacitation Procedures Exist. Current-day procedures are not well designed to address the situation of an incapacitated pilot in single-pilot operation. Although emergency pilot incapacitation procedures do exist, such as lost comm and "request assistance immediately" (Aeronautical Information Manual 6-1-1; 6-1-2), they are not well suited to case of sudden pilot incapacitation. There was a consensus among the participants that there is a gap and areas of improvement needed to address this difficult issue. There was also a consensus among the participants that the primary responsibility of a pilot is to aviate the aircraft and the pilot will only communicate if able. If they are able to communicate, the majority of participants said they would make an emergency call to whichever frequency they are currently dialed into, which would likely be their operator, such as their dispatch or flight follower, or the common traffic advisory frequency. They did note that ATC would be the best entity to announce this emergency to, if able, as ATC can provide separation to other aircraft in the vicinity to stay well clear from the vehicle with the incapacitated pilot and contact emergency services on the ground. Additionally, ATC could advise other traffic in the vicinity to observe and relay back what is happening as long as the observing aircraft are within an appropriate distance and observation would not result in additional disruption to the airspace.

eVTOL Design and Performance Characteristics. Considerations need to be taken into account in the design and performance characteristics (e.g., safety assurance and flight control system redundancies) of the aircraft. A pilot experiencing a medical emergency in-flight can have a significant difference in outcome when it comes to whether the aircraft is fixed-wing or a rotorcraft.

Rotorcraft are known to be less stable than fixed-wing aircraft and have different performance characteristics when it comes to controls. Rotorcraft are unable to glide, even with autorotation capabilities, and require constant pedal input at hover and low forward speed to keep the tail behind the nose. Additionally, the design of the rotocraft flight deck controls can have a significant impact on how the aircraft responds if a pilot suddenly becomes incapacitated and leans over or blocks certain controls. A summary of the responses collected from participants regarding the severity of a pilot's incapacitation in a helicopter can be summarized in the following quote by US journalist Harry Reasoner in 1973 (Flight Safety Australia, 2019) who wrote:

> [A]n airplane by its nature wants to fly and, if not interfered with too strongly by unusual events or by a deliberately incompetent pilot, it will fly. A helicopter does not want to fly. It is maintained in the air by a variety of forces and controls working in opposition to each other, and if there is any disturbance in the delicate balance, the helicopter stops flying, immediately and disastrously. There is no such thing as a gliding helicopter.

All participants agreed that a pilot medical emergency while in-flight is both a medical and aircraft emergency, especially in aircraft resembling modern-day helicopters, such as some versions of envisioned eVTOL aircraft.

eVTOL Altitude and Urban Environment Challenges. Another challenge is the altitude and urban environment in which these operations will take place. Flying close to the ground is nothing new for rotorcraft. Helicopters in current-day operations already fly close to the ground in areas with terrain. Although pilots fly slower when close to terrain, there can often be limited time to react a recovery maneuver, particularly in the event of a sudden in-flight pilot incapacitation. One great advantage to rotorcraft is that they have the potential to land anywhere, such as a parking lot or an open field. However, depending on the nature of the medical emergency, the pilot may not have the ability to maneuver the aircraft to reach that destination and safely land while avoiding ground obstacles, including pedestrians. Additionally, depending on speed and altitude during the medical emergency, the rotorcraft itself may not even be able to autorotate to a safe landing (Prouty, 1986).

Summary

There were many potential solutions discussed during the interview process, including receiving support from onboard and remote resources, such as the Garmin Autoland, onboard passengers, and the idea of having a remote operator with the ability to remotely control the aircraft. Each of these potential solutions pose their own challenges, which ranged from implementation and training to significant security concerns.

More in-depth research is needed to explore the pros and cons of these solutions and others. Important upfront considerations include how eVTOL aircraft are designed and the infrastructure for the urban environment in which these aircraft will be operating in; this is particularly critical for single pilot incapacitation. The NASA UAM sub-project is focused on the research and development needs to help enable future UAM air taxi operations through efforts, such as UAM PIE CAMP study, to identify potential safety concerns and help identify potential gaps and improvements needed to support and sustain near-term UAM operations. These potential gaps and improvements will form the foundation for the development of initial information exchange requirements between the on-board pilot in command and other key entities in order to reduce potential increase of severity and incidences of UAM accidents in the event that the single onboard eVTOL pilot becomes incapacitated while carrying passengers above urban communities.

Acknowledgements

This work was conducted under the NASA Urban Air Mobility Sub-Project under the Air Traffic Management – eXploration (ATM-X) Project as part of the Airspace Procedures and Design (AP&D) sub-element. The support of the sub-project management, Kevin Whitzberger and Ian Levitt, is gratefully appreciated. The unfailing and tireless support and research contributions of Savita Verma, the AP&D technical lead at Ames Research Center, is also gratefully appreciated along with the entire AP&D team. Gratitude is extended to Clay Hubbs, our pilot subject-matter expert, and Jason Prince, the AP&D technical lead at Langley Research Center, for their support and contributions to the success of the UAM PIE CAMP study, and to Bryan Barmore, Taumi Daniels, Kurt Swieringa, David Thippavong, Lindsay Stevens, Heather Arneson, Shivanjli Sharma, and Starr Ginn for their support of this effort at the initial stages of research planning and design. Thanks are also extended to the Crew Systems and Aviation Operations Branch management and review panel committee at Langley Research Center for their continued support of the UAM PIE CAMP research study.

References

Arneson & Thiphavong (2020). "ATM-X and Urban Air Mobility Overview." Retrieved from https://ntrs.nasa.gov/citations/20205000435

Craven et al. (2021). "Preliminary Evaluation of National Campaign Scenarios for Urban Air Mobility." https://ntrs.nasa.gov/citations/20210018334

Federal Aviation Administration (2021). "Charting Aviation's Future: Operations in an Info-Centric National Airspace System," 2021.

Federal Aviation Administration (2020). "Concept of Operations V1.0 Urban Air Mobility (UAM)," Federal Aviation Administration, Washington, D.C., 2020.

Flight Safety Australia (2019). "Human Factors and the Helicopter." April 2, 2019. Retrieved from https://www.flightsafetyaustralia.com/2019/04/human-factors-and-the-helicopter/

Flight Safety Foundation (n.d.). "Pilot Incapacitation." Retrieved from https://www.skybrary.aero/articles/pilot-incapacitation

Google (n.d.). Retrieved 2021, from https://maps.google.com

National Aeronautics and Space Administration, "Sky for All Portal," 2022. Retrieved from https://nari.arc.nasa.gov/skyforall/

MITRE Corporation (2020). "The Future of Aerospace: Interconnected from Surface to Space." FAA Managers Association Managing the Skies. Retrieved from www.faama.org

Price et al. (2020). Urban Air Mobility Operational Concept (OpsCon) Passenger-Carrying Operations. https://ntrs.nasa.gov/citations/20205001587

Prouty, R. (1986). "Helicopter Performance, Stability, and Control." Wadsworth.

Thipphavong et al. (2018). Urban Air Mobility Airspace Integration Concepts and Considerations. Retrieve from https://ntrs.nasa.gov/citations/20180005218

Federal Aviation Administration (2023). Aeronautical Information Manual. Retrieved from https://www.faa.gov/air_traffic/publications/atpubs/aim_html/chap6_section_1.html

CAN YOU HEAR ME? SIMULTANEOUS MASKING BETWEEN THE STARS AIR TRAFFIC CONTROL ALARMS

Corey Hall
General Dynamics Mission Systems, Pittsfield, MA

Elliot Biltekoff
University at Buffalo, Buffalo, NY

Matthew L. Bolton
University of Virginia, Charlottesville, VA

The Standard Terminal Automation Replacement System (STARS) was introduced to eliminate unnecessary Air Traffic Control (ATC) alarms and reduce air traffic controller confusion about alarms. The six STARS alarms are tonal melodies. Because of this, they are susceptible to simultaneous masking: where a tone cannot be heard in the presence of other sounds due to limitations of the human sensory system. This work used a proof-based computational method to analyze the STARS alarms to determine if any masking is possible. Our results found three instances where alarms could be partially masked (have part of their signal made inaudible) by other simultaneous alarms. More importantly, one alarm could be totally masked (rendered completely inaudible) in the presence of three others. In this paper we describe this analysis. We also suggest standardizing the volumes of alarms to prevent the potential for total masking.

Raytheon Systems Company proposed the Standard Terminal Automation Replacement System (STARS) to replace the alarms used in Air Traffic Control (ATC). STARS was created to address several issues facing the industry such as effectiveness, consistency, and discriminability of alarms. Prior to the introduction of STARS, there were three versions of another system being used(Newman & Allendoerfer, 2000). While STARS was previously compared to its predecessors, the specifics of the alarms were not directly addressed. The similarity of the tones utilized by STARS are susceptible to simultaneous masking; the phenomenon where one tone or sound cannot be perceived (total masking) or can only be partially perceived (partial masking) in the presence of another tone or sound due to limitations of the human sensory system (Fastl & Zwicker, 2006).

In work environments at large, auditory masking has been observed and reported on in medicine and aircraft cockpits (Edworthy & Meredith, 1994; Meredith & Edworthy, 1995; Konkani, Oakley, & Bauld, 2012; Edworthy & Hellier, 2006, 2005; Patterson & Mayfield, 1990; Patterson, 1982; Momtahan, Hetu, & Tansley, 1993; Toor, Ryan, & Richard, 2008). It is likely that this information would translate to Air Traffic Control and errors may be able to be contributed to this effect. There is little research connecting aircraft accidents to the simultaneous masking of alarms in ATC. However, there are documented reports, made by controllers, that the discriminability of alarms poses a problem (Newman & Allendoerfer, 2000).

Bolton et al. developed (Bolton, Edworthy, & Boyd, 2018b, 2018a; Hasanain, Boyd, Edworthy,

& Bolton, 2017; Hasanain, Boyd, & Bolton, 2014, 2016; Bolton, Hasanain, Boyd, & Edworthy, 2016) and experimentally validated (Bolton, Zheng, Li, Edworthy, & Boyd, 2020) a computational method that is able to address the complexity issues that have prevented comprehensive masking analyses previously. The method used model checking (Clarke, Grumberg, & Peled, 1999; a method for automatically proving properties about computational modeling) with the psychoacoustics of simultaneous masking (mathematical models of the masking phenomenon) to determine if masking could manifest in a modeled set of alarms. This was ultimately used to evaluate the international medical alarm standard (Bolton, Edworthy, & Boyd, 2022) and inform its update (Edworthy et al., 2018).

For this research, Bolton's computational method is used to evaluate the six STARS alarms, at iterative intervals between 70 dB and 80dB, to determine the level to which the individual alarms are masked by each other. Below, we provide background on material necessary for understanding this research. We then provide a description of the objectives and method that were used to achieve our results. We discuss these results and the implications they pose for the industry moving forward.

Background

The following describes the method utilized in the research and the alarm sounds of STARS.

Method

Bolton's method uses a combination of psychoacoustics and model checking to detect if alarms susceptible to simultaneous masking (Bolton et al., 2018b, 2018a; Hasanain et al., 2017).

The psychoacoustics of simultaneous masking mathematically represent how the physical characteristics of a sound affects its ability to be perceived. The basis of this is derived from a decrease in sensitivity of sensory cells on the basilar membrane when it is exposed to multiple sounds. Because there is an additive effect per the number of sounds, there is a greater chance that a single sound will be masked when there are multiple sounds present (Lutfi, 1983; Bosi & Goldberg, 2003). This additive affect is depicted as a curve and is often referred to as the masking curve.

In formal methods, a model is created to explain a systems behavior and then checked against desirable properties to see if those properties are always true. Dr. Bolton's method uses formal methods, and specifically a technique called model checking, to automatically, mathematically prove properties about masking in models of alarms. When applying Dr. Bolton's method (Bolton et al., 2018b, 2018a; Hasanain et al., 2017) to a given alarm system, formal models can be created for the alarms, along with their specifications, to prove if masking can occur. The method was ultimately extended with a computer software frontend called MAASC (Medical Alarm Audibility System Checker; Bolton, Biltekoff, Boyd, Darget, & Edworthy, 2020), This desktop application enables medical alarms to be modeled and evaluated using simple point-and-click interactions. It is worth noting that one feature of MAASC is its ability to visualize counterexamples, the trace produced by the model checker that shows how a specification violation occurred. This shows exactly which alarm was masked and what timing is required between alarms to produce the masking.

Despite its power, the method has not been applied to alarms in ATC.

STARS

Previous research has found that the alarm systems used in ATC struggled to help controllers identify and differentiate the aural tones used as signals. Raytheon Systems Company developed

Table 1
Standard Terminal Automation Replacement System Low Priority Alarms

Alarm	Frequency	Period
Conflict Alert	1600 Hz	60 ms / 60 ms
Minimum Safe Altitude Warning	1600 Hz to 2000 Hz warble	260 ms / 180 ms
Mode C Intruder	1600 Hz	130 ms / 130 ms
Default	800 Hz	60 ms / 60 ms
Special Transponder Emergency	1400 Hz	600 ms / 250 ms
Critical Subsystem Failure	800 Hz	250 ms / 500 ms

STARS to replace existing systems in effort to relieve this problem. STARS acts to eliminate confusion among the several versions of the previous alarm system and standardize the ATC alarm system at large. This new system utilizes six alarms (Table 1) with frequencies ranging from 800 Hz to 2000 Hz with variable sounding periods. Each alarm is classified as low priority and is comprised of two events. An event is defined as an audible tone at a specific frequency followed by a space of a specific time frame. The alarm name, frequency (in hertz) and period (in milliseconds) are outlined in the following table (Newman & Allendoerfer, 2000).

Methods

While MAASC was originally created for application in the medical field, we modified it to fit the needs of this research. Specifically, we update the mathematical computations used for converting alarm descriptions into formal models. This was necessary due to the different frequency ranges found in STARS alarms. Note that we also manually overrode some of the medical alarm standard values that were incompatible with the STARS alarms (this was something that was supported by MAASC). Thus, MAASC was used to model and analyze all of the alarms listed in Table 1. Note that because no volumes are specified for the STARS alarms (Newman & Allendoerfer, 2000) (volumes of the STARS alarms can be set by individual controllers), we started our analysis by using a fairly standard, listenable volume of 70 dB. This gave us a baseline from which to consider relative volumes between alarms. Then, based on results observed between alarms, we systematically varied alarm volumes (iteratively decreasing each by 1dB down to 60 dB while holding the others at 70 dB) to investigating masking conditions more deeply. In all of these analyses, we stopped descending once we found the masking condition under consideration:

Results

All six alarms and the associated configurations were analyzed for both total and partial masking. MAASC provided results with a mean verification time of 4.4 seconds with a standard deviation of 1.58 seconds. At the 70 dB level, no total masking was found. However, situations where alarms could be partially masked were identified. In particular, the Conflict Alert, Mode C Intruder, and Minimum Safe Altitude Warning are all susceptible to partial masking (Fig. 1(a)–(c)). This masking only occurs in the presence of all three target alarms and the Default alarm. Also, only one event for each of those alarms is masked in a given configuration.

For the Conflict Alert, at 70 dB, the first event is masked by the combination of the first events of the Mode C Intruder, Minimum Safe Altitude Warning, and Default alarms (Fig. 1(a)). For Mode

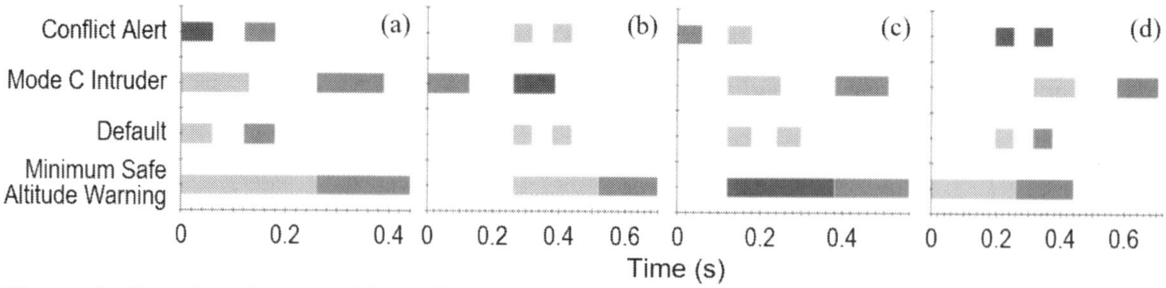

Figure 1. Results of the masking of the STARS alarms from Table 1 using MAASC's (Bolton, Biltekoff, et al., 2020) visualization. Color blocks show the sounding patterns of the two-tone pairs of the alarms. Unmasked tones are green and yellow. Masked tones are red. The masking tones (those that mask the red tone) are yellow. (a) The partial masking of the Conflict Alert alarm (at 70 dB). (b) The partial masking of the Mode C Intruder alarm (at 70 dB). (c) The partial masking of the Minimum Safe Altitude Waring alarm (at 70 dB). (d) The total masking of the Conflict Alert alarm (at 61 dB).

C Intruder, at 70 dB, the second event is masked by the combination of both events from the Conflict Alert and the Default alarms as well as the first event of the Minimum Safe Altitude Warning alarm (Fig. 1(b)). Also at the 70 dB level, the first event of the Minimum Safe Altitude Warning alarm is masked by the combination of the second event of the Conflict Alert, the first event of the Mode C Intruder, and both events of the Default Alarm (Fig. 1(c)). Finally, for the total masking analyses, MAASC identified total masking of the Conflict Alert alarm at 61 dB when the other three alarms remained at 70 dB (Fig. 1(d)).

Discussion

After analyzing the alarm configurations using Bolton et al.'s method (Bolton et al., 2018b, 2018a; Hasanain et al., 2017), the results show that masking is possible and a concern for STARS. Three out of the six alarms analyzed are susceptible to partial masking at our standardized 70 dB level and 1 alarm is susceptible to total masking when there is a drop of 9 dB or more.

There is no current consensus regarding the number of aural tones that can be distinguished between, regardless of masking. However, there is ultimately a maximum number of tones a given individual can distinguish between. It can then be assumed that masking decreases an individual's threshold. Given this, if masking cannot be avoided, the best practice would be to increase the minimum number of alarms necessary before masking has an effect on the distinguishability of the alarms. Therefore, the results for the STARS alarms, at 70 dB, are promising. Each situation where masking is possible requires a minimum of three other alarms to also be present in order for the masking to occur. The masking detected in these configurations is only partial, whereas the total masking of the Conflict Alert poses a larger concern.

The results presented here are encouraging for STARS (and far better than those observed for medical alarms (Bolton et al., 2022)). Only one example of total masking was identified and this required a relatively high number of alarms to manifest. Unfortunately, the fact that combinations of four of the alarms resulted in three being masked at some level poses an issue. If these alarms sound in unison or close proximity, it is likely that one or more could be missed. In the safety-critical environment of air traffic control, even one missed alarm could have potentially costly consequences and could be the difference in life or death.

It is important to note that, while variations in volume can have minor impact on the simultaneous masking effect (Bosi & Goldberg, 2003), air traffic controllers have the ability to manipulate the volume of the alarms. This does pose a concern, illustrated by the total masking of the Conflict Alert. A standardized volume set for STARS would help to further eliminate any situations in which masking could occur.

Future Research

The presented research isolates the alarms and models configurations of them. Ultimately, no other variables, such as ambient or transient noise, are accounted for. Future research should account for these in masking analyses.

Next, it is important to understand these alarms and air traffic control at a more specific level. Variables such as number of STARS terminals that a controller is exposed to in a given environment. Accounting for these, would make our analyses more complete. It is possible that temporal spacing of the alarms is sufficient to eliminate total masking altogether (as was shown for medical alarms (Bolton, Zheng, et al., 2020). However, the frequency of unnecessary or redundant alarms remains high (Newman & Allendoerfer, 2000).

Finally, there may exist combination of frequencies and spacing that would completely eliminate masking. For air traffic control alarms and specifically STARS, such a combination could be identified and implemented to provide controllers with the most clear and discriminable alarms.

Ultimately, the full extent of this problem is unknown. Any effort to expand the understanding of an air traffic controller's interaction with the alarm system will aid in interpreting these results and work to make air traffic control more effective. The ability of MAASC to be implemented and used in an effective and timely manner suggests that is practical to apply the application to further research on this topic and in the field at large.

References

Bolton, M. L., Biltekoff, E., Boyd, A., Darget, T., & Edworthy, J. (2020). Medical alarm audibility system checker (MAASC): A computational tool for checking medical alarm configurations for simultaneous masking. In *Proceedings of the international symposium on human factors and ergonomics in health care* (Vol. 9, pp. 302–303).

Bolton, M. L., Edworthy, J., & Boyd, A. D. (2018a). A computationally efficient formal method for discovering simultaneous masking in medical alarms. *Applied Accoustics*, *141*, 403–415.

Bolton, M. L., Edworthy, J., & Boyd, A. D. (2018b). A formal analysis of masking between reserved alarm sounds of the IEC 60601-1-8 international medical alarm standard. In *Proceedings of the human factors and ergonomics society annual meeting* (pp. 523–527). Los Angeles.

Bolton, M. L., Edworthy, J. R., & Boyd, A. D. (2022). Masking between reserved alarm sounds of the iec 60601-1-8 international medical alarm standard: A systematic, formal analysis. *Human factors*, *64*(5), 835–851.

Bolton, M. L., Hasanain, B., Boyd, A. D., & Edworthy, J. (2016). Using model checking to detect masking in IEC 60601-1-8-compliant alarm configurations. In *Proceedings of the human factors and ergonomics society annual meeting* (pp. 636–640). Los Angeles.

Bolton, M. L., Zheng, X., Li, M., Edworthy, J., & Boyd, A. D. (2020). An experimental validation of masking in IEC 60601-1-8:2006-compliant alarm sounds. *Human Factors*, *62*(6), 954–972.

Bosi, M., & Goldberg, R. E. (2003). *Introduction to digital audio coding and standards*. New York: Springer.

Clarke, E. M., Grumberg, O., & Peled, D. A. (1999). *Model checking*. Cambridge: MIT Press.

Edworthy, J., & Hellier, E. (2005). Fewer but better auditory alarms will improve patient safety. *Quality and Safety in Health Care, 14*(3), 212–215.

Edworthy, J., & Hellier, E. (2006). Alarms and human behaviour: Implications for medical alarms. *British Journal of Anaesthesia, 97*(1), 12–17.

Edworthy, J., McNeer, R. R., Bennett, C. L., Dudaryk, R., McDougall, S. J. P., Schlesinger, J. J., ... Osborn, D. (2018). Getting better hospital alarm sounds into a global standard. *Ergonomics in Design, 26*(4), 4–13.

Edworthy, J., & Meredith, C. S. (1994). Cognitive psychology and the design of alarm sounds. *Medical Engineering & Physics, 16*(6), 445–449.

Fastl, H., & Zwicker, E. (2006). *Psychoacoustics: Facts and models* (Vol. 22). Springer.

Hasanain, B., Boyd, A., & Bolton, M. (2016). Using model checking to detect simultaneous masking in medical alarms. *IEEE Transactions on Human-Machine Systems, 46*(2), 174–185.

Hasanain, B., Boyd, A., & Bolton, M. L. (2014). An approach to model checking the perceptual interactions of medical alarms. In *Proceedings of the 2014 international annual meeting of the human factors and ergonomics society* (pp. 822–826). Santa Monica: HFES.

Hasanain, B., Boyd, A. D., Edworthy, J., & Bolton, M. L. (2017). A formal approach to discovering simultaneous additive masking between auditory medical alarms. *Applied Ergonomics, 58*, 500–514.

Konkani, A., Oakley, B., & Bauld, T. J. (2012). Reducing hospital noise: A review of medical device alarm management. *Biomedical Instrumentation & Technology, 46*(6), 478–487.

Lutfi, R. A. (1983). Additivity of simultaneous masking. *The Journal of the Acoustical Society of America, 73*(1), 262–267.

Meredith, C., & Edworthy, J. (1995). Are there too many alarms in the intensive care unit? An overview of the problems. *Journal of Advanced Nursing, 21*(1), 15–20.

Momtahan, K., Hetu, R., & Tansley, B. (1993). Audibility and identification of auditory alarms in the operating room and intensive care unit. *Ergonomics, 36*(10), 1159–1176.

Newman, R. A., & Allendoerfer, K. (2000). *Assessment of current and proposed audio alarms in terminal air traffic control* (Tech. Rep. No. DOT/FAA/CT-TN00/21). Springfield: William J. Hughes Technical Center (US).

Patterson, R. D. (1982). *Guidelines for auditory warning systems on civil aircraft*. Civil Aviation Authority.

Patterson, R. D., & Mayfield, T. F. (1990). Auditory warning sounds in the work environment. *Philosophical Transactions of the Royal Society of London. B, Biological Sciences, 327*(1241), 485–492.

Toor, O., Ryan, T., & Richard, M. (2008). Auditory masking potential of common operating room sounds: A psychoacoustic analysis. In *Anesthesiology* (Vol. 109, p. A1207). Park Ridge: American Society of Anesthesiologists.

DEFINING NEEDS FOR ENHANCED WEATHER PRODUCTS FOR UAS AND GA STAKEEHOLDERS: A QUALITATIVE STUDY

Scout Hernandez
Oklahoma State University
Stillwater, Oklahoma

Nicoletta Fala
Oklahoma State University
Stillwater, Oklahoma

Weather is a major influencing factor in determining if a pilot can safely fly on any given day. Unfavorable weather conditions can cause accidents and lead to potential injuries or death. In this paper, we use focus groups to gather perspectives of General Aviation (GA) pilots and Uncrewed Aerial Systems (UAS) operators on weather communication products and their influence on weather-related decision-making. GA pilots have used products to make informed decisions for a while, but UAS operators are relatively new to their adoption and may have different methods of usage. Understanding how both groups perceive weather communication and prediction will help us improve future weather products. Novel weather communication products with enhanced features may increase the comfort and confidence levels of all airspace stakeholders by helping them make more informed decisions. We use a qualitative approach to solicit specific needs and provide potential improvements to weather products.

Accurate weather prediction has always been crucial to the success of a flight since the early 1900s (Caldwell, 2017), with technological advancements to accommodate pilot and aircraft needs and comforts (Casner, 2012) improving the accuracy of information (Benjamin, 2010). The evolution from anemometers to current weather products used by modern pilots shows the progressive technological advances. Weather Intelligent Navigation, Data, and Models for Aviation Planning (WINDMAP) is a NASA University Leadership Initiative (ULI) aiming to explore and address weather information demands of crewed and uncrewed aircraft using remote observations and data-driven predictions to improve the safety of pilots and aircraft (Jacob, 2020). With technological progression, pilot needs have adapted as well. We must therefore enhance current weather observation, forecasting, communication methods and accommodate the needs of both General Aviation (GA) and Uncrewed Aerial Systems (UAS) pilots (Thornes, 2001). UAS pilots have weather needs (such as information on wind gusts and turbulence) that are not currently sufficiently addressed by weather products (Campbell, 2017). GA pilots are often limited by their knowledge in weather technology (Blickensderfer, 2015) and would therefore benefit from improvements in weather information availability and communication. Before we design systems, we must focus on understanding the needs of UAS and GA pilots. Their safety and confidence in flight missions are crucial to mission success. In this paper, we designed surveys and focus groups to analyze responses from UAS and GA pilots about weather products they use to guide preplanning and in-flight stages of flights.

In prior research, we conducted a web-based survey to quantitatively question GA pilots and UAS operators on the weather products they use and how the weather products affect their decision making (Fala & Wallace, 2021). Our prior research has concluded that GA pilots and UAS operators are overall satisfied with the weather products they use, but would benefit from modifications (Fala & Wallace, 2021). While surveys can quickly and easily gather information from a large pool of participants, the participants are limited in what they can articulate in their responses (Kamberelis, 2013). In the work presented here, we use focus groups to add context to the responses gathered from the quantitative study. By allowing deeper discussion and explanation from participants qualitative studies provide more thorough communication between the moderator and subjects (Gibbs, 1997; Scheuren, 2004). Leading a focus group is similar to guiding a conversation. It allows participants to express themselves in ways that can be analyzed qualitatively (Cyr, 2016).

This paper evaluates responses provided by UAS and GA pilots in focus groups to assist WINDMAP with accommodating pilots' weather communication needs. In the next section, we present our experiment design. We used a screening survey to recruit participants, collect demographic data, and facilitate logistics. We then designed and tested focus groups that asked participants about how they prepare for flights, products they use in flight preparation, and potential improvements they would prefer to suit their needs. The paper categorically analyzes the pilot responses, answering questions such as "How do you prepare for a flight considering weather?" and "Do you think it is important to modify the weather products you have mentioned?" In the results section, we provide some results collected from the focus groups after analyzing the data. Finally, we discuss conclusions, limitations of the project, and future work. This paper provides context to prior quantitative work, informs the challenges GA and UAS pilots face due to weather and weather-related decision making. This paper also identifies their needs and suggests improvements to the weather products used.

Experiment Design

In this research we conduct UAS and GA pilot focus groups to understand their perspectives on current weather communication products. We asked both groups of participants which products they use, what challenges they face using those products, and how we can improve their confidence in weather-related decision making. We used a two-phase approach to recruit participants and lead focus groups. The experiment was approved by the Oklahoma State University Institutional Review Board (IRB). The two-phase approach begins with a brief survey that identifies qualifying participants as well as information about their backgrounds. The survey allows us to differentiate the pilots through skill, experience, and location. The focus groups allow us to discuss the weather products used by the UAS and GA pilots and describe their experiences. Here, we discuss the screening survey, the design of focus groups, and our participant recruitment effort.

The screening survey was part of the recruitment email and form we sent to several universities to screen interested pilots. The survey is outlined in Table 1. The survey begins with an introduction and consent form which provides prospective participants with a summary of what they can expect from this study. The survey questions begin with the participants' qualifications: "I operate Unmanned Aircraft Systems/Drones" and/or "I am a pilot of a crewed vehicle." If the participant does not have experience with either of these two options, the survey is discontinued. Otherwise, the survey continues to the section about participant's background, where we collect information on location, age, and flight/aviation experience.

Table 1.

Questions posed to participants in the screening survey.

Category	Questions
Qualifications	What is your experience with aviation? {I operate UAS; I am a pilot of a crewed vehicle}
Name	Name
Contact Information	Email Address
Availability	Among the dates, which can you meet for the focus group?
Demographics	How do you identify?
	What is your age?
	What is your highest level of education?
Flight Experience	How many flight hours have you logged?
	On average how frequently do you fly?
	How long have you been a pilot?
Location	Where do you live?
	Based on where you live, are there any seasons you try to avoid flying in?

The screening survey serves as the introduction between our research team and the participants. It helped inform the participants of the experimental process and the privacy measures taken to protect their anonymity. Participation from GA pilots was wider than the UAS operators (Fala & Wallace, 2021): only 26% of the survey responses were from UAS operators. Next, we collected contact information for the focus group participants. The participants were given pseudonyms to maintain anonymity in discussions (Scheuren, 2004). We collected flight hours logged, frequency of flights, and years of experience to see potential effects of experience on the pilots' use of weather products.

The second phase of the experiment informs us of products GA and UAS use for weather prediction, the challenges the pilots face, and their suggestions to improve the products. Our focus group guiding script follows a funnel design format (Morgan, 2012). As seen in Table 2, the funnel design begins with introduction questions which allow the participants and the moderator to open dialog and sets the mood of the focus group for an informal discussion. Questions like "How long has everyone been flying?" and "What types of aircraft have you flown?" allow our participants to identify common interests and experiences. Next, the broad opening questions narrow the discussion topics. Questions in this section shift the participants' focus on our topic. In the broad opening questions section, we ask the participants how they prepare for a flight regarding weather. The questions in this section are purposely open-ended to encourage discussion. We had both a moderator and a notetaker to ensure the questions and answers were expressed and recorded clearly. The moderator aimed to have a thorough discussion with all participants while the notetaker tracked time and progress for the discussion.

Table 2.

Focus group questions outlined using funnel design.

Category	Questions
Introduction Questions	How long have you been flying?
	What type of aircraft have you flown?
Broad Opening Questions	How do you prepare for a flight regarding weather?
	What services do you use apps, websites, new channels?
	Do you use multiple products for weather prediction?
Main Body Questions	Describe the process which you use to prepare for a flight.
	Do you check the weather forecast a week before the flight? The night before the flight? Thirty minutes before flight?
	Do you modify your flight path depending on weather conditions or do you normally cancel your flight?
	Can you share a couple of flights when the weather stressed you during the flight?
	What would you define as "questionable weather"?
Closing Questions	Do you think it is important to modify the weather products you have mentioned?
	How would you prioritize the changes discussed?

In preparation for the focus groups, we conducted a practice focus group. The practice focus group took 39 minutes, as indicated in Table 3. and allowed us to improve our focus group process and structure. Our practice run consisted of two participants (a GA pilot and a GA pilot/UAS operator) and helped clarify the questions in preparation for the actual focus groups. We also used the practice focus group to estimate how long participants would need to answer all questions. The time required for a focus group depends on the number of participants. Therefore, we aimed for up to five participants per focus group, with each focus group scheduled to take up to 90 minutes.

The focus groups followed the funnel format mentioned earlier as closely as possible. We outlined the script to reduce deviation from the topic in Table 2. The moderator would pose the question to the participants as the starting point for the discussion. As the pilots and UAS operators were quite interested in the topic they rarely deviated from the subject.

Results

This study thematically analyzes each participant's response to answer the research question. A response matrix allows us to categorize the responses of the participants. The focus groups included eight pilots and three UAS operators, as summarized in Table 3.

Table 3.
Number of participants per focus group.

Focus Group	UAS	GA	Duration (Minutes)
Practice	1	2	39
1	2	3	70
2	1	2	42
3	0	3	41

Pilots and UAS operators repeatedly stated they rarely felt unsafe or misguided by information from weather products when preparing for a flight. The less experienced pilots tend to cancel flights if the weather products indicate what the participant considers hazardous weather conditions. The pilots did feel misguided if the weather ended up being fair after they had already cancelled a flight. Experienced pilots have a higher tolerance for what they consider hazardous weather. They tend to blame their own hazardous attitudes if they encounter stressful situations. Although pilots and UAS operators did not believe the weather products had led them astray, they did have suggestions for improvements to the products. Three main desires for improvement were gust speed prediction, denser geographical weather information, and cloud base height predictions.

UAS operators and pilots both expressed their frustration with the uncertainty of gusts and the need for improved warnings regarding gusts. UAS operators are cautious of the structural constraints of the drone when planning for flights. All UAS operators mentioned they usually fly in weather conditions within the manufacturer's recommendations. When using weather products, they noted that while the wind speeds predicted are within the manufacturer's limits, the gust speeds, which are not predicted, might be outside the manufacturer's recommendations, putting the aircraft at risk. The unexpected changes due to gusts have startled, and in some cases unnerved, the participants in previous flights. The lack of resources provided for visualizing or warning pilots and UAS operators of potential gusts have led to wariness and concern. Pilots suggested including color codes or markers for the gust speeds highlighting the location and speeds of the gusts.

Participants highlighted a necessity for more dense geographical weather information. The UAS operators expressed concern when relying on weather products away from an airport. When UAS operators conduct flights far away from airports, they are forced to rely on interpolations to predict the weather at their locations. Interpolations, unfortunately, can be inaccurate and starved of crucial details. The pilots also mentioned a need for reduced distance between weather observational sources like weather stations or airports. Estimation to predict weather can mislead the pilots, and in the case of inexperienced pilots prevent them from flying. Both pilots and operators were eager to suggest possibilities that allow for an increase in weather information either by adding more weather service stations in less populous areas or installing weather trackers on aircraft.

Finally, both pilots and operators indicated annoyance with the prediction of cloud heights. Operators depend on weather products predicting low level ceilings being accurate, as they are restricted from flying in clouds. When resources are limited, both operators and pilots described using the height of known mountains to determine the cloud bases. This method, however, is inaccurate, and the stakeholders would prefer more accurate cloud cover predictions. Additionally, many locations across the United States lack mountainous terrain or visual markers pilots can use for reference. The pilots indicated a desire for increased frequency in cloud information and suggested adding better detailed data regarding the beginning layers of the cloud levels.

Conclusions and Future Work

Though many participants claimed that inaccuracies provided by weather products rarely impede their safety, there are at least three improvements pilots and operators identified. Identifying wind gust velocities and locations, increasing the density of geographical weather prediction, and accurate cloud cover predictions are the improvements mentioned most frequently. In future work, we hope to guide designers toward addressing these challenges and adding improvements based on the participants' recommendations, followed by another series of focus groups to determine if enhanced weather products have provided the expected benefits.

Acknowledgements

This research is partially funded by the National Aeronautics and Space Administration (NASA) WINDMAP ULI under Award No. 80NSSC20M0162.

References

Benjamin, S. G., Jamison, B.D., Moninger, W.R., Sahm, S. R., Schwartz, B. E., & Schlatter, T.W. (2010). Relative Short-Range Forecast Impact from Aircraft, Profiler, Radiosonde, VAD, GPS-PW, METAR, and Mesonet Observations via the RUC Hourly Assimilation Cycle. *American Meteorological Society*, 1319 - 1343.

Blickensderfer, E. L., Lanicci, J. M., Vincent, M. J., Thomas, R. L., Smith, M., & Cruit, J. K. (2015). Training general aviation pilots for convective weather situations. *Aerospace medicine and human performance, 86(10)*, 881-888.

Caldwell, B. E., McCarron, E., & Jonas, S. (2017). An abridged history of federal involvement in space weather forecasting. *Space Weather 15.10*, 1222-1237.

Campbell, S. E., Clark, D. A., & Evans, J. E. (2017). Preliminary UAS weather research roadmap. *Lincoln Laboratory, Massachusetts Institute of Technology*, 244.

Casner, S. M., Murphy, M. P., Neville, E. C., & Neville, M. R. (2012). Pilots as weather briefers: The direct use of aviation weather products by general aviation pilots. *The International Journal of Aviation Psychology, 22(4)*, 367-381.

Cyr, J. (2016). The pitfalls and promise of focus groups as a data collection method. *Sociological methods & research*, 231-259.

Fala, N., & Wallace, J. W. (2021). Identification of Potential Gaps and Requirements in Weather Sources for General Aviation and UAS Operations. *American Institute of Aeronautics and Astronautics*.

Gibbs, A. (1997). Focus groups. *Social research update, 19(8)*, 1-8.

Jacob, J., Chilson, P. B., Houston, A. L., Pinto, J. O., & Smith, S. (2020). Real-time Weather Awareness for Enhanced Advanced Aerial Mobility Safety Assurance. *AGU Fall Meeting Abstracts*, A021-03.

Kamberelis, G., & Dimitriadis, G. (2013). *Focus groups.* London: Routledge.

Morgan, D. L. (2012). Focus groups and social interaction. In *The Sage handbook of interview research: The complexity of the craft, 2*, 161-176.

Scheuren, F. (2004). *What is a Survey.* Alexandria: American Statistical Association.

Thornes, J. E., & Stephenson, D. B. (2001). How to judge the quality and value of weather forecast products. *Meteorological Applications, 8(3)*, 307-314.

A NATURAL LANGUAGE PROCESSING MODEL FOR ANALYZING AVIATION SAFETY EVENT REPORTS: A SUBSET OF RESULTS

R. Jordan Hinson, Edward Bynum, Amelia Kinsella, Katherine Berry, Michael Sawyer
Fort Hill Group
Washington, DC, USA

Many civil aviation authorities, operators, and manufacturers utilize voluntary safety reporting programs (VSRPs) to understand risk within their operations. Insights from these first-hand accounts can lead to significant safety and efficiency improvements. Subject matter experts often read and analyze these reports by labeling factors of interest to derive safety insights. The resources required for this analysis can limit the insights an organization can obtain from their VSRP data. A novel machine learning model was developed and trained on over 50,000 rows of manually labeled aviation VSRP data. This model uses machine learning and natural language processing (NLP) to automate the task of labeling aviation safety reporting data and codifying report narratives according to a structured list of human factors topics. This paper presents a subset of interim model results and discusses the implications of using NLP to identify reports citing human factors topics from aviation VSRP data.

Commercial aviation has earned a reputation as a mode of transportation with exceptional safety. To achieve and maintain this high standard of safety the aviation industry has relied on continuously tracking incidents and assessing safety trends. Voluntary safety reporting programs (VSRPs) have been established and used to track and assess safety incidents that occur in the National Airspace System (NAS). One such program is the Aviation Safety Reporting System (ASRS) developed and maintained by the National Aeronautics and Space Administration (NASA; NASA Aviation Safety Reporting System, n.d.). This system encourages the aviation industry to report observed safety problems, discrepancies, or deficiencies. Thousands of reports are submitted, processed, and publicly released each year. For example, 6,428 ASRS records describing events that occurred in 2019 are currently for public download.

While many safety events are reported each year, the process of reading and analyzing reports in a meaningful way can be labor intensive. Drawing safety insights from the reports involves a complex process, and the full potential of these reports is difficult to realize for some organizations. One common analysis approach is utilizing subject matter experts (SMEs) to manually read each report and label all relevant factors using a taxonomy. Applying taxonomies to safety event reports is an effective way to identify trends and gain safety insights across numerous reports. However, this process can be time-consuming and requires SMEs who understand human factors, aviation systems, and nuanced industry jargon.

New machine learning techniques involving natural language processing (NLP) offer opportunities to assess and label factors of interest within safety reports in a more efficient and effective manner. The application of NLP in aviation, and specifically ASRS, has been explored by some researchers. Kierszbaum and Lapasset (2020) used NLP to extract the event date from the free text portion of ASRS reports with relative success. Further, researchers have highlighted the importance of using a pre-trained, aviation model when applying NLP to ASRS reports due to the unique language of aviation (Kierszbaum, Klein, & Lapasset, 2022). NLP has also been utilized in aviation safety reports to examine flight delays in ASRS (Miyanmoto, Bendarker, & Marvis, 2022) and probable cause in National Transportation Safety Board (NTSB) reports (Jonk, et al., 2023). This research, along with other NLP research, emphasizes the potential application of NLP in aviation safety event reporting.

Our team has developed the AVIation Analytic Neural network for Safety events (AVIAN-S) model by incorporating ML and NLP techniques to automate the identification and labeling of a specific human factors (HF) taxonomy within VSRP reports. This model was developed and trained by utilizing publicly available VSRP data that was manually labeled by SMEs. This project is an independent self-funded research effort. Views and results are those of Fort Hill Group and do not represent opinions or views of the Federal Aviation Administration or NASA.

AVIAN-S Model

Model Development and Training Dataset

The AVIAN-S model was developed over a series of iterations. First, a training dataset was established using publicly available ASRS reports. These reports were SME coded utilizing the AirTracs human factors taxonomy (Berry, et al., 2015). AirTracs is a tiered human factors taxonomy that includes specific factors designed to provide insight into human performance. When applying the AirTracs taxonomy to ASRS reports, SMEs identified report narratives, factors, and rationales. These three inputs were utilized to train the current version of the AVIAN-S model.

After establishing the training dataset, model development began. First, full narratives were utilized as model input. However, these narratives were difficult for the model to process as they were long and often contained multiple different factors. It was decided to use SME-identified rationales as the model input. The current version of the model utilizes rationales to identify factors. The output of the model gives AirTracs factors with an associated predicted probability for the factor. These results are compared to the SME-coded factors from the training dataset to determine the accuracy of the model.

Subset of Preliminary Results

Ten percent of the overall SME-analyzed data was randomly retained to validate model accuracy. Accuracy was measured using a metric called "top K score." The model assigns a predicted value for every possible factor to be applied to a report. The top K score takes the top "K" highest probabilities and compares them to the SME-coded factors. In this case, K = 9. If the model correctly predicts a factor within the top 9 values, then it is considered a success. In the current 10% sample, the accuracy measured using the Average Top K Score was 80.9%, where K<=9. This indicates that the SME manually assigned factor was included in the top 9 model predicted factors 80.9% of the time. A subset of the results will be further discussed in this section to highlight model accuracy insights.

Overall Accuracy

Figure 1 shows a scatter plot of the average Top K scores per factor by the number of ASRS records, with a few example factors highlighted. A significant positive correlation was observed, $r(137) = .32, p < .001$, indicating higher factor occurrence is associated with a higher top-K score (model accuracy). This makes sense, as the amount of training data made available to the model increases, the more learning is afforded to the model. However, it is important to note that there are many instances in which the model performed well when there were not many factor occurrences. Conversely, there are a few instances where the model accuracy was poor to moderate even though there were many factor occurrences. This is potentially due to the nature of the safety reports and specific taxonomy factors.

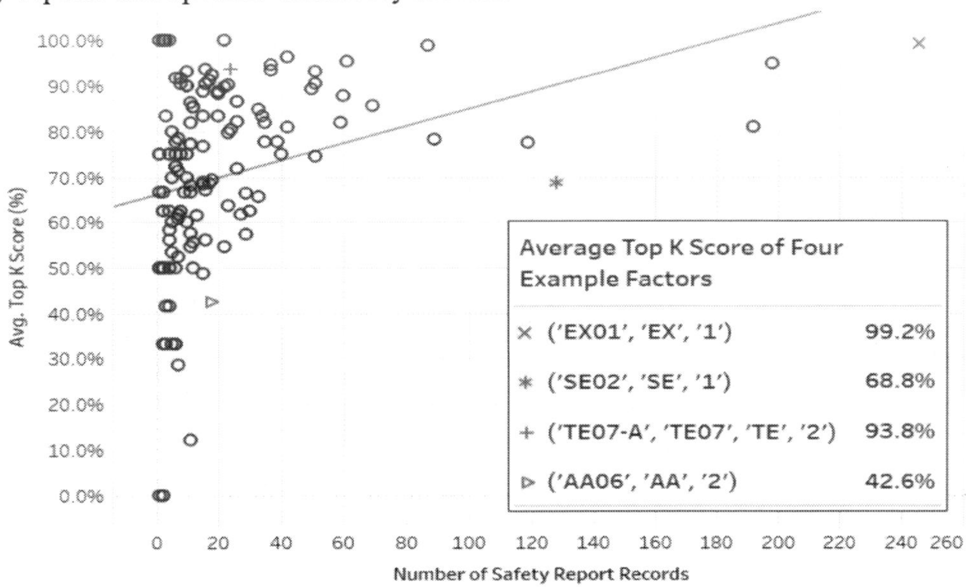

Figure 1. Average Factor Accuracy (Top K Score) by Factor with Examples.

Four Examples

Factor characteristics could be the source of variance in the accuracy of the model. The AirTracs Human Factors Taxonomy has 155 possible factors. As such, there is a range in specificity of factor definitions, as well as occurrences of factors in real-world safety reports. For example, the EX01 Technique Factor can be broadly applied to many different events, while the PE03 Noise Interference Factor is only appropriate when the effect of noise is mentioned in the safety report. Additionally, the DE02 Knowledge/Planning Factor is applied to a broad set of vocabulary describing circumstances involving various types of decision making and planning and is identified frequently in reports. The SP02 Staffing Factor on the other hand is less commonly identified and applies to a more limited scope of circumstances. Therefore, the training dataset has an unproportionate and variable amount of data per factor. This may have an impact on model performance. To better highlight these differences, four example factors were pulled and reviewed. The four factors include an example of 1) a high occurrence, high accuracy factor, 2) a high occurrence, low accuracy factor, 3) a low occurrence, high accuracy factor, and 4) a low occurrence low accuracy factor.

Figure 2 shows a density graph of the top K scores for four example factors. The top portion of Figure 2 shows the data for the most granular level of the taxonomy that was coded.

The bottom portion shows the same density graph across four levels of granularity. Here the most granular level is called “F1” and the least granular level “F4.” The grey-shaded portions of the graph represent the K value of 9 for this analysis. Any points that fall within the gray shaded areas are considered successes. In other words, when compared to the SME-labeled factors, the model correctly predicted these factors in the top 9 ranked factors. In general, the average top K scores increase as the factor level becomes less granular. It makes sense that the model does a better job at predicting factors in the broader sense compared to the more granular levels.

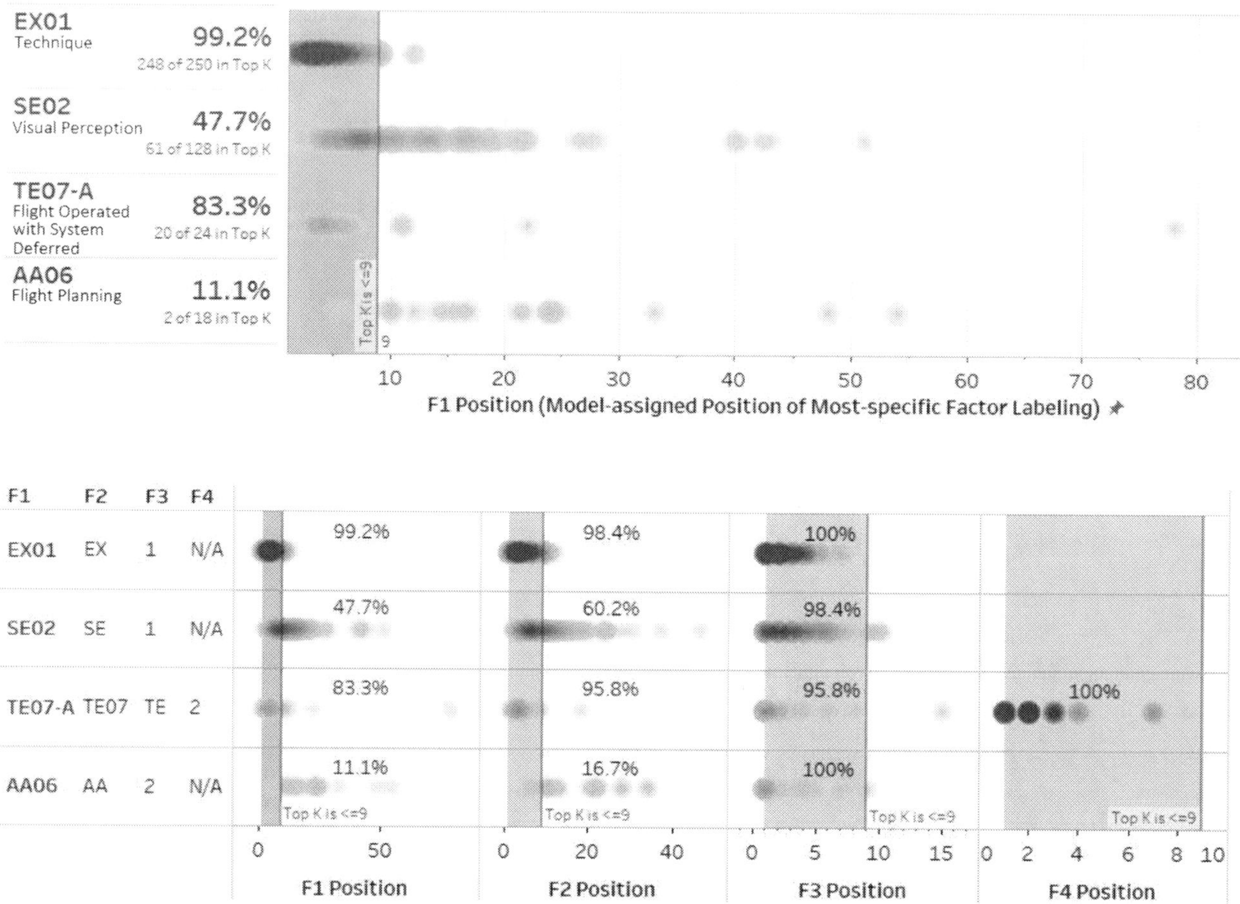

Figure 2. Top K Accuracy Score for four example factors at each level of specificity, F1 position individually (top) and positions of F1, F2, F3, and F4 combined (bottom). Note: The two density charts show stacked data points. Darker blue shaded areas contain more data points than lighter blue areas. Blue marks inside of the gray bands indicate accurate model assignments.

High occurrence, high accuracy factor example. EX01, or “Technique”, occurred 250 times in the retained 10% validation data set. This factor is nested under the “Operator Acts” tier of the taxonomy and can be applied when “a controller performs a task or job with an inadequate technique or uses an inadequate sequence” (Berry, et al., 2015). EX01 had an average top K score of 0.992 indicating the model correctly predicted this factor 99.2% of the time when compared to SME-coded reports.

This was one of the highest occurring factors with one of the highest success rates. Based on the correlation between factor occurrence and accuracy, it makes sense that the model performed well with this factor.

High occurrence, low accuracy factor example. SE02, or "Visual Perception," occurred 128 times in the 10% validation data set. This factor is nested under the "Operator Acts" tier of the taxonomy and is appropriately applied when "a controller's perception of visual information differs from the actual visual information" (Berry, et al., 2015). In the current 10% data pull, SE02 has an average top K score of 0.688 indicating the model correctly predicted this factor 68.8% of the time when compared to SME-coded safety reports.

While this factor occurred frequently in the current validation data set, the model accuracy was relatively low. Despite the frequency of this factor, the model still struggled with how to correctly apply it to safety reports. One reason for this may be the broad nature of "visual perception." To correctly apply this factor, the rationale needs to include a perception discrepancy between what the operator perceives and what is occurring in their environment. This discrepancy may be difficult for the model because of its nuanced application. It is possible there is an especially diverse range of explanations reporters could use to describe an event with this factor.

Low occurrence, high accuracy factor example. TE07-A, or "Flight Operated with System Deferred," occurred 24 times in the current 10% validation data set. This factor is nested under the "Operator Context" tier of the taxonomy and can be applied to safety reports when "a flight is operated with inoperative equipment legally in accordance with an approved Minimum Equipment List (MEL) or under the provisions of FAR 91.213 which allows for the equipment to be non-functional if it is not required by the aircraft equipment list, type of operation requirements (FAR 91.205), by airworthiness directive, or by other governing regulations as specified by the reporter" (Berry, et al., 2015). TE07-A has an average top K score of 0.938 indicating that the model correctly predicted this factor 93.8% of the time when compared to SME-coded safety reports.

While only 24 instances of this factor were observed in the current validation data set, the model predicted this factor successfully. This is likely due to the specificity of the definition and the common use of keywords associated with the factor. For example, words like "MEL" and "deferred" are often associated with this factor and not with other factors, so the model knows to assign TE07-A when these words appear in the rationale. For future model iterations, or when applying this model to other domains, incorporating specific taxonomy definitions similar to TE07-A may be beneficial in the model's predictive success.

Low occurrence, low accuracy factor example. AA06, or "Flight Planning" occurred 18 times in the current 10% validation data set. This factor is nested under the "Operator Context" tier of the taxonomy and can be appropriately applied to a report when "a pilot's preparations or planning for a flight impacts operations" (Berry, et al., 2015). AA06 has an average top K score of 0.426 indicating the model correctly predicted this factor 42.6% of the time when compared to SME-coded safety reports.

AA06 only occurred 18 times in the current validation data set and had a less than 50% success rate. While it is impossible to determine the exact causal reason for the low success rate, it is likely a combination of low factor occurrences and the taxonomy factor definition. To correctly apply this factor, the pilot must relay their flight planning process. This sentiment may not seem important to the reporter of the safety report when in conjunction with a specific incident that is being reported. Additionally, this sentiment may take many forms and is not always associated with specific keywords. It is possible the low success rate may be mitigated by incorporating more factor instances into the training data. This would give the model more information from which to learn how to apply this factor.

Conclusions and Next Steps

Overall, the AVIAN-S model does a good job of predicting AirTracs factors to be applied to aviation safety event reports. However, there is considerable variance in performance between some factors. A positive correlation was found between model performance and factor occurrence in the training dataset. This indicates the model tends to perform better when the training data consists of more factor instances. To improve future model performance, researchers may need to increase the training dataset, specifically in factor areas that were not as prominent in the current training dataset.

Another key finding is the model is generally better at predicting factors in the higher tiers of the taxonomy – meaning the less-specific levels. While certain factors perform well at the most granular level, some factors were more difficult for the model to correctly predict at that level of specificity. Future analyses can be conducted to better identify factors that do not perform as well at the granular level. Once identified, this knowledge will be helpful when utilizing the model in real-world analyses of safety event reports. For example, the model could be used to reliably identify one or two levels of the taxonomy for those identified factors, and SMEs could then finish the task by manually labeling the lower levels of granularity. While this might still be time consuming for the SMEs, having the model identify the appropriate top tiers of the taxonomy will markedly decrease the time spent on each report.

References

Berry, K., Sawyer, M., Hinson, J., & Rohde, R. (2015). Understanding Human Performance: The Air Traffic Analysis and Classification System.

Jonk, P., de Vries, V., Wever, R., Sidiropoulos, G., & Kanoulas, E. (2023). Natural Language Processing of Aviation Occurrence Reports for Safety Management. Proceedings of the 32nd European Safety and Reliability Conference.

Kierszbaum, S., & Lapasset, L. (2020, November). Applying distilled BERT for question answering on ASRS reports. In 2020 New Trends in Civil Aviation (NTCA) (pp. 33-38). IEEE

Kierszbaum, S., Klein, T., & Lapasset, L. (2022). ASRS-CMFS vs. RoBERTa: Comparing Two Pre-Trained Language Models to Predict Anomalies in Aviation Occurrence Reports with a Low Volume of In-Domain Data Available. Aerospace, 9(10), 591.

Miyamoto, A., Bendarkar, M. V., & Mavris, D. N. (2022). Natural Language Processing of Aviation Safety Reports to Identify Inefficient Operational Patterns. Aerospace, 9(8), 450.

NASA Aviation Safety Reporting System (n.d.). *Aviation Safety Reporting System*. Aviation Safety Reporting System. Retrieved April 28, 2023, from https://asrs.arc.nasa.gov/

LEARNING ABOUT ROUTINE SUCCESSFUL PILOT TECHNIQUES USING A CUED RETROSPECTIVE THINK-ALOUD TASK

Jon Holbrook, Chad Stephens, Lawrence Prinzel III, Sepehr Bastami
National Aeronautics and Space Administration
Hampton, VA

Daniel Kiggins
San Jose State University Research Foundation
Moffett Field, CA

Self-report can be a valuable method for collecting data about people's goals and perceived motivations – data about aspects of crew thinking that are not otherwise readily observable. One of the challenges associated with collecting self-report data on routine successful performance, however, is that details may go unreported, be deemed unimportant, or may not be recalled. We report a study in which commercial airline flight crews participated in a video-cued retrospective think aloud after flying a high-fidelity simulated arrival into Charlotte airport. One day after flying the simulated arrival, crews were shown a video recording of their flight. The video was paused after each minute, and crew members were each asked to describe what they were doing and thinking during that interval. Reported data analysis focused on aspects of performance that are often ambiguously described as "pilot technique" or "airmanship," in an attempt to provide more detail around these types of behaviors.

Today's commercial air transport industry collects aviation safety data through many mechanisms, including system-generated data (e.g., Flight Operational Quality Assurance [FOQA]), observer-generated data (e.g., Line Operations Safety Audits [LOSA]), and self-reported data (e.g., Aviation Safety Action Program [ASAP] reports). Analysis and reporting of collected data are frequently triggered by undesired events, such as operational exceedances, failures, and errors. While understanding and mitigating undesired outcomes is an important part of aviation safety, analysis of flight crew operational performance suggests that pilots intervene to keep flights safe over 157,000 times for every time that pilot error contributes to an accident (Holbrook, 2021). Understanding what pilots routinely do to *produce* safety, not just what they rarely do to *reduce* safety, should represent a significant additional source of aviation safety data.

A critical challenge to learning from pilots' contributions to safety is understanding how to measure them. This challenge is the subject of recent efforts by NASA, which has created a data testbed to enable exploration of methods and metrics for pilot contributions to safety and mission success. Details of the full data collection plan and flight simulation scenarios are described in Stephens et al. (2021). Data from 24 pilots (12 Captains [CA] and 12 First Officers [FO]) from a major US air carrier are included in the testbed. Six scenarios were designed to include a range of challenging but manageable disturbances, inspired by previously conducted pilot interviews (Holbrook et al., 2019) and event reports submitted to NASA's Aviation Safety Reporting System (Billings et al., 1976). All scenarios involved Area Navigation (RNAV) arrivals into Charlotte Douglas International Airport (CLT) and were flown in NASA Langley's Integrated Flight Deck motion-base Boeing 737-NG simulator. Participating pilots were all Boeing 737 type-rated. Air traffic control (ATC) was provided in real time by a recently-retired CLT Terminal Radar Approach Control (TRACON) controller confederate who also participated in the design of the scenarios. The intent was to create an environment realistic enough to leverage the expertise of the participating pilots and flexible enough to enable a range of possible behaviors in response to the pressures encountered across the scenarios. The data collected as part of the testbed were intended to capture the processes by which flight crews managed those pressures in addition to performance outcomes. The testbed includes (simulated) flight data; multiple psychophysiological measures; "over the shoulder" video recordings; pilot-generated event narratives, similar to ASRS reports; verbal account of "what they were thinking," provided during scenario video playback; and results from a range of surveys and subjective questionnaires intended to capture information about pilots' workload, situation awareness, resilient performance behaviors, and organizational support for resilient performance.

Self-report can be a valuable method for collecting data about people's goals and perceived motivations – data about aspects of crew thinking that are not otherwise readily observable. One of the challenges associated with collecting self-report data on routine successful performance, however, is that details may go unreported, be deemed unimportant, or may not be recalled at the time of reporting. The current study is specifically focused on pilots' verbal accounts of what they were thinking during one of the testbed scenarios. Think-aloud protocols have been used extensively in fields from cognitive psychology to usability testing to address a range of questions, resulting in many different specific methodologies (see Boren & Ramey, 2000, for a review). To ensure that the task of thinking aloud did not impact pilots' performance or data collected during scenarios, a retrospective think-aloud method was used (Kuusela & Paul, 2000). Details of the method employed are described here, along with illustrations of the types of insights into crew performance that this method can provide.

Method

The day after flying the simulated scenarios described above, each of 12 flight crews, consisting of one CA and one FO, were told that they would be shown a video of their performance for one of the arrivals they flew the previous day. Participants were told that the video would be paused after each minute, and they were asked to verbally describe what they were doing and, more importantly, thinking, during that one-minute interval. Participants were informed that they would both have an opportunity to speak, with the FO speaking first in each case. After both crew members described what they were thinking, they were asked to write down a personal workload rating, on a scale from 1 (very low) to 7 (very high) for that one-minute interval.

The scenario used for the retrospective think-aloud involved flying the FILPZ3 RNAV standard terminal arrival (STAR). For this scenario, the FO was the pilot flying, and the CA served as the pilot monitoring. The challenges present in this scenario included the following (see Figure 1A):

- Dynamically evolving convective weather and associated turbulence near and along the route of flight, including strong weather cells over the final approach fix and just off the departure end of the planned landing runway
- Hearing other aircraft deviating left and right of course on the party-line ATC frequency
- ATC reporting a microburst alert at the airfield
- Receiving a call from the cabin that a passenger has barricaded themselves in the aft lavatory

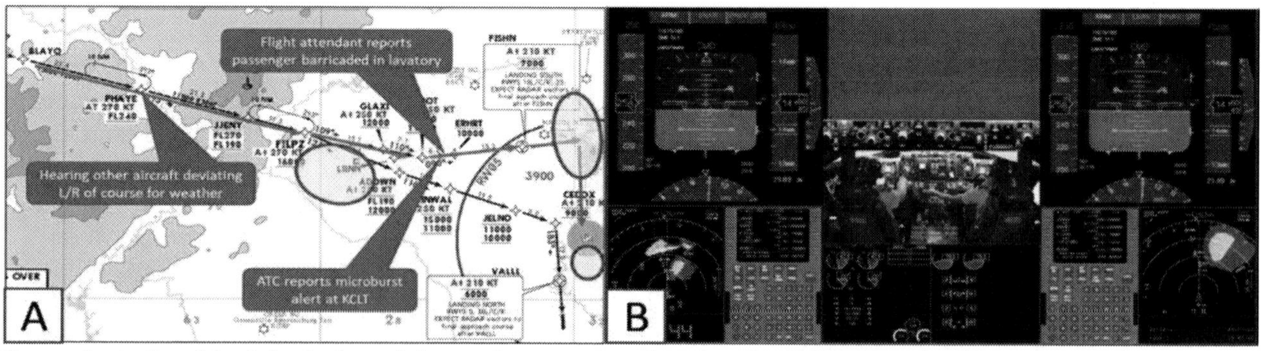

Figure 1. A. Graphical depiction of the study scenario, showing the FILPZ3 arrival into CLT. The planned arrival route is shown in green, and the approximate location of convective weather cells are shown as red-bordered yellow ovals. *B.* Screen capture of a representative study video, showing over-the-shoulder camera view (center), CA's instruments (left), FO's instruments (right), engine indicating and crew alerting display (EICAS, bottom center), and out-the-window forward view (top center).

The approximately 20-minute video (see Figure 1B) was played on a 55-inch high-definition television, and participants were seated between 6-8 ft from the display. The experimenter present in the room was responsible for pausing the recording after each minute. If a minute ended in the middle of a statement by the crew or ATC, the video was not paused until that sentence concluded, therefore, the boundaries of each "minute" were approximate. A separate audio recording captured pilots' think-aloud responses. Audio recordings averaged 68 minutes in length, with a range of 48 to 100 minutes. That time included the playback of the video, during which crews were silently watching, as well as the video pauses, during which crews described what they were thinking during the previously played minute and recorded their written workload ratings.

Results

Participants' think-aloud responses were manually transcribed by one of the authors from the audio recordings using an "intelligent verbatim transcription" technique. In this approach, vocal disfluencies (e.g., "um," "ah," "like"), laughter, pauses, etc., were omitted, and light editing to correct grammar and eliminate irrelevant statements was performed. This transcription approach preserves and focuses on the meaning of what was said, but does not capture behaviors or reactions of the participants outside of their spoken words. The transcription was parsed into distinct statements based on the topic discussed, and each statement was labeled by speaker (FO or CA) and the minute (numbered sequentially) during which the statement was made. Statements were each independently coded by two authors and discussed to consensus when differences in individual coding occurred. Coding comprised the following:

- Most-applicable American Airlines Learning and Improvement Team (LIT) code (American Airlines, 2021, Appendix A). LIT describes a process for capturing resilient behavior data by trained flightdeck observers. Twenty-seven codes are used to describe observable flightdeck behaviors that reflect crews' capability to plan, adapt, coordinate, and learn.
- Most-applicable macrocognitive function and process, as described by Klein et al., (2003). Macrocognition is a term used to describe the mental activities to successfully perform a task or achieve a goal in naturalistic or real-world settings. Macrocognitive functions (decision making, sensemaking/situation assessment, planning, adaptation, problem detection, coordination) describe the goal the person is trying to achieve, and macrocognitive processes (managing attention, identifying opportunities, managing uncertainty/risk, mental simulation, developing mental models, maintaining common ground) describe means for achieving those functions.
- Whether the described thinking was something that could be reasonably identified by a trained and attentive observer. For example, in some cases, pilots verbalized what they were thinking to the other pilot during the scenario, or took an action that clearly revealed what they were thinking. Instances such as these would be coded as "readily observable".
- Whether the described thinking captures something that is specifically covered in formal training or is something that reflects informal knowledge (e.g., picked up during operations, developed through personal experience, etc.). This determination was confirmed through discussion with a current pilot from the same company as the participant pilots.
- Whether a statement made by the CA was semantically paired (i.e., related to the same specific event) with a statement by the FO.

The focus of this paper is on insights that can be gained using this approach – in particular, insights into behaviors that might otherwise go unreported or unobserved. As an illustration, data from analysis of one representative crew are discussed. Using the procedure described above, 100 statements were coded for the crew in question. Of those statements, 55% were made by the FO, and 45% were made by the CA. In total, 33% were coded as "not readily observable," and 20% of statements described positive actions not directly covered in formal training. Some insights derived from the analysis are provided below, along with supporting statements.

Insight 1. Pilots used informal body language cues and gestures to communicate.

Pilots used body language cues to quickly convey their thoughts at times when verbal communication was not a good or convenient option. Gestures were used to convey simple messages, such as "I trust you to handle this" or to quickly establish real-time agreement with a verbal back-and-forth.

> CA: *Because I was on ATIS [Automated Terminal Information Service], I gestured to let the FO know that I was aware and trusting him to respond to ATC clearances. Because I was distracted by the ATIS, you can't be certain you're getting the full information from ATC, so I was letting the FO know that I was trusting him on that.*

> FO: *During ATC's call providing an update on weather conditions at the airport, CA gives a thumbs-down, making it clear we're not going to mess with that, and it's time to start changing gears.*

Insight 2. Pilots generalized application of formally learned techniques to other situations.

Independent verification by each crew member is part of formal training to ensure arrival procedures are loaded correctly in the flight management system (FMS) and that charts match the aircraft database. This training occurs in the simulator and during initial operating experience (IOE), but can be generalized to other tasks that are potentially vulnerable to single-fault failures.

> CA: *One thing I'm always concerned about when briefing the approach. I never want to set his mins [minimums]. I'll set other things – course and frequency – but I want to make sure 2 pilots are looking at the approach plate. I'll say "244, you got it?" and he'll say "yep, those are the mins I've got". Versus possibly making a mistake without a secondary chance to correct if I'm wrong.*

Strategic offloading of tasks from the other pilot to enable focus on a priority task is formalized during simulator training for emergency and abnormal conditions requiring complex error-intolerant procedures, but it can be generalized to other situations with potential distractions.

> CA: *This is a technique I use a lot. When we're doing something different or things are kind of off, I want the guy flying the airplane to focus on flying the airplane. By taking the briefing, it allows him the extra space to focus on where the airplane's going.*

Order of operations are formalized for emergencies, but the principles used to guide prioritization can be generalized to support other tasks that also require prioritization.

> CA: *I decided I would get back to the cabin [flight attendant] later. Right now, we needed to fly the airplane and decide what we're going to do initially to get away from this weather. That's all I wanted to focus on – just fly the airplane and get away from the weather.*

Insight 3. Pilots described instances in which they were trying to ascertain the "right time" or a "good time" to perform a task.

> CA: *I was kind of letting the controller do with us what he wanted, but there is a point during the approach where we have to be proactive and say what we need to do. I wanted to make sure we had the basics out of the way, to minimize distractions down the line as we get closer to the ground. Decision-making strategy about "what's important when?"*

> FO: *Now that we had briefed the approach, it was a good time to circle back and say "this is what I'm thinking". Maybe a good time to ask if we can go direct to one of these points.*

Insight 4. Pilots used tactile cues to maintain situational awareness.

FO: *I rest my arm on the thrust levers, so I can feel if they change. Are they pushing up? Is that something I need to happen? Oh, I probably don't need the spreedbrakes any more if the thrust levers are coming back up.*

Insight 5. Pilots adapted how they communicated to support shared situational awareness.

FO: *I decided to really verbalize what I was thinking because I knew the CA was just coming back from talking to the cabin, so wanted to make sure he knew where we were at, and that I wasn't going to fly through the weather.*

Insight 6. Pilots gauged the competency of their copilots, and this determination impacted their decision making.

CA: *I thought it was great that the FO let me know what he was thinking, because there are FOs that would just go "well, that's our clearance, so I'm just going to drive ahead". It really showed his competency. Because he showed he was highly competent, I was like "oh good, now I can go deal with this cabin issue".*

Insight 7. CAs thought about their role as "mentor" to the FO.

CA: *As a CA, you're always trying to adjust your leadership style and intervention strategy – what you're doing either for or with the FO to make it work.*

CA: *I like it when the FO asks me "what do you think about doing this?". If I like it, I'll go ahead and request it. If I don't, I'll say "I don't know about that" and we'll talk about it. In this case, I totally agreed and made the call to ATC.*

CA: *I commented "we've got a tailwind" to prompt him [the FO] to be a little more aggressive on the speed control, which he did by using more speedbrakes.*

Insight 8. Paired statements by the FO and the CA do not always reflect paired thinking.

In some instances, although the FO and CA were clearly talking about the same event or decision, they sometimes approached that decision differently. In the example below, the crew is briefing the landing. The CA brought up that it would likely be wet on the runway, and suggested a different autobrake setting, to which the FO concurred. The following statements pick up immediately afterward, and highlight a difference between the FO's rule-based thought process and the CA's context-based thinking.

FO: *I asked the CA about the wind, because typically when we talk about brake setting we are also talking about flap setting. Depending on wind speed and direction, I was considering between flaps 30 and flaps 40.*

CA: *I wasn't as worried about the winds, but more concerned about dynamic weather – if everything is going to change. That's what was going through my brain.*

Insight 9. Automating a procedure does not necessarily reduce crew workload.

For some tasks that have been automated, crewmembers still mentally perform the tasks themselves as an independent verification of the automation and to support building or maintaining their own mental model of the situation. Backing up the automation requires many if not all of the same mental resources used to perform the task without automation, and given that analysis of LOSA data has indicated that pilots must intervene to manage aircraft malfunctions on 20% of normal flights (PARC/CAST, 2013), this represents an important crew responsibility.

FO: *I performed my own personal check of the upcoming points on the arrival while checking the chart and what we had in the FMS to make sure those point were going to be met.*

Discussion

The analysis reported here just begins to scratch the surface of what could be learned from these retrospective think aloud data. For example, planned future analyses will compare the results of structured video observations, using approaches based on LOSA and LIT, against pilots' introspective accounts from the think-aloud task. Furthermore, patterns across macrocognitive functions and processes will be explored within the context of other data from the testbed. Understanding the routine resilient performance of flight crews has the potential to massively expand the pool of safety-relevant data, but depends upon development and application of techniques to collect and analyze those data. NASA has created a data testbed to enable exploration of these techniques. Cued retrospective think aloud protocols represent a technique widely used in fields from cognitive psychology to usability testing, and their application here shows promise for revealing insights into how pilots contribute to safety that are otherwise difficult or impossible to obtain. A more complete understanding of what pilots do and think about could inform the design of automated tools intended to support or supplement pilot performance.

Acknowledgements

This work was funded by NASA's System-Wide Safety Project, part of the Aeronautics Research Mission Directorate's Aviation Operations and Safety Program.

References

American Airlines Department of Flight Safety (2021). Charting a New Approach: What Goes Well and Why at American Airlines, A white paper outlining the second phase of AA's Learning and Improvement Team (LIT). Retrieved from https://www.skybrary.aero/articles/trailblazers-safety-ii-american-airlines-learning-and-improvement-team

Billings, C. E., Lauber, J. K., Funkhouser, H., Lyman, E. G., & Huff, E. M. (1976). NASA Aviation Safety Reporting System quarterly report number 76-1. NASA Technical Memorandum X-3445.

Boren, M. T. & Ramey, J. (2000). Thinking aloud: Reconciling theory and practice. *IEEE Transactions on Professional Communication, 43*(3), 261-278.

Holbrook, J. (2021). Exploring methods to collect and analyze data on human contributions to aviation safety. *Proceedings of the 21st International Symposium on Aviation Psychology*, 110-115.

Holbrook, J., Prinzel III, L. J., Stewart, M. J., Smith, B. E., & Matthews, B. L. (2019). Resilience and safety for in-time monitoring, prediction, and mitigation of emergent risks in commercial aviation. *Proceedings of the 20th International Symposium on Aviation Psychology*, 109-114.

Klein. G., Ross, K. G., Moon, B. M., Klein, D. E., Hoffman, R. R., & Hollnagel, E. (2003). Macrocognition. *IEEE Intelligent Systems. DOI: 10.1109/MIS.2003.1200735.*

Kuusela, H. & Paul, P. (2000). A comparison of concurrent and retrospective verbal protocol analysis. *American Journal of Psychology, 113*(3), 387-404.

PARC/CAST Flight Deck Automation Working Group. (2013). Operational use of flight path management systems. Final Report of the Performance-based operations Aviation Rulemaking Committee/Commercial Aviation Safety Team Flight Deck Automation Working Group. Washington, DC: Federal Aviation Administration.

Stephens C., Prinzel L., Kiggins D., Ballard K., & Holbrook J. (2021). Evaluating the use of high-fidelity simulator research methods to study airline flight crew resilience. Proceedings of the 21st International Symposium on Aviation Psychology, 140-145.

FLIGHT DECK PROCEDURES FOR A NEW GENERATION OF PILOTS

Erik-Jan A.M. Huijbrechts
Airline Pilot and Independent Aviation Safety Researcher
Sassenheim, the Netherlands
M.M. (René) van Paassen
Associate Professor Delft University of Technology
Delft, the Netherlands

This paper considers the combined effect of two trends in commercial aviation. On the one hand, there is a continuing demand for pilots, implying that a new generation of pilots, will soon be flying our aircraft. On the other hand, legal aspects have had an adverse effect on innovation in the safety level of established procedures, leading to a trend for aircraft operating companies to adopt manufacturer's recommended flight deck procedures rather than reviewing these and adapting these where appropriate to local needs. However, with the influx of new pilots to the workforce, the lack of innovation and adaptation of flight deck procedures poses a safety treat. The safety level of the aviation industry relies on the experience of the individual operator (pilot), the demands of the tasks and the available support tools.
If the experience is not in the operator (pilot) we have to put the experience in the procedures to maintain the existing safety level in aviation.
This conference paper is a plea to develop type, and operation, specific flightdeck procedures, adapted to present day operation and usable by the future pilot population.

Suitable flight deck procedures are an essential component in the safety of airline operations. They serve to achieve Work-as-Imagined by legislators, manufacturers and aircraft operating companies and are an integral part of the 4p's (Philosophy, Policy, Procedures and Practices) (Degani A. & Wiener E. 1994) that can be used to model aircraft operation. Clear guidance from procedures is particularly required for less experienced operators (pilots), who cannot yet rely on experience to guide their actions. Manufacturers of aircraft and aircraft equipment provide procedures for operation, supplied as a package with the purchase or lease of the aircraft. These procedures can however often still be improved on the basis of feedback from operational experience, tailoring these to the operational needs that may have changed from the time when the aircraft and instrumentations were designed, or that are specific to localized use that was not imagined by manufacturers.

Sources of Flight Deck Procedures

Flight deck procedures are presented by aircraft operating companies to their operators. They are composed of rules and procedures originating from legislation, manufacturers and commercial incentives. Procedures in general are presented to operators in a Basic Operations Manual (BOM) and operational procedures in an, aircraft type specific, company Flight Crew Operations Manual (FCOM).

The organisational model presented in (Huijbrechts & van Paassen 2021), can be used to illustrate the different influences on procedures. The model in Fig.1 shows the blunt and sharp end in aircraft operation and the way in which flight deck procedures arrive in operation.

Although flight deck procedures preferably must be tailored to the circumstances within which the airline company works (Barshi I. et al., 2016), many aircraft operating companies nowadays choose to present manufacturers' procedures without much adaptation to their operators (pilots). In smaller companies, the knowledge or assets to adapt procedures may not be available. In bigger companies, a fear for liability issues often hinders initiatives to improve safety by tailoring procedures to (changed) circumstances or company's specific needs.

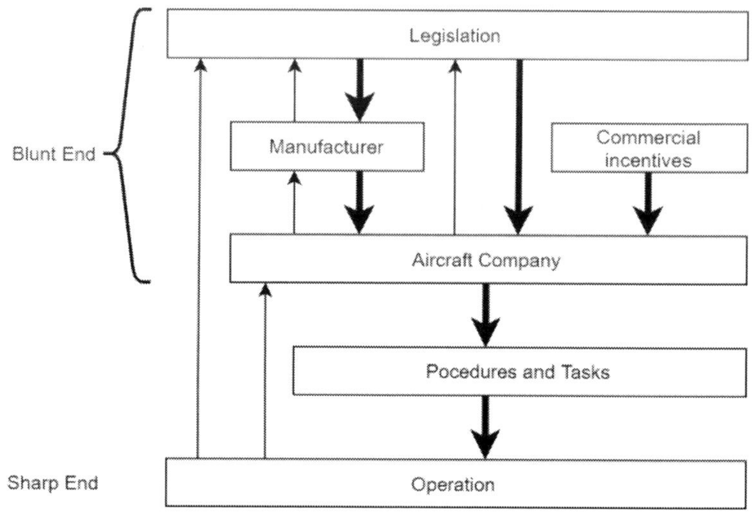

Fig.1 Blunt End and Sharp End in aircraft operation (Huijbrechts & van Paassen 2021)

On the work floor (at the sharp end), safety related issues with systems and procedures and flaws in legislation may emerge that are not obvious to the blunt end. The thin, upward flowing, lines in fig. 1 indicate that the feedback to the blunt end is not as strong as the instructions from the blunt end downwards. From the work floor perspective, there are several observations that may help to explain why present flight deck procedures must be improved to serve less experienced pilots.

The focus of management has shifted from operation and product oriented to process and legal oriented.

Until about the turn of the century, in many companies, management supported efforts to tailor flight deck procedures for ease of use and safety. Companies developed their own procedures and management took responsibility for the operation on the work floor. This led to remarkable differences in operating procedures between different companies (Degani A. & Wiener E., 1994).

But management policy has changed. Instead of assuming responsibility for tailored company procedures, management nowadays prefers to use manufacturers' procedures. Where these procedures are not suitable in a specific operational context, adapting and deviating from these is left to the responsibility of individual operators. This same observation implies that safety related efforts of management often are focused on (and limited to) compliance with legislation.

Manufacturers' procedures are not always tailored for operational use.

A manufacturer is not an operator. (Barshi I. et al., 2016 p. 6) This observation requires adaptation of procedures on the work floor. Manufacturers procedures are sometimes made up by non-operational experts or in a simulator. As an example the Boeing remote de-icing procedure will be considered in this paper.

Manufacturers' procedures do not cover all operational situations.

This observation shows the need to add specific procedures. Companies sometimes add structured guidance on how to handle non-normal situations. Operators may identify the need for a type specific checklist to guard for omissions in case of a late runway change or return to gate.

The effect of legal aspects on Flight Deck Procedures

Certification is a legal process that was intended to provide a good safety level in aircraft operation. Examples can however be found where the certified status of procedures and equipment has hindered further safety innovations (Huijbrechts & van Paassen 2021).

A specific example is found in a major European company, that decided to revert to using manufacturers' procedures discarding the operational experience that was enclosed in their own, adapted, procedures. This decision was influenced by an accident in a daughter company that showed the vulnerability of companies to liability claims after an accident. This change did not provide an improvement in operational safety but shifted the responsibility for safe operation to individual operators, mitigating the risk for the company for being held liable for company procedures.

Authorities require companies to install incident reporting systems that include questions on how procedures can be improved. The aim of such systems is to transform companies into learning organizations. Results of the incident reporting are, in general, shared by company management with operators, adding to their experience. But even here, with an explicit structure in place to improve safety, manufacturers' procedures are seldomly adapted by companies in response to incident reports, because the fear for liability issues outweighs the drive for safety. Thus liability can be seen as a barrier for a learning organization.

Certification and liability can be considered as legal aspects that have an adverse effect on safety innovation in flight deck procedures.

Types of Operators

The SRK framework (Rasmussen 1983) can be used to distinguish between operators that can rely on experience and less experienced operators. Operators on a flight deck (pilots) in general are smart people that are selected and employed because they can show, or develop, a high level of knowledge-based behaviour. Combining this with the 4P's Knowledge based behaviour can be seen as the Practice to use, select and adapt Procedures to the situation according to a Policy within the Philosophy of the aircraft operating company.
Based on experience and 'modus operandi' we propose here a distinction between operators in craftsman and rulesman.

Craftsman:

Craftsman in general do not need formalized rules and procedures to perform their task. They rely on their experience to guide their actions. Their rule-based behaviour (in Rasmussen's definition) is experience-based and effortless, rather than recipe-driven, and their high level of skill-based behaviour provides them with time and resources to display knowledge-based behaviour in reflecting on circumstances and adapting procedures to the actual situation.
Rigid rules and procedures can hinder craftsman in achieving their goals.

Rulesman:

Rulesman strongly rely on formalized rules and procedures to perform their tasks. Inexperienced operators often need a clear set of rules and procedures to guide them. Rule-based behaviour prevails over skill and knowledge-based behaviour. Over time a rulesman can turn into a craftsman by gaining experience.

Example: The Boeing remote De-Icing procedure

Before an aircraft can start its flight it has to be clear of contaminants on critical surfaces. If snow or ice is present this has to be removed through de-icing. Guidance can be found in FCOM. Through the years the practice of de-icing has changed. In the sixties and seventies it was common practice to have the aircraft de-iced at the gate before engine start. In the eighties remote de-icing on an apron platform became in use. This offered a more efficient use of de-icing equipment, better use of Hold Over Times and the spilled de-icing fluids could better be collected to prevent damage to the environment.

The Boeing de-icing procedure was originally designed for gate de-icing but later adapted for remote de-icing. The Boeing remote de-icing procedure shows evidence that it is not made up based on operational experience. For example, it requires operators to move controls and flaps before de-icing (The Boeing Company FCOM B737). The major European company that reverted to using manufacturers procedures was confronted by operators (pilots) that refused to perform the procedure by the book as this could damage control surfaces covered with a layer of snow or ice. In response the company did not change the procedure but added notes in their company FCOM stating that it is Subject to Captains Discretion (SCD) to delay control and flaps movement to after being de-iced (KLM FCOM B737). Examining the procedures of a major American company shows that they simply omitted publishing a detailed procedure for de-icing (Continental FCOM B737).

In such cases the responsibility to perform a safe procedure is shifted to the crew.

For most operators (pilots) de-icing is not a daily procedure and, in particular, inexperienced pilots would likely benefit from having clear guidance on what actions have to be performed.

Note: When I (the first author) was involved in company procedure development a simple solution was found to cover the remote de-icing situation. The before taxi procedure was performed twice; once before taxiing to the de-icing station omitting moving controls and flaps and once after de-icing, including the controls check and flaps setting (The latter may be postponed to just before take-off if prolonged taxi is required through precipitation after de-icing). This principle can be used for all aircraft types including those with electronic checklists.

Collecting information to improve Flight Deck Procedures

Manufacturers cannot foresee changes in common practices and only have a limited knowledge of operational practice. In order to improve procedures operational information has to be collected from aircraft operating companies and operators (pilots).

Collect information from aircraft operating companies.

Aircraft operating companies may have adapted manufacturers procedures based on incident reports or their own operational experience. Companies may have collected information from their operators that procedures can be improved without acting on this. Companies may have added structured guidance to operators e.g. on how to handle non-normal situations.

Collect information from operators.

Operators (pilots) use their own tricks to assure a safe operation. E.g. leave the Aircraft Maintenance Log on the glareshield as long as not all technical issues are resolved or use trigger events to check if all necessary actions are performed. Collecting and sharing these tricks may help new pilots to develop their own way of working. Operators may have good reasons to divert from company or manufacturers procedures based on their own experience. If diverting from Procedures becomes common Practice the former may have to be changed. Operators may recognise flaws in legislation that can be improved. Operators may also recognize situations that are not covered by manufacturers procedures that would benefit from better guidance.

Developing Flight Deck Procedures for a new generation of Pilots

If a significant number of companies and operators are willing to share information, this can be used to develop Procedures that contain both the manufacturers technical knowledge and the operational experience. The aim must be to combine these, to develop procedures that can be used by both craftsman and rulesman. In this practice the operational safety level has to prevail over legal correctness. In our opinion, this works best if procedures are developed, at the sharp end on the work floor and that this working method is supported and approved by the authorities.

Ecological Interface Design

Ecological interface design (Burns, C.M. & Hajdukiewicz, J., 2004) can be used to improve safety on the work floor. Improvement is already being made in flight deck instrumentation and e.g. performance software with graphical displays that may help in recognizing invalid input values and offer a check against gross input errors compared to the predicted load and fuel figures. Preferably procedures will be designed to offer enough flexibility that they will not hinder craftsman in their skill & knowledge-based behaviour but can still be used by rulesman as a do-list (Rantanen & Huijbrechts, 2021).

Improving Manufacturers Procedures

The biggest challenge is to break through the legal barriers that nowadays prevent improvement on manufacturers' procedures. This means that the responsibility for published procedures cannot be shifted to manufacturers, companies or legislators. An independent party, e.g. the Flight Safety Foundation (FSF) or NASA, can take the initiative to provide a recommended format to include manufacturers, company and operators input in a, type specific, Flight Crew Operations Manual (FCOM) that can be used as a standard for many companies with approval/validation of legislators and other stake holders. This will however require a big effort comparable to the Global Action Plan to Prevent Runway Excursions (GAPPRE) (FSF & Eurocontrol 2021).

Evaluation

A system has evolved in aviation where manufacturers try to shift responsibilities to legislators by means of certification (Huijbrechts & van Paassen 2021). Legislators are not always effective in promoting companies as learning organizations because of the barriers posed by liability issues. The ultimate responsibility for a safe operation thus greatly rests on the work floor that sometimes has to cope with unpractical procedures. Collecting operational feedback by an independent party may reveal flaws in legislation and (certified) procedures and equipment that can be improved. If operational experience can be included in flight deck procedures the loss of safety level through the change of focus in management and resulting effect on organizational safety (Rantanen & Huijbrechts, 2021[2]) can be partly recovered.

Conclusions

There are several barriers to further improving safety in the aviation system, one of these is the certified status of procedures and equipment as a barrier for safety innovations. Another is the liability issues faced by aircraft operating companies, which act as a barrier for a learning organization, by limiting innovation in procedures.

For operators (pilots) new to the aviation system, who did not yet have the opportunity to collect a wide experience, we have to put previously collected experience in the procedures to maintain the existing safety level in aviation.

To improve flight deck procedures the knowledge of manufacturers has to be combined with the operational experience of aircraft operating companies and operators (pilots).

Acknowledgements

I want to thank Maartje Huijbrechts for critical reading, hints and tips. This report is not initiated by, nor does it reflect, the views of my employer.

References

Barshi, I., Mauro, R., Degani, A., and Loukopoulou, L., *Designing Flightdeck Procedures,* NASA/TM-2016-219421, October 2016

Burns, C.M. & Hajdukiewicz, J. *Ecological Interface Design.* CRC Press. 2004

Continental Airlines *Flight Crew Operations Manual B737*

Degani A. & Wiener E. *On the Design of Flight-Deck Procedures* NASA Contracter Report 177642, June 1994

Flight Safety Foundation & Eurocontrol, *Global Action Plan to Prevent Runway Excursions (GAPPRE)* 2021.

Huijbrechts & van Paassen *Is our Current Certification Process a Threat to Safety Innovation?* Conference Paper, ISAP 21, may 2021

KLM Royal Dutch Airlines *Flight Crew Operations Manual B737*

Rantanen, E. & Huijbrechts E-J.A.M., *Procedures as an Ecological Interface* Conference Paper, ISAP 21, may 2021

Rantanen, E. & Huijbrechts E-J.A.M. [2], *Organizational Safety in Airline Operations* Conference Paper, ISAP 21, may 2021

Rasmussen, J. *Skills, Rules, and Knowledge; signals, signs and symbols and other distinctions in human performance models.* IEEE transactions on Systems, Man and Cybernetics, SMC-13(3) 257-266

The Boeing Company *Flight Crew Operations Manual B737*

A NON-TECHNICAL SKILLS TRAINING CONCEPT FROM THE INITIAL FLIGHT TRAINING STAGE TO AIRLINE OPERATION

IKEBA Hiroshi
Tokyo, Japan

TSUDA Hiroka
FUNABIKI Kohei
Japan Aerospace Explorlation Agency
Tokyo, Japan

This study proposed a framework of Non-Techical Skills (NTS) that integrates existing NTS frameworks, such as CRM(Crew Resource Management), SRM(Single-pilot Resource Management), and TEM(Threat and Error Management). First, CRM and SRM were compared, and most of the elements of CRM and SRM were found to be commonly useful in multi-crew and single-pilot operations. Second, Risk Management in SRM was compared with TEM, and these were integrated into a single framework called Unified Risk Management. Third, DODAR model, which is widely used as a checklist for Decision Making process, was modified and extended to cover all the processes of Risk Management and proposed VNS/DRODAR model.

In Japan, there are two ways to become an airline pilot, namely via CPL(Commercial Pilot License) course and MPL(Multicrew Pilot License) course. In the CPL course, although the importance of Non-Technical Skills (NTS) is emphasized in the training and education materials, few practical lessons are conducted before graduating from the course. As Japanese airlines do not require kinds of MCC (Multi-Crew Cooperation) training at the entry of FO (First Officer) training, most of the pilots from CPL course experience the NTS training at the time of beginning the training of transport category airplane. On the other hand, in the MPL course, the NTS is integrated with the course from the initial stage of the training.

There is continuing discussion on whether and how NTS should be taught and trained in the early phase of the training. Although the majority of instructors argue that learning technical skills is more important than learning NTS because the Technical Skills would be the bases of all competencies. One can say that the NTS is better to be taught after beginning the FO training, because the company policies of each airline are different, on the other hand, another says that most of the basics of the NTS are common among the airlines. Experiencing and learning from several accidents that happened in flight training, the authors concluded that it is important to provide NTS education from the early phase of training. The next challenge is to define NTS be required for those who want to become airline pilots but be exercised during the training for acquiring competencies for single-pilot operation. It is clear that the most straightforward approach is combining CRM and SRM. In this paper, relationships between SRM, CRM, and TEM are analyzed and integrated into a new framework named Unified Risk Management.

Analysis of Existing NTS

In this chapter, major frameworks and associated elements of NTS are explained and analyzed. Figure 1 shows the overview of the NTS discussed in this paper. In the NTS, there are two major frameworks of CRM/TEM and SRM, where the CRM/TEM framework is comprised of CRM Skills framework and TEM framework. CRM Skills and SRM include some elements (blue letter) and tools (green letter) that look similar to each other. The goal of this study is to compare those frameworks and elements and to construct a single framework that can be used in both single-pilot and multi-crew operations.

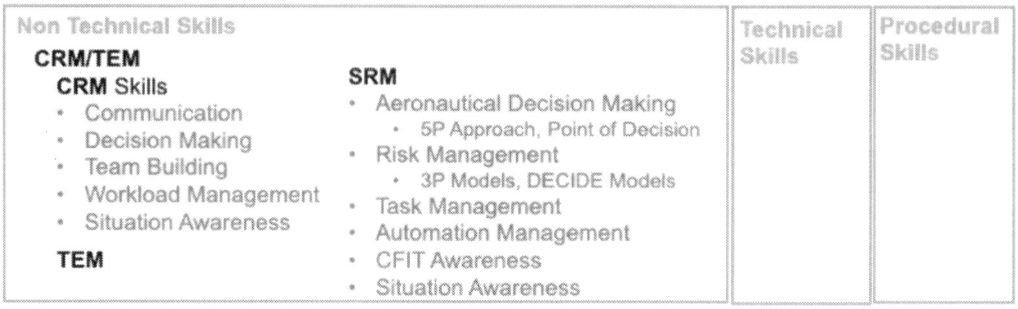

Figure 1. Overview of NTS.

CRM

There are several frameworks of non-technical skills, but the most widely known of them is CRM. CRM is defined as "using all available resources, information, equipment, and people to achieve safe and efficient flight operations" by Lauber (1984). In the early stage of CRM, the focus was on specific skills and behaviors that would enable pilots to perform their tasks more effectively. CRM skills, developed and introduced in the late 1980s, are specific examples of how to act in order to practice CRM in actual flight operations. JAXA(Japan Aerospace Exploration Agency), with the cooperation of Japanese airlines, developed a set of CRM skills for practical operation and pilot resources in Japan by Iijima(2003) shown in Figure 2. Based on the idea that CRM is based on the team concept, it has been regarded as being applied mainly to multi-crew operations, but CRM skills have many elements that can be applied to single-pilot operations. However, at least in Japan, it is considered that CRM and SRM are different and that CRM is not for single-pilot operation.

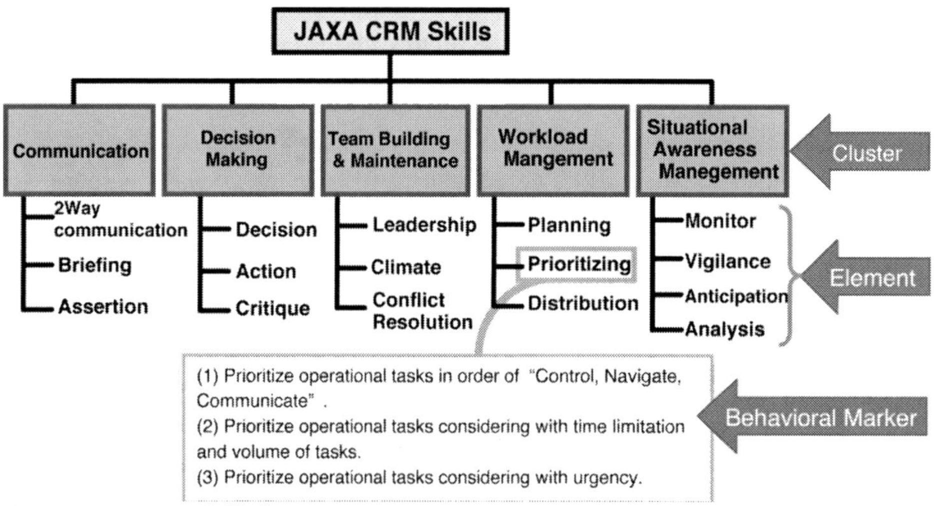

Figure 2. CRM Skills by JAXA.

TEM

TEM, as defned by ICAO(2015), is an approach or a framework that focuses not only on the prevention and early detection of errors but also attempts to manage threats, which are factors that cause errors. TEM is now widely accepted by airline pilots as a practical framework to carry out the concept of

CRM in actual operation. It is said that "Practicing CRM means implementing the concept of TEM using CRM skills". On the other hand, among pilots of smaller aircraft, such recognition is not so widespread.

SRM

SRM is defined as "the art and science of managing all the resources, both on-board the aircraft and from outside sources, available to a single pilot before and during flight, to ensure the successful outcome of the flight". It may be natural to say that the SRM is CRM for single-pilot operation. SRM includes the concepts of Aeronautical Decision Making, Risk Management, Task Management, Automation Management, CFIT Awareness, and Situational Awareness. SRM practices include the 3P model, the 5Ps approach, and the DECIDE model developed by the FAA Aviation Safety Program(2016). In the following sections, essential and typical elements of SRM are explained.

Aeronautical Decision Making(ADM). ADM is one of the skills included in SRM and is defined as "a systematic approach to the mental process in which pilots consistently determine the best course of action in response to a specific situation." Important point related to ADM is the appropriate timing of making decisions. Pilots encounter various expected and unexpected events during flight, especially in VFR. Under high workload or time pressures, humans do not realize that they are standing at a Point of Decision, described by Craig(1997), even if they were well prepared. Whether in single-piloted or multi-crew operation, there is no difference in the importance of the Go/No Go decision at the Point of Decision.

Risk Management. Risk Management is one of the elements of SRM, but at the same time, it is a component of ADM. The FAA defines Risk Management as "a part of the decision-making process that relies on situational awareness, problem awareness, and good deciion to mitigate the risks associated with each flight," and FAA(2016) explains that "the goal of Risk Management is to proactively identify safety-related hazards and mitigate the associated risks".

3P Model. The 3P model is positioned as a method for practicing Risk Management, especially among SRM. It consists of three steps: PERCEIVE, PROCESS, and PERFORM, associated with three checklists, namely PAVE, CARE, and TEAM for each. The FAA recommends the use of the 3P model when introducing the concepts of Aeronautical Decision Making and Risk Management to training sites.

DECIDE and DORDAR. As with the 5Ps approach, one of the tools for implementing SRM is the DECIDE model. There is another model called DODAR, introduced by CAA (2016), which is very similar to the DECIDE model. Those two models are compared in Figure 3.

DECIDE	DODAR
Detect fact	
Estimate the need to respond	Diagnose
Choose	Options
I dentify the solution	
Do	Decide
	Assign tasks (to crew members)
Evaluate the effect of actions	Review

Figure 3. Comparison of DECIDE and DODAR Models

Both the DECIDE model and the DODAR model indicate actions that should be taken after a pilot discovers a non-normal condition such as equipment failure. The DODAR model is considered to be more specific and superior to the DECIDE model in terms of the ease of understanding the actions to be taken and taking into account of multi-crew concept. However, from the perspective of Risk Management, it has been pointed out that neither the DECIDE model nor the DODAR model includes the term indicating Hazrd Identification and Risk Assessment.

Unification of NTS

Comparion of CRM Skills and SRM Skills

As shown in Table 1, the three common skills required for CRM and SRM are Situational Awareness, Workload Management, and Decision Making. Differences are Communication and Team Building in CRM, Risk Management, Automation Management, and CFIT Awareness in SRM.

Table 1.

Comparison CRM Skills and SRM Skills.

CRM/multi-crew	**SRM**/single-pilot	Discussions
Communication	→ commonly used	Also be useful in single, especially in teaming with non-flight crew.
Team Building	→ commonly used	
Decision Making	Aeronautical Decision Making	common
TEM commonly used ←	Risk Management	It is more important in single, but also be useful in multicrew. TEM is sharing the same goal.
Workload Management	Task Management	common
Situation Awareness	Situation Awareness	common
procedural skills	CFIT Awareness	They are included in procedural skills in multicrew.
procedural skills	Automation Management	

The two skills of Communication and Team Building that appear only as CRM skills, seem to be also useful in single pilot operations. Communication and Team Building are undoubtedly essential not only between flight crews, but also between pilots and ATC controllers, mechanics, and non-flight crew members such as air medics.

The two skills of Automation Management and CFIT Awareness are said to be particularly necessary for the single pilot operation, as mentioned by JCAB(2020). The automation Management refers to the skill of understanding and mastering the automated systems of TAA (Technically Advanced Aircraft), which are increasing in recent years. In addition, CFIT Awareness skill is required from the viewpoint of accident prevention for small aircraft that often fly at low altitudes, without the equipment of TAWS (Terrain Awareness and Warning System). Although Automation Management and CFIT awareness are not explicitly defined in CRM, they are incorporated into standard operating proceduresfor large aircraft. Therefore, it can be said that Communication, Team Building, Automation Management, and CFIT Awareness are commonly required for both multi-crew and single-pilot operations.

As a result of the above comparison of CRM and SRM, the difference remains in the positioning of Risk Management in the multi-crew operation. In the next section, we will discuss Risk Management in the operation of multi-crew and single-pilot operations, and whether it can be regarded as a common item.

Importance of Risk Management in SRM

In the operation of a small aircraft, it is not possible to obtain systematic operation support like an airline, and the pilot must collect, evaluate, and make decisions by himself. The FAA(2016) stated, "Single Pilots without other crew members to consult must contend with intangible elements that place them at risk. Single Pilots are therefore more vulnerable than multi-crew operations." For this reason, in single pilot operations, risk management regarding whether to depart, whether to continue the flight, and whether to divert is an important skill. On the other hand, in ailine operation, a Go/No-Go decision is supported by not only other crewmember but by many ground personnel and onboard equipment. However, that does not mean that the airline pilots themselves do not need risk management. Risk Management element in SRM is implemented as part of Decision Making and Situation Awareness of CRM, and of TEM. In the next section, Risk Management and TEM are compared.

Unification of Risk Management and TEM

Figure 4 shows the comparison of TEM and Risk Management processes, from the viewpoint of how minor threats or hazards lead to accidents. In the risk management model, Hazards are considered to be the same as Threats in the TEM model. The Risk of UAS is then calculated as the product of hazard and probability of failure in managing Threat and Error. Although it is possible to say that the start and the goal of the TEM model and the Risk Management model are common, the Risk Management model can provide more specific tools and frameworks as derivatives of SRM. We call the Risk Management model which is comparable with the TEM model, as the "Unified Risk Management" model.

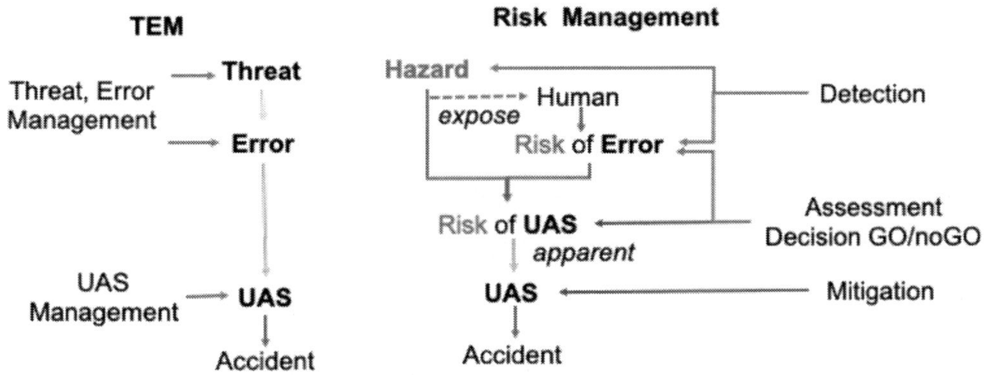

Figure 4. Comparing TEM and Risk Management.

Introduction of VNS/DRODAR Model

In the broad sense, there are a lot of frameworks or tools to support risk management, such as 3P Model, DECIDE, DODAR and TEM. As mentioned earlier, all of the tools have some week points. Therefore, we propose a new frame work of risk management, called VNS/DRODAR. The VNS/DRODAR is based on DODAR model, and including three elements for situation awareness and one element for risk assessment. As shown in Table XX, VNS/DRODAR framework can be used as a replacement of 3P model. It is supposed to be used as a checklist at the checkpoints, may be together with tools like 5P Approach.

3P Model	VNS/DRODAR	CRM/SRM Skills/Elements
Perceive/ PAVEc'k: - Pilot - Aircraft - enVironment - External pressures	Vigilance Notice Stay Calm	Situation Awareness - Vigilance - Anticipation - Monitor - Analysis
Process/ CARE c'k - Consequences - Alternatives - Reality - External factors	Diagnose Risk Assessment Options Decide	Risk Management Decision Making - Decision
Perform/ TEAM c'k - Transfer - Eliminate - Accept - Mitigate	Assign Task	Workload Management - Planning - Prioritizing - Task Distribution Decision Making - Action
	Review	Decision Making - Critique

Figure 5. VNS/DRODAR Model

Conclusions

As a result of analyzing the existing NTS frameworks, it was found that CRM and SRM do not differ greatly at skill levels, and can be integrated as complementary to each other. In addition, Risk Management, which is emphasized in SRM, is not included in CRM, suggesting the possibility of unifying Risk Management with TEM. We proposed VNS/DRODAR as a model applicable to practice Unified Risk Management. Using these integrated frameworks and models will enable consistent NTS education and training from Single-Pilot to Multi-crew.

References

Lauber, J. Resource management in the cockpit. Air Line Pilot 1984, 53, 20–35.

Iijima, T.,Noda F., Sudo K., Muraoka K. and Funabiki K.(2003). Development of CRM skills behavioral markers, TR-1465, National Aerospace Laboratory Report.

ICAO Doc 9868 (2015), PANS Training, Chapter 1, Attachment C to Chapter 1 (TEM), Second Edition 2.

Federal Aviation Administration (2016); Pilot's Handbook of Aeronautical Knowledge (FAA-H-8083-25B), pp. 2-19,

Craig, P. A.(1997), BE A BETTER PILOT Making the Right Decisions, Tab Books, 107.

Civil Aviation Authority (CAA)(2016),Rational (classical) decision making, Flight-crew human factors handbook (CAP 737), 83.

Japan Civil Aviation Bureau (JCAB)(2018), YouTube MLIT Channel, https://www.youtube.com/watch?v=oHF692-4Qkk, viewed 2020.

Federal Aviation Administration (FAA) (2022), Risk Management Handbook (FAA-H-8083-2A).

IMPROVING GENERAL AVIATION SAFETY THROUGH HUMAN FACTORS RESEARCH

Ian Johnson, Ph.D.
Federal Aviation Administration
Washington, DC.
Mary E. Johnson, Ph.D., Brandon J. Pitts, Ph.D., & Shantanu Gupta, M.S.
Purdue University
West Lafayette, IN
Beth Blickensderfer, Ph.D., John Kleber, M.S., Cassandra Domingo, M.S., Robert Thomas, Ph.D., & Thomas Guinn, Ph.D.
Embry-Riddle Aeronautical University
Daytona Beach, FL
Lori Brown, FRAeS, MSc
Western Michigan University
Kalamazoo, MI
Debbie Carstens, Ph.D., PMP, Tianhua Li, Ph.D., Michael E. Splitt, & M. Harwin, J.D., M.S.
Florida Institute of Technology
Melbourne, FL
Barrett Caldwell, Ph.D., & Mel Futrell
Purdue University
West Lafayette, IN

Weather continues to play a significant role in general aviation (GA) events. GA pilots use various technologies to access and view weather information in the cockpit. These technologies range from handheld devices to installed displays. A contributing factor in many weather-related events was the pilots' failure to correctly interpret the displayed weather information compared to what was observed out the window. This session will highlight ongoing Human Factors (HF) research aimed at understanding and addressing this problem. Topics include an overview of the FAA's Weather Technology in the Cockpit (WTIC) program; barriers to automating and implementing a speech-to-coded Pilot Report (PIREP) system; effect of weather briefing strategy on a pilot's understanding of weather; pilot perception of hands-minimized PIREP submittal tools; use of Virtual Reality (VR) in aviation training; and pilot use of weather information in low altitude and special operations. This paper provides an abstract for each of these topics.

Despite enhancements in weather information and the proliferation of weather-related cockpit displays, mobile technology, and applications by industry, weather-related accidents continue to account for the majority of general aviation (GA) fatal accidents. Previous research has shown that in many instances a contributing factor in many of these accidents was the pilots' failure to correctly interpret the weather information being depicted inside and viewed outside the cockpit, and inadvertently entering instrument meteorological conditions. (Pearson, 2002; Aarons, 2014). Fortunately, a body of ongoing, human factors research exists aimed at understanding and addressing this problem.

The purpose of this session is to highlight a body of ongoing Human Factors (HF) research aimed at understanding and addressing this problem. Topics include an overview of the FAA's Weather Technology in the Cockpit (WTIC) program; barriers to automating and implementing a speech-to-coded Pilot Report (PIREP) system; effect of weather briefing strategy on a pilot's understanding of weather; pilot perception of hands-minimized PIREP submittal tools; use of Virtual Reality (VR) Human Interfaces in aviation training; and pilot use of weather information in low altitude and special operations. This session is designed to foster a discussion about the complexity of interpreting aviation weather, the hazards of weather in GA operations, and the research underway to mitigate the hazards and improve GA safety.

WTIC Program Research Impacts

The Weather Technology in the Cockpit (WTIC) Program is an FAA NextGen weather research program that sponsors research to develop, verify, and validate Minimum Weather Service (MinWxSvc) recommendations for incorporation into standards, guidance documents, and training materials, and for technical transfer to Government agencies and industry for implementation. The WTIC program sponsors research to identify and resolve cockpit weather-related gaps in information and technology and then explore ways to improve the following: cockpit weather information and its rendering; pilot understanding and interpretation of cockpit weather information and technologies; weather information training; and operational efficiency and safety of commercial, business, and general aviation operations. A vital part of the WTIC program research is performing Human Factors analyses on the rendering of meteorological information presented in the cockpit. The topics presented in this paper represent examples of WTIC-sponsored Human Factors research.

Barriers to Automating and Implementing a Speech-to-Coded PIREP System

Weather information is important to pilots so that they may better understand conditions enroute and at airports that may affect the safe operation of their aircraft. Currently, weather conditions such as icing and turbulence, are only available from airborne reports. Pilot Reports (PIREPs) are reports of the actual weather conditions encountered by an aircraft in flight and are used to notify other pilots of these conditions and modify or inform weather forecasts and forecasting models (NTSB, 2017). Most of the PIREPs submitted are from commercial airliners across the globe. General Aviation (GA) pilots use PIREPS but do not submit a proportionate number of them. Recent developments in speech-to-text technology can make the process of submitting PIREPs easier and ultimately increase the total number of weather reports submitted. However, GA pilots may have differing levels of automation and communication capabilities in the reporting and receiving of airborne weather information. In this study, some of the technological and cultural barriers to the automation of the submittal of PIREPs are identified and discussed. This presentation provides the background, rationale, and methodology for a large-scale study of the GA PIREP submittal process, the sources of PIREPs used by GA pilots, data science approaches to automating PIREP submissions, and barriers involved in the automation of PIREP submittal.

Weather Self-briefings in General Aviation: A Laboratory Study

Over the past two decades, a shift has been underway regarding how General Aviation (GA) pilots obtain weather information prior to flight. Traditionally, GA pilots obtained weather information using Flight Services. With the advent of the internet, GA pilots are now able to obtain weather information from their own computers and mobile devices and perform weather pre-flight planning independently (also known as a "self-briefing"). Research indicates that GA pilots are increasingly conducting weather self-briefings during preflight (Duke et al., 2019). The purpose of this presentation is to describe an ongoing laboratory study of GA pilots' preflight planning for weather. The 2x2x2 experimental design includes three independent variables: self-briefing (yes vs. no), flight services briefing (yes vs. no), and weather scenario (fog vs. icing). The dependent variable is the pilots' understanding of the weather in relation to the planned flight path. Measures of the dependent variable include mental model assessments and interview data. The presentation will include the preliminary results that indicate an effect of briefing strategy on pilots' understanding of weather. Implications for research, product design, and pilot training will be discussed.

Virtual Reality Human Interfaces for the Next Generation of Aviation Professionals

The Next Generation of Aviation Professionals (NGAP) entering the aviation industry today represents a new generation of learners that requires us to look beyond our traditional training and evaluation methods (Brown, 2017; Felder & Brent, 2005; Tulis, 2018). Specific aviation tasks require an understanding of several interrelated human and machine components requiring practice and immersion. To meet these challenges, we can harness three-dimensional (3D) simulated environments using Virtual Reality (VR) human interfaces to provide adaptive learning methodologies to meet the learning styles of changing generations. We have seen how VR has started to transform aviation training and now we can optimize real-world training to enhance aviation weather understanding with digital tools such as spatial computing. In spatial computing training environments, the avatar serves various purposes for virtual training sessions while connecting with other digital objects and the ability to create virtual worlds. In practice, this means an instructor can present a digital version of an aircraft, aircraft cockpit, or system to multiple students in various locations around the world at the same time. Students are mirrored as avatars and can walk around the aircraft, learn cockpit procedures, practice two-crew operations with multiple people and examine components up-close, ask questions, and share opinions, just as if they were together in a traditional classroom, simulator, or aircraft setting. With spatial computing, you can run a training or outreach session from anywhere in the world in a way that is cost-effective, interactive, and engaging. The potential of using 3D graphically rendered content to communicate and illustrate objects for aviation training and outreach is limitless.

This approach has several practical advantages, and we can enable students to demonstrate proficiency, enhance retention, and engagement with 3D learning outcomes while meeting future training and operational demands (Holcomb, 2018). As the aviation industry is challenged by changing demographics, growing demand, and innovative technologies with far-reaching potential, it becomes increasingly urgent to evaluate the human-machine interfaces associated with such technologies (Bellotti, et al., 2013; Nazir, et al., 2005; Rupasinghe, et al., 2011).

Pilot Perception of Hands-Minimized PIREP Submittal Tools

Flying into hazardous weather can lead to aviation incidents and accidents. Pilot Reports (PIREPs) can increase the accuracy and timeliness of current and forecasted weather conditions and are an essential tool used by pilots to avoid flying into hazardous weather as well as meteorologists to develop and update aviation forecasts. This study administered a descriptive survey to inquire about how likely pilots would be to use a Speech Recognition System (SRS) to transcribe and submit PIREPs automatically while flying in three distinct flight regimes: instrument flight rules (IFR), visual flight rules (VFR) with flight following, and VFR without flight following. The survey employed a cross-section design and included Likert scale questions. For each flight regime, additional information was obtained through an open-ended follow-up question. The Likert scale responses indicated that pilots were neutral about using an SRS to transcribe and submit PIREPs in each flight regime. Spradley's (1979) domain analysis was used to identify common themes and patterns from the open-ended responses. Major findings from flying IFR were that pilots found it easier to speak directly to air traffic control, or pilots were too busy to submit PIREPs while flying IFR. Major findings from flying VFR with flight following were that pilots thought it was easier to report PIREPs directly to air traffic control or a flight service station, and it was more accurate to report PIREPs directly to an aviation professional. However, they were willing to try an SRS. Major findings from flying VFR without flight following were that pilots wanted the opportunity to review a PIREP submission for accuracy and were willing to try the system. Significant differences were determined by making a comparison between the three groups.

Pilot Use of Weather Information in Low-Altitude and Special Operations Missions

Research conducted by the authors as part of the FAA Partnership to Enhance General Aviation Safety, Accessibility, and Sustainability (PEGASAS) has focused on issues of weather information presentation to GA pilots, including questions of how well weather information from one site can represent conditions at another site (without FAA certified weather information available). An important consideration for both fixed-wing and rotorcraft pilots operating in low-altitude operations (LAO) missions is that of information uncertainty associated with dynamic weather conditions and terrain variations. Even expert meteorologists have found it difficult to appropriately assess weather conditions in these types of areas. Our research has identified that other meteorological variables, such as climate zones, can help provide information to support machine learning approaches to assist in weather-related decision-making. Previous research has demonstrated that areas, where distinct climate zone models are in conflict, can help identify areas of increased weather-related decision risk and uncertainty. An initial "climate zone matching index" prototype has been demonstrated to show promise in identifying LAO areas in the Los Angeles basis with increased risk of weather information uncertainty. Planned PEGASAS research will investigate the impact of climate zone matching evaluation in other areas of the United States. This presentation will include results of prior survey data collection, climate zone matching approaches, and characteristics of important LAO features influencing climate zone / machine learning systems.

Acknowledgment

The research presented in this paper was sponsored by the FAA Weather Technology in the Cockpit (WTIC) program.

References

Aircraft Owners and Pilots Association (AOPA). (2016). *2016 Pilot report survey.* http://download.aopa.org/advocacy/0417_2016_pilot_report_survey_final_report.pdf

Bellotti, B., Kapralos, K., Moreno-Ger, P. (2013). User Assessment in Serious Games and Technology-Enhanced Learning. *Advances in Human-Computer Interaction, 2013*, 2. https://doi.org/10.1155/2013/120791.

Blickensderfer, B. L., Guinn, T. A., Laniccci, J. M., Ortiz, Y., King, J. M., Thomas, R. L., & DeFilippis, N. (2020). Interpretability of aviation weather information displays for general aviation. *Aerospace Medicine and Human Performance, 91*(4), 318-325. https://doi.org/10.3357/AMHP.5245.2020

Blickensderfer, B. (2021). Weather Self-briefings in General Aviation: A Human Factors Perspective.

Brown, L. (2017). Augmenting the Next Generation of Aviation Training with Holograms. *ICAO Training Report*, 7, 22-25. https://www.icao.int/publications/journalsreports/2017/icao_training_report_vol7_No3.pdf

Di Serio, Á., Ibáñez, M. B., & Kloos, C. D. (2013). Impact of an augmented reality system on students' motivation for a visual art course. *Computers & Education, 68*, 586-596.

Duke, R., George, T., Davis, K., & Bell, E. (2019). AOPA 2019 Weather Survey. http://download.aopa.org/advocacy/2019/190820_weather_survey.pdf?_ga=2.109147461.1913807553.1566334445-279783132.1506440972

Felder, R. M., & Brent, R. (2005). Understanding student differences. *Journal of Engineering Education, 94*(1), 57-72.

Holcomb, K. (January 12, 2018). Researchers test virtual reality Adaptive Flight Training Study. 14th Flying Training Wing Public Affairs, US Air Force Space Command. https://www.afspc.af.mil/News/Article-Display/Article/1414771/researchers-test-virtual-reality-adaptive-flight-training-study/

National Transportation Safety Board (2017). *NTSB/SIR-17/02: Improving pilot weather report submission and dissemination to benefit safety in the National Airspace System.* https://www.ntsb.gov/safety/safety-studies/Documents/SIR1702.pdf

Nazir, S., Sorensen, L. J., Overgård, K. I., Manca, D. (2015). Impact of training methods on Distributed Situation Awareness of industrial operators. *Safety Science, 73*, 136-145.

Pearson, D. C. (2002). VFR flight not recommended: A study of weather-related fatal aviation accidents. Technical Attachment SR SSD, 18, 2443-2448.

Rupasinghe, T. D., Kurz, M. E., Washburn, C., Gramopadhye, A. K. (2011). Virtual reality training integrated curriculum: An aircraft maintenance technology (AMT) education perspective. *International Journal of Engineering Education, 27*(4), 778.

Spradley, J. P. (1979). The ethnographic interview. New York: Holt, Rhinehart & Winston. *LeCompte, MD (2000). Analyzing Qualitative Data. Theory into Practice, 39*(3), 146-156.

Tulis, D. (March 12, 2018). Aviation Professor Bridges Real, Artificial Worlds: JetXplore Holographic Learning Enhances Training. *AOPA e-pilot, Flight Training Edition.* https://www.aopa.org/news-and-media/all-news/2018/march/12/aviation-professor-bridges-real-artificial-worlds.

GET THE JOB DONE OR SAFETY ABOVE ALL? HOW TRAINING BACKGROUND AFFECTS SAFETY IN HELICOPTER PILOTS

Anna Kaminska, Amy Irwin, Devin Ray
University of Aberdeen
Aberdeen, UK
Rhona Flin
Robert Gordon University
Aberdeen, UK

Culture has been identified as one of the main input factors impacting flight safety and team performance. Diverse methodologies were used to examine how professional culture influences helicopter pilots' safety-related behaviours. Study 1 (mixed-methods survey) showed that the main difference between civilian- and military-trained pilots can be put down to 'safety vs. efficiency', with pilots mentioning that what is perceived to be a threat seems to differ between military-and civilian-trained helicopter pilots. Additionally, having a multi-professional crew (military- and civilian-trained pilots together in a cockpit) was seen as having a positive effect on all non-technical skills, especially on situation awareness. Study 2 examined implicit risk perception of military- and civilian-trained pilots. The results indicated that all participants perceived the risk associated with flying in adverse weather. Interestingly, no differences between military-/civilian-trained pilots were observed. The studies presented provide an original, in-depth look at how helicopter pilots perceive their professional culture.

Culture has been identified as one of the main input factors impacting flight safety and team performance (Helmreich, 2000). Professional culture, specifically, is based on job role and training background. In aviation, this often relates to civilian vs. military training background and the shared norms and behaviours embedded via this initial training. Kaminska et al. (2021) conducted in-depth interviews with helicopter pilots to determine which aspects of culture are perceived as factors influencing safety behaviours and performance during flight. One of the most prominent findings was the profound impact of training background on pilots for the rest of their careers, especially in relation to the approach to flying. Most mentioned that military-trained pilots approach flight with the 'must get the job done' attitude, whereas civilian-trained pilots with 'safety above all'.

This finding leads to the question of whether military-trained pilots perceive less risk associated with flight? Risk perception is considered to be inherently subjective, dependant on such factors as the perceiver's past experiences, perceived control over the risk, and consideration of how the risk is likely to impact them personally (Slovic, 1987). A person's risk-taking is thus linked to their individual stance on these factors as well as organisational influence and social factors (Harris et al., 2022). This could potentially explain the risk-taking of ex-military pilots, as they are also described by other helicopter pilots to have better, more in-depth training and might have broader flight experience than civilian-trained pilots (Kaminska et al., 2021).

The current package of studies aims to replicate the findings of Kaminska et al. (2021) in a larger, multi-national sample and further determine the effect of professional culture on helicopter pilots' risk-perception. A mixed-methods survey (Study 1) was chosen to assess how each non-technical skill is affected by professional culture, while allowing participants to comment on their opinions. The Implicit Association Test (IAT; Study 2) was chosen for its superior capacity to predict individuals' behaviour, when compared to measures of explicit attitudes in the context of safety-related behaviour (Hatfield et al., 2008; Marquardt et al., 2012).

Study 1

Methods

Participants. A sample of 128 (3 female, 1 preferred not to say) helicopter pilots completed the study. Participants came various training backgrounds (79 civilian-, 47 military-trained) and ranged in their total time in aviation: from <5 years since beginning of training to >35 years of experience. Majority of the pilots surveyed most often flew in the role of a commander (9 co-pilots, 2 preferred not to say, 20 missing data). The pilots worked in a wide variety of operation types (offshore transport, search and rescue, air ambulance, police, etc.).

Materials and Procedure. A survey assessing how various non-technical skills are used in flight was adapted from Hamlet (2021). Participants had to rate how often (from 1 *never* to 7 *every time*) they use each element during flight. The second questionnaire was developed specifically for this study. Participants were asked what impact does either a multi-national or a multi-professional flight crew have on a non-technical skill during flight. Participants were asked these questions for each NTS (situation awareness, decision making, workload management, communication, teamwork, leadership, cognitive readiness). The participants were asked to answer these questions as 'no effect' or on a scale from -3 (*completely negative: 100% negative, 0% positive*) to 3 (*completely positive: 0% negative, 100% positive*). Participants were also asked to briefly explain their rating. This was included to allow participants to not be constrained by ratings, but also motivate their chosen rating. Finally, pilots filled out Attitudes towards a Multi-National Cockpit (MNCA) Questionnaire (adapted from Peksatici, 2018). For full materials, please contact first author.

Results

Quantitative analyses. It was aimed to (1) compare self-rated NTS performance across professional training background; and (2) assess the extent to which a multi-professional (crew of military- and civilian-trained pilots) flight team influences NTS.

Impact of training (type / country) on self-rated NTS performance. Firstly, the difference in how military-trained vs. civilian-trained helicopter pilots assessed their own non-technical skills was examined. The higher the score, the more often the pilot engaged in a certain activity.

Each NTS average score was submitted to one-way (*training type*: military vs. civilian) between-subjects ANOVA. The ANOVAs revealed a main effect of training on pilots' self-assessment of leadership ($F(1, 126) = 4.812$, $p = .030$, $\eta_p^2 = .037$) and communication ($F(1, 126) = 11.472$, $p = .001$, $\eta_p^2 = .083$). Military-trained pilots rated their own communication skills ($M = 5.95$, $SD = 0.58$) and leadership ($M = 5.97$, $SD = 0.53$) significantly higher than civilian-trained pilots assess their communication skills ($M = 5.52$, $SD = 0.75$) and leadership ($M = 5.75$, $SD = 0.56$), respectively. No other differences observed between military vs. civilian-trained pilots' self-assessment of their non-technical skills were observed.

Effect of multi-professional crew on crew NTS. The second aim was to examine whether pilots perceive having a multi-professional cockpit as affecting crew non-technical skills. To test this, a repeated measured ANOVA (*7 multi-professional NTS: decision making, workload management, situation awareness, teamwork, communication, leadership, cognitive readiness*) was conducted. Mauchly's test indicated that the assumption of sphericity had been violated, $\chi2$ (20) = 57.242, $p < 0.001$, therefore, the degrees of freedom were corrected using Greenhouse-Geisser estimates of sphericity ($\varepsilon = 0.834$). The results show that there was a significant effect of a multi-professional crew on non-technical skills, $F(5.004, 445.400) = 1.944$, $p = 0.010$, $\eta_p^2 = .033$.

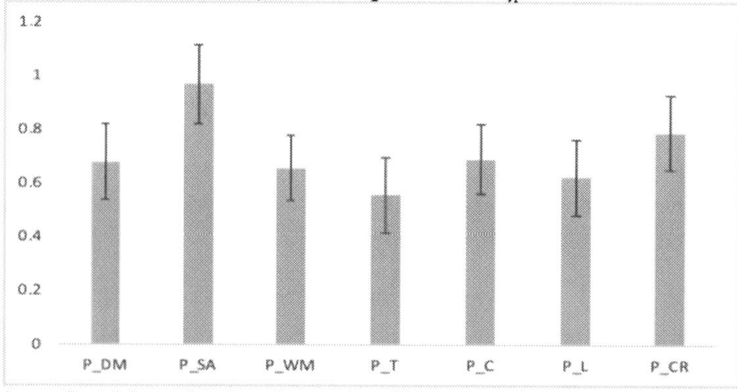

Figure 1. Mean rating of multi-professional crew's effect on each NTS (error bars represent 1SE). 'P' in labels refers to 'multi-professional' cockpit. DM, SA, WM, T, C, L and CR are non-technical skills

(i.e., decision making, situation awareness, workload management, teamwork, communication, leadership, and cognitive readiness, respectively).

Pairwise comparisons revealed that situation awareness was affected significantly differently from all other non-technical skills ($ps < 0.012$), apart from cognitive readiness ($p = 0.155$), and cognitive readiness was affected significantly differently from teamwork ($p = 0.05$). There were no significant differences between any other NTS. This suggests that a multi-professional team was viewed as enhancing all NTS, with Situation Awareness being enhanced the most (see Figure 1).

Qualitative Analysis. The conventional content analysis (Hsieh & Shannon, 2005) generated four categories (Power Distance Contrasts, Flight Approach Contrasts, Mixed Crews, Ex-military) with eight sub-categories related to the impact of professional culture on crew non-technical skills.

Power Distance Contrasts. One of the most described differences and sources of issues in a mixed-professional crew was related to Power Distance in military vs. civilian settings, as well as the continued manifestation of this in ex-military pilots.

Ex-military Rank & PD Issues. Pilots described ex-military pilots as clinging to 'perceived' rank ('*if paired with a civilian who does not care about [hierarchy], there could be conflicts inside the cockpit.*' #47), struggling to accept civilian leadership and less cooperative ('*Ex military more like one man show and less teamship*' #82). It was mentioned that these behaviours might be due to military training being leadership-enforcing and military training idealism. The often-harsh language used to describe ex-military pilots (e.g., overbearing, not open to criticism, condescending, etc.) is suggestive of the tension in the cockpit between military- and civilian-trained pilots.

Civilian Low PD = 'A Team of Equals'. Civilian pilots were described as better at teamwork ('*Civilian pilots are usually more a team player.*' #100) and more relaxed/flexible. However, it was also noted that civilian captains are not confident enough ('*Pilots from a civilian background tend to be less confident in assuming a leadership role in-flight.*' #123) and fail to 'take charge'.

Flight Approach Contrast. A difference in approaches to flight was another often-described disparity between pilots trained through the military versus civilian route.

Military 'Getting the Job Done' Attitude. Military-trained pilots were described as 'pushing further' ('*military pilots will accept greater risks*' #89), reverting back to military-style decision making and perhaps having a different opinion of what is a threat. However, pilots also noted that military-trained colleagues are always ready, more resilient and have better performance under pressure ('*The ability of the military trained pilot handle standard and non-standard emergency situations is much higher*' #88).

Civilian 'Safety Above Efficiency'. Civilian-trained pilots were considered to be more cautious ('*Civ mentality is mostly cautious*' #55), falling back on SOPs and making decisions based on revenue and commercial pressure ('*Civilian flight crew are more exposed to the Commercial pressures of Contracts and Clients.*' #13).

Mixed Crews. Pilots also discussed benefits and drawbacks of having a mixed crew on NTS and flight safety.

Benefits. Some of the benefits described related to better recognition of unique aspects of a situation and having a complete picture ('*Both might look at things a different way but will give a more complete picture to the crew.*' #32), as well as bringing more options. Pilots also mentioned that a military-trained co-pilot can help create a level cockpit ('*Someone from a military background may be more assertive, leading to a steep cockpit gradient if he/she is commander, and reverse gradient if copilot*' #83) and having assigned roles helps to minimise differences.

Drawbacks. On the other hand, it was mentioned that military-trained pilots occasionally reduce the efficiency of a crew and can force their opinion ('*Might be an issue of the military person to «force» his/her opinion on the other.*' #12). It was also mentioned that communication can suffer in a mixed crew due to terminological differences ('*Different and unclear terms and abbreviations might block understanding of what is said*' #57) and non-standard communication issues, which can sometimes lead to misunderstandings and conflict.

Ex-military. The final category related to descriptions of military training and in-groups.

Military Training 'Quality Assurance'. Military training was described as superior and more in-depth than civilian training ('*the military pilots training might have been superior*' #103) and that there is a difference in standards between military and civilian training (*'Military pilots has different standards, and sometimes this missmatching can be a problem.'* #93). This also manifested in military-trained pilots being described as more experienced, and superior in all NTS and related behaviours (e.g., multitasking, quick reactions, etc.).

Ex-military In-group. Some pilots also mentioned an ex-military in-group formation (*'Those that served I see in general as my brothers and sisters regardless of nationality.'* #116).

Study 2

Study 2 used the IAT to examine the role of culture in implicit risk perception by measuring helicopter pilots' implicit associations between adverse weather and risk. Pilots were asked to sort photos of excellent/marginal weather and risky/safe words. Firstly, due to the ingrained awareness of riskiness of adverse weather throughout training, it was hypothesised that helicopter pilots will have a stronger implicit association between congruent pairings (IMC/risky and VMC/safe) than incongruent ones (IMC/safe and VMC/risky). Secondly, given the difference in threat perception of military-trained helicopter pilots discussed in the previous studies, it was hypothesized that military-trained pilots will have an overall weaker implicit association between risk and bad weather than civilian-trained pilots. Observing a weaker association would suggest decreased risk perception and, in turn, an increased potential for risk taking.

Methods

Participants. A sample of 109 (4 female, 1 preferred not to say) helicopter pilots was recruited. Participants were from a wide range of countries (e.g., Netherlands, UK, USA, Canada, Spain, Italy). The participants were primarily civilian-trained (n = 80) versus military-trained (n = 29). Participants ranged in their total time in aviation: from <5 years since beginning of training to >35 years of experience. There was also a range in pilots' usual role in the cockpit: 48 pilots most often flew in the role of a commander, 19 as co-pilot, 40 single-pilot, two preferred not to say. Notably, there was a marked difference in years of experience and cockpit role by the training type (military-trained were more experienced and more senior). All pilots worked in a wide variety of operation types (e.g., offshore transport, search and rescue, air ambulance, police, various aerial work).

Materials and Procedure. Participants completed the study on their own device. The IAT was used to measure implicit associations between depiction on VMC (excellent) and IMC (marginal) weather conditions and sets of words meaning safe (*protected, secure, home, reliable, sure*) and risky (*danger, threatened, harm, lethal, hazard*). The words were taken from a similar study by Pauley et al. (2008), described as having been pre-validated as relevant to risk perception. Colour screenshots (n=10) from Microsoft Flight depicting excellent and marginal weather (5 each) were used. All photos were at altitude, taken from various locations in Europe (selected for not having any major geographical features that could pose a risk to flight: mostly flat, no buildings, no tall trees), with no identifying features of where they were taken). For excellent weather conditions, photos with either clear skies or with few clouds were used. For marginal weather conditions, photos were set to be overcast, have 100% cloud coverage and 0m ceiling level. Precipitation level was varied between 1 and 4.

The standard IAT procedure (as described in Carpenter et al., 2019) was followed. The task was divided into 7 blocks (B1 photo sorting practice, B2 word practice, B3 combined practice, B4 combined test, B5 reversed word practice, B6 reversed combined practice, B7 reversed combined test). The order of the blocks (congruent vs. incongruent versions) was counterbalanced. Participants were randomly allocated to one of the versions. Participants sorted the items using letters 'd' and 'k' on their keyboard. An interstimulus presentation interval of 250ms was used. If a participant made a mistake (sorted an item wrong), they were shown a red X in the middle of the screen for 300ms. Before the start of each block, participants saw a screen with a message of what items they were going to sort in the next block (e.g., 'Part 1: Pictures) and a reminder which keys they will use to sort those items. They had to click 'next' when ready to begin. The presentation of items in each block was randomized across participants.

Results

IAT effects. To calculate the IAT effect for each participant, the recommended procedure by Greenwald et al. (2003) was used. Implicit attitudes were assessed using a difference score (*D*), which is calculated as: (mean reaction time congruent – mean reaction time incongruent)/standard deviation of all latencies. It is assumed that a stronger association requires a shorter response time, thus, if a participant's score is below 0: the lower the participant's *D* score, the stronger is the association between congruent pairings. If a participant's *D* score is above 0, they have a stronger association between incongruent parings: the higher the score, the stronger the association. Data from both combined practice and test blocks was used (B3, B4, B6 & B7), with only trials of $RT > 10000$ms being excluded.

The *D* scores ranged from -1.55 to 0.14 ($M = -0.65$, $SD = 0.32$). The mean reaction times to the congruent pairings of IMC/risky and VMC/safe ($M = 1061.77$, $SD = 284.14$) were significantly faster than the incongruent reverse pairings of IMC/safe and VMC/risky ($M = 1621.57$, $SD = 534.36$), $t(108) = -14.224$, $p < 0.001$. Bias-corrected and accelerated bootstrapping with 10000 samples indicated a Cohen's *d* of -1.362 with 95%CI [-1.622; 1.100]. Thus, Hypothesis 1 that pilots will have a stronger implicit association between IMC/risky and VMC/safe than IMC/safe and VMC/risky was supported.

Effect of training background. It was hypothesised that military-trained pilots will have an overall weaker implicit association between risk and bad weather than civilian-trained pilots (Hypothesis 2).

An independent samples *t*-test showed no difference between military-trained ($M = -0.70$, $SD = 0.28$) and civilian-trained ($M = -0.63$, $SD = 0.34$) pilots, $t(107) = -0.934$, $p = 0.352$. Bias-corrected and accelerated bootstrapping with 10000 samples indicated a Cohen's *d* of -.203 with 95%CI [-.628; .224]. This suggests that there are no large differences in implicit risk perception between military- and civilian-trained pilots, however small or even medium size effect of training background might have been missed. Thus, support for hypothesis 2 was not found.

General discussion

The results of Study 1 showed that, similarly to findings of Kaminska et al. (2021), the main difference between civilian- and military-trained pilots can be put down to 'safety vs. efficiency'. Pilots also mentioned that what is perceived to be a threat seems to differ between military-and civilian-trained helicopter pilots. Having a multi-professional crew (military- and civilian-trained pilots together in a cockpit) was seen as having a positive effect on all non-technical skills, with a significantly more positive effect on situation awareness than the other NTS.

In Study 2 helicopter pilots were found to have a stronger implicit association for congruent weather/word pairings than incongruent ones. Recognising the inherent risk involved in flying through IMC suggests that pilots are more cautious about approaching IMC and thus might be more likely to abort or change course if faced with that. Despite previous findings of flight approach differences between military-/civilian-trained helicopter pilots, no large differences in implicit associations between weather and risk were found, with any effect being small or medium. This suggests that the previously observed effect of training background is unlikely to be due to a difference in risk perception. All pilots being able to perceive the associated risks with IMC, however, does not necessarily mean that their behaviour will be the same.

As such, being able to perceive the risk posed by adverse weather (risk perception) is only one part of risk management. Pilots need to be able to manage risks by recognising hazards (such as weather), understanding the risks involved, and making appropriate decisions based on this assessment (Pauley et al., 2008). Risk tolerance and subsequent decision making, thus, are equally important. Interestingly, risk perception and risk tolerance are only slightly related to one another (Hunter, 2002). Both can contribute to engagement in risky behaviour to various extents.

Thus, despite the finding of Study 3 that there was no large difference between military- and civilian-trained pilots in their risk perception, it is hard to predict how pilots would actually behave if they were faced with these weather conditions. Thus, examining risk tolerance and subsequent decision making of pilots is the next logical step. If the previously reported risk-taking behaviour by military-trained pilots is in fact true, it might be due to their higher risk tolerance, rather than lower ability to perceive risk. On the other hand, it is also possible that the reported risk-taking of military-trained pilots and risk aversion of civilian-trained pilots can be attributed to stereotyping of the out-group. As reported in Study 1, military-trained pilots were described as having their own in-group, thus, potentially creating this in-group/out-group view with civilian-trained pilots. Perhaps, there are no quantifiable differences between the two groups in their risk taking, and the previously observed effect is purely down to stereotyping occurring between the two.

The current package of studies provides an original, in-depth look at how helicopter pilots perceive their own and other's professional culture, as well as how it affects their day-to-day work performance. The work reported here builds a foundation for further research in the field of professional culture in aviation, with its' novel findings of flight approach differences between civilian- and military-trained pilots.

Acknowledgements

This work was supported by the Economic and Social Research Council through a collaborative studentship (ES/P000681/1). The research was conducted as part of a doctoral dissertation.

References

Carpenter, T. P., Pogacar, R., Pullig, C., Kouril, M., Aguilar, S., LaBouff, J., ... & Chakroff, A. (2019). Survey-software implicit association tests: A methodological and empirical analysis. *Behavior Research Methods*, *51*(5), 2194-2208. https://doi.org/10.3758/s13428-019-01293-3

Greenwald, A. G., Nosek, B. A., & Banaji, M. R. (2003). Understanding and using the implicit association test: I. An improved scoring algorithm. *Journal of Personality and Social Psychology*, *85*(2), 197-216. https://doi.org/10.1037/0022-3514.85.2.197

Hamlet, O. E. D. (2021). *It's all about the mission: assessing the utilisation of non-technical skills across offshore transport and search and rescue helicopter crews.* [Unpublished doctoral dissertation/master's thesis]. University of Aberdeen.

Harris, M. R., Fein, E. C., & Machin, M. A. (2022). A systematic review of multilevel influenced risk-taking in helicopter and small airplane normal operations. *Frontiers in Public Health*, *10*, 823276. doi: 10.3389/fpubh.2022.823276

Hatfield, J., Fernandes, R., Faunce, G., & Job, R. S. (2008). An implicit non-self-report measure of attitudes to speeding: Development and validation. *Accident Analysis & Prevention*, *40*(2), 616-627. https://doi.org/10.1016/j.aap.2007.08.020

Helmreich, R. L. (2000). Culture and error in space: Implications from analog environments. *Aviation Space and Environmental Medicine*, *71*(9-11), 133-139.

Hunter, D. R. (2002). Risk perception and risk tolerance in aircraft pilots (Report No. DOT/FAA/AM-02/17). Washington DC: Office of Aerospace Medicine Federal Aviation Administration.

Kaminska, A., Irwin, A., Ray, D., & Flin, R. (2021, June). Pilot is a Pilot is a Pilot?: Exploration of Effects of Professional Culture in Helicopter Pilots. In *Congress of the International Ergonomics Association* (pp. 682-690). Springer, Cham.

Marquardt, N., Gades, R., & Robelski, S. (2012). Implicit social cognition and safety culture. *Human Factors and Ergonomics in Manufacturing & Service Industries*, *22*(3), 213-234. https://doi.org/10.1002/hfm.20264

Pauley, K. A., O'Hare, D., Mullen, N. W., & Wiggins, M. (2008). Implicit perceptions of risk and anxiety and pilot involvement in hazardous events. *Human Factors*, *50*(5), 723-733. https://doi.org/10.1518/001872008X312350.

Peksatici, O. (2018). Crew resource management (CRM) and cultural differences among cockpit crew-the case of Turkey. *Journal of Aviation/Aerospace Education & Research*, *27*(2). https://doi.org/ 10.15394/jaaer.2018.1742

Slovic, P. (1987). Perception of risk. *Science, 236*(4799), 280–285. https://doi.org/10.1126/science.3563507

UAS VOICE COMMUNICATIONS INTELLIGIBILITY TESTING

Alex Konkel, Ph.D.
Federal Aviation Administration
Atlantic City International Airport, NJ

Unmanned Aircraft Systems (UAS) operations are increasing rapidly. UAS would like to operate similarly to current aircraft in the National Airspace System (NAS), including communicating with air traffic control and possibly each other. The Federal Aviation Administration (FAA) is evaluating a potential voice communications system that would allow this to see if the system's latency and voice intelligibility are sufficient to meet FAA requirements. This paper describes the results of the first phase of voice intelligibility testing. Participants completed two tests, a Message Completion Test developed by the Human Factors Branch and the Modified Rhyme Test, with the audio being sent through the FAA's voice switch test bed. Participants also completed a short questionnaire to rate the audio for intelligibility and acceptability. We found that intelligibility varied across switches but was acceptable, with no statistical impact of the UAS voice communications system.

Unmanned Aircraft Systems (UAS) operations are rapidly increasing, with nearly 1 million UAS registered with the Federal Aviation Administration (FAA). UAS operations are highly dependent on reliable signals as the aircraft have no pilot on board to control the craft or talk with air traffic control (ATC). For example, if the Command and Control (C2) link is lost or disrupted (called a lost link), the UAS must follow pre-programmed commands instead of being flown by the pilot. A lost link situation can be dangerous as there is an uncontrolled aircraft and ATC is unsure what it will do. If the operator were able to communicate with ATC, they would be able to convey the lost link information and coordinate on how to handle the aircraft.

Besides emergency situations, an infrastructure for normal UAS – ATC communications would be beneficial. For example, transport companies might want to have unmanned aircraft carry cargo. Aircraft of that size would need to fly in controlled airspace and be able to maintain communications. Such a communications system is subject to various requirements such as National Airspace System Requirements Document (NAS-RD) 2013 (Federal Aviation Administration, 2013), which places limits on the voice communication latency between users and specialists (3.3.2.0-5.0 1 through 3). Ensuring that voice communications are clear and understandable is also important; a timely but non-comprehensible message is as bad if not worse than a delayed message. This report describes an assessment of the intelligibility of speech sent through a UAS voice communications system. This effort also provides the opportunity to collect fresh data on intelligibility in the five FAA voice switches themselves. Notably, this is the first phase of testing and as such is using a test bed set up at the William J. Hughes Technical Center (WJHTC). The test bed emulates a UAS voice system that could potentially be in use but no UAS were used during this phase of testing.

The UAS voice test bed is integrated into the FAA's Voice Communications Laboratory voice switch test bed at the WJHTC. It sends audio input through an aviation audio panel, vocoder, and internet protocol (IP) interface before going to a UAS base station. The base station relays the signal via Ethernet cable to an Unmanned Aerial Vehicle (UAV) remote station where it again goes through an IP interface and vocoder before being converted to an ATC VHF radio signal. Given the various processing steps, there is the potential for degradation or other changes to the typical ATC radio signal that could affect how well speech is understood.

Speech intelligibility can be evaluated in a number of ways. For example, Perceptual Objective Listening Quality Analysis (POLQA; POLQA, n.d.) is an algorithm that compares the input and output signals of a digital speech system (such as voice over internet protocol (VoIP)) to predict speech quality. While such algorithms are objective, they are not a direct measure of speech intelligibility. They are models based on subjective ratings of speech intelligibility from previous datasets. For this study we chose to collect direct measures of intelligibility. The primary goals of the study are to address:

- Baseline levels of intelligibility on the five FAA voice switches
- If intelligibility is affected by including the UAS communications system in the loop, including if there are any differences across the five voice switches

Following on previous research, we chose two tests. The first is the Modified Rhyme Test (MRT) (American National Standards Institute, 2009). This test uses sets of six rhyming (e.g., went, sent, bent, dent, tent, rent) or alliterative (e.g., pat, pad, pan, path, pack, pass) words. A single word is presented auditorily to the participant during each trial (e.g., "please select the word pad") and the participant chooses it from the set of six. Thus, the MRT evaluates voice intelligibility by ensuring that listeners can distinguish between similar-sounding words. While the MRT is an established speech intelligibility test, it is limited. The key words are all monosyllabic and intentionally confusable, and they are presented with no context. As such, it may not be representative of speech in ATC situations.

The second test is the Message Completion Test (MCT) used by Friedman-Berg, Allendoerfer, and Deshmukh (2009). This test uses ATC phrases and asks participants to repeat key pieces of information from the phrase. For example, the participant may hear "United 748, turn right heading 270, runway 28, cleared for takeoff" and be asked to report the call sign, turn direction, heading, and runway. Speech in the Message Completion Test is longer and more complicated than the MRT speech but has the benefit of being ATC-relevant. With the complementary features of the two tests, we believe the results will provide a good overall measure of speech intelligibility.

Method

Participants

We recruited 17 participants from the WJHTC community. We asked participants if they have normal, uncorrected hearing but otherwise there were no requirements to participate. Participation was voluntary and uncompensated, and the study was approved by the WJHTC local IRB. The participants consisted of four women and 13 men. Their ages ranged from 23 to 63 with an average of 48.7. Four participants completed the testing in two sessions. Two participants reported having pilot experience, although they did not perform differently than the other participants, and none reported any air traffic control experience. Being employees at the WJHTC, the participants had varying levels of general familiarity with air traffic control, air traffic control phraseology, and the voice switch systems.

Materials

The FAA and AURA Network Systems, Inc. entered a Cooperative Research And Development Agreement (CRADA) to use a proposed UAS communications system developed by AURA. The proposed system, in the long-term, will use ground-based cell stations to enable UAS operators to communicate with their aircraft as well as ATC via the standard push-to-talk (PTT) radio system. A test bed version of the system was installed in the WJHTC Voice Communications laboratory and configured to interface with the various FAA voice switch test beds. The FAA's Voice Communications laboratory houses five voice switches: the ETVS, IVSR, RDVS, STVS, and VSCS. A single set of radio equipment was used with all switches. Participants listened to audio files over a Plantronics headset when audio was injected into the pilot side of the communications system (i.e., they used an ATC-style headset when listening as ATC). When listening as a pilot, participant used a Radioshack headset. The headset had two earcups but was set to mono output to match the single-ear style of the ATC headset. It also had volume control on the cord, which researchers attempted to keep at a single location throughout testing.

The Modified Rhyme Test consists of 50 sets of six words. Each set of six words differs from one another only in their initial or ending phoneme. We downloaded the source audio files from https://www.nist.gov/ctl/pscr/pscr-audio-source-files, which consists of nine different voices saying the entire set. We recreated the MRT in PsychoPy. Participants saw a given word set on each trial with the options appearing simultaneously with the audio file playing over the headset. The audio only played once but the options remained on the screen until the participant made a response. The next trial did not begin until the participant pressed a button to proceed.

We adapted the Message Completion Test used by Friedman-Berg et al. (2009). We expanded the set to 12 sentence frames each from the controller and pilot perspective (24 sentences total). Each sentence had five answer sets for a total of 120 messages. To create multiple voices for the MCT as in the MRT, we used text-to-speech software to create audio files of the sentences. Murf (murf.ai) uses artificial intelligence to generate audio files with different voices and voice characteristics (e.g. 'general' or 'excited'). We selected five voices that were fairly generic (not 'excited', for example) and suitable to stand in as a controller or pilot. While the MRT was straightforward, the MCT required more instruction. Researchers gave a verbal description of the test prior to the first run, and written instructions also appeared at the beginning of each run. Participants then saw an example trial with the prompt and audio from a potential trial as well as the expected answer for that trial. The example allowed the researcher to better describe what the participant might hear and how they should type it in. In particular, participants were encouraged to use shorthand while listening to the audio and then go back to fill in the message (e.g., for the example audio they might begin by typing "den, l8975, 8, f" then go back and expand to the answer of Denver, Lindbergh8975, 8000, foxtrot). Researchers encouraged this system based on participant feedback from preliminary testing to emphasize the listening aspect of the test over trying to hold the message in memory and then typing it out. On each trial, participants saw a prompt that told them what information to enter from the message they were going to hear. The audio played one second after the prompt appeared. The audio only played once but the prompt remained on the screen until the participant pressed the enter key. The next trial did not begin until the participant pressed a button to proceed.

In addition to the accuracy data generated by the voice intelligibility tests, we collected subjective ratings of intelligibility and audio quality via a questionnaire. We based the questionnaire on that used by Friedman-Berg et al. (2009). Their questionnaire consisted of two ratings questions, asking participants to respond on a Likert scale from one to seven as to the intelligibility and acceptability of the audio they heard during a test. There was also an open-ended question for the participants to provide other feedback. We also asked participants to fill out a basic background questionnaire for demographic purposes.

The voice intelligibility tests were administered on a standard PC laptop. The experiments were programmed using the PsychoPy package (version 2022.2.4; https://www.psychopy.org) for Python software (version 3.8; https://www.python.org). The experiment code played the appropriate audio file on each trial and recorded the participant's response. It also administered the questionnaire. The FAA voice switch was configured to allow for continuous audio transmission. The UAS communications system, however, was set to squelch audio after 20 seconds (as is typical to avoid 'stuck mic' situations). Thus for test cycles where the participant was tested with the UAS system integrated and audio was injected into the pilot side (the participant was listening at the ATC station), it was necessary to push-to-talk on at the beginning of each trial and off at the end. The researcher did this to better allow the participant to focus on the test itself. The experiment code displayed screens before and after a trial with reminders to toggle the push-to-talk on or off as appropriate. Testing occurred at the voice switch test bed in the Voice Communications laboratory. This is an open-air area with other laboratories and equipment nearby. Thus there was consistent background noise during testing, typically fan noise from the computer racks and other equipment in the area. Occasionally people would walk by having a conversation, or there was construction noise. The latter two examples were rare, but performance was very likely affected by these extraneous sources.

Procedure

Each participant was tested individually. A complete session of 10 test runs lasted 3-4 hours; some participants chose to complete them in two sessions. Those participants completed five runs in one session and then returned to complete the other five a different time, typically a week later. Researchers set the UAS system configuration prior to the participant arriving, and set the voice switch and computer equipment configuration prior to each run. Each run consisted of the MRT and MCT in that order, and the configuration order (combination of voice switch and station) across runs was set randomly. Participants began by receiving a brief introduction to the study from the researchers and then completed the background questionnaire. The WJHTC IRB approved the study and determined it to be exempt, so no informed consent was necessary. Participants then went through the testing procedure.

Prior to the first time completing each test, the researcher gave a more detailed description of the test to the participant. The researcher also described the push-to-talk system. When it was not necessary to push-to-talk, the participant was allowed to progress themselves through trials at their own pace since they did not have to coordinate with the researcher toggling the switch. To conduct a test, the researcher used a laptop to run the PsychoPy experiment files, which played the appropriate audio files directly into the voice system, and which the participant heard via headphones (one of two sets, as described previously). The participant followed prompts on the laptop to either select the word they heard (for the MRT) or type a response based on what they heard (for the MCT). After each test was finished, the participant completed a questionnaire on the laptop and was offered a short break while the voice switch configuration and headphones were changed as necessary. Participants completed 20 voice intelligibility tests in total, the MRT and MCT 10 times each. Each MRT session lasted five to 10 minutes and each Message Completion Test lasted 10 to 15 minutes. At the end of a session the participant was reminded of their next testing appointment or, if it was their last or only testing session, debriefed and thanked for their participation.

Data Analysis

Our general plan for data analysis was to use a Bayesian approach with generalized linear regression models. The simplest model that statistically fit as well or better than a larger model was chosen as the final model. The model also included a multilevel (also sometimes called hierarchical or random effects) component such that participants could have varying intercepts and varying slopes for runs and trials (i.e., different learning curves). The varying slopes were also removed and tested as part of the model choosing procedure. Main effects were always kept in the final model to allow for inspection even if non-significant.

The primary outcome from the MRT was accuracy on each trial. Due to technical problems (such as a fault with the voice switch) or experimenter error, two runs from two participants were not analyzed, and two trials from two other participants. The analyzed data set consisted of 98.8% of the possible full data set. The covariate predictors were gender, age, whether the participant split the test session or not, run, trial, the voice heard on that trial, and the word set for the trial. The final best-fitting model contained only main effects, with no interactions between predictors, and only a random intercept across participants (no random effects for run or trial). Overall accuracy was 79%, with performance across participants ranging from 71% to 84%. The results are shown in Table 1 with asterisks denoting reliable differences according to the model results. Notably, accuracy differed across the voice switches and the station, but not with UAS system integration.

Table 1.
Accuracy Results from the MRT

Voice Switch	Accuracy	Station	Accuracy	UAS System	Accuracy
VSCS	82%	ATC	83%	In the loop	78%
ETVS	81%				
IVSR	79%	Pilot **	75%	Out of loop	80%
STVS **	79%				
RDVS **	74%				

Note. STVS was statistically different from the VSCS, and RDVS from the STVS.

The dependent measure for the MCT was accuracy at the 'element' level. If a message included a call sign, altitude, and heading, each of these elements were scored and independently marked as correct or incorrect. Scoring was done by two researchers and all discrepancies were resolved before analysis. Due to experimenter error, twelve trials were removed from the analysis. In addition, one participant did not follow instructions in regard to filling out abbreviations in their initial answers, which made scoring difficult. Coupled with very low performance in general, we decided to remove the participant's data as unrepresentative. The analyzed data set consisted of 93.4% of the possible full data set which was 1,584 trials consisting of 4,447 responses at the element level. The covariate predictors were gender, age, whether the participant split the test session or not, run, trial, the voice heard on that trial, the message for that trial, the type of element, the element position, and the total number of elements in the message. The final best-fitting model contained only main effects, with no interactions between predictors, and only a random intercept across participants (no random effects for run or trial). Overall accuracy was 75%, with large performance differences across participants. Ignoring the participant who was excluded from the analysis, participants ranged in accuracy from 55.9% to 93.9%. The results are shown in Table 2 with asterisks denoting reliable differences according to the model results. Notably, accuracy differed across the voice switches but not station or with UAS system integration. There were also differences across materials, such as accuracy varying with message length and with element type (e.g. call sign or speed).

Table 2.
Accuracy Results from the MCT

Voice Switch	Accuracy	Station	Accuracy	UAS System	Accuracy
VSCS	78%	ATC	76%	In the loop	73%
ETVS	73%				
IVSR	76%	Pilot **	74%	Out of loop	78%
STVS	75%				
RDVS **	73%				

Note. STVS was statistically different from the VSCS, with no other reliable switch effects.

Results and Conclusions

The most notable result is that accuracy varied across the five FAA voice switches. Accuracy was highest on the VSCS and lowest on the RDVS, with performance on the other three switches falling in between. The differences, while statistically significant, only covered a range of a few percentage points. Accuracy was notably lower at the pilot station on the MRT but was essentially equivalent to the ATC station on the MCT, suggesting at best a small effect of station on intelligibility. Accuracy was also numerically lower with the UAS communications system in the loop, but this effect did not reach statistical significance for either test perhaps due in part to being a between-subjects comparison. There was also no statistical interaction between voice switch and UAS integration, suggesting that the UAS system works fairly equivalently with each switch. Performance on the switches in general was around 80% for the MRT (although again lower on the RDVS) and 75% for the MCT. For the MRT, this would correspond to 'minimally acceptable intelligibility' according to the FAA Human Factors Standard (HF-STD-001B; Ahlstrom, 2016). Based on the results, we make the following recommendations:

- Intelligibility levels should be verified through other means, such as the on-going objective measurement effort or additional tests.
- Intelligibility levels should be tested in a higher-fidelity environment, given that both the FAA voice switches and UAS communications system used in this study were test bed versions.
- Higher-fidelity testing could also include air traffic controllers and pilots who are more accustomed to the audio characteristics and ATC phraseology used in this test.
- Future users of the Message Completion Test should consider alternative means of administration to reduce the impact of memory and typing ability on performance.
- Further research should look into the potential impact, both objectively and subjectively, of using synthetic voices in intelligibility testing.

Acknowledgements

We appreciate the contract support provided by Eve Perchanok, DSoft Technology, and support from our CRADA partner AURA Network Systems. We also appreciate the technical work done by the Voice Communications Laboratory, ANG-E153. This work is sponsored by the UAS Integration Office, AUS-300, and managed by ANG-E64, the ATC Voice Communications Branch, and ANG-C21, the UAS R&D Portfolio Branch. The U.S. Government assumes no liability for the contents or use of this document. The U.S. Government does not endorse products or manufacturers. Trade or manufacturers' names appear herein solely because they are considered essential to the objective of this report. The findings and conclusions in this report are those of the author(s) and do not necessarily represent the views of the funding agency. This document does not constitute FAA policy.

References

Ahlstrom, V. (2016). Human Factors Design Standard (DOT/FAA/HF-STD-001B). Atlantic City International Airport, NJ: Federal Aviation Administration William J. Hughes Technical Center.

American National Standards Institute (2009). Method for measuring the intelligibility of speech over communication systems. *ANSI/ASA S3.2*. Acoustical Society of America. Retrieved from https://www.tc.faa.gov/its/worldpac/Standards/ansi/ANSI-ASA_S3.2.pdf

Federal Aviation Administration (2013). National Airspace System Requirements Document. Retrieved from https://employees.faa.gov/org/linebusiness/ato/operations/technical_operations/neo/issp/authorization/media/NAS_Requirements_Document_2013.pdf

Friedman-Berg, F., Allendoerfer, K., & Deshmukh, A. (2009). Voice over Internet Protocol: Speech intelligibility assessment (DOT/FAA/TC-TN09/4). Atlatnic City International Airport, NJ; Federal Aviation Administration. Retrieved from https://tc.faa.gov/its/worldpac/techrpt/tctn094.pdf

POLQA (n.d.). *The next-generation mobile voice quality testing standard*. https://polqa.info

DEVELOPMENT OF A PROTOTYPE VERTIPORT HUMAN AUTOMATION TEAMING TOOLBOX FOR URBAN AIR MOBILITY

Paul Krois, Ph.D.
Joseph Block
Paul Cobb
Gano Chatterji, Ph.D.
Cherie Kurian
Crown Consulting, Inc.
Arlington, VA

Peng Wei, Ph.D.
Shulu Chen
George Washington University
Washington, D.C.

The future airspace system is envisioned to include urban air mobility enabled by new types of electric vertical takeoff and landing aircraft for transporting passengers and cargo quickly and safely. Success will depend, in part, on the design and operation of vertiports, that like airports and heliports, will enable these aircraft to transfer passengers and cargo, land, recharge and takeoff. Human factors need to be considered in these designs with humans in the role of vertiport operators. Requirements for arrival, surface, and departure traffic and the interaction of the human operators with increasingly autonomous aircraft and decision support systems have to be determined. A proof-of-concept simulation with a prototype workstation called the Vertiport-Human Automation Teaming Toolbox, employing arrival scheduling automation, highlights human-system interaction considerations. Needs for further research are identified for improving the understanding of human teaming with machine agents for integrated arrival, surface, and departure management.

The Federal Aviation Administration (FAA) in its Concept of Operations (ConOps) for an information-centric NAS (ICN) presents a vision for the National Airspace System (NAS) circa 2035 that includes Urban Air Mobility (UAM) based on a foundation of operations, supporting infrastructure, and integrated safety management (FAA, 2022a).The National Aeronautics and Space Administration (NASA) in its Sky-for-All vision of the NAS circa 2045 foresees highly automated aircraft operating in dense, complex urban airspace (NASA, 2022, 2023). Success of UAM will depend, in part, on the design and operation of vertiports for enabling quick and safe transport of passengers and cargo.

Purpose

The thesis of this paper is that vertiport operations will rely on the human Vertiport Operator (VO) interacting with human-centered automation for acquiring the traffic data, processing the data for decision support, and displaying the information for enabling safe and efficient operations. Depending on their complexity, high density vertiports may share similarities in the design and operation of today's high tempo heliports and airports with multiple takeoff and landing areas and taxiways for surface movements. This paradigm shift drives the need to understand the information requirements of the VO and the interactions between the VO and vertiport automation to manage high volumes of traffic.

The prototype Vertiport-Human Automation Teaming Toolbox (V-HATT) was developed to assess VO information and performance requirements for vertiport design and operations, test assumptions, and evaluate off-nominal scenarios (Crown, 2023). V-HATT has been used to study terminal airspace management (Chen et al., 2023).

Vertiports are seen as the bottleneck in future UAM transportation networks, limiting traffic flow throughput and therefore impacting business outcomes. Past studies include addressing route network design, vertiport operational capacity, and vertiport surface topology (e.g., Zelinski, 2020).

Vertiport Design

A conceptual vertiport automation system from the Northeast Unmanned Aircraft System (UAS) Airspace Integration Research Alliance is shown in Figure 1 (NUAIR, 2021). This figure shows the layout of a vertiport with arrivals on the left (shown in green) with final approach and takeoff (FATO) airspace flowing to touchdown and liftoff areas (TLOFs). On the right is a departure FATO and a missed approach (shown in red). Other vertiport features include parking stands and passenger movement areas.

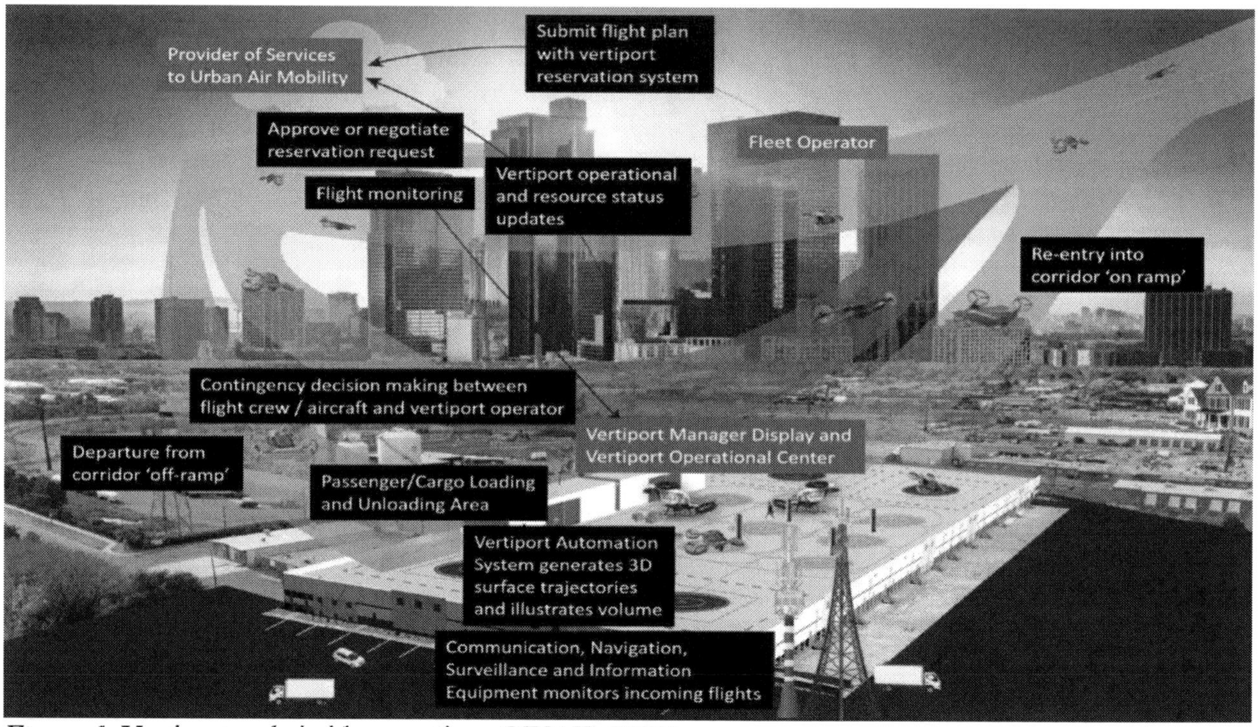

Figure 1. Vertiport and airside operations (NUAIR 2021).

The FAA provides specifications and guidance for vertiport design such as for TLOF and FATO design, VFR approach, and charging/electronic infrastructure (2022b). In addition, industry input on vertiport design included operational integration and safety considerations (Mendonca et al., 2022).

Technical Approach

The technical approach consisted of a series of steps necessary to understand the envelope of VO responsibilities, information requirements, and human-machine interface (HMI) capabilities.

Human-Autonomy Teaming Knowledge Elicitation

Human-Autonomy Teaming (HAT) knowledge elicitation involved identifying operational needs with stakeholders including heliport and airport operators. The operational needs were categorized to provide context and support insights that were leveraged in subsequent steps to derive user stories and functional capability requirements for V-HATT. Key points included that general understanding, expectations, and assumptions about vertiport throughput exceed the limits posed by practical concerns associated with off-nominal scenarios, and the airspace navigation services expected to be provided by Providers of Services to UAM (PSU) may not be sufficient for complexities such as helicopters and electric vertical takeoff and landing (eVTOL) aircraft arriving unannounced, requesting landing and parking without significant advance pre-coordination.

Vertiport Operator User Interface Requirements Development

The operational needs were used to develop a set of user stories. A user story was in the form of "As a [ ], I would like to [ ] so that [ ]" with mission phase assigned to each use story. For example, "As a Vertiport Operator, I would like to specify and configure the type of schedule (scheduled, on-demand, hybrid) of operations so that I may simulate a specific type of vertiport schedule approach." The user stories were construed to be an acceptable starting point for further analysis.

The user stories were used to develop a set of V-HATT functional capabilities broken into Pre-Mission, Mission, and Post-Mission phases. Some user stories were based on a single functional capability and other stories on multiple capabilities. Some stories uncovered additional capabilities. The analysis brought forth assumptions about these capabilities including that there are no locations in the terminal airspace where hovering will be required due to energy management concerns, air traffic control or other air navigation manager may provide en route handoff to the VO, and taxiing capabilities involve use of powered ground taxi and hover taxi but not use of tugs.

The Pre-Mission Phase involved the Surface Resource Management Design with surface objects such as the TLOF, FATO, taxiway, and parking stand. V-HATT capabilities include creating different areas on the vertiport surface, adjusting object position and spacing, and assigning aircraft performance attributes. The Arrival and Departure Airspace Design concerned approach and departure fixes, obstructions, approach decision point, and holding pattern. V-HATT capabilities included visualizing the local airspace, ground environment, and weather data. The Operational Parameter Configuration involved settings such as for weather, types and probabilities of off-nominal situations, and the type of vertiport operating model (scheduled, on-demand, or hybrid approach). Pre-Mission is the simulation design.

The Mission Phase involved actions taken by automation or the human VO. Surface Resource Management capabilities included providing clearance to taxi, introducing delay, designating a resource as unavailable, and assigning aircraft to an arrival TLOF or parking stand. Arrival and Departure Management capabilities included actions for scheduling and sequencing, resolving schedule conflicts, and providing situational awareness such as aircraft position on a terminal airspace map and displaying the density of traffic along a current fix or holding pattern. Mission is the simulation execution.

The Post-Mission Phase consisted of a human-in-the-loop simulation for the proof-of-concept with human factors analysis. V-HATT was designed to collect all HMI interactions, data exchanges, and data from the simulation. Measures included instantaneous subjective workload every two to three minutes using a 5-point rating scale, post-scenario measures using NASA Task Load Index for average and peak workload, and activity measures including counts of data inputs using the keyboard or mouse. Post-Mission is the simulation performance and human factors analysis.

VO-Automation Workflow

The V-HATT prototype demonstrated actions and interactions of the VO and vertiport automation, as shown in Figure 2. Automated scheduling algorithms were developed to calculate a schedule of operations that sufficiently meets the throughput operations as well as pre-specified separation criteria. If there is a conflict in the vertiport, the VO uses the vertiport scheduling service to change the throughput. Throughput is then propagated to the automated arrival scheduling algorithms, which then recalculates a new set of required time of arrivals (RTAs) for all aircraft. The algorithms will then maneuver the aircraft (e.g., speed up, slow down, enter holding pattern) to meet the new set of RTAs.

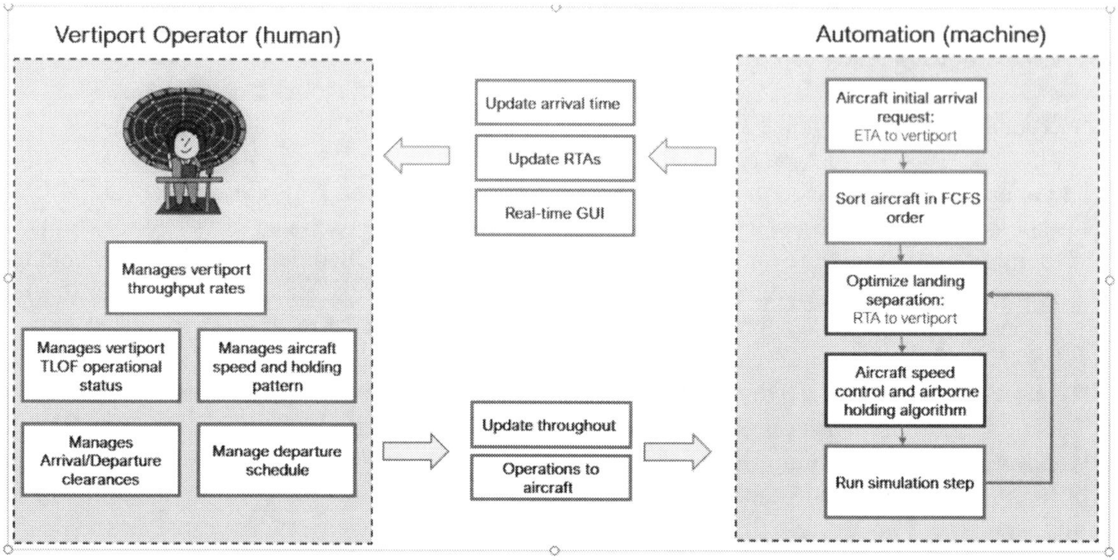

Figure 2. Vertiport VO display design.

On the human operator's side, a set of vertiport display management interfaces was designed for vertiport operators to monitor and direct the aircraft. For example, the VO can change the current vertiport throughput rate, or directly issue maneuvers to the in-air aircraft for safety separation or emergency situations, like leave space for medical helicopter. On the automation side, the centralized system would firstly collect the landing requests from all aircraft, then sort the aircraft in first-come-first-serve order. Then, an optimization method is used to compute the required time of arrival (RTA) to the vertiport, which is based on the current traffic density and vertiport required throughput. After getting the RTA, an aircraft speed control and airborne holding algorithm is used to compute the desired speed and holding time for each aircraft. During the operation, the automation system will keep listening to the vertiport. If the throughput changes, the system could reschedule and issue the new RTAs to aircraft. On the other hand, automation will also keep posting messages like aircraft RTA and actual arrival time to the vertiport, to help human operators make the decision.

Design of HMI Configurable Interfaces

The display design for the vertiport operator is shown in Figure 3. The top-left area is a Surface Situational Awareness Display showing the locations of TLOFs, parking stands, and real-time locations of aircraft. The top-right area is an airside traffic situational awareness display showing arrivals starting, for purposes of this simulation, three miles out from the vertiport. Traffic was shown against a background of geo-located rings marking operational flow areas. Along the bottom area several arrival and departure flow ribbon displays showed the sequencing and spacing of traffic based on scheduler automation. It was assumed the VO would issue the RTA to the pilot and automation would handle holding and reroutes. The display design was evaluated through a walkthrough of an off-nominal scenario involving closure of an arrival TLOF to assess how the VO would interact with the arrival scheduler automation to re-assign arrivals to another TLOF.

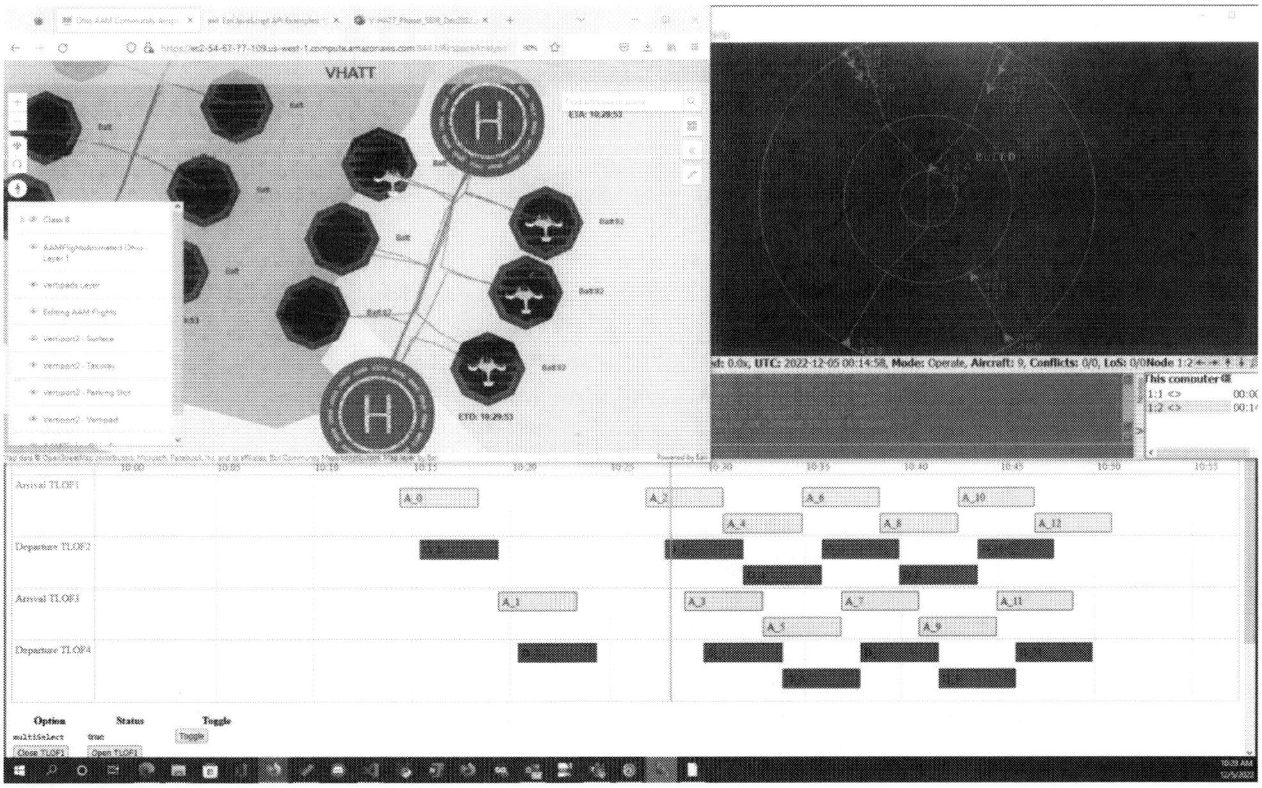

Figure 3. Vertiport VO display design.

Proof-of-Concept Simulation

The proof-of-concept simulation demonstrated the importance of the teaming of human operator actions with arrival traffic. This teaming would extend to actions with a fully capable prototype that includes surface and departure management. A simulation provides valuable understanding and insight into HMI design based on VO performance and workload data. A post-simulation questionnaire provides information about VO concerns about HMI ease of use and areas for improvement.

Discussion

Further Vertiport-Human Automation Teaming Toolbox development will explore the relationship of the VO managing vertiport operations by collaborating with automation acting as a machine teaming agent. The human-machine teaming component of proposed toolbox and the simulation capability will be designed to employ various teaming strategies with differing degrees of automation to examine the performance of the teaming relationship. This entails configuring the vertiport to a specific set of circumstances that impact vertiport throughput in a specific manner. For example, a vertiport operator may need to designate a touchdown and liftoff area for helicopter traffic at certain times so all eVTOLs could be redirected to other touchdown and liftoff areas. The operator could load a 'playbook' operation for this that would automatically reconfigure the vertiport for this operation signaling the scheduling services for a different throughput from normal operations (NASA, 2023). Also, the VO or automation could provide the RTA, holding, or reroutes to the pilot or aircraft.

Additional human factors considerations include that the HMI design should follow the FAA Human Factors Design Standard, HF-STD-001B (2016). VO information requirements related to real-time vertiport surface and airside operations could be supported through use of remote cameras including during low visibility conditions. The complexity of HMI parallels changes in the balance between humans and automation. The design of algorithms and the processes for their use provide a context for potential issues with automation involving "use, misuse, disuse, and abuse" (Parasuraman & Riley, 1997). Issues shaping the use of automation include trust, over-reliance on automation to detect problems, reduced attentiveness to deal with false alarms, and degradation of skills (Smith & Baumann, 2019).

In conclusion, vertiports have a critical role in future visions of UAM. The V-HATT prototype provides a significant tool for designing the HMI for vertiports of different sizes and operational complexity. Further development will integrate arrival, surface, and departure capabilities.

Acknowledgements

This work was completed under NASA Small Business Innovative Research Phase I contract 80NSSC22PB003 titled, "Vertiport Human Automation Teaming Toolbox (V-HATT)." The authors thank the NASA contract monitor, Ms. Savita Verma, for her guidance and enthusiastic support.

References

Chen, S., Wei, P., Krois, P., Block, J., Cobb, P., Chatterji, G., & Kurian, C. (2023). Arrival Management for High-density Vertiport and Terminal Airspace Operations. Air Traffic Control Association Technical Symposium.

Crown Consulting. (2023). Vertiport Human Automation Teaming Toolbox (V-HATT) Phase I Final Report. Alexandria, VA.

Northeast UAS Airspace Integration Research Alliance, Inc. (2021). High-Density Automated Vertiport Concept of Operations. Rome, NY.

Federal Aviation Administration. (2016). Human Factors Design Standard HF-STD-001B. Washington, DC: FAA.

Federal Aviation Administration (2022a). Charting Aviation's Future: Operations in an Info-Centric National Airspace System. Washington, DC: FAA.

Federal Aviation Administration. (2022b). Engineering Brief No 105, Vertiport Design. Washington, DC.

Mendonca, N., Murphy, J., Patterson, J., Alexander, R., Juarez, G., & Harper, C. (2022). Advanced Air Mobility Vertiport Considerations: A List and Overview. AIAA Aviation Forum.

National Aeronautics and Space Administration. (2022). Sky for All Portal. https://nari.arc.nasa.gov/skyforall/

National Aeronautics and Space Administration. (2023). NASA is Creating an Advanced Air Mobility Playbook. https://www.nasa.gov/feature/nasa-is-creating-an-advanced-air-mobility-playbook

Parasuraman, R., & Riley, V. (1997). Humans and automation: Use, misuse, disuse, abuse. *Human Factors*, 39, 230-253.

Smith, P.J. & Baumann, E. (2020). Human-automation teaming: Unintended consequences of automation on user performance. Proceedings of the 2020 Digital Avionics Systems Conference (DASC).

Zelinski, S. (2020). Operational Analysis of Vertiport Surface Topology. Proceedings of the 2020 Digital Avionics Systems Conference (DASC).

CONCEPT OF A GOAL AND PLAN RECOGNITION SYSTEM FOR ADAPTIVE PILOT ASSISTANCE IN HELICOPTER OPERATIONS

Dominik Künzel & Axel Schulte
Institute of Flight Systems
University of the Bundeswehr Munich
Neubiberg, Germany

This article presents a first concept of a pilot assistant system that adapts its support to the current intent of the pilot during Manned-Unmanned-Teaming (MUM-T) helicopter missions. Assistant systems often depend on a pre-defined plan. Due to unpredicted situational changes, the plan can deteriorate, and the system is not able to assist anymore. We envisage a system design that will infer the pilot's intent by using a domain theory approach (plan recognition as planning). To compose a possible plan, a sequence of decisions about the relevant actions is necessary. Thus, we formulate sequential planning problems using Partially Observable Markov Decision Processes (POMDP). POMDP enables us to consider the uncertainty of the mission's course and environment. To perform human-in-the-loop experiments, the next steps are to develop the functions of the designed assistant system and integrate them into our mission and cockpit simulation environment.

Assistant systems are getting more and more attention in an increasing number of domains, in which the human operator has to operate along the system or is to be supported. The application ranges from assistive technologies for elderly people (Hoey et al., 2011), and robotic control (e.g. Pushp et al., 2017), up to complex military flight missions (e.g. Brand & Schulte, 2021; Schwerd & Schulte, 2021). Especially when the human has to perform multiple tasks simultaneously, as it is the case in military helicopter operations, assistance can be a beneficial contribution to efficiency as well as safety. There is a great aspiration, that these systems behave cooperatively and assist the human operator adaptively. For this reason, the capabilities of the system are constantly being expanded to consider the current situation and the operator's state.

Currently, adaptive assistant systems, as being subject to research studies in the field of military domain, often depend on a plan representing discrete tasks during the mission. The plan is usually predefined by the pilot. However, we are looking into highly dynamic missions, where a mission plan can quickly deteriorate due to unpredicted situational changes. If the pilot rapidly adapts to the new mission constraints before re-planning, the assistant system, which still derives its interventions based upon the obsolete plan, is then no longer effective or could even disturb the pilot. For adapted support, the system must therefore be able to understand the pilot's pursued goal and plan.

Related work

Pilot assistant systems

Usually, pilot assistant systems are used to support the pilot with focus on flight deck operations (e.g. Onken & Prévot, 1994, Russwinkel et al., 2020, Estes et al., 2016, Suck & Fortmann, 2016) or the mission management and on-board planning (e.g. Brand & Schulte, 2021; Schwerd & Schulte, 2021) for improving the pilot's performance and safety. Those systems are usually described as cooperative co-pilots with a heterarchical relation to the pilot (Schulte et al., 2016). Hence, they are capable in understanding the situation, environment, system, and the pilot. Based on these analyses the assisting interventions are generated. These can be simple hints or warnings (based on the current situation) given to the pilot, task simplifications, or even the adoption of tasks by automation. The triggers can be vastly different and usually relate to the pilot (cognitive processing) or the task and plan situation. Brand and Schulte (2021) used the mental workload (current and projected) to avoid high workload task situations. Another approach is to use the pilot's awareness of relevant situational information (Schwerd & Schulte, 2021). In contrast, Estes et al. (2016) provide the pilot with relevant information, related to the current flight phase.

There are already various approaches in which the intent of the pilot is considered for assistance. Thus, support of the pilot is possible, even if the intention is not explicitly communicated beforehand. The assistance and intent are usually related to flight plan changes (Strohal & Onken, 1998) or phases and tasks concerning a current flight plan (Estes et al., 2016). However, this assistance mostly refers to flight execution or concrete flight planning in rather structured domains such as instrument flight. Moreover, looking at assistance within missions that are not just about a specific phase or plan, more research needs to be done.

Plan and Goal Recognition

To adapt its support to the pilot's momentary plan, the assistant system shall infer the pursued goal and plan. Goal recognition (also referred to as intent) is defined "[…] as the problem of inferring an agent's intention through its actions and their effects on the environment" (Han & Pereira, 2013). Additionally, plan recognition is the problem of understanding "the set of actions that have been or will be performed […] to reach that goal" (Van-Horenbeke & Peer, 2021). According to the review of Van-Horenbeke and Peer (2021), there are two approaches to infer the goal and plan: plan library based and domain theory based. Since we are following the domain theory approach, the plan library will not be considered further. Defining the recognition problem over a domain theory (also known as plan recognition as planning), off-the-shelf-planning algorithms are used to create candidate plans of the observed agent. For that, the planning problem can be modeled as a Partially Observable Markov Decision Process (POMDP). Ramirez and Geffner (2011) infer the agent's goal over POMDPs in a daily environment like an office, kitchen, and drawers. An approach of using plan recognition over domain theory for assistance is given by Oh et al. (2014). Here, the authors want to infer the user's goal to proactively assist with tedious and time-consuming tasks (e.g., anticipating information needs).

To our knowledge, there is currently no research working in the field of goal and plan recognition systems for adaptive assistance for military helicopter missions, which are highly dynamic due to their dependence on the tactical situation and without a well-structured pre-defined plan. Thus, we want to use the inferred goal and plan to continuously adapt the assistance and support the pilot on mission management and a tactical level.

Assistant system design

Architecture of the adaptive assistance

To assist the pilot in pursuing his plan, the system shall be enabled to define the goal and plan hypotheses based on the mission, environment (tactical situation), and system. Each hypothesis describes one goal and a sequence of actions to reach that goal. Through a sequence of observations, the pursued hypothesis can be inferred. Then, the intervention is generated based on the inferred hypothesis. Figure 1 shows the essential functional modules of the adaptive assistant system. The system is subdivided into four information processing stages: situation detection, assessment, diagnosis, and intervention generation. Each stage is explained below.

Situation detection

The situation detection describes the collection of all available and necessary data for the recognition. Four types of information are considered: system (aircraft capabilities, durations, positions, UAV status), environment (tactical situation, sensor information), mission (goal and task), and pilot's state. The

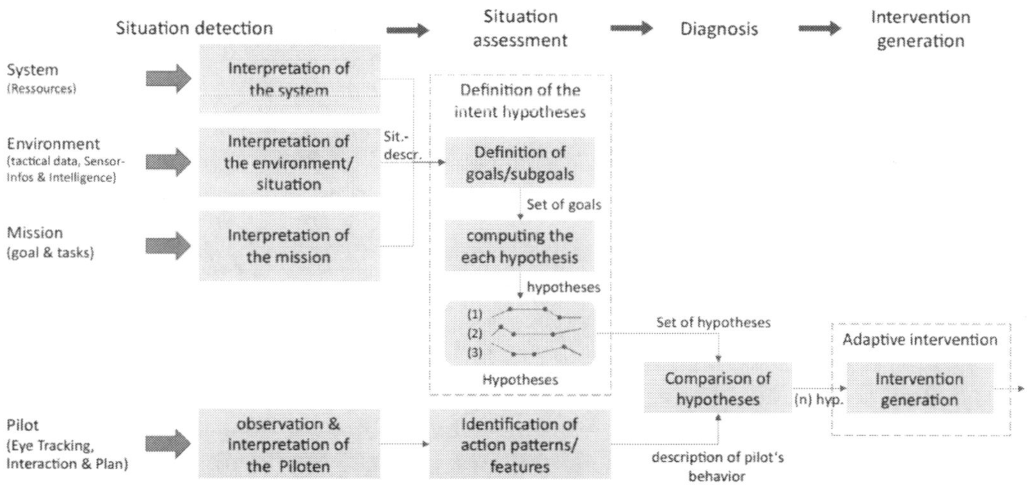

Figure 1. The modular architecture of the assistant system

collected data are used to infer the pilot goals, respectively to generate the hypothesis/candidate plans. For that, the collected data has to be interpreted to create a machine description of the situation.

To infer the pursued plan, the observation of the pilot is essential. The pilot can be observed related to the usage of the available tools (and their resulting actions; which tasks are given to the assets), the manual system control interactions (e.g. pressing buttons), and by use of psycho-physiological sensors (i.e., eye tracking, movements, etc.). In this particular case, we focus mainly on interaction and gazes. Concerning the system control interactions, we provide a cockpit interface with a tactical map and a task scheduling page (cf. Figure 2). Via these interfaces, the pilot can add relevant areas/points and assign tasks related to objects. For locating and classifying the reported vehicle, the pilot has to assign the necessary tasks to his helicopter and/or the UAV(s).

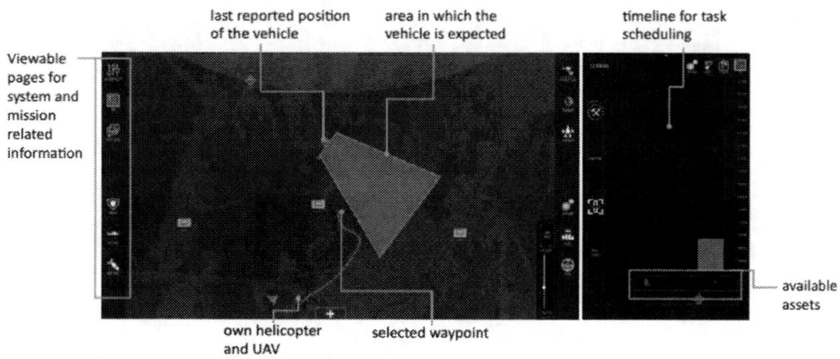

Figure 2. Cockpit interface providing a tactical map (left) and a timeline for task scheduling (right). The map shows the current situation and the selected waypoint (including the trajectory).

Situation assessment and diagnosis

The situation assessment and diagnosis describe the generation of the hypotheses and the inference of the pursued one. For that, the available information about the situation is used to generate the hypotheses based on a POMDP model. For this, it is necessary to specify the states, the initial state, and possible goals. To simplify the problem, we assume, that the pilot starts from an idle state (a not related task concerning an upcoming pursued goal). A hypothesis is determined for each target state.

For generating the intent hypotheses, we use a POMDP, which is a tuple $\langle S, A, T, R, \Omega, O, \gamma \rangle$. Here, the **states S** are defined as $S = S_1 \cup S_2 \cup S_3$, $S_1 = \{(i, e, p) | \, i = selected, e \in E, p \in P_{selected})\}$, $S_2 = \{(i, e, p) | \, i = assigned, e \in E, p \in P_{assigned})\}$, and $S_3 = \{(i, e, p) | \, i = classified, \; e \in E, p \in P_{classified})\}$. I refers to the type of interaction $I = \{selected, assigned, classified\}$, E refers to the related elements and P to the properties. P_n are defined as $P_{selected} = \{done, pending\}$, $P_{assigned} = \{go\ to, locate, engage, assess, classify\}$ and $P_{classified} = \{neutral_{good}, neutral_{destroyed}, foe_{good}, foe_{destroyed}\}$. The **set of actions** $\boldsymbol{A} = \{find\ e\ on\ map, select\ e, assign\ task\ to\ e, classify\ e\}$ is closely related to the states and describe the interaction of the pilot with the system to achieve a certain (mission) state. $\boldsymbol{\Omega}$ is the observation distribution function (describing the probability of observing $o \in O$ from state $s \in S$ after taking the action $a \in A$. The **set of observations O** consist of the perceived interactions with the user interface as well as the gaze obtained by the eye-tracking system. Based on the object with which the pilot interacts and the tactical situation, the **transition probability T** enfolds the most likely transition between the states. Currently, the transition data is based on observations from experiments but can be extended with the help of collecting data during operations. The **reward function R** defines the reward for reaching a state after performing an action. Hence, there is a positive reward for reaching the goal state and a negative reward for each necessary action. We define the **discount factor** $\boldsymbol{\gamma} = 0.9$. As a result, the optimal sequence of actions, called policy π, is computed for each possible goal state (which results from the situations that had occurred) by using the policy iteration algorithm. Finally, the policies are compared to the sequence of observations (interactions and gazes) to infer the pursued hypothesis.

Intervention generation

Decision to intervene. Figure 3 shows the modules of the intervention generation. First, a decision to intervene must be made. The necessary information can be derived from the inferred plan, system state, pilot state in relation to the inferred plan (i.e., what kind of task is the pilot currently performing; what kind of information is available to the pilot), mission, and environment. The following criteria are used to decide, whether an intervention is necessary: *Which actions need to be done (to reach the goal)? Which actions are already done? Are there any constraints to be considered?*

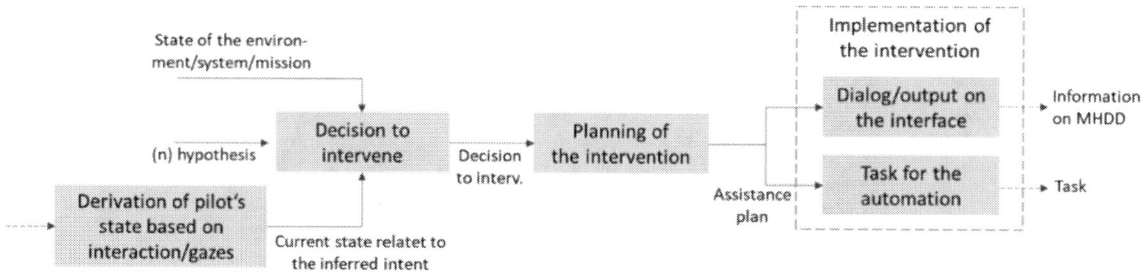

Figure 3. Modular description of the intervention generation.

Planning of the intervention. Based on the decision, the planning of the intervention needs to be performed to obtain the assistance plan. The assistance plan describes the approach for intervening and contains the type and strategy of intervention (e.g., text-based information, tactical overlays, etc.) as well as the time (for multiple interventions also the sequence). Generally, four different types of intervention can be implemented by the assistant system (based on Onken & Schulte, 2010): *No intervention, suggestions for planning, suggestions for planning and preparing the system, assigning and execution of actions/tasks.* As part of our investigation, we also want to consider how communicating the intent affects the transparency of the system.

Description of the design challenge. Figure 4 illustrates a possible situation during a military helicopter reconnaissance mission, which we use to design the intervention of our adaptive pilot assistant system. The overall mission objective is to locate and identify vehicles in a designated area. To fulfill this mission objective, the pilot has to scan the area and needs to find the most southern vehicle and identify it (report the appearance as well as the direction of movement) as quickly as possible. To add urgency, the pilot is presented with a situation update. During the situation update, three simultaneous events at different locations are reported (see Figure 4a). The pilot must therefore decide which event should be dealt with first. Figure 4b presents the pilot following the scenario shown in the middle of the three simultaneously occurring events. Here, the pilot just receives an update on the vehicle position, which is already a few minutes old, and the direction of movement. Based on that, the pilot must locate the vehicle (scan the area where the vehicle is suspected) before the identification of the vehicle can be done.

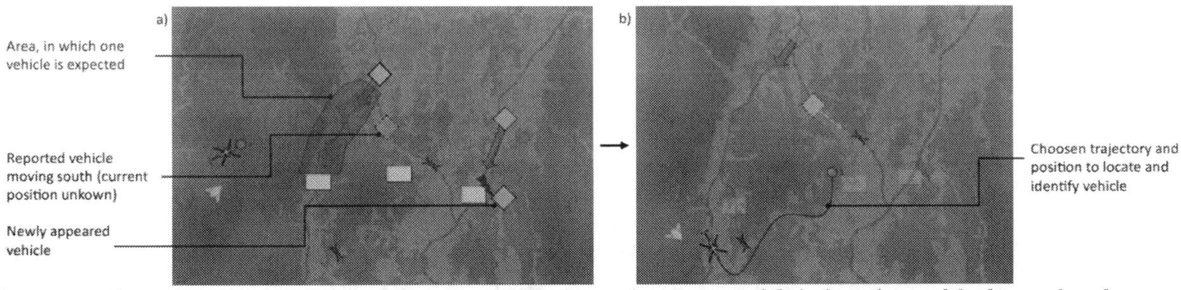

Figure 4. The pilot pursues one of three possible intentions. (a) Initial situation with three simultaneous occurring events. (b) The pilot defines the position to locate the reported vehicle heading to the south. Based on the set position and the tactical situation, the assistant system can infer the intent to locate the approaching vehicle.

Intervention implementation. With the help of the assistance plan, the intervention implementation is the next step. There are two ways in which the interventions are implemented. One way is the output for the pilot via the interfaces, and the other way is to transfer an associated task to the automation. The output via the interface is shown schematically in Figure 5. This is done as an entry in the tactical map or as a text notification on the interface. Figure 5 illustrates an example of the approach route as well as possible positions of the helicopter and UAV to locate the reported vehicle and the considered area of the suspected vehicle.

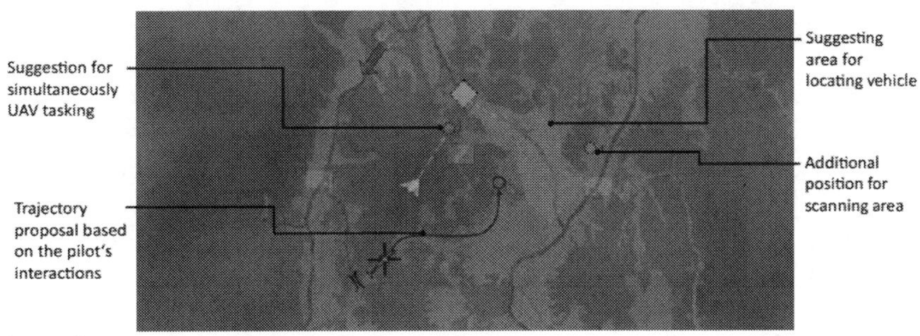

Figure 5. Possible assistance based on the inferred intent.

Conclusion and Future Work

In our contribution, we presented an approach to an assistant system that supports the pilot of a future military helicopter in highly unpredictable missions by determining their plan and providing adapted assistance. We addressed the basic system design, the goal and plan recognition, as well as the assistance behavior. The concept will be integrated into our mission and cockpit simulation environment at the Institute of Flight Systems. We plan to investigate by means of human-in-the-loop experimentation whether our system approach can help to improve the flexibility and thus the applicability of plan-based assistant systems in highly dynamic domains, where pre-planning, replanning, and configuring the system is not easily possible.

References

Brand, Y., & Schulte, A. (2021). Workload-adaptive and task-specific support for cockpit crews: design and evaluation of an adaptive associate system. *Human-Intelligent Systems Integration, 3*(2), 187–199. https://doi.org/10.1007/s42454-020-00018-8

Estes, S. L., Burns, K. J., Helleberg, J. R., Long, Kevin M., Pollack, Matthew E., & Stein, J. L. (2016). Digital copilot: Cognitive assistance for pilots. In *2016 AAAI Fall Symposium Series* (Cognitive Assistance in Government and Public Sector Applications).

Han, T. A., & Pereira, L. M. (2013). State-of-the-art of intention recognition and its use in decision making. *AI Communications, 26*, 237–246. https://doi.org/10.3233/AIC-130559

Hoey, J., Poupart, P., Boutilier, C., & Mihailidis, A. (2011). POMDP models for assistive technology. *Decision Theory Models for Applications in Artificial Intelligence: Concepts and Solutions.* Advance online publication. https://doi.org/10.4018/978-1-60960-165-2.ch013

Oh, J., Meneguzzi, F., & Sycara, K. (2014). Probabilistic Plan Recognition for Proactive Assistant Agents. In *Plan, Activity, and Intent Recognition* (pp. 275–288). Elsevier. https://doi.org/10.1016/b978-0-12-398532-3.00011-7

Onken, R., & Prévot, T. (1994). CASSY - Cockpit Assistant System for IFR Operation. In *ICAS proceedings 1994. 19th Congress of the International Council of the Aeronautical Sciences.*

Onken, R., & Schulte, A. (2010). *System-ergonomic design of cognitive automation: Dual-mode cognitive design of vehicle guidance and control work systems ; [with DVD]. Studies in computational intelligence: Vol. 235.* Springer.

Pushp, S., Bhardwaj, B., & Hazarika, S. M. (2017). Cognitive Decision Making for Navigation Assistance Based on Intent Recognition. In A. Ghosh, R. Pal, & R. Prasath (Eds.), *Mining Intelligence and Knowledge Exploration* (pp. 81–89). Springer International Publishing.

Ramirez, M., & Geffner, H. (2011). Goal recognition over POMDPs: Inferring the intention of a POMDP agent. *IJCAI International Joint Conference on Artificial Intelligence.* Advance online publication. https://doi.org/10.5591/978-1-57735-516-8/IJCAI11-335

Russwinkel, N., Vernaleken, C., & Klaproth, O. W. (2020). Towards Cognitive Assistance and Teaming in Aviation by Inferring Pilot's Mental State. In T. Ahram (Ed.), *Advances in Intelligent Systems and Computing Ser. Intelligent Human Systems Integration 2020* (Vol. 1131, pp. 1021–1027). Springer International Publishing AG. https://doi.org/10.1007/978-3-030-39512-4_155

Schulte, A., Donath, D., & Lange, D. S. (2016). *Design Patterns for Human-Cognitive Agent Teaming: In: Harris, D. (eds) Engineering Psychology and Cognitive Ergonomics. EPCE 2016. Lecture Notes in Computer Science, vol 9736. Springer, Cham.* https://doi.org/10.1007/978-3-319-40030-3_24

Schwerd, S., & Schulte, A. (2021). Operator State Estimation to Enable Adaptive Assistance in Manned-Unmanned-Teaming. *Cognitive Systems Research, 67*, 73–83. https://doi.org/10.1016/j.cogsys.2021.01.002

Strohal, M., & Onken, R. (1998). Intent and error recognition as part of a knowledge-based cockpit assistant. In S. K. Rogers, D. B. Fogel, J. C. Bezdek, & B. Bosacchi (Eds.), *SPIE Proceedings, Applications and Science of Computational Intelligence* (p. 287). SPIE. https://doi.org/10.1117/12.304818

Suck, S., & Fortmann, F. (2016). Aircraft Pilot Intention Recognition for Advanced Cockpit Assistance Systems. In D. D. Schmorrow & C. M. Fidopiastis (Eds.), *Foundations of Augmented Cognition: Neuroergonomics and Operational Neuroscience* (pp. 231–240). Springer International Publishing.

Van-Horenbeke, F. A., & Peer, A. (2021). Activity, Plan, and Goal Recognition: A Review. *Frontiers in Robotics and AI, 8*, 1-18. https://doi.org/10.3389/frobt.2021.643010

ACQUIRING MANUAL FLYING SKILLS IN A VIRTUAL REALITY FLIGHT SIMULATOR

Wietse D. Ledegang
Erik van der Burg
Ivo V. Stuldreher
Mark M.J. Houben
Eric. L. Groen
TNO, Human Performance, The Netherlands
Danny van der Horst
Erik A.M. Starmans
Guido Almekinders
Royal Netherlands Air Force, The Netherlands

In this study, we explored the possibility of objectively assessing the progress in manual flying skills by student pilots using Virtual Reality (VR). Using a VR flight simulator of the Pilatus PC-7 training aircraft, fifteen participants without flying experience practiced basic flight maneuvers based on self-study and without receiving feedback. Relevant flight performance measures were normalized and a learning curve was fitted, representing learning speed and end-level. During some runs an N-back task was included as a secondary task to quantify the participants' cognitive capacity. Interestingly, performance on the N-back was not a good predictor of someone's learning curve. The correlation between performance measures and flight instructor gradings confirmed that, for a limited set of maneuvers, we were able to objectify the students' learning behavior of acquiring a set of manual flying skills in a VR flight simulator. The results of this study show the potential of measuring learning performance in VR.

Aspiring military pilots within the Royal Netherlands Air Force (RNLAF) undergo Elementary Military Pilot Training (in Dutch: Elementaire Militaire Vlieger Opleiding, EMVO) in the Pilatus PC-7 turboprop training aircraft. Because of its lifetime, the PC-7 aircraft will be replaced by a new training capacity in 2026. In addition to a new aircraft and high-fidelity simulation training, Virtual Reality (VR) is identified as a potential training means to accomplish part of the training objectives in the future training syllabus. VR has already been introduced in initial pilot training within the Royal Air Force (RAF), United States Air Force (USAF) and Royal Australian Air Force (RAAF) (Pope, 2019; Air Education and Training Command, 2020; Lewis & Livingston, 2018; Pennington et al., 2019). Ross (2022) concludes that, based on the results of eighteen studies performed between 2018 and 2021, student pilots trained in VR performed at least as well as students trained with traditional means. Furthermore, a combination of VR and traditional flight training can decrease the training time required (Lewis & Livingston, 2018; McCoy-Fisher et al., 2019; Pope, 2019; Sheets & Elmore, 2018; Pennington et al., 2019; Mishler et al., 2022).

In this study, we explored the possibility of using a VR flight simulator to objectively measure the learning performance of student pilots while acquiring manual flying skills and associated visual behavior. In VR, objective performance measures can be recorded, which may be helpful for monitoring the student's progression. These measures may be derived from control inputs, flight performance, and the gaze behavior as recorded by a built-in eyetracker.

Learning theory shows that that during learning the ability to execute tasks evolves from slow and effortful controlled processing to fast and less effortful, or automatic, processing (Tinga et al., 2019; Schneider en Chein, 2003). In this way, learning improves task proficiency while cognitive demands decrease. We therefore hypothesized that an increase in so-called 'cognitive spare capacity' can indicate learning. These considerations led us to define two research questions: 1) Can we determine an overall learning curve for the acquisition of technical flying skills, based on various performance measures obtained across a limited set of basic flight maneuvers, and 2) does the cognitive spare capacity of student pilots correlate with their ability to learn these basic flying skills? Note that the learning of associated visual behavior is described in a separate paper, see Stuldreher et al. (*in press*).

Method

Participants

Fifteen military cadets (12 males and 3 females) of the Royal Military Academy participated in this study (mean age: 23.7 years, ± standard deviation of 2.4 years). They had an average of 3.6 ±7.8 hours of flight experience on powered- and glider aircraft and 2.4 ±7.7 hours on flight simulators. Prior to the experiment, all pilots signed an informed consent, stating that the details of the experiment had been sufficiently explained and that they participated voluntarily. The experiment was conducted with the approval of the institutional ethics committee and was in accordance with the (revised) Helsinki Declaration.

Materials

The simulator environment (see Figure 1), developed by multiSIM BV, consisted of a fixed-base cockpit (front-seat) of a Pilatus PC-7 turboprop trainer aircraft and control devices with control loading. A VARJO-Aero VR headset with built-in eye-tracker was used to present the cockpit and virtual environment near Woensdrecht Air Force Base, The Netherlands, rendering at 90Hz. The flight model characteristics were comparable to the PC-7 aircraft and were validated by EMVO flight instructors.

Procedure

The participants repeatedly practiced three flight maneuvers in a fixed order: Straight-and-Level flight (SAL); Speed Change (SC); and Level Turn (LT). Each maneuver was performed three times during runs of 210 seconds each, followed by a fourth run in which the same maneuver was performed while simultaneously executing an additional memory task as a measure of cognitive spare capacity. Each block of four consecutive runs was repeated three times, spread over two days, thus cumulating to twelve runs per maneuver (i.e., 36 runs overall).

The primary task consisted of manual aircraft control, including the instrument scan and lookout. The secondary task during each fourth run consisted of an auditory 2-back memory task (Kirchner, 1958), which required the participant to continuously update their working memory (i.e., remembering the last two letters of an auditory sequence of continuously changing letters at a fixed 3-seconds interval with a 25% repetition probability). The participants were instructed to respond self-paced by pressing a button on the throttle if the letter heard was identical to the letter two trials back and to withhold a response if the letter was different.

Prior to each block, the participants studied standardized instruction material that included a video of each flight maneuver in which a flight instructor explained the task in the same simulation environment. During the experiment participants received no feedback on their performance.

Figure 1. Setup of the VR simulator during the experiment, with the participant inside the cockpit mock-up and the experimental test leader behind the instructor station.

Measurements

For the analysis of the SAL maneuver the parameters during an entire run were used, while for SC and LT maneuvers the parameters were extracted during phases of deceleration and turning, respectively. To compare errors in performance (i.e., the deviation of a parameter from a target value) across different flight parameters, these measures were normalized in relation to the largest error observed across all participants and combined into an overall performance measure (ranging from 0: worst performance to 1: perfect performance).

Although very simplified, learning curves were estimated by fitting two linear functions representing a 'learning part' and an 'end-level' on the runs without the N-back task. Learning speed is quantified by the number of runs needed to reach end-level performance. It is assumed that the combination of a high end-level and fast learning resembles a high learning performance.

For the N-back task the percentage of errors (i.e., miss or false hit) was calculated. A baseline-corrected number of errors was calculated by subtracting the number of errors that were made in the N-back task prior to the experiment (i.e., without flying).

After the experiment, a flight instructor graded a semi-random selection of 27 runs based on a video replay of each recording. For each of these runs the instructor rated Overall performance, Basic aircraft control, and Multi-tasking according to EMVO grading categories (i.e., Unsatisfactory 1-3, Fair 4-6, and Good 7-9).

Statistical analysis

For each maneuver, an explorative analysis was conducted to examine which performance measures varied significantly over the twelve runs. This was done in separate repeated-measures Analyses of Variance (ANOVAs) with run (1-12) as within-subjects variable. The normalized performance in runs with the N-back task (runs 4, 8, and 12) was compared to the preceding runs without the N-back task (3, 7, and 11, respectively) by means of three separate two-tailed t-tests. A repeated-measures ANOVA was conducted on the overall normalized performance, averaged over the SAL, SC and LT maneuvers together, as function of the different runs for all participants. In all analyses, alpha was set to .05.

Results

The results show that learning to fly SAL is related to the errors in airspeed, roll, altitude and heading as these measures show significant effects as function of run. Learning to execute the SC is related to the errors in altitude, airspeed and heading, while learning of the LT is related to the errors in altitude, roll and side slip. See Table 1 for an overview.

After normalizing the flight performance measures and combining these into an overall mean normalized performance, repeated measures ANOVA shows significant main effects of run for SAL ($F(11, 154) = 17.556, p < .001$), LT ($F(11, 154) = 8.042, p < .001$) and SC ($F(11, 154) = 12.827, p < .001$). This indicates that for all maneuvers the participants were able to improve their performance with more repetitions.

Table 1.
Relevant performance measures and learning curve fit details per flight maneuver.

	Straight-and-Level (SAL)		Speed Change (SC)		Level Turn (LT)	
Performance errors	***p***	***F*(1,11)**	***p***	***F*(1,11)**	***p***	***F*(1,11)**
Airspeed	<.001	7.06	.004	2.68		
Roll	<.001	3.81			.022	2.12
Altitude	<.001	6.25	<.001	8.45	<.001	7.61
Heading	<.001	9.68	<.001	7.42		
Side slip					.042	1.91
Learning curve fit	Mean (Std)	Range	Mean (Std)	Range	Mean (Std)	Range
R^2	.93		.83		.69	
Start level	.56 (.16)	.21-.79	.59 (.21)	.19-.82	.69 (.17)	.30-.87
End level	.80 (.08)	.58-.89	.86 (.09)	.63-.94	.86 (.08)	.67-.94
Time to end level	8.36 (1.78)	5.2-11.0	5.86 (2.89)	1.96-11.0	6.78 (3.25)	2.04-11.0
Learning speed	.69 (.09)	.58-.89	.73 (.11)	.63-.94	.74 (.09)	.67-.94
Learning performance	1.06 (.06)	.94-1.19	1.14 (.06)	.99-1.25	1.14 (.09)	.95-1.30

Even though it is very simplified to fit a linear learning curve on the normalized performance measures, the fits show good results (i.e., R^2 varies from .69 to .93). Participants were able to improve their manual flying skills up to a normalized end-level of .80, .86 and .86 within, on average, 8.36, 5.86 and 6.78 runs for the SAL, SC and LT maneuvers respectively. Due to the normalization procedure, the impression can be given that SAL was the most difficult to learn (i.e., most runs needed to achieve end-level). However, because SAL was quite easy, only small performance improvements could be achieved with each repetition, which took longer to reach end-level. See Table 1 and Figure 2 for more details.

Comparing flight performance between runs with the N-back task (i.e., the fourth run of a session) and their preceding run (i.e., the third run) yielded no significant effect for the SAL and LT maneuver, indicating that flight performance did not improve when the participants performed the additional N-back task. Since flight performance did improve across the first three runs without the N-back task, it appears that the additional N-back task interfered with learning. For the SC maneuver the analysis even yielded a significant drop in performance between run 3 and 4, $t(14) = 2.205$, $p = .045$, and between run 7 and 8, $t(14) = 3.385$, $p = .004$. There was no difference between runs 11 and 12, $t(14) = 0.790$, $p = .443$. During the SC the N-back task thus not only caused 'stagnation' of the learning, but even a performance decline.

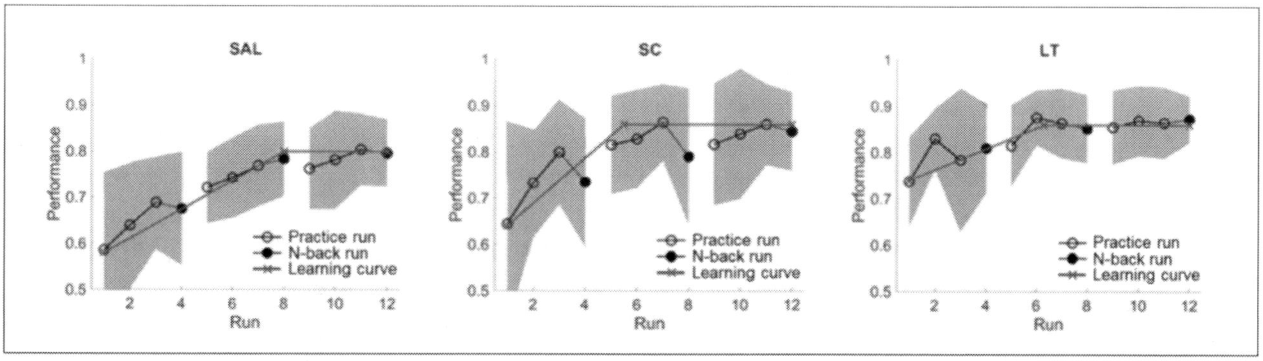

Figure 2. Normalized flight performance as function of run for the Straight-and-Level (SAL), Speed Change (SC) and Level Turn (LT) maneuvers. The filled symbols correspond to the runs with N-back task, while the shading reflects the standard deviation. The optimal fit of a learning curve, in terms of a 'learning part' and 'end-level', is shown with a red line.

The ANOVA on the baseline corrected N-back task with run and flight maneuver as within-subject variables showed a two-way interaction, $F(4, 56) = 2.842$, $p = .032$, as well as main effects of flight maneuver, $F(2, 28) = 22.757$, $p < .001$, and run, $F(2, 28) = 5.599$, $p = .009$. Participants made more errors in SC (7.7%) than in SAL (3.7%) and LT (4.1%) maneuvers, $t(14) = 5.231$, $p < .001$, and $t(14) = 5.309$, $p < .001$, respectively. The two-way interaction was further examined by separate ANOVAs for each flight maneuver. This yielded a significant effect of run for SAL, $F(2, 28) = 5.989$, $p = .007$, but not for LT, $F(2, 28) = 2.370$, $p = .112$, and SC, $F(2, 28) = 2.687$, $p = .086$. For the SAL maneuver, the baseline-corrected N-back performance improved over runs. Separate two-tailed t-tests showed that the baseline-corrected value significantly differed from zero in run 4, $t(14) = 3.129$, $p = .007$, but not for run 8 and 12 (p values $\geq .074$). This indicates that the participants were able to perform the N-back task while flying SAL after a few runs, confirming that SAL allowed some degree of cognitive spare capacity.

Computing correlations between N-back task performance and learning curve metrics yielded only one significant negative correlation ($r = -.52$, $p = .045$) in the LT maneuver, which is driven by the 'end-level' component ($p = .051$). There are no significant correlations between the N-back task performance and learning curve metrics in the other flight maneuvers or when averaging over the three flight maneuvers.

Finally, normalized performance measures showed significant positive correlations with the instructor ratings for Overall performance ($r = .76$, $p < .001$), Basic aircraft control ($r = .70$, $p < .001$) and Multi-tasking ($r = .59$, $p < .005$).

Discussion

Our primary interest was how the progress in performance (i.e., the learning ability) of student pilots could be measured using objective measures extracted from a VR flight simulator. Because comparing flight performance across various flight parameters in different maneuvers is not trivial, our data analysis had a strong exploratory character, in particular when estimating the participants' learning performance. The extent to which the learned skills, on the limited set of maneuvers, transfer to actual flying still needs to be investigated.

The data shows that the performance of the participants during the three maneuvers could be described with a limited set of objective performance measures, which were normalized and combined into an overall performance measure on which a learning curve was fitted. Although fitting a learning curve by a linear function is an over-simplification, we obtained good fit coefficients by fitting two separate linear functions to the 'learning part' and 'end-level'.

The additional N-back task hindered the progress on flight performance, indicating that the additional memory task drew cognitive capacity away from the primary task. Vice versa, the N-back performance dropped below baseline scores when it was performed in combination with the flight task. We did not find a statistical correlation between the N-back performance and the learning curve parameters. Hence, we did not find evidence for our hypothesis that the learning performance is related to the student's cognitive spare capacity as measured by the N-back task.

While only one instructor performed the post-experiment grading for a limited set of recordings of the performed maneuvers, the results showed strong correlations between the normalized performance measures and the instructor gradings.

Conclusions

Using a VR flight simulator, fifteen participants without flying experience practiced basic flight maneuvers based on self-study and without receiving feedback. Learning performance was extracted from relevant flight parameters, which were normalized and combined into an overall measure. This measure was fitted with a learning curve representing learning speed and end-level. The high correlation with instructor gradings suggests that, for the limited set of maneuvers, the student's progress in manual flying skills could objectively be assessed in the VR flight simulator. Addition of the N-back task hampered the students' flight performance and their learning progression, indicating that the additional task absorbed cognitive capacity. However, the performance on the N-back was not a good predictor of someone's learning curve. The results of this study show the potential of measuring learning performance in a VR simulator, whereas the transfer of training from VR to the real aircraft has yet to be explored.

Acknowledgements

This research was supported by the Defence Research and Development Programmes V1917 and V1903, and RNLAF AIR. The authors acknowledge all RNLAF subject matter experts from EMVO, FCL and CMA for their guidance and advise in the preparation of the experiment and interpretation of the results. Furthermore, we thank multiSIM BV for their VR simulator developments and all participants for their enthusiastic contribution.

References

Air Education and Training Command. (2020). About PTN

Lewis, J., & Livingston, J. (2018). Pilot Training Next: Breaking institutional paradigms using student centered multimodal learning. Interservice/Industry Training, Simulation, and Education Conference (I/ITSEC), Orlando, FL.

McCoy-Fisher, C., Mishler, A., Bush, D., Severe-Valsaint, G., Natali, M., & Riner, B. (2019). Student naval aviation extended reality device capability evaluation.

Mishler, A., Severe-Valsaint, G., Natali, M., Seech, T., McCoy-Fisher, C., Cooper, T., & Astwood, R. (2022). Project Avenger Training Effectiveness Evaluation. Naval Air Warfare Center Training Systems DivisionChief of Naval Air Training.

Pennington, E., Hafer, R., Nistler, E., Seech, T., & Tossell, C. (2019, 26/04). Integration of advanced technology in initial flight training. Systems and Information Engineering Design Symposium (SIEDS), University of Virginia, USA.

Pope, T. (2019). A cost-benefit analysis of Pilot Training Next Air University. Wright-Patterson Air Force Base, OH, USA.

Ross, G.A. (2022) Extended Reality Flight Simulators as an Adjunct to Traditional Flight Training Methods, Thesis, Massey University, 2022.

Schneider, W., and Chein, J. (2003) Controlled & automatic processing: behavior, theory, and biological mechanisms, Cognitive Science.

Sheets, T. H., & Elmore, M. P. (2018). Abstract to action: Targeted learning system theory applied to adaptive flight training.

Stuldreher, I.V., Van der Burg, E., Ledegang, W. D., Houben, M.M.J., Groen, E. L., Van der Horst, D., Starmans, E.A.M., Almekinders, G. (in press). Measuring the Lookout Behaviour of Student Pilots in a Virtual Reality Flight Simulator. International Symposium on Aviation Psychology.

Tinga, A.M., de Back, T.T., and Louwerse, M.M. (2019) Non-invasive neurophysiological measures of learning: A meta-analysis, Neuroscience & Biobehavioral reviews.

DEVELOPMENT OF A TEST SCENARIO TO ASSESS KINETOSIS RISK IN MILITARY FLIGHT TRAINING

Simone Vera Löhlein
Institute of Flight Systems,
University of the Bundeswehr Munich
Munich, Germany

Mara Kaufeld
Fraunhofer Institute FKIE
Wachtberg, Germany

Andreas Seefried
German Aerospace Center DLR
Munich, Germany

Axel Schulte
Institute of Flight Systems,
University of the Bundeswehr Munich
Munich, Germany

In order to develop a comprehensive test scenario to detect the risk of early motion induced kinetosis, in aviation more commonly known as airsickness (AS), we conducted a detailed survey on the experiences of jet-aircraft student pilots, flight instructors, and flight physicians to discover AS-related flight maneuvers and other non-motion-induced triggers. Subsequently, we use these findings to design test scenarios that simulate the relevant stimuli in a controlled laboratory setting. Additionally, we propose how gaze tracking can be used to get further information about the pilot's behaviour. For instance, it gives information about head-down times related to secondary tasks and the use of landmarks for orientation in space. Finally, we suggest machine-learning algorithms that combine those parameters with psycho-physiological measures to estimate the AS risk.

Kinetosis, or airsickness (AS) in the context of aviation, is a physiological response of individuals to motion cues. Previous theories attribute inter- and intra-sensory conflicts in the form of contradictory or uncorrelated information of the visual and vestibular system as causes to AS (Griffin, 1991). AS manifests in a symptom pattern of nausea, disorientation, and oculomotor difficulties (Keshavarz & Hecht, 2011). The time course of kinetosis symptoms depends on the intensity of the stimulus and the sensitivity of the person (Cheung, 2008). The symptom pattern not only leads to personal discomfort and reduced performance, but can also result in errors and accidents in safety-critical settings such as aviation (Hixson, Guedry, & Holtzman, 1980). Therefore, the identification of causes and the prediction of pilots' vulnerability is an important concern, especially in military aviation where training slots are expensive and very limited.

To identify causes, initial attempts started to explore AS in the military aviation environment. Hixson et al. (1980) analyzed AS in 79 naval flight officer students. 83% of them reported being airsick on one or more hops, 47% vomited, and 48% admitted to worse in-flight performance due to AS. In a follow-up study they highlighted differences in AS depending on the training phase and associated them with the flight itinerary. Finally, they showed that the extent of AS was significantly lower in the second training stage than in the first (Hixson, Guedry, Lentz, & Holtzman, 1983). Lucertini, Lugli, Casagrande, and Trivelloni (2008) collected AS data over four years from air force student pilots and confirmed an habituation effect over time. However, they also showed that this adaptation is lost during breaks between flights. Of the 63 students, two developed AS in late stages of their education and were classified as incapable of flying. In addition to field studies, attempts have also been made to analyze single features of AS triggers in a laboratory setting. Lawther and Griffin (1987) investigated the influence of magnitude, frequency, and duration of vertical oscillation. This resulted in the important findings that the AS risk increases with acceleration magnitude as well as over time and reaches its maximum at a frequency of about 0.2 Hz. Although these findings provide a basis for understanding the development of AS, they do not yet allow person-specific predictions to be made. This aspect of predictability has been poorly addressed so far. However, an important foundation is provided by the meta-analysis of Kennedy, Dunlap, and Fowlkes (1990), which identified the combination of an AS history questionnaire, physiological variables (measuring motion sickness), and standardized provocative laboratory tests to be the best multivariate predictor. Currently, there is a lack of studies investigating AS considering the variety of triggers and combining it with modern machine learning approaches. As a result, it is not yet possible to predict AS risk in military flight training. Our hypothesis is that by creating an test scenario mimicking a variety of AS-triggers from military flight training, we will be able to record psycho-physiological as well as behavioral measures that will allow us to predict the AS risk with high accuracy. By early AS-risk detection, preventive desensitization training can avert high costs for interruption or discontinuation of training.

An extended survey of military personnel in the field of aviation training for jet pilots was conducted as the basis for the presented experimental concept. In the first section, the findings are analyzed and set into context with literature to define aviation missions as well as side tasks. Then, the feasibility of the implementation of the selected maneuver in a seven degree-of-freedom (DOF) motion simulator will be discussed. Finally, the complete test scenario including previous findings, as well as the intended psycho-physiological and behavior analysis methods, will be presented.

Evaluation of Surveys

The basis for the development of the test scenario was a survey of six students within their practical jet pilot training, two instructor pilots (IP), two flight training directors, and two medical advisors responsible for the anti-airsickness training program (AATP) of the German Air Force. The semi-structured interviews comprised questions regarding the main topics ‘Description of flight training’, ‘Airsickness – triggers and symptoms’, and ‘Approach of the AATP’ and were conducted either in person or by phone. Each interview lasted about one and a half hours and the responses were written down immediately.

In the interviews, several maneuvers were mentioned by both instructors and students in which AS symptoms occurred. These include lazy eights, stalls, and passive flights. Furthermore, spins, cuban eight, unusual attitude recovery (UAR) were mentioned by the students. Passive flight phases, during which the pilot is not in control of the aircraft controls, particularly in combination with head-down times were described as critical triggers. It was highlighted that symptoms often occurred at the end of high-performance maneuver (HPM) sets. In addition to the physical stimuli, other influencing side factors may play an important role in the development of AS. Flight instructors reported the occurrence of AS symptoms in connection with turbulence, heat, improper nutrition, and lack of sleep. Additionally, there were situations in which orientation in space was impaired. These include visibility limitations due to cloud cover as well as loss of orientation during dynamic maneuvers when looking at a procedure checklist. Finally, stress factors, such as family problems, general stress and especially pressure to perform, were mentioned numerous times as triggers for AS. According to their own statements, the psychological stress of the students results from the pressure exerted by the IPs in exam situations as well as self-induced stress from their own performance expectations. Based on years of experience, flight-training instructors categorized students with AS based on the triggers (e.g., first solo flight or first aerobatic lesson) as well as the time course of symptoms. Most problematic are the cases in which AS symptoms recur at nonspecific times and cannot be attributed to a particular trigger. Interestingly, several students reported that the symptoms did not appear until after the maneuvers. According to psychologists at the AATP, this could be due to some sort of internal filter that causes students to self-suppress symptoms during the stressful situation. Although the AS symptoms are already there beforehand, they are only consciously noticed after the situation has ended. This assumption would be well evaluated in the context of our study through a comparison between the measured physiological response and the perceived symptomatology.

Aviation Maneuvers Selection and Simulation Restrictions

For the flight simulation, the DLR Robotic Motion Simulator (Bellmann, 2014) was chosen. It is based on an industrial robot (see Figure 1 left). Due to a linear axis and serial arrangement of joints, it allows for a large translational and rotational workspace, which is particularly advantageous for maneuvers outside of normal flight conditions. The base functionality of motion simulation is provided by the so-called Motion-cueing Algorithm: The accelerations and rates of the simulated aircraft are filtered in a washout filter in order to scale down the large-scale motions to the limited workspace of the simulator. The resulting reference position and orientation is then converted into the seven joint axes trajectories of the robotic motion platform. The motion cueing system has been tuned and tested in initial preliminary trials by experienced Air Force jet pilots and found to be very realistic. However, as the lack of strong G-forces during overhead maneuvers was criticized, these were excluded from the mission planning.

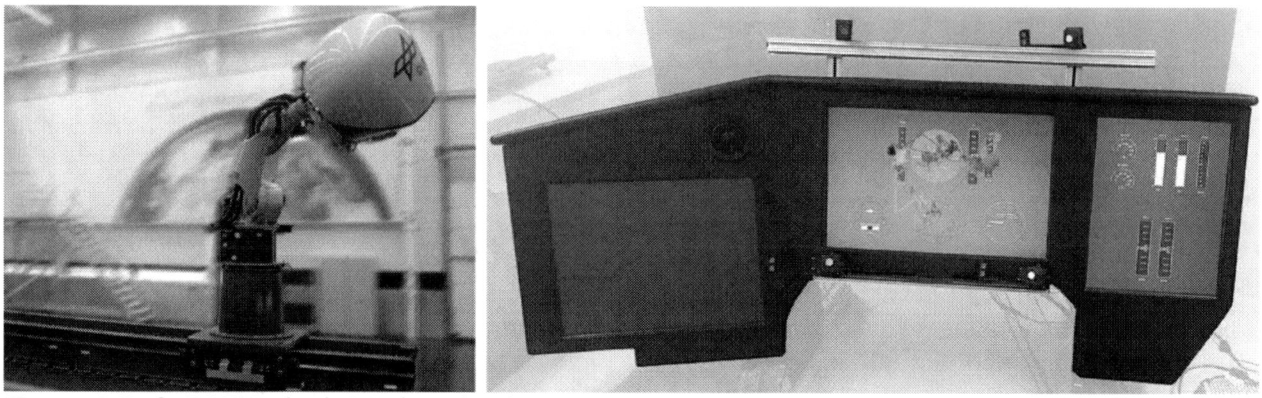

Figure 1. Left. DLR Robotic Motion Simulator (Bellmann, 2014), covering seven degrees of freedom. Right: Exemplary eye-tracking recording for standard cross check (yellow) in a generic cockpit dashboard.

For the selection of the maneuvers, findings from literature on triggers of motion sickness in various environments were considered in addition to the analysis of the surveys. For instance, Lucertini et al. (2008) found that the specific maneuvers experienced as well as the flight duration have an influence on AS. Additionally, their results show that even individuals who are already experienced at flying develop AS symptoms in response to certain aerobatic maneuvers. It has long been known that oscillations in slow frequencies between 0.1 - 0.3 Hz lead to an increased risk of motion sickness (Yen Pik Sang, Billar, Gresty, & Golding, 2005). The maximum risk is observed at a frequency of 0.2 Hz (Lawther & Griffin, 1987). Especially in the context of the development of seasickness, they also emphasize the importance of vertical movements. However, other studies cast doubt on this and attribute a significant role to pitching and rolling movements (Wertheim, Bos, & Bles, 1998). In other words, these movements do not trigger nausea on their own, but they can increase motion sickness if they are combined in a nonlinear manner with vertical heave movements. Based on these indications from the literature as well as the findings from the interviews, the following HPMs were selected for the test scenario: S-turns along the road, steep turns, lazy eights, and chandelles. This selection combines rotations around the vertical, longitudinal, and lateral axes.

Since students as well as IPs reported AS cases during passive flight phases as well as in response to unusual attitude recoveries, these were also included. Literature supports this findings, as false expectations about movement direction can promote discomfort (Bles, Bos, & Graaf, 1998). In the passive flight phase, subjects are moved along a previously recorded motion trajectory which mimics a HPM set. During the recovery task, the subjects are asked to close their eyes while they are moved into an unusual aircraft attitude. They then have to react within a short time to recover the plane into straight level flight. Finally, random components such as wind gusts and vents complete the test scenario design. They increase immersion and add another factor of uncertainty.

Selection of Mentally Demanding Side Tasks

Since both high workload and stress, in combination with head movements, were mentioned as AS triggers in the interviews, secondary tasks were included in the test scenario to elicit such responses. As a positive side effect, tasks based on the standard flight instruction program also served to increase participants' sense of presence. Although research on motion sickness is sparse, literature on visually induced sickness confirms a positive correlation between high workload and symptoms. According to Svensson, Angelborg-Thanderz, Sjoberg, and Olsson (1997), durations and frequencies of eye fixation, and thus head up and down movements, change as a function of workload. In space research, head movements, especially in tilt, were found to be particularly potent in triggering motion sickness (Lackner & Dizio, 2006).

Based on these findings, two types of tasks were selected. The first variety will be referred to as the communication task. Here, the subject listens to an audio stream of radio announcements through headphones. As soon as a certain callsign is mentioned, the subject has to memorize the announced radio frequency and subsequently manually change the setting and check the number on a screen lateral to the standard gaze direction. This task combines general manual, auditory, and visual activities and is intended to provoke an overload through multiple-resource conflicts with the parallel flight control (Wickens, 2008). The second variety of side tasks are in-flight checks. According to the practical training instructions, subjects are supposed to perform so-called operational safety checks during the flight before and after HPMs. These include checking the engine instruments as well as the crew alerting system. In addition, fuel quantity and distribution should be rechecked in straight and level flight. This task requires focusing on the instrument panel within the cockpit, as well as a downward head movement while referencing notes on a knee board. In addition, the visual clearing process, which has to be done before HPM also requires 180° lateral head movements to scan the environment for obstacles and planes.

Test Scenario

In the test scenario, the previously obtained findings are combined to present a comprehensive picture of the AS triggers in the military-jet-training environment (see Figure 2). In preparation for the experiment, the student pilots must familiarize themselves with the maneuver parameters, such as angles and reference speeds. On site, they are then equipped with the physiological sensors (electrodes for dermal activity (EDA), heart rate, etc.). Similar to the official qualification test, the mission is structured according to the scheme "Demonstration-Practice-Able to perform". First, there is a passive flight in which the subjects sit in the simulator while recorded flight trajectories are traversed. In this approximately 20-minute section, they get an impression of the approaching maneuvers on the one hand, and on the other hand, head movements are provoked by secondary tasks, which could trigger AS. This section is followed by a flight training in which the participants have to fly the given (traffic pattern - s- turn along the road - chandelle - lazy eight). The phase lasts a maximum of 45 minutes or until each maneuver has been flown once within the tolerance values. Afterwards the test phase begins, in which the same maneuvers are flown, now in a given order. At the same time, both types of side task, in-flight checks, and the communication task, also have to be completed.

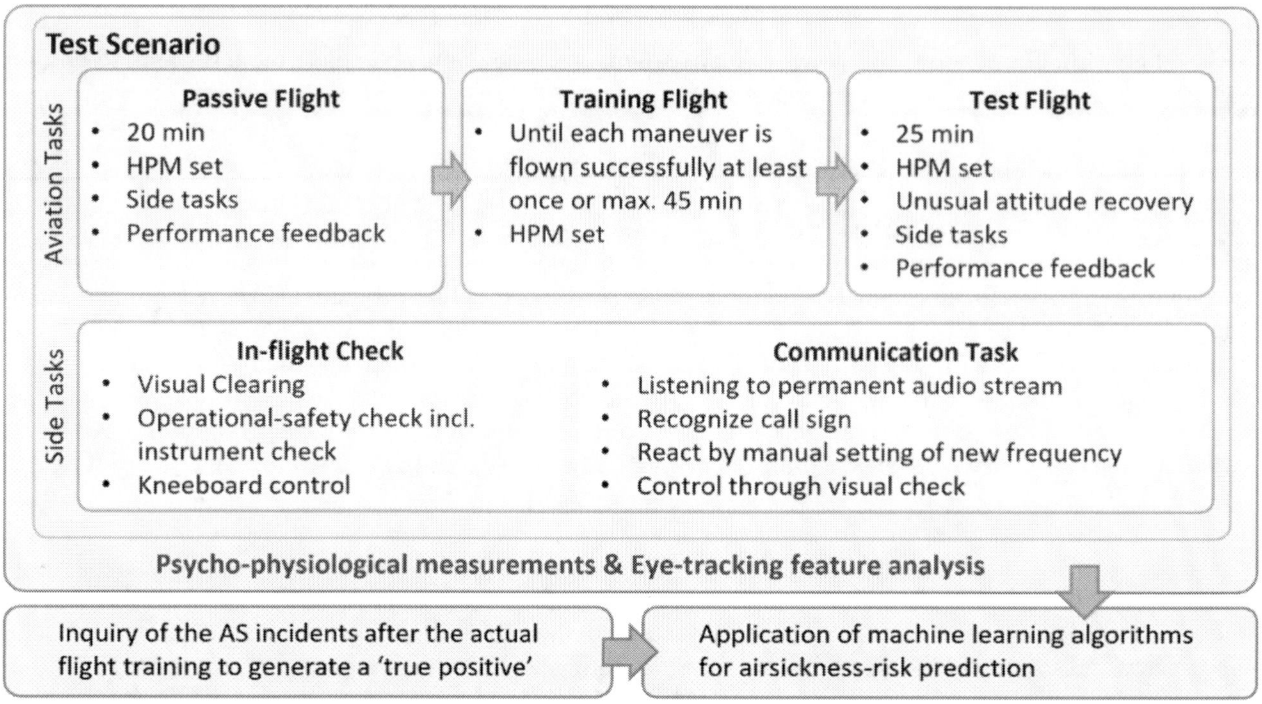

Figure 2. The main pillars of the test scenario are the flight maneuvers, together with the secondary tasks. Psycho-physiological data and eye-tracking parameters are recorded throughout the experiment. After the military flight training, information about actual AS incidences is collected and correlated with the experimental data using machine learning algorithms.

Data Collection and Evaluation

Psycho-physiological data such as EDA and heart rate are recorded throughout the experiment and are further analyzed e.g., by calculating the heart rate variability. Additionally, subjects are asked to complete a Fast Motion Sickness Scale (FMS) questionnaire at regular intervals, which provides information about their current motion sickness symptoms (Keshavarz & Hecht, 2011). Furthermore, the eye-tracking system SmartEye© Pro will be used to provide information about head position, gaze direction, pupil diameter, saccade length, fixation times as well as about areas of interest in which the gaze vector intersects with objects in the environment (see Figure 1 right). Based on this data, behavioral patterns are to be extracted. Since changes of behavior are an indicator for high workload, it can serve as another feature for the machine learning algorithm (Harris, Tole, Stephens, & Ephrath, 1982; Sperandio, 1978). Finally, after the students have finished their first and second practical flight training, the ground truth data of AS incidences will be collected and used to train machine learning algorithms. The selection of algorithms depends on the number of final data sets. The estimated runtime for data collection is two years resulting in approximately 45 datasets from the initial flight training program, which lasts about three month, and an additional 30 data sets from subjects who complete the second practical flight training program, which takes another 15 months.

Conclusion and Future Directions

The presented concept is the first time a model has been presented that entails such a variety of airsickness triggers in military flight simulation. Due to the profound insight on jet flight training provided by the initial interviews, and the realistic movement simulation in the seven-DOF motion simulator, we expect to be able to assign the occurrence of symptoms to specific combinations of stimuli. By recording psycho-physiological as well as behavior-indicating eye movement data, the reaction of the student pilots can be tracked with high precision and in the end may be used to assign certain AS risk classes to these characteristics. By early risk detection, students could undergo preventive desensitization training, which can avert high costs for interruption or discontinuation of training.

References

Bellmann, T. (2014). *Optimierungsbasierte Bahnplanung für interaktive robotische Bewegungssimulatoren* (Doctoral dissertation). Universität der Bundeswehr München.

Bles, W., Bos, J. E., & Graaf, B. (1998). *Motion sickness: only one provocative conflict?* United States: Elsevier Inc.

Cheung, B. (2008). Seasickness: Guidelines for All Operators of Marine Vessels Marine Helicopters and Offshore Oil Installations: Survival at Sea for Mariners, Aviators and Search and Rescue Personnel. Retrieved from https://pdfs.semanticscholar.org/50a9/5ed5ec25c1315dd16e021fe056d825e3831c.pdf

Griffin, M. J. (1991). Physical characteristics of stimuli provoking motion sickness, *175*, 3-1–3-32.

Harris, R. L., Tole, J. R., Stephens, A. T., & Ephrath, A. R. (1982). Visual scanning behavior and pilot workload. *Aviation, Space, and Environmental Medicine*, *53*(11), 1067–1072.

Hixson, W. C., Guedry, F. E., JR, & Holtzman, G. L. (1980). *Airsickness during Naval Flight Officer Training: Advanced Squadron VT86-RIO.*

Hixson, W. C., Guedry, F. E., JR, Lentz, J. M., & Holtzman, G. L. (1983). *Airsickness during naval flight officer training: fleet readiness squadrons*. Retrieved from https://apps.dtic.mil/sti/citations/ada138973

Kennedy, R. S., Dunlap, W. P., & Fowlkes, J. E. (1990). Prediction of motion sickness susceptibility. *G. H. Crampton (Hrsg.), Motion and Space Sickness*, 179–216.

Keshavarz, B., & Hecht, H. (2011). Validating an efficient method to quantify motion sickness. *Human Factors*, *53*(4), 415–426. https://doi.org/10.1177/0018720811403736

Lackner, J. R., & Dizio, P. (2006). Space motion sickness. *Experimental Brain Research*, *175*(3), 377–399. https://doi.org/10.1007/s00221-006-0697-y

Lawther, A., & Griffin, M. J. (1987). Prediction of the incidence of motion sickness from the magnitude, frequency, and duration of vertical oscillation. *The Journal of the Acoustical Society of America*, *82*(3), 957–966. https://doi.org/10.1121/1.395295

Lucertini, M., Lugli, V., Casagrande, M., & Trivelloni, P. (2008). Effects of airsickness in male and female student pilots: Adaptation rates and 4-year outcomes. *Aviation, Space, and Environmental Medicine*, *79*(7), 677–684. https://doi.org/10.3357/asem.2146.2008

Sperandio, J. C. (1978). The regulation of working methods as a function of work-load among air traffic controllers. *ERGONOMICS*, *21*(3), 195–202. https://doi.org/10.1080/00140137808931713

Svensson, E., Angelborg-Thanderz, M., Sjoberg, L., & Olsson, S. (1997). Information complexity--mental workload and performance in combat aircraft. *ERGONOMICS*, *40*(3), 362–380. https://doi.org/10.1080/001401397188206

Wertheim, A. H., Bos, J. E., & Bles, W. (1998). Contributions of roll and pitch to sea sickness. *Brain Research Bulletin*, *47*(5), 517–524. https://doi.org/10.1016/s0361-9230(98)00098-7

Wickens, C. D. (2008). Multiple resources and mental workload. *Human Factors: The Journal of the Human Factors and Ergonomics Society*, *50*(3), 449–455. https://doi.org/10.1518/001872008x288394

Yen Pik Sang, F., Billar, J., Gresty, M. A., & Golding, J. F. (2005). Effect of a novel motion desensitization training regime and controlled breathing on habituation to motion sickness. *Perceptual and Motor Skills*, *101*(1), 29–34. https://doi.org/10.2466/pms.101.1.244–256

EXPERIMENTAL EVALUATION OF CLOUD-BASED SYNCHRONOUS MULT-PILOT MULTI-UAV MISSION PLAN GENERATION IN A MUM-T ENVIRONMENT

Siegfried Maier & Axel Schulte
University of the Bundeswehr Munich
Neubiberg, 85577, Germany

In this study, we compares two approaches for creating mission plans in manned-unmanned teams (MUM-T). In traditional military MUM-T air operations, one human pilot commands multiple UAVs in package-based planning. In situations where multiple teams are working together, it could be helpful to provide all human pilots with equal and simultaneous access to all available UAVs and remove hierarchical boundaries at the team level through a cloud-based approach. The experimental study involved 10 teams of 2 participants each to compare the two approaches. After each mission, participants completed a NASA-TLX questionnaire to assess their workload and rated their perceptions of the two approaches after the second mission. With modifications, cloud-based planning could be a viable option for creating mission plans with multiple pilots and UAVs in MUM-T environments.

MUM-T missions involve teams of manned and unmanned systems, with at least one manned and one unmanned vehicle controlled by human cockpit crew (Strenzke et al., 2011). Humans provide cognitive abilities such as problem solving, mission planning, and decision-making while unmanned vehicles reduce risk to the manned command fighter (Schulte & Donath, 2019). Large-scale combined military air operations (COMAO) use a strict hierarchy and responsibility among participants, including Mission Commanders, Flight Leads, and Wingmen (Fredriksen, 2018) (the latter replaced by unmanned aerial vehicles (UAVs) in MUM-T operations (Figure 1 (left)). We use our task-based guidance approach (Uhrmann & Schulte, 2011) and a mixed-initiative mission planner (Heilemann, Schmitt, & Schulte, 2019) to enable users to delegate tasks to UAVs and create mission plans. Our previous studies focused on one user and multiple UAVs. In (Maier & Schulte, 2021), we proposed an approach that allows users to delegate tasks to other packages and soften the hierarchical structures at the team level, we call this package-based-planning (PBP). Here the users only have direct access to the UAVs in their own dedicated team. A cloud-based approach (cloud-based planning, CBP) (Maier & Schulte, 2022) aims to dissolve the COMAO at the team-level and give all human users direct, equal and simultaneous access to all available UAVs (Figure 1 (right)).

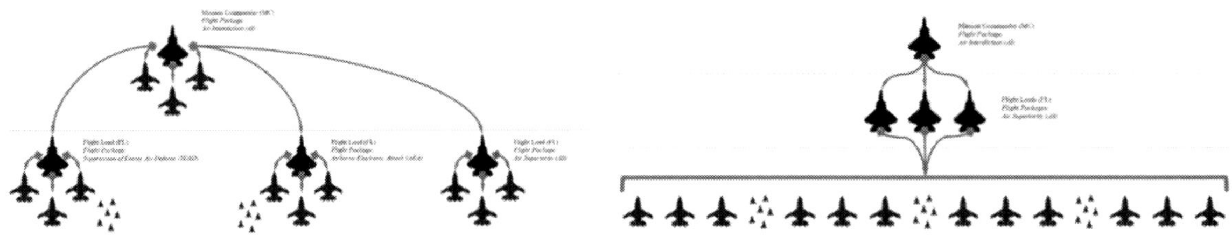

Figure 1. COMAO with a MUM-T PBP approach (left) and a CBP approach (right)

In this contribution we present the results of the experimental evaluation in which we compared PBP with CBP in the form of a usability study. We performed human-in-the-loop experiments using our MUM-T fighter simulator and obtained subjective measures through questionnaires. To evaluate the results we used the NASA Task Load Index (NASA-TLX) (Hart & Staveland, 1988), self-created questionnaires using the Likert Scale (Joshi, Kale, Chandel, & Pal, 2015) and semi-structured interviews.

Experimental Setup

We conducted 10 experiments, with two participants as a team in each experiment. The participants' ages ranged from 19 to 55 years, with three of the 20 participants being female. Two of the participants were civilians, and the others had a military background with ranks ranging from Second Lieutenant to Lieutenant Colonel.

Experimental process

At the start, each team was introduced to the simulator and trained in handling the basics of flying, tasking, generating and modifying mission plans in PBP and CBP, and interacting with objects on the tactical map. After completing their training and addressing any remaining questions, two experimental missions were conducted. After the first mission, the subjects completed a NASA-TLX questionnaire. The second mission was then executed, followed by a second NASA-TLX and additional questions aimed at gathering further information to compare PBP and CBP.

Experimental missions

Three experimental missions were developed to investigate the usability of the two mission planning variants. The primary mission objective was to conduct reconnaissance and engage hostile targets while suppressing enemy air defense systems, with a requirement for at least one engagement by a manned aircraft. Pre-existing surface-to-air missile (SAM) sites (Bolkcom, 2005) were taken into consideration, as well as unknown SAM-sites that would appear when an UAV or fighter came near them. Figure 2 shows all three experimental scenarios. By varying the number of targets and known/unknown SAM sites, the mission difficulty and complexity matched the proficiency of the pilot pairs.

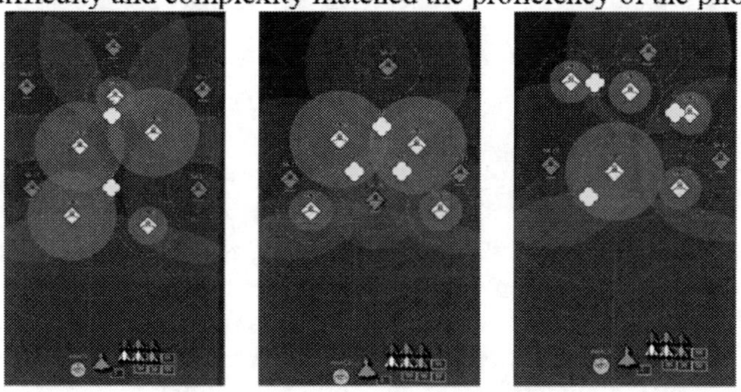

Figure 2. Experimental scenarios

Red diamonds represent known enemy SAM sites, while yellow diamonds represent unknown SAM sites. In their cockpits, the pilots' visibility is limited to known SAM sites. Unknown SAM sites change their status by approaching UAVs or fighter. Yellow clouds represent the mission objectives. The colored objects at the bottom of the image are the available UAVs, while the gray object represents one of the fighter aircraft controlled by the pilots, with the second one, controlled by the second pilot, in the same position at the mission start.

Experimental evaluation

NASA-TLX questionnaire

Before evaluating the data obtained through the NASA-TLX questionnaire, it is necessary to check for normal distribution. This can be done using histograms and the Shapiro-Wilk Test (Shapiro &

Wilk, 1965). By examining the histograms and also by performing the Shapiro-Wilk Test, it has been determined that the assumption of normal distribution for the data must be rejected. Therefore, the median and median absolute deviation (MAD) were calculated and are shown in Table 1. Figure 3 shows the NASA-TLX results for each participant. In Figure 4 the average unweighted demands for each dimension of the NASA-TLX is shown. Participants 1 – 6 and 11 – 14 used PBP in their first mission and participants 7 – 10 and 15 – 20 CBP.

Table 1.
Median and MAD of the NASA-TLX Score for PBP and CBP.

Planning Variant	Median	MAD
PBP	50	7
CBP	46	6

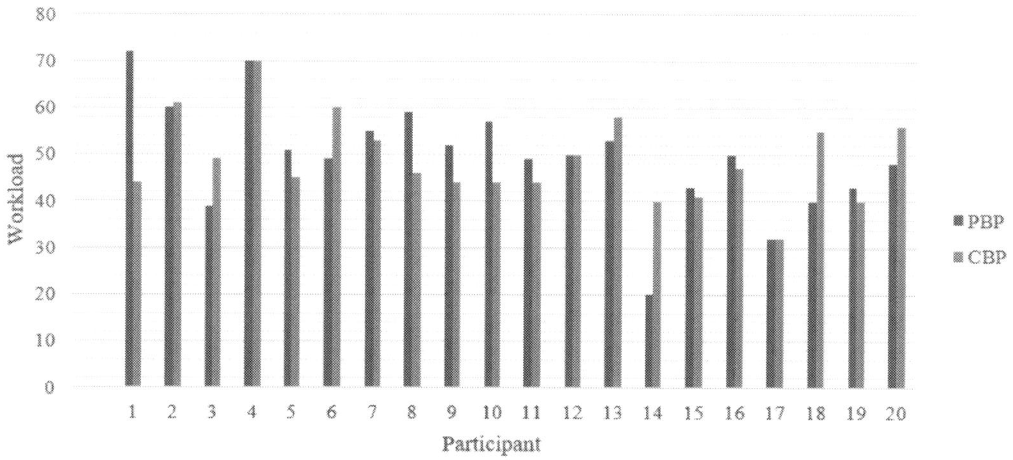

Figure 3. Task Loads in PBP and CBP for each participant

The medians suggest that PBP was slightly more demanding than CBP overall. Figure 3 shows that 6 participants found PBP more demanding and a further 6 perceived CBP as more demanding. 8 participants found both variants almost equally demanding. When PBP was used in the first mission, 4 out of 10 participants found CBP more demanding. However, when CBP was used in the first mission, only 2 out of 10 participants found CBP more demanding than PBP.

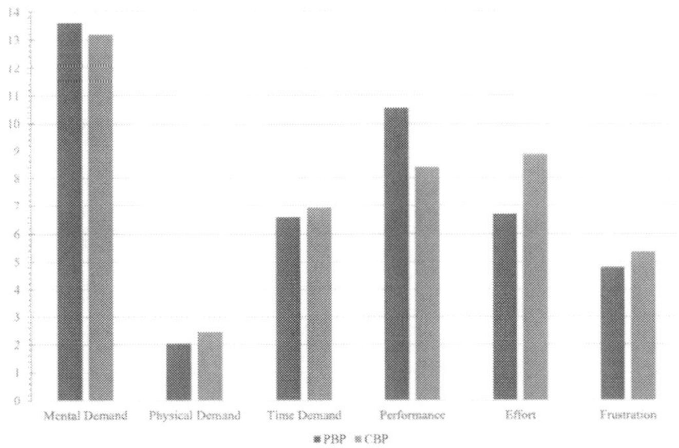

Figure 4. Average unweighted demands

Figure 4 does not allow a clear statement to be made as to which variants the participants found more demanding. The only exception was that participants perceived their performance to be better in CBP. It could be argued that the difference in performance and effort cancelled each other out in favor of PBP. However, the average weight for PBP (performance: 3.7 and effort: 2.25) and CBP (performance: 3.75 and effort: 2.7) is quite similar, with the weight for performance being higher than for effort.

Likert Scale evaluation

Likert scales measure ordinal data and visualize values as categories. Therefore, it is not appropriate to use average values for data rating. Here we will show the graphical representation of the data. There were a total of 12 questions, with 5 questions specifically focusing on conflicts that can arise when using CBP (Maier & Schulte, 2022). The remaining 7 questions were specifically aimed at CBP in general. All questions had to be answered on scale from 1 = strongly disagree to 5 = strongly agree. Figure 5 shows the Likert-Scales for conflict resolution questions, and Figure 6 displays them for specific CBP questions.

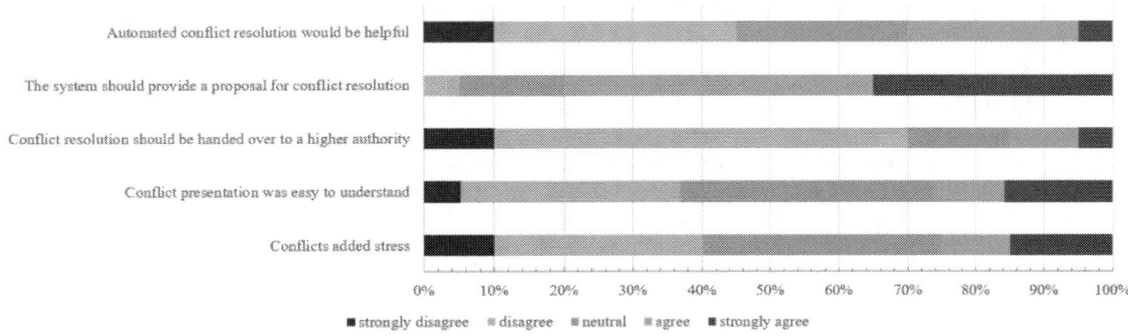

Figure 5. Likert-Scales for conflict resolution questions

Figure 5 shows that for some of the participants, automatic conflict resolution would have been helpful, while the other half see no benefit in it. This could be because the participants who disagree with this point have had little to no conflicts. However, almost all participants agree that the system should offer suggestions for conflict resolution. It also shows that the participants want to handle conflict resolution themselves and that this should not be left to a higher level of the hierarchy. Most participants were reasonably satisfied with the conflict representation. Conflicts that arose did not lead to a noticeable increase in stress for a large proportion of the participants.

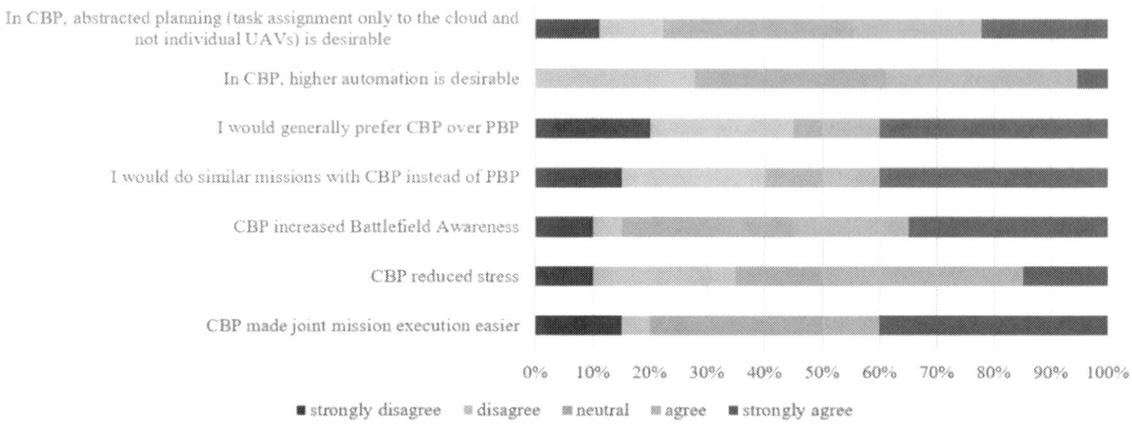

Figure 6. Likert-Scales for CBP specific questions

Figure 6 does not provide a clear conclusion as to whether CBP is preferred or improves mission performance, as there are similar levels of agreement and disagreement on all aspects except for a slightly higher level of agreement that CBP increases Battlefield Awareness.

Additional remarks

In order to gather further insights on the usability of CBP, an open-ended feedback component was incorporated into the survey. This allowed participants to provide additional qualitative data and remarks after each section of questions, thus providing a more comprehensive understanding of their attitudes and perceptions.

How would you adjust the conflict visualization? The majority of participants provided feedback to improve conflict visualization by utilizing stronger colors and effects to make it more salient. Several participants also suggested the use of additional acoustic signals to alert them when conflicts arise. A commonly noted issue was that multiple conflicts were presented simultaneously, leading to difficulties in managing and resolving them. This feedback provides valuable insights on how to improve the visual and auditory design of the system to enhance its usability.

How would you present an automated conflict representation? The majority of participants provided feedback that the system should have the capability of proposing a solution for conflict resolution, with the option for human pilots to accept or decline the proposal. This feedback highlights the participants' preference for a system that can assist in conflict resolution, while still having the decision-making authority.

General remarks regarding CBP vs. PBP. These remarks reflect the results of Figure 6. Participants' subjective evaluations were conflicting. Some found CBP to improve situational awareness by facilitating a more comprehensive view of tasks, while others found its cognitive demands to be excessive, requiring more mental effort and attentional resources.

Conclusion and Outlook

We conducted experiments to compare two methods of mission planning in MUM-T fighter operations: PBP and CBP. In PBP, pilots had access only to their own dedicated UAVs, while in CBP, both pilots had equal access to all available UAVs. After training, each team completed two missions, one with PBP and one with CBP. NASA-TLX and Likert-Scale questionnaires were administered after each mission, with additional free-response questions after the second mission. The results indicate that CBP was slightly less demanding, but further analysis shows that PBP was easier to use. In principle, however, it is not possible to make a clear statement about which variant was preferred by the participants, since the feedback was at both ends of the Likert scales. The data in Figure 6 illustrates the conflicting opinions.

The results show that it might be reasonable to investigate CBP further and also what a tasking interface might look like in this type of mission plan generation. We will improve the current implementation and also make the plan generation process more abstract. In addition, we will add mixed-initiative planning capabilities to the agent that assists pilots in plan generation.

References

Bolkcom, C. (2005). Military Suppression of Enemy Air Defenses (SEAD): Assessing Future Needs. *Congressional Research Service*. Retrieved from https://apps.dtic.mil/sti/pdfs/ADA445462.pdf

Fredriksen, P. K. (2018). Interaction in Aerial Warfare: The Role of the Mission Commander in Composite Air Operations (COMAO). In *Interaction_'Samhandling' Under Risk: A Step Ahead of the Unforeseen* (pp. 481–500). Cappelen Damm Akademisk/NOASP. https://doi.org/10.23865/noasp.36.ch26

Hart, S. G., & Staveland, L. E. (1988). Development of NASA-TLX (Task Load Index): Results of Empirical and Theoretical Research. *Advances in Psychology*, *52*, 139–183. https://doi.org/10.1016/S0166-4115(08)62386-9

Heilemann, F., Schmitt, F., & Schulte, A. [A.] (2019). Mixed-Initiative Mission Planning of Multiple UCAVs from Aboard a Single Seat Fighter Aircraft.

Joshi, A., Kale, S., Chandel, S., & Pal, D. (2015). Likert Scale: Explored and Explained. *British Journal of Applied Science & Technology*, *7*(4), 396–403. https://doi.org/10.9734/BJAST/2015/14975

Maier, S., & Schulte, A. [Axel] (2021). Concept for Cross-platform Delegation of Heterogeneous UAVs in a MUM-T Environment. In (pp. 3–9). Springer, Cham. https://doi.org/10.1007/978-3-030-79997-7_1

Maier, S., & Schulte, A. [Axel] (2022). A Cloud-based approach for synchronous multi-pilot multi-UAV mission plan generation in a MUM-T environment. In *AIAA SCITECH 2022 Forum*. Reston, Virginia: American Institute of Aeronautics and Astronautics. https://doi.org/10.2514/6.2022-2345

Schulte, A. [A.], & Donath, D. (2019). Cognitive engineering approach to human-autonomy teaming (HAT). *20th International Symposium on Aviation Psychology*. Retrieved from https://corescholar.libraries.wright.edu/cgi/viewcontent.cgi?article=1072&context=isap_2019

Shapiro, S. S., & Wilk, M. B. (1965). An Analysis of Variance Test for Normality (Complete Samples). *Biometrika*, *52*(3/4), 591. https://doi.org/10.2307/2333709

Strenzke, R., Uhrmann, J., Benzler, A., Maiwald, F., Rauschert, A., & Schulte A. (2011). Managing Cockpit Crew Excess Task Load in Military Manned-Unmanned Teaming Missions by Dual-Mode Cognitive Automation Approaches. *AIAA Guidance, Navigation, and Control Conference*, 6237–6260.

Uhrmann, J., & Schulte, A. [A.] (Eds.) (2011). *Task-based Guidance of Multiple UAV Using Cognitive Automation: The Third International Conference on Advanced Cognitive Technologies and Applications : September 25-30, 2011, Rome, Italy*. Wilmington, DE, USA: IARIA.

INTERVIEW OF PILOTS' INFLIGHT LOSS OF CONTROL EXPERIENCES

Neelakshi Majumdar
Purdue University
West Lafayette, IN

Karen Marais
Purdue University
West Lafayette, IN

Inflight loss of control (LOC-I) continues to be a significant cause of General Aviation fixed-wing aircraft accidents. Nearly 20% of fixed-wing aircraft accidents in the last two decades involved LOC-I, and approximately 45% of LOC-I accidents are fatal. Previous studies suggest that the leading factors in aviation accidents are human factors-related. One approach to better understand the causes of LOC-I accidents is to analyze accidents using sources such as the National Transportation Safety Board (NTSB) database. However, the NTSB General Aviation accident reports include limited detail on human factors, specifically pilot actions and conditions. We interviewed pilots about their LOC-I experiences and the training in practice to provide a deeper understanding of LOC-I causation. This paper discusses the initial findings from the interview study and the causes of the LOC-I experiences. The findings from the study may help improve training methods and operating procedures for General Aviation pilots.

The International Civil Aviation Organization (ICAO) defines General Aviation (GA) as a category of aircraft operations, exclusive of all military operations, scheduled air services, and non-scheduled air transport operations for remuneration or hire (ICAO, 2009). GA accidents comprise approximately 94% of all aviation accidents in the United States annually (NTSB, 2023). Inflight loss of control (LOC-I) continues to be a significant cause of GA fixed-wing aircraft accidents each year. Loss of control is "a hazardous condition that involves an unintended departure of an aircraft from controlled flight regime" (FAA, 2019). Nearly 20% of fixed-wing GA accidents in the last two decades in the United States (U.S.) involved LOC-I (NTSB, 2023). In 2020, 19% of total fixed-wing GA accidents involved LOC-I; for fatal accidents, this percentage increases to 52%. 49.2% of all LOC-I accidents in 2010–2020 were fatal, with an average of 163 fatalities per year (NTSB, 2023).

Most LOC-I incidents involve some kind of pilot error (Belcastro et al., 2014). In 2019, 65% of LOC-I accidents and 51% of fatal LOC-I accidents were pilot-related (NTSB, 2023). Two of the most cited pilot-related triggers in 2009–2019 LOC-I accidents were improper inflight planning (12.91%) and improper action performance (7.08%) (Majumdar et al., 2021). These pilot-related NTSB codes tend to be broad and vague. The accident reports do not explain specifically what lacked in inflight planning or pilot actions. One approach to better understand the causes of LOC-I accidents is to analyze accidents using historical data such as the National Transportation Safety Board (NTSB) database. Unfortunately, the NTSB General Aviation accident reports include limited detail on human factor-related causes, specifically pilot actions and conditions. GA accident reports, in particular, include limited detail on human factor (HF) related causes compared to other causes such as mechanical faults and weather.

Another way to identify LOC-I accident causation is by conducting surveys and interviewing pilots to reveal insights into aviation risk. Most previous studies focused on military pilots to understand pilot fatigue and spatial disorientation (e.g., Lewkowicz & Biernacki, 2020 and Dawson et al., 2017). General Aviation related studies investigated pilots' experiences with hypoxia, pilots' flight planning, and decision-making processes (Holt et al., 2019; Keller et al., 2019; and Psyllou et al., 2017). To the best of our knowledge, no published survey or interview studies to date have investigated pilots' LOC-I experiences and related human factors. Understanding specific pilot actions and conditions may help better focus GA training methods to prevent LOC-I accidents. To investigate how pilot actions and other human factors lead to LOC-I, we conducted a two-fold study by (1) surveying pilots who had experienced or prevented an inadvertent LOC-I; and (2) interviewing pilots and flight instructors about their LOC-I experiences and their perspectives about LOC-I training.

Building on the survey findings, we interviewed pilots to delve deeper into human factors issues and the training in practice (Majumdar & Marais, 2022). In this paper, we discuss the initial findings from the interview responses and the causes of the LOC-I experiences as shared by pilots.

Research Instrument and Procedures

We conducted virtual semi-structured interviews with pilots, asking them about their perspectives on LOC-I, their LOC-I experiences, if any, and the training they received or provide as an instructor to prevent LOC-I. Participants were eligible to participate in the study if they were at least 18 years old, and had either experienced an LOC-I or were certified flight instructors. We asked a total of 58 standard questions, including five demographic questions. Each interview lasted about an hour.

After the study was approved by Purdue's Institutional Review Board, we contacted sixteen pilots. Out of the sixteen pilots, nine chose to participate in the study. We shared the interview study consent form with participants before scheduling an interview via Zoom. We recorded the audio in the calls with the participant's verbal consent to participate in the study.

Data Analysis

We first transcribed the interview recordings to text and then destroyed the interview recordings. We used NVivo12, a qualitative data analysis software, to analyze the responses.

Demographic Data

Of the nine pilots we interviewed, there were four airline transport pilots (ATP), four commercial pilots, and one private pilot. Seven pilots were certified flight instructors. Pilots' ages ranged from 44 to 73 years. There were seven male pilots and two female pilots. Pilots had multiple ratings and endorsements. All pilots had single-engine land, instrument, and complex rating. Pilots had flying experience ranging from 603 to 10,200 hours and 20 to 48 years. Eight pilots had flown for more than 1,500 hours.

A Summary of the LOC-I Experiences

Six pilots had experienced or prevented an inadvertent LOC-I. One pilot shared two of their incidents. At the time of their LOC-I incident, three pilots had a private certificate, two were commercial pilots, one was an ATP, and another was a student pilot. Two of the seven incidents resulted from an intentional attempted maneuver (a stall maneuver going wrong). Table 1 shows the hours and years of flying experience, flying frequency, pilot certificate, aircraft make and model, and who the pilots were flying with at the time of their LOC-I.

Table 1. *Details of LOC-I events. The pilot number corresponds to the pilots who shared their LOC-I related incidents. Pilots 4, 6, and 7 did not have LOC-I experiences.*

Pilot No.	Flying hours	Flying years	Flying frequency (average days a month)	Grade of pilot certificate	Aircraft make and model	Flying solo or with someone
Pilot 1	2000	25	6	ATP	Cessna 175	Solo
Pilot 2	700	25	2	Private	Cessna 172	Solo
Pilot 3	30	1	20	Student	Cessna 172	Solo
Pilot 5	60	1.5	2	Private	Cessna 172	Two passengers
	750	24	1	Commercial	Cessna R182	One passenger
Pilot 8	1000	10	25	Commercial	Piper Cherokee	Student (who was acting as a passenger)
Pilot 9	60	1	5	Private	Cessna 172	Instructor (for currency)

What caused the LOC-I incidents?

We identified causes in the LOC-I experiences, as mentioned by the pilots, and classified them into different categories. Figure 1 shows a tree map of probable causes in the LOC-I experiences. The size of the boxes represents the relative frequency of the causes mentioned in the events.

1. Pilot Conditions

Pilots underwent spatial disorientation and vertigo during instrumental meteorological conditions (IMC). Improper practices and unusual flight attitude caused their disorientation. A pilot mentioned that they could not rely on the instruments in night IMC because the cockpit lights were not working, making them disoriented for a while before regaining aircraft control. Another pilot mentioned that they did not perceive the severity of the situation due to their overconfidence. Some pilots mentioned feeling pressured during or before the flight. A pilot said they had self-imposed pressure to initiate a go-around during their flight, leading to an aircraft upset during the initial climb. Another pilot mentioned that they felt pressure to fly on the day even when they felt that they should not because of the weather conditions.

2. Perceptual Error

Perceptual errors happen when sensory input is either degraded or unusual such as during visual illusions, spatial disorientation, or vertigo (e.g., not being able to recognize hazardous conditions). Pilots mentioned that they were not mentally prepared for the threat and did not recognize the severity of the situation.

3. Weather Conditions

Weather contributed to a LOC-I in four of the seven incidents. One pilot experienced an upset during approach, and another experienced LOC-I during landing in gusty weather conditions. The third pilot experienced spatial disorientation in day IMC and the fourth experienced LOC-I in night IMC.

4. Improper Action

Pilots performed incorrect actions, such as improper procedures, pilot-induced oscillations, and instrument fixation that caused a LOC-I. Pilots used improper procedures, such as using the wrong aircraft configuration. Improper use of flight controls, such as using the opposite aileron instead of the opposite rudder during a spin recovery, caused a LOC-I. A pilot mentioned that they focused on the airspeed and altitude alone, fixating on the airspeed indicator and the altimeter. Thus, they ignored other instruments, such as the turn and heading indicator, which caused the aircraft upset.

5. Lack of Experience

A pilot mentioned that they were an inexperienced student pilot, possibly on their third solo flight when the LOC-I happened. They did not remember their procedures well while performing the go-around maneuver. Another pilot mentioned that they took an instinctive wrong action to recover from a stall because they were inexperienced. Some pilots said they could not manually maintain aircraft control in conditions such as IMC and strong crosswind because they were not used to flying without an autopilot.

6. Organizational Factors

Issues with the organization's culture were mentioned in one incident. A pilot said they felt uncomfortable discussing the LOC-I incident with their organization because they feared losing their job. New pilots feel pressured to maintain a clean record since they cannot afford scrutiny on their record, that might result in not getting the certifications they need to further their careers. A pilot stated that they received an incorrect weather reporting (VFR conditions) before the flight. But when they took off, they inadvertently entered IMC at night.

7. Aircraft System Issue

During a flight at night IMC, the lights in the cockpit panel went off, and the pilot became disoriented and felt they could not rely on their instruments. The pilot hence entered a spiral and lost a lot of altitude before they finally recovered.

8. ATC (Controller's) Communication

A pilot mentioned feeling stressed when communicating with the controller during the incident. The controller sounded stressed and did not communicate calmly, further exacerbating the situation. The pilot recommended controllers to communicate in a more empathetic way in emergencies that could help pilots get out of the threat more easily without creating a panic.

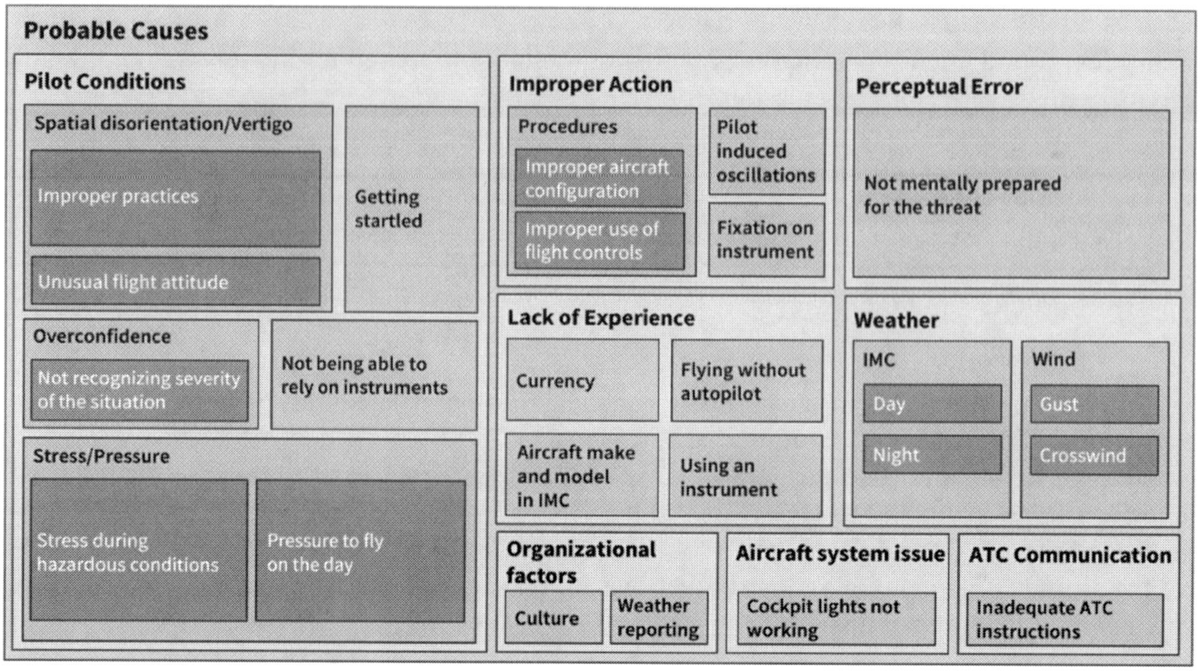

Figure 1. Tree map of hierarchical categorization of probable causes as found in the LOC-I experiences. The size of the boxes represents the relative frequency of the causes mentioned in the events.

Conclusions and Future Work

The initial findings from the study provided a deeper perspective into LOC-I causation. We found LOC-I causes ranging from pilot conditions to organizational factors. Weather contributed to four of the seven LOC-I incidents. Lack of experience or familiarity with specific instruments was another common factor that contributed to LOC-I. Issues with the organizational culture suggest possible reasons why new pilots tend not to share their incidents and experiences. Additionally, ATC's communication style may contribute to pilots' stress in emergencies. In the next phase of this study, we will evaluate pilots' responses regarding their recovery actions and training in practice to identify any gaps in training methods. The findings from this study may help pilots improve their operating procedures and focus training methods to prevent LOC-I in the future.

Acknowledgements

This research was partially funded by the US Department of Transportation/Federal Aviation Administration PEGASAS Center of Excellence under Award No 12-C-GA-PU AM44, 55. The project was managed by Michael Vu and Cliff Johnson. The views expressed in this paper are those of the authors and do not necessarily reflect those of the FAA. The information in this research does not constitute FAA Flight Standards or FAA Aircraft Certification policy.

References

Belcastro, C. M., Foster, J., Newman, R. L., Groff, L., Crider, D. A., & Klyde, D. H. (2014). Preliminary analysis of aircraft loss of control accidents: Worst case precursor combinations and temporal sequencing. In AIAA guidance, navigation, and control conference (p. 0612). https://doi.org/10.2514/6.2014-0612

Dawson, D., Cleggett, C., Thompson, K., & Thomas, M. J. (2017). Fatigue proofing: the role of protective behaviours in mediating fatigue-related risk in a defence aviation environment. Accident Analysis & Prevention, 99, 465–468. https://doi.org/10.1016/j.aap.2015.10.011

Federal Aviation Administration (FAA) (2019). Fly Safe: Prevent Loss of Control Accidents. Retrieved from https://www.faa.gov/newsroom/fly-safe-prevent-loss-control-accidents-34?newsId=94566

Holt, T., Perry, J., Carr, R., Ward, P., Luedtke, J., Hight, M., & Schindler, C. (2019). General Aviation Hypoxia and Reporting Statistics. Journal of Aviation Technology and Engineering, 8(2), 2. https://doi.org/10.7771/2159-6670.1176

International Civil Aviation Organization (2009). Review of The Classification and Definitions Used for Civil Aviation Activities. Retrieved from https://www.icao.int/Meetings/STA10/Documents/Sta10_Wp007_en.pdf

Keller, J., Mendonca MR, F. C., & Cutter, J. E. (2019). Collegiate aviation pilots: Analyses of fatigue related decision-making scenarios. International Journal of Aviation, Aeronautics, and Aerospace, 6(4), 9. https://doi.org/10.15394/ijaaa.2019.1360

Lewkowicz, R., & Biernacki, M. P. (2020). A survey of spatial disorientation incidence in Polish military pilots. International Journal of Occupational Medicine and Environmental Health, 33(6), 791–810. https://doi.org/10.13075/ijomeh.1896.01621

Majumdar, N., & Marais, K. (2022). A Survey of Pilots' Experiences of Inflight Loss of Control Incidents and Training. In AIAA AVIATION 2022 Forum (p. 3778). https://doi.org/10.2514/6.2022-3778

Majumdar, N., Marais, K., & Rao, A. (2021). Analysis of General Aviation fixed-wing aircraft accidents involving inflight loss of control using a state-based approach. Aviation, 25(4), 283-294. https://doi.org/10.3846/aviation.2021.15837

National Transportation Safety Board (NTSB). (2023). Aviation Accident Database & Synopses. Retrieved from https://www.ntsb.gov/_layouts/ntsb.aviation/index.aspx

Psyllou, E., Majumdar, A., & Ochieng, W. (2017). Planning of general aviation pilots using interviews. In International Conference on Applied Human Factors and Ergonomics (pp. 48–57). Springer, Cham. https://doi.org/10.1007/978-3-319-60441-1_5

A MULTI-METHOD APPROACH TO WORK DESIGN FOR CREW IN FUTURE REMOTELY PILOTED AIRCRAFT OPERATIONS

Kayler Marshall
Penelope Sanderson
Andrew Neal
The University of Queensland
Brisbane, Australia

Globally, most safety regulators only allow crew to operate one remotely piloted aircraft (RPA) at a time due to workload concerns. More sophisticated automation is anticipated to alleviate operator workload, allowing crew to simultaneously operate more than one RPA. However, how work should be distributed amongst crew is still unknown. We employ a complementary set of methods for work design in a future system of RPA operation: Cognitive Work Analysis, computational modelling, and human-in-the-loop experiments. In this paper we describe each method, outlining the unique insights gained and how these are applied in the evaluation of work in a future RPA system. We also identify the limitations of each method and outline how these are addressed in successive methods. Our approach provides rich evaluations of alternative work designs and—once established—can be used to efficiently identify designs that maximise safety and efficiency of operations in a future RPA system.

The remotely piloted aircraft (RPA) industry is rapidly expanding, with market predictions anticipating the value of the RPA sector to reach approximately USD$54.2 billion by 2027 (IndustryARC, 2022). Improvements in automation, artificial intelligence (AI), and other technologies on board RPA have been credited for the industry's growth, enabling RPA to be used as a more efficient and economic means of executing previously labour-intensive operations (Civil Aviation Safety Authority [CASA], 2022; IndustryARC, 2022). Furthermore, the growing capability and accessibility of RPA is expected to greatly increase adoption of RPA technology (CASA, 2022). However, if the RPA industry is to expand as predicted, revisions to current airspace regulations will be required.

Most airspace regulators only allow a small team of operators to control one RPA at a time due to concerns about operator workload and system safety. Within the foreseeable future, advances in automation and technology on board RPA will likely enable a small team of operators to safely manage more than RPA at a time. This is known as 'one-to-many operations'. However, even with reduced crew involvement, human input will remain a critical aspect of successful RPA operations, especially during unforeseen events. Therefore, work must be designed in a way that supports the crew and allows them to work cohesively with automation. There are many different possibilities for how work roles could be designed. However, we do not yet know what the safest, and most effective allocation of responsibilities and tasking is, for a small crew operating more than one RPA at a time.

To address this problem, we employ a multi-method approach for developing and evaluating work designs for one-to-many RPA operations. Our approach consists of three complementary methods: Cognitive Work Analysis (CWA), computational modelling, and human-in-the-loop experiments. CWA provides a framework for understanding the nature of the work and establishing work designs that maximise performance in a future work system. Computational modelling provides a means of evaluating the efficacy of different work designs under a range of operational scenarios, and human-in-the-loop experiments allow model predictions to be validated against behavioural data. In the following sections, we summarise each method, outlining the insights provided, noting the shortcomings or limitations, and stating how these are addressed by subsequent methods.

Cognitive Work Analysis

CWA is a framework for designing and evaluating work in complex sociotechnical systems (Rasmussen et al., 1994; Vicente, 1999). CWA consists of a series of analytic tools that can be used to identify constraints within the work system. Once the boundaries or constraints imposed on work have been identified, what remains are the various ways in which work can occur. Therefore, rather than establishing designs based on how work is currently done (descriptive), or how work should be done (normative), CWA establishes designs based on how work *could* be done (formative) (Rasmussen et al., 1994; Vicente, 1999).

Establishing work designs in a formative manner has two primary advantages for the current research. First, designs can be established before the existence of the physical work system, as designs are not limited to existing procedures, processes, and physical attributes of the environment (Naikar et al., 2006). Second, designs that maximise possibilities for how work can occur enable greater adaptability in how work is conducted. In principle, this should support better overall performance, as the system will be better able to adapt and cope with unforeseen events (Naikar, 2011). This is especially important in complex sociotechnical systems such as RPA operation, which often experience high levels of uncertainty and unpredictability.

We applied CWA to the design of work for future one-to-many operations using a phased approach, broadly following a procedure introduced by Naikar et al. (2003). In Phase 1, two analytic tools were developed in collaboration with industry experts, academics specialising in similar fields and five subject-matter experts (SMEs). Phase 2 involved using the analytic tools to develop a scenario for the preliminary evaluation of work designs with SMEs.

In Phase 1, the first analytic tool developed was a Work Domain Analysis (WDA). A WDA represents the constraints imposed on work by the functional structure of the environment or work domain (Naikar, 2013). A common modelling tool for conducting a WDA is the abstraction hierarchy. The abstraction hierarchy represents the entire work system at different levels of abstraction, ranging from the purposes and priorities of the work system to the physical resources available (Naikar, 2013) (see Figure 1). In this way, an abstraction hierarchy represents the constraints or conditions within the work domain that must be respected, regardless of the situation (Naikar, 2013). Applied to the current research, the abstraction hierarchy provides a useful framework for evaluating different work designs, according to whether each design satisfies the conditions of the work domain.

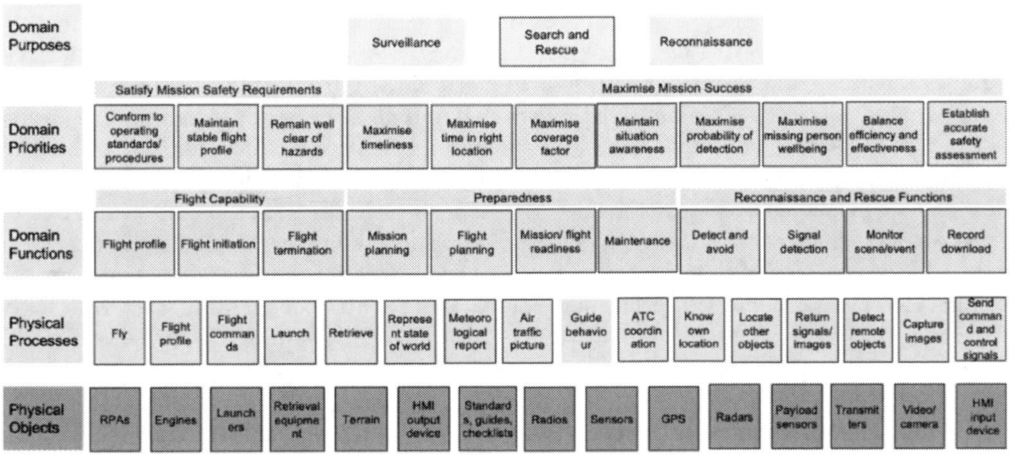

Figure 1. An abstraction hierarchy of RPA operations representing the domain purposes, priorities, functions, processes, and physical objects of the work domain. Elements of the abstraction hierarchy relevant to search and rescue are outlined.

In the current research, we identified the conditions to be satisfied as the domain purposes and priorities of the RPA system. The system's purpose was deemed to be performing three types of RPA mission. However, for present purposes, we focus on the search and rescue mission context. Further, we categorised the domain priorities of search and rescue as being broadly related to satisfying mission safety requirements and maximising mission success (see Figure 1, above).

The second analytic tool developed in Phase 1 was a Control Task Analysis (ConTA). A ConTA identifies the constraints associated with the activity necessary to obtain the systems' objectives, as identified in the WDA (Naikar, 2013). We developed a Contextual Activity Template (CAT) to represent required activity in the context of a search and rescue mission (see Figure 2). The CAT represents required activity as a series of work functions that occur across a sequence of work situations (Naikar, 2013; Vicente, 1999). The CAT identifies required activity independently of who is responsible for the activity or how the activity is conducted (Naikar et al., 2006). In this way, the CAT provides a template for evaluating how RPA activity is managed under different work designs.

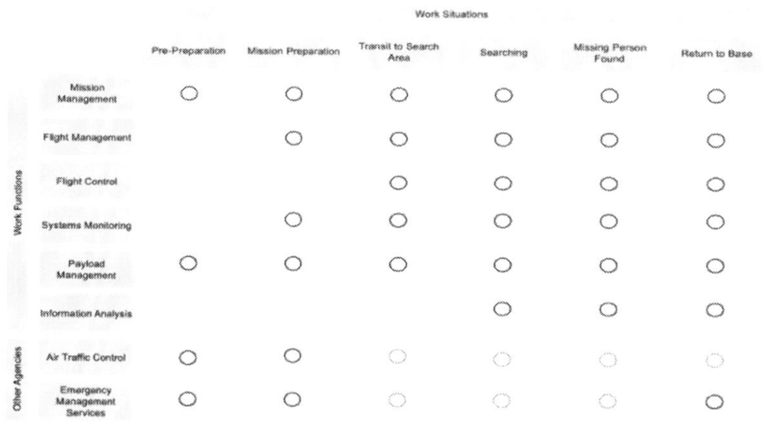

Figure 2. A simplified CAT of a search and rescue mission, representing required activity for each (work) function across different phases of the mission (work situation). In the full version, required activity is listed in cells currently occupied by circles. Activity in grey indicates 'as required'.

In Phase 2, we combine elements of the WDA and ConTA to develop a framework for the preliminary evaluation of work designs. The ConTA describes the activity required across different phases of the search and rescue mission and the WDA provides the priorities and values of the work system that need to be upheld to achieve the system's objectives. To evaluate work designs, we present the framework to SMEs as a scenario. Work designs included in this analysis consist of functional, structural, and dynamic allocations of work ranging from a ratio of one operator, two RPA to four operators, nine RPA. Before each scenario we outline a series of assumptions regarding the capabilities of the RPA in a future system of operation. Namely, RPA are equipped with an autopilot, an automated conflict detection system and an AI-assisted detection system for identifying points of interest that are inspected by humans. Each SME steps through the scenario contrasting two work designs. As the scenario unfolds, we ask SMEs to describe which work design best satisfies the domain priorities of the search and rescue mission. Data collection for Phase 2 is ongoing at the time of writing. However, preliminary results suggest that there are advantages and trade-offs for each work design. Once completed, this process will provide us with preliminary evaluations of alternative work designs and identify designs that best satisfy the criteria of the work system.

Although evaluating work designs using CWA is useful in establishing insight into the activity that needs to be done and possible ways the work could be allocated amongst crew, there are limitations to this approach. First, it is infeasible to use this approach to evaluate multiple designs under various operating conditions. Second, it is impossible to generate quantitative evaluations of the efficacy of different work designs. Therefore, to overcome these shortcomings, we develop and apply a computational model to evaluate alternative work designs for future one-to-many operations.

Computational Modelling

Computational modelling involves developing precise mathematical models to make sense of behavioural data and to generate predictions (Wilson & Collins, 2019). The current research employs a queuing model as a test bed for evaluating alternative work designs (for a similar approach, see: Hannah & Neal, 2014; IJtsma et al., 2019; Mekdeci & Cummings, 2009).

In the computational model, pending work is added to queues that are serviced by agents (human or automation) who carry out the tasks. The work design is coded in the form of an allocation policy that determines how the tasks are allocated to agents. Each agent chooses tasks from their queue based on the task priority and deadline, and the time it takes an agent to complete a task is represented as a rate of progress within the work model. New tasks are added to the queue when the preconditions of that task are met, and tasks are removed from the queue once completed. The model environment captures momentary changes in the environment across time based on the completion of tasks and scripted events. See Figure 3 for an overview of the model architecture.

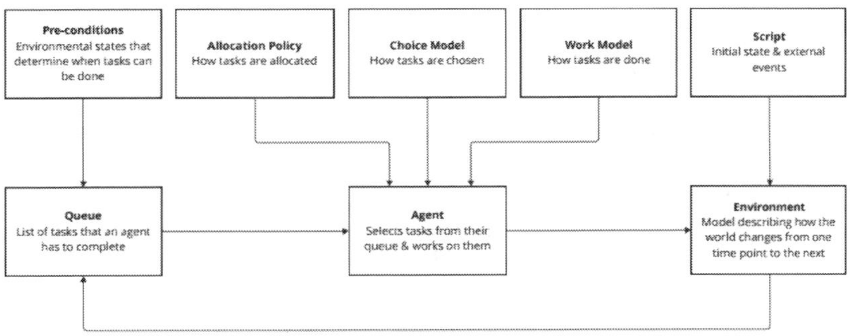

Figure 3. Diagram of the model architecture outlining all key elements.

Using the model, we evaluate how different work designs influence task load, coordination load, and system performance. System-level performance measures are adapted from the domain priorities in the WDA and include measures of safety, effectiveness, and efficiency of operations.

The computational model provides unique insights regarding the effectiveness of different work designs. Analysts can use the model to explore the logical consequences of different work designs, given a set of assumptions regarding the work that has to be done (e.g., the conditions under which different tasks can be done). The model provides a quantitative evaluation of each design, enabling predictions regarding the efficacy of work designs across various operating contexts to be generated.

Although computational modelling enables a more efficient and quantifiable means of evaluating alternative work designs, it is not without limitations. The major limitation of computational modelling is that the predictions of the model are dependent upon the assumptions that are made. Key assumptions include the number and type of tasks to be completed, the priority of those tasks, and how long they take to complete. The model's sensitivity to these assumptions can be evaluated by running a simulation study and assessing how the behaviour of the model changes under different parameter settings. However, work may be done in a fundamentally different way to that envisaged by the model, producing misleading results. For this reason, it is important to validate the model using real behavioural data (Wilson & Collins, 2019).

Human In-the-Loop Experiments

To validate the model, we conduct a series of human-in-the-loop experiments using a desktop simulation. We have developed a simulated microworld that is representative of a future system used for one-to-many operations. The microworld simulates a search and rescue mission in uncontrolled airspace, where crew a of up to four can manage up to 12 RPA simultaneously. Each RPA is equipped with an autopilot that flies the RPA according to the flight plan, an automated conflict detection system that identifies aircraft that will breach the separation standard and an AI-assisted point of interest detection system that identifies points of interest to be inspected by the operators (see Figure 4 for simulation interface).

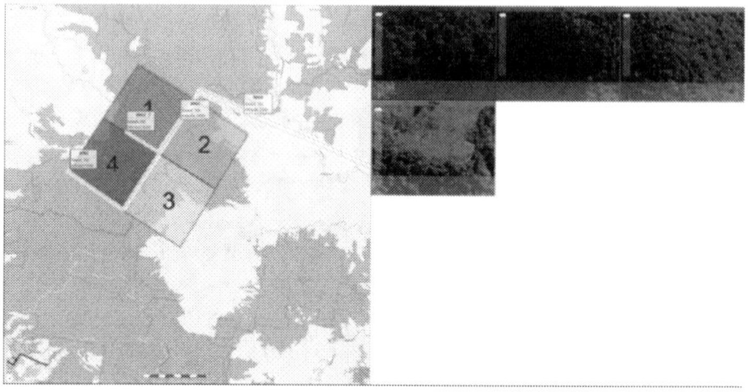

Figure 4. Simulation interface used for operating RPA and performing the search and rescue mission.

We are running a series of experiments in which teams of participants conduct search and rescue missions using RPA in the simulation environment. Each mission requires the crew to coordinate and carry out flight, payload, and mission management functions. The experiments vary the size of the crew, the number of RPA that they control, and the way that the roles are designed.

The efficacy of alternative work designs are evaluated using measures of system-level performance, individual, and team-level processes, similar to the computational model. The primary difference is that individual and team-level processes include measures of participant communication load as well as subjective workload and response time to a vibrotactile detection response task which has been found to accurately capture fluctuations in cognitive load in experimental contexts (Innes et al., 2021).

Investigating work designs for a future system of RPA operation using human-in-the-loop experiments allows us to evaluate designs using real behavioural data. Data gathered in human-in-the-loop experiments is then applied to validate the model. To do this, we examine whether the model can reproduce the observed trends in the data. If there are findings that the model cannot account for, the assumptions embedded within the model are revised accordingly. This may involve revising the number and type of tasks to be completed, the conditions under which those tasks are done, the priority of those tasks, or how long they take to complete. Once the model provides an adequate account of the behavioural data, we then cross-validate the model using new data to establish that it generalises to a new set of conditions.

The multi-method approach employed in this research provides a cost-effective means of evaluating work designs for future operational systems. CWA allows the system requirements and range of possible work designs to be identified, as well as providing a framework for the evaluation of a small number of designs. Computational modelling affords a systematic evaluation of the full range of possible designs under a wide range of scenarios, and human-in-the-loop experiments provide the behavioural data needed to verify and improve the model. However, the models and experiments can only provide a simplified representation of a future operational system. Whilst they can identify a set of promising work design options during the early phases of the design process, the effectiveness of these designs needs to be evaluated as the system is developed. In this way, manufacturers and operators can use this multi-method approach to establish designs that maximise safety and efficiency of operations and generate the body of evidence required to mount a safety case for the introduction of one-to-many operations within civil airspace.

Acknowledgements

This research is funded by the Australian Research Council Linkage Project (LP190199188), with Boeing Research and Technology-Australia as the industry partner.

References

Civil Aviation Safety Authority.(2022). *The RPAS and AAM strategic regulatory roadmap*. Retrieved from https://www.casa.gov.au/sites/default/files/2022-06/the-rpas-and-aam-roadmap.pdf

Hannah, S. D., & Neal, A. (2014). On-the-fly scheduling as a manifestation of partial-order planning and dynamic task values. *Human Factors*, *56*, 1093-1112. doi: 10.1177/0018720814525629

IJtsma, M., Lanssie M. Ma, Pritchett, A. R., & Feigh, K. M. (2019). Computational methodology for the allocation of work and interaction in human-robot teams. *Journal of Cognitive Engineering and Decision Making*, *13*, 221-241. doi: 10.1177/1555343419869484

IndustryARC. (2022). *Unmanned aircraft systems market – forecast (2023 - 2028)*. Retrieved from https://www.industryarc.com/Report/15014/unmanned-aircraft-systems-market.html

Innes, R. J., Evans, N. J., Howard, Z. L., Eidels, A., & Brown, S. D. (2021). A Broader application of the detection response task to cognitive tasks and online environments. *Human Factors*, *63*, 896-909. doi: 10.1177/0018720820936800

Mekdeci, B., & Cummings, M. (2009). Modeling multiple human operators in the supervisory control of heterogeneous unmanned vehicles. *Proceedings of the 9th Workshop on Performance Metrics for Intelligent Systems*, 1-8. doi: 10.1145/1865909.1865911

Naikar, N. (2011). *Cognitive Work Analysis: Foundations, extensions, and challenges*. Retrieved from https://apps.dtic.mil/sti/pdfs/ADA564221.pdf

Naikar, N., Pearce, B., Drumm, D., & Sanderson, P. M. (2003). Designing teams for first-of-a-kind, complex systems using the initial phases of cognitive work analysis: Case study. *Human Factors*, *45*, 202-217. doi: 10.1518/hfes.45.2.202.27236

Naikar, N.(2013). *Work domain analysis concepts, guidelines, and cases*. CRC Press.

Naikar, N., Moylan, A., & Pearce, B. (2006). Analysing activity in complex systems with cognitive work analysis: concepts, guidelines and case study for control task analysis. *Theoretical Issues in Ergonomics Science*, *7*, 371-394. doi: 10.1080/14639220500098821

Rasmussen, J., Pejtersen, A., & Goodstein, L. (1994). *Cognitive systems engineering*. Wiley.

Vicente, K. J. (1999). *Cognitive work analysis : toward safe, productive, and healthy computer-based work*. Lawrence Erlbaum Associates.

Wilson, R. C., & Collins, A. G. E. (2019). Ten simple rules for the computational modeling of behavioral data. *eLife*, *8*,1-33. doi: 10.7554/eLife.49547

INITIAL TESTING OF THE UNCREWED AERIAL SYSTEM PILOT KIT (UASP-kit) IN OPERATIONAL SETTINGS

Lynne Martin[1], Lauren Roberts[1], Joey Mercer[1], Yasmin Arbab[1], Charles Walter[2], William McCarty[2], Charles Sheehe III[3], David Fuller[3]

[1] NASA Ames Research Center, Moffett Field, CA
[2] ASRC Federal Data Solutions, Moffett Field, CA
[3] NASA Glenn Research Center, Cleveland, OH

Pilots for small uncrewed aerial systems (sUAS) are at a disadvantage for building situation awareness of the remote airspace in which they are flying, simply because they are distant from their vehicles. A tool to provide increased air traffic situation awareness for an sUAS pilot is being developed. The UAS pilot kit, "UASP-kit," is small and self-contained, with its chief capability being to collect and display Automatic Dependent Surveillance-Broadcast reports from local aircraft. UASP-kits were taken into the field, introduced to users during a training course, and then left with them for use throughout the summer fire season. sUAS pilots used the prototypes when it was appropriate during the summer. The UASP-kits were operational for a total of 79 flight-days. Users reported that the UASP-kits supported their situation awareness but also identified several usability issues. The findings contribute to the validation of the UASP-kit, and support continuing the work to improve the tool and develop additional functionality.

The use of uncrewed aerial systems (UAS) is proliferating through many domains, particularly within disaster and emergency response, as the capabilities of these aircraft and recognition of their versatility increases. One example of the use of UAS within disaster and emergency response is combatting wildland fire. The increasing number and severity of wildland fires over the last two decades (Hoover & Hanson, 2023; NIFC, 2022) have emphasized that new methods need to be explored to provide greater assistance to the firefighters working in wildland areas. One way to achieve this is to take advantage of technological developments to provide firefighters with strategic tools in addition to improved physical tools. Strategic tools, designed to assist awareness and decision making, could provide more, and better-organized, information to assist operational personnel to identify and select the most effective strategies and methods for fighting a fire.

The use of UAS by disaster and emergency response services is growing rapidly because, as their name describes, they remove the operator from the vehicle and thereby do not expose the remote pilot to the same risks as the aircraft. The remote operator or UAS pilot (UASP) is still subjected to the environmental hazards around a wildland fire, e.g., smoke, and must be aware of additional ground hazards, such as ground equipment. One tradeoff for a remote UAS pilot, however, is that they no longer have a wider view of aerial operations because they can only view the airspace from the ground. UASPs have to build airspace situation awareness (SA) from the information shared over the radio (and through briefings). In addition, if the UASP is operating a small UAS (sUAS), e.g., for Infrared (IR) imaging or controlled burn missions, the other aviators in crewed vehicles are unlikely to be able to see their sUAS vehicle. The burden is therefore on the UASP to stay clear of crewed aircraft.

NASA's Scalable Traffic Management for Emergency Response Operations (STEReO) research activity investigated developing a prototype tool that would assist sUAS pilots to maintain an awareness of the airspace in which their vehicle is operating. The initial ideas for such a tool were formulated in collaboration with the U.S. Forest Service (USFS) and CAL FIRE, during 2020, through two demonstrations and a series of discussions; for more details see Martin, et al. (2022). The necessary properties for a tool that supports UASP situation awareness are both physical and informational. The tool needs to operate in a communications-denied environment (without Wi-Fi or cellular connections) and be small and light enough for a person to transport it. It needs to be easy to use, to provide information about the airspace around the sUAS, and draw the UASP's attention to potential hazards in the airspace. The UAS Pilot-kit (UASP-kit) was designed through these discussions and is intended to meet these prerequisites.

Description of the First Prototype UASP-kit

As the aim of the UASP-kit is to provide increased air traffic situation awareness for one sUAS pilot, it is designed to be self-contained and portable. The components include a display and a communications infrastructure that collects Automatic Dependent Surveillance-Broadcast (ADS-B) messages to give the user a view of crewed vehicles in the surrounding airspace, especially when they are in areas of low connectivity with poor cell service. The first prototype was designed and built during the summer of 2021. It consists of an ADS-B data link receiver with a power over ethernet (POE) switch, a server, a power source, and a display. These are housed in a 21" by 32" by 13" ruggedized case. The view of traffic in the airspace is generated by receiving ADS-B messages from airborne traffic that are broadcasting their enhanced Global Positioning System (GPS) position to other traffic and to the ground (FAA, 2022). The ADS-B receiver (uAvionix, 2020) listens for and receives messages reported on the 978MHz and 1090MHz frequency bands. The messages are interpreted and displayed as icons on the UASP-kit's graphical user interface (GUI) (Figure 1), which is a touchscreen tablet. The JavaScript browser-based GUI application uses a base map, e.g., a satellite image, as a canvas on which the aircraft traffic is displayed. This interface has features to assist with interpretation of the display including aircraft icons to distinguish between types of aircraft, and a filter that allows the user to reduce the range of the ADS-B traffic shown. In addition, the UASP-kit can import and display a fire operations map onto the base map display and allow users to define an operational volume for an sUAS, which includes area, height, and location. The user can control when the UASP-kit notifies them of a situation that requires their attention based on the proximity between ADS-B tracks and the operational volume of the sUAS.

After the UASP-kit prototype was built, and reviewed by Subject Matter Experts, it underwent two phases of field assessments to evaluate its performance in real-world settings and collect user feedback to direct future development. These two phases are described below.

Method for Field Data Collection

The first user-testing data were collected during a two-week spring sUAS prescribed burn (PB) training session that was hosted by the USFS. The second set of data was collected during the summer fire season of 2022 when the USFS and CAL FIRE used sUAS to help with their efforts to combat wildland fires.

Prescribed Burn Data Collection

During the spring of 2022, researchers from the STEReO team shadowed three units of sUAS prescribed burn instructors and trainees as they traveled throughout the south-eastern U.S., conducting prescribed burns with sUAS as part of their hands-on training to become qualified for aerial ignition. Each unit was comprised of six trainees and two instructors (18 UASP-kit users in total). The units set up the UASP-kits as they prepared their equipment (sUAS and Ground Control Stations) for the day's flights, setting the operational volume dimensions and alerting dimensions to the sizes that they determined would be most useful each day.

While active, each UASP-kit recorded logs of the ADS-B messages received and the users' interactions with the display, i.e., those to set up the operational volume and the alerting. Feedback from users was gathered in an intentionally ad-hoc manner. Each research team had a list of prepared questions and topics of interest, e.g., questions asking about constructing situation awareness, usability of the UASP-kit and communications between team members. Researchers solicited feedback from the users when there was an opportunity and asked a selection of these questions to prompt conversation. User responses were hand-written by the research team and transcribed into a common spreadsheet.

Summer Fire Season Data Collection

Five UASP-kits were supplied to sUAS crews for their use during the summer fire season of 2022. sUAS crews (usually two to three people) set up the UASP-kits when they considered it appropriate, as they were on missions to fly sUAS to assist with control of wildland fires, mainly in the western U.S. While it was turned on, each UASP-kit recorded logs of the ADS-B messages received from crewed aircraft and the users' interactions with the display to set up the operational volume and the alerting. Twice during the summer fire season, feedback about the UASP-kit's usability was solicited from the UASPs – once via phone conversations and again at a second point in person as the logs were retrieved from the UASP-kits. During these conversations, the usability of the UASP-kit was the focus of the questions.

Comparison and Discussion of UASP-kit Settings and Usability

Logs from the three UASP-kits used during the spring prescribed burn training event revealed that UASP-kit-1 was active for the most flight-days (14) and UASP-kit-2 showed the most alerts (76 total or approximately 60% of all collected alerts during that event). From the five UASP-kits in the field and operational during the summer fire season, UASP-kit-5 was switched on for the most flight-days (26) with UASP-kit-4 showing the most alerts (72 or approximately 45% of the summer's alerts).

During the prescribed burn data collection, the three UASP-kits were used for 27 flight-days. Over the summer fire season, the UASP-kits were not active for all sUAS missions but were used on 52 flight-days. Because the number of flight-days differed a good amount between the prescribed burn and

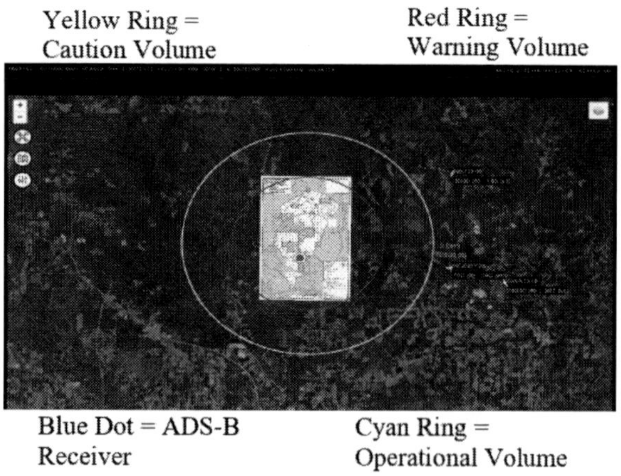

Figure 1. Image of the UASP-kit display showing caution and warning rings.

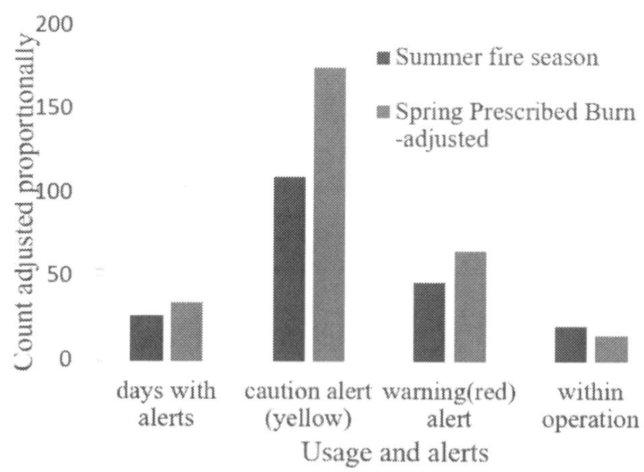

Figure 2. Proportionally adjusted UASP-kit usage by flight-day, and alerts generated to nearby aircraft.

summer season, the data from the prescribed burn were adjusted, in proportion with the difference, to allow for a comparison between the two data collection periods (see Figure 2). The UASP-kit traffic alerting was triggered on 66% of the flight-days during prescribed burns, alerting 91 times. During summer fire season, the UASP-kit alerting was triggered on 52% of the flight-days and alerted 110 times. All these alerts announced a crewed aircraft flying into the caution (yellow) alert volume defined by the user. During prescribed burns, 37% of the alerted aircraft continued to move closer to the sUAS operational volume, flying into the warning (red) volume, while a similar percentage of the alerted aircraft (43%) continued into the warning volume during summer fire season. Further, 9% of the aircraft tracked by the UASP-kits during prescribed burn flight-days remained on their approaching trajectories to fly into the operational volume defined by the sUAS pilot, the nearest of these coming as close as 0.11 nautical miles (nmi) to the center of the operation. Of the alerted aircraft tracked during the summer, 19% flew into the operational volume of the sUAS, and the closest approach was within 0.05nmi (304ft) of the operation's center.

Users created cylinder-shaped operational volumes 91.5% of the time during the spring prescribed burn operations, while over the summer fire season they selected cylindrical operational volumes only 58.2% of the time (the alternative was a cube). Most often users selected operational volumes that had 1nmi radii and a 700ft Mean Sea Level (MSL) ceiling (Figure 3a and b). Both the width and the height of operations varied more widely during the summer data collection (from 0.17nmi to 6.73nmi laterally (radius) and 100ft MSL to 8500ft MSL vertically) than during prescribed burns (0.86nmi to 3nmi laterally (radius) and 700ft MSL to 2107ft MSL vertically).

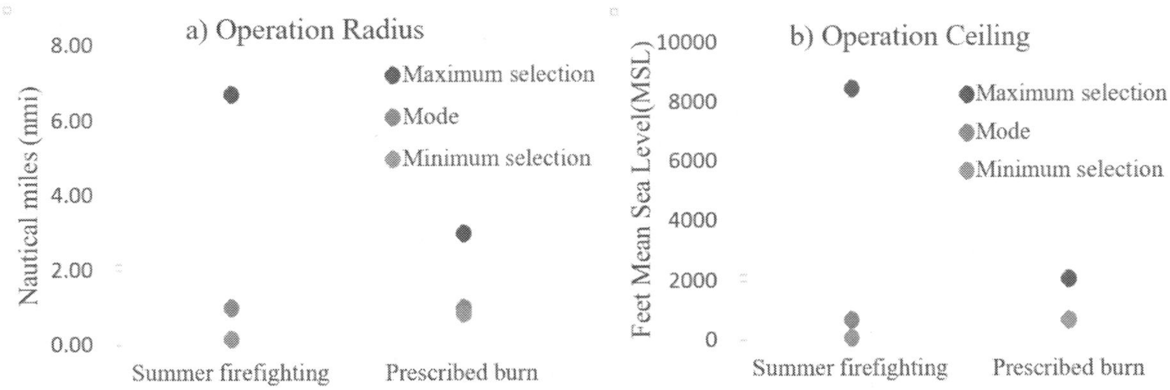

Figure 3. User-selected operational volume dimensions with a) showing the range of radius selections, and b) showing the range of height selections.

Although there was no significant difference in the most commonly chosen operation sizes between types of mission (both having a mode of 1nmi lateral radius and 700ft MSL vertical), the range for both dimensions during the summer firefighting season was larger. During prescribed burns, the largest operation volumes had a 3nmi lateral radius and were 2100ft MSL high, during the summer the largest operation volumes were more than twice that, with a 6.7nmi radius and an 8500ft MSL profile although, when compared using a Mann-Whitney U-test, these differences were non-significant.

Regarding the user-selected alerting volumes, most often users chose caution alerts that had 5nmi radii and warning alerts that had 2nmi radii with a 12,000ft MSL ceiling (Figure 4a and b). While the width of alerting volumes was almost the same across both data collection periods, the height of the volumes varied more widely during the prescribed burn data collection (from 2,100ft MSL to 30,000ft MSL) than during the summer season (2,000ft MSL to 12,000ft MSL vertically), see Figure 3b. Although it could be argued that the prescribed burn alerting volumes were substantially taller than the summer volumes, these higher values were only selected 6% of the time and both the mode and median alerting ceiling height was 12,000ft MSL.

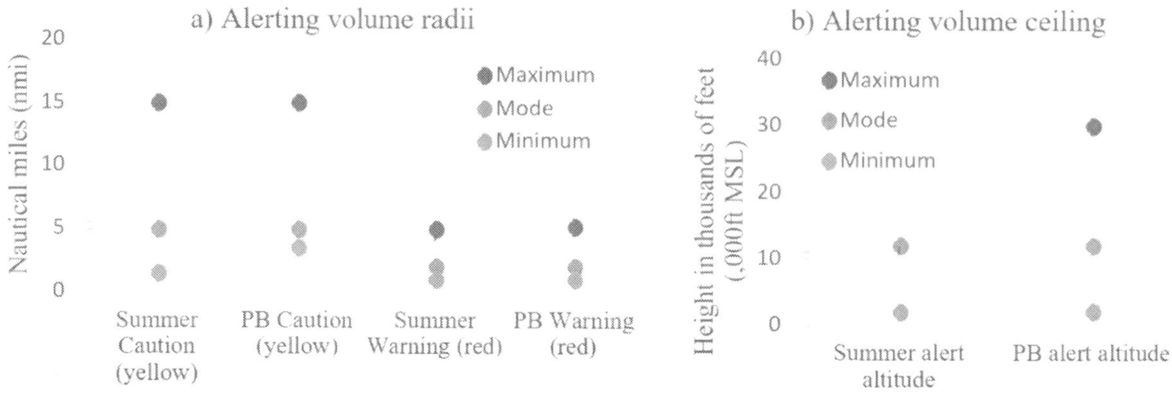

Figure 4. User-selected alerting volume dimensions with a) showing the range of radius selections for both the caution and warning alerting and b) showing the range of height selections.

Given that the alerting volumes were usually similar sizes across the two data collection periods, the number of caution and warning alerts were informally compared. Although proportionally more aircraft flew through the caution (yellow) airspace during prescribed burns than the summer fire season data collection flight-days, that difference was far smaller for the aircraft that flew through the warning area (Figure 2). Operational volumes were constructed to be a good amount larger during the summer fire season (Figure 3a & b), making a comparison uneven. Nevertheless, the closest point of approach of a crewed aircraft to the center of the operation was during the summer fire season (0.05nmi). From these descriptions a hypothesis could be presented that users' choices when setting up the UASP-kit are consistent with the different environments in which they were conducting their missions. During a prescribed burn, because they are flying in class G airspace, a UASP only sometimes has advanced knowledge of the aircraft that could fly close to him/her, often the first-time s/he becomes aware of an aircraft is when s/he hears a radio transmission or physically sees/hears the aircraft. There could be many of these aircraft transiting the airspace or none. In this environment, it may help to set alerting distances farther from your operation to give earlier warnings and a longer time to react. During daytime aerial operations over a wildland fire, the number of aircraft and their flight paths will have been discussed in the morning briefing and so, although the UASP still has to watch and listen to radio transmissions, when s/he is notified by the UASP-kit of an aircraft, s/he knows approximately in which direction to expect to look and what type of vehicle to expect. In this environment, it may help to set alerting distances closer to the operation to reduce repeated alerts as other aircraft fly many passes over the fire. Specific user feedback on their reasons for the way they set up the UASP-kit should be solicited from UASPs to support or refute this hypothesis.

For both data collection periods, the research team asked users about the usability of the UASP-kit. Questions were asked face-to-face during the prescribed burn about which functions UASPs found useful and which new functions users felt would increase the tool's usefulness. Users liked the UASP-kit alerting function, especially the audio alert, saying alerting was "what [they] cared most about." They found configuring the alerting rings straightforward, and tried different combinations of ADS-B filters and alerting dimensions to explore how they could change their view of the airspace. If a UASP-kit alert sounded, a crew member viewed the display to track the aircraft in case there was a need for deconfliction. Crew members also used the map to track crewed aircraft over time for general awareness, as well as to anticipate potential interactions with other airspace users. Suggestions for improvements included having more information announced in the aural alert, e.g., the aircraft callsign, altitude and speed. Yet, other users commented that the UASP-kit adds complexity to operations and will take time to integrate into the workflow. Overall, the UASP-kit "took too long to get going" and would be improved if it could be activated with fewer steps.

One discussion concerned the need to build strategies for how best to use the alerting rings. There is a tradeoff between too many alerts and alerting volumes that are too small. The UASP-kit was noted to have useful functions for the prescribed burn setting but some users debated whether the airspace complexities associated with a busy wildland fire bring unique challenges that these early versions of the UASP-kit (like the version used in this data collection and described above) cannot support.

During telephone interviews conducted over the summer, questions focused on usability issues reported about the UASP-kits and new functions or features to mitigate these issues. Users described difficulties setting up the UASP-kits, commenting that the user guide was difficult to follow unaided, and that the logic of the initial location showed by the UASP-kit as the GUI was brought up was confusing, with some crews not ever moving past this initial set up step. Users reported frustration with the length of the startup process. A key suggestion was to streamline this, including having the UASP-kit automatically display a graphical indication of its current location. Users also faced challenges with the physical UASP-kit itself. Many users removed the power supply from the box. This made the UASP-kit much lighter but also allowed the components to shift. Some users found on opening the UASP-kit at their work sites, the contents were jumbled, and they were not sure whether set up issues they experienced were because not all the components were firmly connected.

Conclusions

The UASP-kit showed promise as a tool to support sUAS pilots' situation awareness. Having tried the UASP-kit in the field, pilots reported that the tool was useful, and they offered many ideas for expanding the functionality of the prototype. There were no significant differences in the way users set up the UASP-kits, but it also became apparent that, without ongoing support from the research team, users found the UASP-kit more difficult to use than expected. These findings indicate there is more work needed to improve training and the usability of the UASP-kits, including reworking the user guide and simplifying the start-up procedures.

Acknowledgements

The authors thank the members of the USFS UAS Program, CAL FIRE's Tactical Air Operations Program and Global UAS Solutions for all their help and guidance. The UASP-kit evaluation would also not have been possible without the support and invaluable feedback of the participants in the 2022 Aerial Ignition Academy who provided invaluable feedback.

References

Federal Aviation Administration (2022). *Equip ADS-B*, FAA, Washington DC, https://www.faa.gov/air_traffic/technology/equipadsb, March 2022.

Hoover, K. & Hanson, L. (2023). *Wildfire Statistics: In Focus*, IF10244, Version 66, March 1st, Congressional Research Service, Library of Congress, Washington, DC, https://crsreports.congress.gov/product/pdf/IF/IF10244

Martin, L., Arbab, Y., Roberts, L., Mercer, J., Walter, C., McCarty, W. & Sheehe III, C. (2022). Developing an Unmanned Aircraft System Pilot Kit (UASP-kit) for Wildland Fire UAS Operators, *AIAA Aviation Forum*, American Institute of Aeronautics, Chicago, IL, 27 June-1 July.

National Interagency Fire Center (NIFC) (2022). *Total wildland fires and acres (1983–2022)*, www.nifc.gov/fireInfo/fireInfo_stats_totalFires.html

uAvionix (2020) *pingStation User and Installation Guide, Revision L, UAV-1001358-00,1,* uAvionix Corporation, Bigfork, MT.

THE VIABILITY OF SEE-AND-AVOID FOR URBAN AIR MOBILITY OPERATIONS

Richard Mogford
NASA
Moffett Field, CA
Walter Johnson
NORCAL Pacific Consultants
Mountain View, CA

Urban Air Mobility (UAM) is an emerging aviation concept that could supplement today's ground and air transportation systems. For UAM, it is generally assumed that the private sector will manage separation and not rely on the U.S. Federal Aviation Administration air traffic control system. To date, discussions of initial operations focus on using the visual abilities of the pilot to see-and-avoid (SAA) other aircraft. Decades of research on SAA has demonstrated that it is inadequate for reliable detection of aircraft that might pose a collision risk. The literature on multi-object tracking is also reviewed for findings on how well humans can visually track objects. This research shows that observers have limited resources for tracking and that this may be affected by object characteristics and cognitive resources. The conclusion is that SAA is a risky method for avoiding midair collisions. It is recommended that flight deck displays and automated collision avoidance systems be implemented in UAM aircraft at the outset of their introduction.

Urban Air Mobility (UAM) will transport passengers and cargo in urban areas using new types of aircraft (Mueller et al., 2017; Uber Elevate, 2017). Electric vertical takeoff and landing (eVTOL) vehicles are being developed that have sufficient capacity and range to efficiently move people, particularly between urban vertiports and airports. UAM is expected to improve mobility, decongest road traffic, reduce trip time, and mitigate strain on existing transportation networks (Thipphavong et al., 2018).

To support UAM, and as an alternative to the Federal Aviation Administration's (FAA) current publicly managed air traffic management system, the FAA has proposed allotting responsibility for tactical UAM separation services to the private sector. The UAM system would employ multiple dedicated flight corridors, servicing urban vertiports with the responsibility for conformance and tactical separation residing with UAM ground operators, onboard pilots, or with an independent Provider of Services for UAM (in a future mature system) (FAA, 2020). Flights would operate under Visual Flight Rules in Visual Meteorological Conditions. Most eVTOL aircraft are expected to have a single pilot with out-of-the window visibility similar to current helicopters and general aviation (GA) aircraft. A critical issue for UAM flights will be the use of see-and-avoid (SAA) for tactical separation and collision avoidance as is currently the practice with aircraft in uncontrolled airspace. SAA is defined as the detection and avoidance of other aircraft using the unaided perceptual and cognitive abilities of the pilot.

See and Avoid Process

When using SAA as a collision avoidance strategy, a series of functions is needed for any given encounter with another aircraft. These are:

1. Detect intruder
2. Track intruder
3. Evaluate collision potential
5. Calculate an avoidance maneuver
6. Execute the avoidance maneuver
7. Return to course

Figure 1 is a timeline of the SAA process. It begins with the pilot's detection of the possible threat and ends with an avoidance maneuver prior to a return to course. Between the two endpoints the intruding aircraft must be tracked and evaluated for collision potential. If a collision is predicted, an avoidance maneuver must be formulated and executed. These activities are performed in the context of the ongoing pilot's tasks of operating the aircraft, communicating by radio, scanning for other aircraft, responding to passengers, etc.

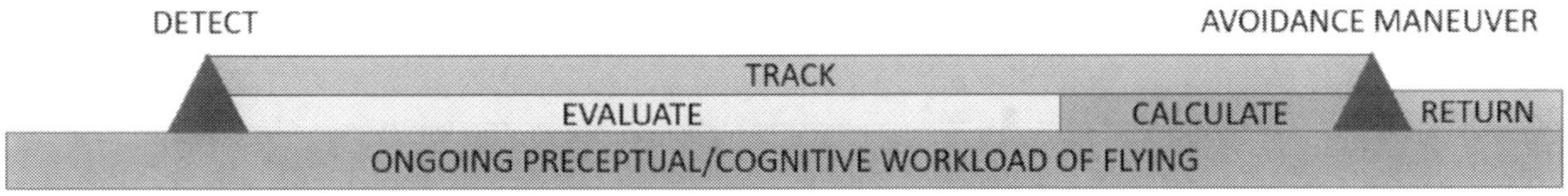

Figure 1. The SAA process.

There is considerable research on the effectiveness of SAA. This work generally addresses the detection stage of SAA. There are also psychological investigations on the perceptual and cognitive aspects of locating and tracking objects. Relevant research will be reviewed and the discussion section will focus on how SAA might be used for UAM.

SAA Literature Review

Graham (1989) surveyed publications on visual detection for SAA. Collision risk (or the probability of a collision if no action is taken) increases in proportion to the number of proximate aircraft pairs and approximately as the square of the number of aircraft. He also analyzed 649 near midair collision reports from 1968-69. The results covered several closing speed intervals for air carrier, general aviation, and military aircraft. See and avoid effectiveness probability was 0.97 from 101 to 199 knots closing speeds but was reduced to 0.47 at 400 knots or more closing speeds.

Graham (1989) noted that the failure of SAA is mostly due to the failure to see as opposed to avoid. Target detection is affected by many factors such as pilot visual acuity, air-to-air visibility, target size and aspect, target contrast, background complexity, crew workload, visual search patterns, and sun position. He also reported that the conspicuity of aircraft (paint color, etc.) does not have much effect on visual detection. Lights also have little influence on target detection in the daytime.

In a detailed review, Morris (2005) analyzed data on midair collisions in the U.S. between 1991 and 2000. There was total of 156 midair collisions for an average of 15.6 collisions per year with failure of SAA accounting for 94% of the incidents. Most collisions occurred during daylight hours. In 87.5% of the cases, at least one aircraft was maneuvering and for 69.7% both were. For 66.9% of the incidents, weather conditions were clear. Over half of the collisions occurred over or on a runway. Of the total of 156 collisions, 23.1% were head-to-tail on final approach or over the runway. The approach geometry of aircraft on final can make it impossible for the pilots to see the other aircraft. Morris concluded, "The see-and-avoid concept has physical and behavioral limitations such that pilots cannot reliably see-and-avoid conflicting aircraft. Pilots can find it physically impossible to see converging aircraft, especially when climbing or descending in an airport traffic pattern." (Morris, 2005, p. 364).

Hobbs (1991) thoroughly reviewed previous research on the use of SAA for collision avoidance, discussing the characteristics of visual search that affect the detectability of aircraft as well as other factors including workload, diffusion of responsibility, cockpit obstructions, glare, and limitations of the human visual system that impact SAA.[1] Hobbs concluded that, "The see-and-avoid principle in the absence of traffic alerts is subject to serious limitations." and "The most effective response to the many flaws of see-and-avoid is to minimize the reliance on see-and-avoid in Australian airspace." (Hobbs, 1991, p. 23).

Further buttressing Hobbs' cautions, a Canadian Transportation Safety Board report (Transportation Safety Board of Canada, 2016) on a midair collision in 2012 concluded that, "This accident has demonstrated yet again that relying solely on the see-and-avoid principle to avoid collisions between aircraft operating under visual flight rules in congested airspace is inadequate." In another review of the literature on SAA, Williams and Gildea (2014) stated that, "The majority of this [SAA] research has found a consistent inability on the part of a pilot to see other aircraft with a high degree of probability (e.g., Hobbs, 1991). Limitations of see-and-avoid have been shown in both actual flight tests (Andrews, 1977, 1984, 1991) and simulation studies (Wickens, Helleberg, Kroft, Talleur, & Xu, 2001; Colvin, Dodhia, & Dismukes, 2005; Morris, 2005)." (Williams and Gildea, 2014, p. 6).[2]

A paper by Andrews (1989) is particularly instructive. Following a midair collision between a Piper Archer and DC-9 in Southern California that resulted in 83 deaths, the National Transportation Safety Board contacted MIT Lincoln Laboratory for assistance with the analysis of the accident using their mathematical model of visual acquisition. In previous work, Lincoln Laboratories created estimates of unalerted and altered visual acquisition using the Traffic Collision and Avoidance System (TCAS). It was clear that alerted acquisition improved pilot performance. "… the presence of the TCAS traffic advisory increased search effectiveness by a factor of 8. In other words, one second of search with the TCAS advisory was as effective as eight seconds of search with no alert" (Andrews, 1989, p. 480). It was concluded that, where SAA failed, the DC-9 flight crew would have had a 95% chance of seeing the Piper Archer in time to avoid it had they been equipped with TCAS.

The above research suggests that the ability of a human, either pilot or observer, to see another aircraft is problematic even under ideal conditions. A recent review by Cianciolo (2022) notes that from 2016 to 2021, there were 43 reports of midair collisions involving GA operations in the United States, resulting in 79 fatalities, 43/6 = 7.2 per year. The literature is clear that using SAA to prevent midair collisions is a risky approach.

Multi-Object Tracking Research

While the research examined above has looked primarily at the detection problem, it is also relevant to consider what takes place following a detection and how this may explain the pilot's ability to avoid a collision. As shown in Figure 1, once detected, a nearby aircraft must be tracked to determine if it is a threat and continue to be tracked in case it becomes a threat. Furthermore, multiple aircraft may need to be tracked at any one time, particularly in dense or crowded airspace such as is envisioned for mature UAM. The literature on SAA does not generally consider this issue. However, the ability of the pilot to track another aircraft, once detected, is essential for determining if it is problem and, if so, to initiate a plan for an avoidance maneuver.

There is an extensive literature in cognitive psychology on multi-object tracking (MOT) that is useful to review regarding the stages of SAA that follow detection. These studies are focused on laboratory research where stimuli are presented on computer displays to investigate the perceptual and cognitive aspects of MOT.

---

[1] Refer to Hobbs (1991) for details on human visual and cognitive systems as they relate to SAA performance.
[2] Refer to Williams and Gildea (2014) for the references cited in the quotation.

In one experiment Tripathy et al. (2007) found that "The effective number of tracked trajectories varied between one and four, depending on the magnitude of the angle of deviation of the target trajectories" (Tripathy et al., p. 17). However, other researchers have argued that this limit may not be valid. Holcombe noted that "…it is incorrect to say that people can track about four moving objects, or even that once some number of targets is reached, performance declines very rapidly with additional targets. The number that can be tracked is quite specific to the display arrangement, object spacing, and object speeds" (Holcombe, 2022, p. 17).

It can be assumed from MOT research that there is a finite (and relatively small) number of objects a human observer can track concurrently. This means that, once aircraft have been visually detected, there will be a limited number that the pilot can track while evaluating collision potential. Other tasks that demand perceptual and cognitive resources (such as flying the aircraft) will limit tracking ability.

Multiple factors affect the ability to detect potential collisions during MOT. Some may be beneficial for SAA. For example, Lin et al. (2008) reported that during a visual search experiment, items that loom or grow larger abruptly capture attention more strongly when they approach from the visual periphery rather than from near the center of gaze. Also, objects are more likely to be attended to when they are on a collision path with the observer rather than on a near-miss path. Their findings suggest that the human visual system prioritizes events that are likely to require a behaviorally urgent response as is the case with detecting an aircraft that may be on a collision course.

However, there are factors which negatively impact performance. Tombu and Seiffert (2008) manipulated the visual aspects of an MOT experiment using a dual-task paradigm. The results showed that unrelated demands on perceptual and cognitive resources can have a negative effect on object tracking. Engaging in radio communications and manipulating flight displays and controls are some of the activities a UAM pilot would be engaged in addition to SAA. Performance decrements in detecting and tracking intruder aircraft would most certainly occur if these tasks occurred concurrently.

Airspace Structure

Airspace structure and operating procedures could improve the performance of SAA. It is expected that UAM aircraft will use well-defined corridors when operating in controlled airspace (FAA, 2020), The structure provided within the corridor may improve the performance of SAA by providing predictability. For example, vertically and horizontally fixed, one-way tracks inside the corridors would ensure that most other proximate aircraft should be either behind or in front of own ship, while other aircraft are confined to different corridors, thus decreasing the likelihood of collisions. On the other hand, pilots on tracks in corridors might be less likely to detect intruders coming from unanticipated directions. The chances of failing to detect an aircraft being overtaken are low since closure rates are low although aircraft ahead will appear smaller than those at other intersecting angles.

It may be impractical to use a corridor structure outside controlled airspace (Class B/C/D). As operations increase, there would be a proliferation of intersecting corridors, making traffic management difficult. Thus, UAM aircraft will, like conventional GA traffic, use SAA in uncontrolled airspace.

Discussion

The aviation literature is consistent in stating that unaided SAA is a risky method for avoiding midair collisions. Each step in the SAA sequence requires perceptual and cognitive resources in addition to those needed to aviate, navigate, and communicate and has its own probability of success. Detection of and tracking other aircraft is negatively affected by perceptual and cognitive limitations and competing demands. Then, once detected, a pilot must track the aircraft - and humans can only track a limited number of targets - while evaluating the collision threat and planning any needed avoidance maneuvers.

This paints a gloomy picture for the effectiveness of unaided SAA for UAM. From 2016 to 2021, there were 7.2 midair collisions per year involving GA operations (which use SAA) in the United States. While these numbers are not high, even one or two accidents involving UAM, passenger-carrying aircraft could be catastrophic for the burgeoning UAM industry.

What are the prospects for using SAA for initial UAM operations? The conservative approach is that unaided SAA outside of airspace corridors is unsafe at any traffic density. However, research has shown that detection probability is improved by a factor of eight if a cockpit display of traffic information is used to aid visual search. Such a display could, at a minimum, also assist with tracking the target by showing a history trail and predictor line as found on air traffic control screens. This would augment the human visual, out of the window visual search and tracking skills of the pilot. If a surveillance system locates, tracks, and predicts the intruding aircraft's trajectory and displays this to the pilot, a conflict detection and resolution algorithm could complete the evaluate and calculate phases of the SAA process. Thus, a strong case can be made for flight deck systems to provide location information and collision avoidance for the pilot (Chamberlain et al., 2017).

Conclusions

The use of SAA for UAM operations is risky. The performance of SAA can be improved by using airspace structure and supportive flight deck technologies. As UAM vehicle and airspace designs evolve, a detailed analysis of collision avoidance risk using SAA and other approaches needs to be conducted. Although SAA is generally accepted for today's operations, this does not mean it should be carried forward for the new industry. An accident rate of 7.2 midair collisions per year may be implicitly accepted as a reasonable risk for GA flights. This would never be tolerated for large, passenger-carrying aircraft and should be not acceptable for UAM. These kinds of accidents would deter the advent and growth of the UAM industry.

References

Andrews, J. W. (1989). Modeling of air-to-air visual acquisition. The Lincoln Laboratory Journal, 2(3), 475-481.

Chamberlain, J., Consiglio, M, & Munoz, C. (2017). DANTi: Detect and avoid in the cockpit. 17th AIAA Aviation Technology, Integration, and Operations Conference, Denver, CO.

Cianciolo, P. (2022, October). Mid-air collision report. https://www.gajsc.org/gajsc/press-release/#:~:text=Midair%20collisions%20are%20a%20persistent%20and%20deadly%20threat,in%20the%20United%20States%2C%20resulting%20in%2079%20fatalities.

Federal Aviation Administration (2020). Concept of operations V1.0 urban air mobility (UAM), Washington, D.C.

Graham, W. (1989). See and avoid/cockpit visibility. FAA Technical Note, DOT/FAA/CT-TN/89/18.

Hobbs, A. (1991). Limitations of the see-and-avoid principle. Australian Transport Safety Bureau.

Holcombe, A. O. (2022). Attending to moving objects. [Unpublished manuscript.]

Lin, Y. L., Franconeri, S., & Enns, J. T. (2008). Objects on a collision path with the observer demand attention. Psychological Science, 19(7) 686-692.

Morris, C. (2005). Midair collisions: Limitations of the see-and-avoid concept in civil aviation. Aviation, Space, and Environmental Medicine, 76(4) 357-365.

Mueller, E., Kopardekar, P., & Goodrich, K., (2017, June 5-9). Enabling airspace integration for high-density on-demand mobility operations. 17th AIAA Aviation Technology, Integration, and Operations Conference, Paper 2017-3086.

Thipphavong, D., Apaza, R., Barmore, B., Battiste, V., Burian, B., Dao, Q., Feary, M., Go, S., Goodrich, K., Homola, J., & Idris, H. (2018, June 25-29). Urban Air Mobility Airspace Integration Concepts and Considerations. 18th AIAA Aviation Technology, Integration, and Operations Conference. Paper 2018-3676.

Tombu, M., & Seiffert, A. E. (2008). Attentional costs in multiple-object tracking. Cognition, 108, 1–25.

Transportation Safety Board of Canada (2017). Mid-air collision between W.M.K. Holding Ltd. (DBA McMurry Aviation), Cessna 172P, C-GJSE and Cessna AE185E, C-FAXO, Fort McMurray Alberta, 21 NM NE, 21 June 2015. Aviation Investigation Report A15W0087.

Tripathy, S. P., Narasimhan, S., & Barrett, B. T. (2007) On the effective number of tracked trajectories in normal human vision. Journal of Vision, 7(6), 1–18.

Uber Elevate. (2017). Fast-forwarding to a future of on-demand urban air transportation. https://www.uber.com/elevate.pdf.

Williams, K.W., & Gildea, K.M. (2014). *A review of research related to unmanned aircraft system visual observers*. FAA/CAMI Report: DOT/FAA/AM-14/9.

WHAT CAN WE LEARN FROM RESILIENT PILOT BEHAVIORS? THE CASE OF ENERGY MANAGEMENT WHILE FLYING A STAR

Randall J. Mumaw
San Jose State University Foundation
San Jose, CA

Dorrit Billman
NASA Ames Research Center
Moffett Field, CA

Barth Baron, Jr.
San Jose State University Foundation
San Jose, CA

Recently, there has been increased interest in documenting flightcrew behaviors that contribute to safe operations. Instead of only capturing errors, new efforts are attempting to understand how pilots manage complexity and variability in the operational environment to ensure a safe mission. This approach highlights pilot responses to events and conditions that fall outside typical TEM threats; e.g., revised ATC clearances. This approach presents a two-sided coin: characterize flightcrew resilience /or/ generate insights regarding complexity in the operational environment that is not adequately managed by current flight deck interface designs, procedures, and training. To capture operational complexity, we have been analyzing flight path management tied to flying an RNAV STAR. Because ATC often requests revisions—e.g., descend late—and because RNAV STARs may not align with airplane performance limits, flightcrews need to monitor, anticipate threats to RNAV STAR compliance, and devise ways to accommodate unexpected challenges. In this paper, we identify general strategies that can support response adaptation and explore methods to facilitate training these strategies.

The Emergence of Safety II and Resilience

Operational safety in aviation (and other domains) has long been framed in terms of probabilistic risk assessment (PRA) in which risk is associated with airplane system failures, upsets, or erroneous flightcrew actions that need to be managed or mitigated. In this framework, flightcrew performance is judged by the flightcrew's ability to recognize and manage failures and upsets. When an unsafe outcome occurs, the event is typically described in terms of a flightcrew failure; e.g., loss of situation awareness, misdiagnosis, inappropriate control actions. Thus, the primary markers of safety within the PRA framework—accidents and incidents—are described as events in which flightcrew performance falls short of prescribed decisions and actions. This approach has led to an investment in error classification schemes (e.g., Wiegmann & Shappell, 1997) to capture and understand the types of errors that flightcrews are most likely to make.

In the last 15 years, however, a complementary perspective on flightcrew performance and operational safety has emerged that focuses on the flightcrew's ability to manage the normal variability and complexity in the operational environment that is not adequately managed by current flight deck interface designs, procedures, and training. This perspective has been referred to as Safety II (Hollnagel, 2014).

Similarly, there has long been an emphasis in aviation on adherence to standard operating procedures (SOPs) for the flightcrew, but careful analysis reveals that SOPs fall short in describing the full range of necessary flightcrew actions. According to this perspective, to understand operational safety, it is important to capture how operators identify and respond to unexpected or atypical operational demands. It is rare that a commercial transport flight, especially in the US, proceeds exactly as specified in the flight plan. Air Traffic Control (ATC) responds to weather and traffic patterns and other disruptions in the National Airspace System (NAS) by revising the flight path of aircraft in the NAS; examples are changes to routing, airspeed, or altitude. For example, as an airplane is descending to an airport and cleared to land on a specific runway, the winds shift considerably, and ATC asks approaching airplanes to re-route to a different approach and runway. The flightcrew makes changes to flight plan restrictions to force the airplane down earlier, which requires a bit of creative problem-solving.

Hollnagel and others (e.g., Hollnagel et al., 2006) have developed a language around these system behaviors that focuses on "resilience." According to Hollnagel (2019), "*A system is resilient if it can adjust its functioning prior to, during, or following events (changes, disturbances, and opportunities), and thereby sustain required operations under both expected and unexpected conditions.*"

Thus, as a complement to traditional safety practices that attempt to reduce or mitigate flightcrew errors and establish strong SOPs, the idea behind Safety II and resilience is acknowledging that an effective flightcrew plays a significant role in anticipating and managing the variability and complexity in the operational environment.

Two Views: Resilience vs Operational Complexity

This emerging perspective on positive flightcrew contributions to operational safety has generated a strong interest in capturing and documenting resilient flightcrew behaviors. Analysts have largely borrowed the Hollnagel framework—monitor, anticipate, respond, and learn—for categorizing these behaviors. From Hollnagel (2011):

- The ability to anticipate. Knowing what to expect or being able to anticipate developments further into the future, such as potential disruptions, novel demands or constraints, new opportunities, or changing operating conditions.
- The ability to monitor. Knowing what to look for or being able to monitor that which is or could seriously affect the system's performance in the near term – positively or negatively. The monitoring must cover the system's own performance as well as what happens in the environment.
- The ability to respond. Knowing what to do or being able to respond to regular and irregular changes, disturbances, and opportunities by activating prepared actions or by adjusting current mode of functioning.

- The ability to learn. Knowing what has happened, or being able to learn from experience, in particular to learn the right lessons from the right experience.

Clearly, there is value in documenting that flightcrew behaviors frequently reveal resilience, which furthers our understanding of SOP limitations (broadly defined). On the other side of the coin—opposite resilient behaviors—is the variability and complexity in the operational environment. We believe that there is equal (if not greater) value in understanding the drivers of resilient behaviors. Specifically, how is the operational environment creating situations that require the flightcrew to adapt and use resources outside SOPs and training? Understanding complexity in the operational environment is important because it forces us to acknowledge that the larger system—airplane design, ATC procedures and clearances, pilot training, etc.—needs to evolve to reduce the need for unsupported flightcrew performance. These insights can potentially lead to changes in interface design, training, or other system characteristics.

Case Study: Monitoring for Flight Path Management

In an exploration of monitoring (Mumaw et al., 2020), we documented knowledge, skills, and strategies that experienced pilots use for flight path management during descents; specifically, in flying Area Navigation (RNAV) Standard Arrival Routes (STARs) to the approach. While RNAV STARs are designed to account for a certain degree of adverse conditions, such as a tailwind, there can be considerable complexity and variability introduced by ATC revisions. There is a range of conditions in which a flight can be forced off the RNAV STAR. For example, ATC's traffic load can require them to slow a flight earlier than planned or to take it off its planned lateral path. The revised ATC clearance may still require that the flightcrew meet waypoint airspeed and altitude restrictions, and the flightcrew needs to understand how to revise some element of the clearance and still comply with waypoint restrictions. In these cases, the flight management system (FMS) predictions likely become invalid, and the flightcrew is required to reason through the changes to intervene effectively.

When we discussed these situations with experienced pilots, it became clear that there is no formal/explicit training on how to
- anticipate potential threats to flight path compliance,
- monitor indications to determine how likely it is that the airplane will comply with the clearance,
- respond/intervene through FMS flight plan modifications or actions on the flight controls.

However, despite the lack of explicit training, these pilots had developed methods for dealing with the "normal" variability and complexity that can be encountered on most flights. These methods offer a clear illustration of resilient performance.

Having uncovered this demand for adaptive responding from the operational environment, the challenge then becomes how to improve flightcrew performance, especially for less-experienced pilots. We chose to develop targeted training to fill the current gap. Although the situations that flightcrews might face can vary considerably—across RNAV STARs, airports, wind conditions, and air frames (to name a few factors)—we were able to articulate the knowledge, skills, and strategies for managing descents (discovered in our work) and convert them into a training module to support resilient performance. An initial question is how to select a level for describing and training this resilient performance. At one end of the continuum, training could focus on general energy-management principles for all airplanes, or could even attempt to introduce principles of "resilient responding" more abstractly. At the other end, training could separate out the specifics of airplane performance, runway layouts, local airport customs, etc. We chose instead a middle ground that would give pilots a set of general problem-solving skills around a small number of problem types. We believe this formulation can serve as a model for training skills foundational to resilient flightcrew performance.

The Problem Space and Training Approach

The problem space, as we first encountered it, was large: managing compliance to an RNAV STAR in the face of shifting winds, weather, ATC needs, airplane performance limits, etc. To identify anchors for training, we sought representative problem types. By filtering Aviation Safety Reporting System (ASRS) reports for missed crossing restrictions during descent and discussing operational practices with experienced line pilots, we

- collected a set of cases where crews reported violations of altitude or airspeed constraints along RNAV STARs, and
- identified knowledge, skills, and strategies pilots used to manage these successfully.

From these data, we noticed that poor outcomes had these two characteristics:

- crews found themselves higher than originally planned (too much energy), often violating the constraints and,
- the situation could have been anticipated and prevented through early control action or FMC flight plan changes

We were able to also identify three problem types, which are connected to three types of ATC clearance revisions[1]:

1. Held high, meaning prevented from descending at the anticipated point along the arrival,
2. Slowed early, causing the crew to shallow their descent gradient to accomplish the deceleration, and
3. Loss of track miles; a change that substantially reduces required track miles.

Further, each of these problem types can occur for different reasons; for example, loss of track miles can occur when ATC gives a "direct to" clearance that eliminates intermediate waypoints, or when there is a need to land on a closer runway. Together, these problem types capture the range of energy-management / flight path management situations, and each type represents common ATC practices for managing traffic and environmental conditions. More importantly, these types of clearance revisions commonly lead to flight path management difficulties in line operations, and they are issues that SOPs may not directly address.

Our approach to training attempts to use a range of operational scenarios to illustrate strategies for anticipating, monitoring, and responding to manage each problem type. We believe it is possible to use this approach to aid pilots in seeing specific operational cues for action and to

[1] A few other problem types could be called out, but we believe the three specified here provide adequate grounding for training the necessary knowledge and skills.

provide problem-solving skills for each problem type. The goal is to both support performance on specific problems and to facilitate transfer across a wide range of operational variability and complexity.

We are also combining training on flight path management skills—anticipating, monitoring, responding/controlling—with important principles about flightcrew communication. When one pilot becomes concerned about potential threats or inadequate performance, it is critical to share those concerns/expectations with the other pilot. More specifically, we are emphasizing several types of communication: identifying potential concerns and jointly planning how to monitor for them; sharing expectations about flight path and how it will be managed; and updating information about status. Updates may include positive information (e.g., potential threats resolved), as well as notification of developing concerns. A subtext of this material is an elevation of the role of the Pilot Monitoring (PM). Traditionally, the PM is trained (and assessed) largely to identify and call out deviations from current flight path targets; e.g., airspeed is 6 kts too fast. In our training module, the PM is given broader responsibilities to develop a view of downstream flight path constraints to aid in anticipating potential threats to compliance[2]. Finally, we believe that the identification and training of "resilience" needs to be grounded in specific operations; that is, solving operational problems within a specific domain. This grounding allows pilots to understand that the operational environment demands working "beyond SOPs" and highlights specific knowledge and skills to address that need.

We are currently planning a flight simulator-based study to determine if this training can improve flightcrew performance when flying challenging RNAV STARs. In early reviews of our training module, reviewers recognized the relevance of the skills for addressing the current gaps in training. Looking forward, another potential use of this training is to facilitate transfer to the full range of operational situations. Indeed, we have discussed whether these flight path management skills could be generalized to operational problems around fuel management or other aspects of mission monitoring. Our approach is to start with a grounding in one operational area and then create awareness of the applicability of these skills to other operational needs.

Summary and Conclusions

We identified an area—flight path management along RNAV STARs—in which experienced pilots revealed knowledge, skills, and strategies that allowed them to successfully manage variability and complexity in the operational environment; these were not formally trained but acquired through experience. We developed a training module intended to improve flightcrew performance. The basic tenets behind our training module are that training content should:

- present common features of the variability and complexity in the operational environment to identify
    - strategies relevant to monitoring and assessing flight path
    - the relevant classes of operational situations that can be addressed using these strategies

[2] Note that this framing is different from current Threat and Error Management (TEM) notions. For our training, a potential threat to fight path management can be a reduction in track miles, which is unlikely to be treated as a threat in TEM.

- features and cues that can be used to recognize these classes
- build supporting skills, such as crew communications, for integration back into the flight deck setting
- provide principles that support generalization to novel situations

Further, with respect to the learning process, our training module emphasizes that training should provide:

- frequent opportunities for realistic problem solving and trainee interaction with the content
- opportunities to practice sub-skills
- opportunities to reason with foundational concepts
- opportunities for reflection and relating new material to prior experience

An upcoming evaluation study will assess the impact of our training module on understanding and performing in both practiced and novel flight path management situations. It will also inform us of strengths and weakness of the module and about directions for improvement.

Acknowledgments

The work reported here was funded by NASA's System-Wide Safety Project, part of the Aeronautics Research Mission Directorate's Aviation Operations and Safety Program.

References

Hollnagel, E. (2011). RAG – The resilience analysis grid. In E. Hollnagel, J. Paries & J. Wreathall (eds), *Resilience engineering in practice*, London, UK: CRC Press.

Hollnagel, E. (2014). Safety-I and Safety-II: The Past and Future of Safety Management. Farnham, UK: Ashgate.

Hollnagel, E. (2019). https://www.resilience-engineering-association.org/blog/2019/11/09/what-is-resilience-engineering/

Hollnagel, E., Woods, D.D. & Leveson, N.G. (2006). Resilience engineering: Concepts and precepts. Aldershot, UK: Ashgate.

Mumaw, R.J., Billman, D., & Feary, M. (2020). Analysis of pilot monitoring skills and a review of training effectiveness. NASA/TM-20210000047.

Wiegmann, D. & Shappell, S. (1997). Human factors analysis of post-accident data: Applying theoretical taxonomies of human error. *The International Journal of Aviation Psychology, 7*, pp. 67-81.

EVALUATING ENVISIONED AIR MOBILITY ARCHITECTURES USING COMPUTATIONAL SIMULATIONS OF WORK

Abhinay Paladugu[1], Alicia Fernandes[2], Stuart Wilson[2], Thomas J. Davis[3], Jarrod Lichty[3], Martijn IJtsma[1]
[1]Department of Integrated Systems Engineering, The Ohio State University, Columbus, OH, United States
[2]Mosaic ATM and [3]Aerial Vantage, Leesburg, VA, United States

Urban Air Mobility (UAM) is an envisioned concept of operation for managing uncrewed and crewed flights for urban, regional, and interregional air transportation. One element of further development of this envisioned system is to specify architectures in terms of roles and procedures for managing contingencies. Contingency management is a highly distributed function involving coordination between multiple system actors. In this study, a computational model is applied to analyze envisioned procedures and identify architectural solutions to improve the robustness of the contingency response. The simulation framework Work Models that Compute (WMC) is used to analyze a proposed UAM lost link procedure in the Dallas-Fort Worth (DFW) airspace while varying task design and control authority of operators, service providers, pilots, and vertiport operators. The simulations provide insights into how candidate designs align or misalign with the dynamics of the contingency. This approach can improve the design and verification of procedures in similar envisioned operations.

Urban Air Mobility (UAM) and Uncrewed Air Systems (UAS) Traffic Management (UTM) are vision concepts gathered from over 100 stakeholder organizations. The envisioned nature of these concepts of operations (ConOps) creates challenges in designing, verifying and validating these concepts (National Aeronautics and Space Administration (NASA), 2020). UAM concepts envision highly automated future air transportation and describe procedures for addressing contingencies that implicitly assume automated services will be able to coordinate well enough with little or no human input, if procedures are pre-defined. However, if traditional air traffic operations are to be a guide, humans must be involved in coordinated contingency planning for UAM and UTM operations.

The present study aims to identify and evaluate alternative architecture and procedure designs for future operations to support efficient and robust contingency management between human and automated systems. This can be accomplished by developing and simulating computational models of the work involved in envisioned UAM contingency management.

Background

Although the UAM and UTM ConOps recognize the need for contingency management and include relevant procedures, there has been limited opportunity to validate the procedures, including the appropriate distribution of functions between organizations in the UAM and UTM ecosystem and between humans and automated systems. There is some disagreement among the different concept documents regarding sharing of contingency planning information. For example, the FAA's UTM Evaluation 4 assumed that "Operators submit information on their contingency procedures to the FAA when obtaining operational approvals to conduct flights" (Mosaic ATM, 2021). However, the progressing draft standard for UTM (ASTM, 2021) does not include contingency plans as part of a vehicle's operational intent except as a supplement when the aircraft is in the Nonconforming or Contingent states. The justification is that an operator does not intend for the aircraft to enter the Nonconforming or Contingent states, and so it should not include off-nominal volumes in its operational intent. Rather, the expectation is that off-nominal volumes encompass where the aircraft may travel to support situation awareness for other operators.

Several candidate procedures for coordinated contingency planning are conceivable. For example, contingency planning could be centralized in a Centralized Contingency Planning Service or distributed across providers of services to UAM (PSUs), UAS service suppliers (USS), and other organizations. Procedures for responding to contingencies could range from local to system-level adjustments. There are fundamental trade-offs associated with these alternatives that remain relatively underexplored in existing ConOps; for example, how coordination overhead are incurred for different allocations of control authority.

To address this gap in the UAM ConOps, this study implements a selected, detailed contingency planning scenario in the Work Models that Compute (WMC) fast-time simulation framework for comparing candidate architectures on their relative merits (e.g., coherence, ensuring access to the needed information, coordination overhead). WMC is a computational modeling and simulation framework for analyzing situated work (Pritchett et al., 2014), used before to study air traffic management (Pritchett et al., 2016) and space operations (IJtsma et al., 2019), among other applications. WMC simulates the detailed interaction between actions and the work environment (as captured in resources), including how activity of actors in the system is interconnected through dynamics and information exchanges. WMC provides quantitative data on the dynamics of activity, such as when and how often actions are performed, and frequency and content of information sharing amongst actors.

Computational Model of Contingency Planning Functions

Scenario. The UTM ConOps lists multiple off-nominal scenarios, including a loss of command and control (C2) link event (NASA, 2019) in which a single vehicle loses a C2 link. The present study modeled a scenario with two UAM aircraft: one remotely piloted using the C2 link and another with an onboard pilot. Figure 1 shows the original filed flight plan for both aircraft at the start of the scenario. Aircraft with "tail number" N12345 takes off from the Frisco vertiport and flies through six waypoints before reaching the Dallas vertiport. Aircraft N54321 travels from the Dallas vertiport to the T57 Garland vertiport. Midflight, the remotely piloted N12345 loses its link with the remote pilot in command (RPIC). Aircraft N12345 has a pre-determined contingency plan loaded for a loss of the C2 link, as described in Uncrewed UAM ConOps (Boeing, 2022), which is automatically activated by the vehicle. This alternate plan is to land at one of three pre-coordinated vertiports, selected based on the current aircraft position.

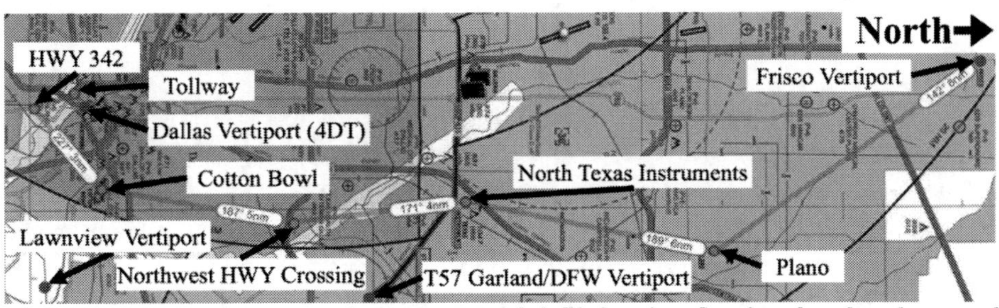

Figure 1. Airspace layout and N12345 Initial flight plan for the simulated scenario.

System Actors and Envisioned Procedure. A subset of system actors involved in responding to a loss of C2 link contingency were identified: the UAM aircraft (N12345 and N54321), one pilot in command of the crewed vehicle (PIC2), one remote pilot in command for the uncrewed vehicle (RPIC1), two fleet operators (FleetOperator1 and FleetOperator2), two agents representing vehicle automation (N12345Automation and N54321Automation), one PSU, and one vertiport agent (Vertiport1). Following detection and confirmation of the lost C2 link event, the procedure is envisioned as the fleet operator or RPIC alerting the PSU and air traffic control (ATC). The PSU network then distributes the aircraft's pre-determined contingency plan to impacted UAM actors (PSU(s), vertiport(s), fleet operator(s)). The affected traffic alters its trajectories as needed to avoid conflicts with the pre-coordinated contingency flight trajectory of the lost C2 link aircraft. This replanning involves coordination and negotiation between PSU(s), vertiport(s) and fleet operator(s)) to finalize new operational plans.

WMC Modeling. The full list of WMC actions is given in Table 1. The flight dynamics of the aircraft are an essential driver of the dynamics of these functions, determining much of the system actors' timing of activity to keep pace with disturbances. Thus, the computational work model includes a model of the flight dynamics for a generic UAS, with parameters that can be changed to simulate a variety of vehicle classes (e.g., a small quadrotor UAS or a large package delivery drone). The actions model includes code that describes the interaction between the actions and the flight dynamics.

Table 1.
List of actions modeled in the WMC simulation framework.

Detect loss of link	Update flight plan	Divert and land at alternate vertiport
Confirm loss of link	Update route trajectory	Avoid loss of separation
Communicate lost link	Request new operations plan	Respond to request for new ops plan
Distribute loss of link	Request new route trajectory	Accept or deny operations plan
Update arrival time	Request new arrival time	Clear landing pad lost link aircraft

Task Design and Architecture Candidates. Table 2 outlines alternative architectures that differ in what actions are performed and which system actor has the authority to conduct each action. Thus, they define each system actor's role in managing the contingency, with each alternative requiring a unique set of information sharing and coordination requirements. The baseline architecture (Case 0) was simulated with alternate ways of responding to the contingency: not diverting whatsoever (i.e., all aircraft continuing their originally planned trajectory), only diverting the lost C2 link aircraft (i.e., no change in trajectory for other traffic), and full system replanning (i.e., all impacted aircraft rerouting). Alternate roles (allocations of authority) were also explored for the full system replanning response. While theoretically there are $x^n = 15^{10}$ alternative ways of distributing the actions between the agents, with x the number of functions and n the number of agents, a subset was defined based on subject-matter expertise, shown in Table 2. The enumerated cases were simulated with the loss of link occurring at 160 seconds and at 240 seconds to evaluate the sensitivity with respect to timing of the contingency event, resulting in a total of 16x2=32 simulation runs. From the results, the time at which the link was lost was not an important factor.

Results

Figure 2 shows various metrics relative to the "No Diversion" baseline (except "Delay of Flight," which is measured relative to longest recorded delay) to allow comparison. The results show distinct differences between alternate ways of responding to the contingency. In the "No Diversion" case, the lost C2 link vehicle (N12345) flies without a C2 datalink for a significant amount of time (more than 16 minutes). All other responses that have N12345 divert to an alternate vertiport show half the flight time without datalink but do cause the secondary aircraft to incur a delay due to its reroute. Flying without a datalink has inherent risks, in that there is a vehicle in the airspace that is uncontrollable and possibly unpredictable. Thus, while the "No Diversion" case shows less impact to the system (i.e., improved performance on flight delays), the risk and uncertainty will be higher (and possibly unacceptable).

The "No Diversion of N54321" case has the lost C2 link aircraft diverting to the alternate vertiport, but the secondary aircraft continues its original flight plan. This is a local response to the contingency, without a system-level response of rerouting other impacted traffic. This case resulted in a loss of separation and avoidance maneuver to prevent a collision.

Table 2.

Different architectures evaluated through WMC simulations

Case	Description	Detect Loss of Link, Confirm Loss of Link, Communicate Loss of Link Alert	Distribute Lost Link	Request Updated Arrival Time, Request Updated Route Trajectory	Accept or Deny Updated Operations Plan	Update Flight Plan
0	**No Diversion**	FO1	PSU	--		
0a	**No Diversion of N54321**					
1a	**System Response + Baseline Design**	FO1	PSU	PIC2	PIC2	PSU
1b	RPIC Detects	RPIC1				
1c	RPIC, FO Detect	RPIC1 & FO1				
2a	FO Detects + Distributes Alert	FO1	FO1	PIC2	PIC2	PSU
2b	RPIC Detects + FO Distributes Alert	RPIC1				
2c	RPIC, FO Detect + FO Distributes Alert	RPIC1 & FO1				
3b	FO Requests Update	FO1	PSU	FO2	PIC2	PSU
3c	PSU Requests Update			PSU		
4a	PIC Requests + FO Accepts Update	FO1	PSU	PIC2	FO2	PSU
4b	FO Requests + Accepts Update			FO2		
4c	PSU Requests + FO Accepts Update			PSU		
5a	FO Updates Plan	FO1	PSU	PIC2	PIC2	FO2

Note. FO = Fleet Operator. Gray cells note differences w.r.t. baseline for system-wide response (Case 1a).

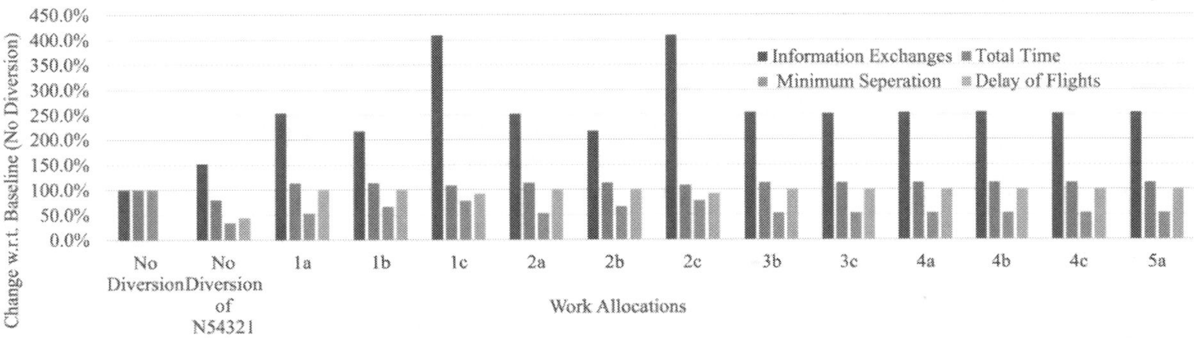

Figure 2. System Metrics with loss of C2 link at 240s

Three alternate strategies were tested for allocating detection tasks for lost C2 links to fleet operators alone, pilots alone, or both fleet operators and pilots. The simulation results showed that a lost C2 link is detected earlier when both actors monitor for this event (both RPIC1 and FleetOperator1 perform the "Detect Loss of Link" function). Comparing cases 1a/2a with 1b/2b and 1c/2c shows that earlier detection results in a higher minimum distance between the vehicles but comes at an increased coordination overhead for information sharing. The increase in information exchange load is because both agents share information with each other regarding the last time each agent received a ping.

Figure 3 shows the total taskwork duration for each alternative test condition. Different architectures distributed taskload differently among the agents. Taskload data shows that total taskwork duration increases moving from cases 1a, 1b to 1c and 2a, 2b to 2c. Thus, while an improved (faster) detection of a lost C2 link allows the vehicles to maintain a higher minimum separation, it increases the information exchange requirements and the overall taskload.

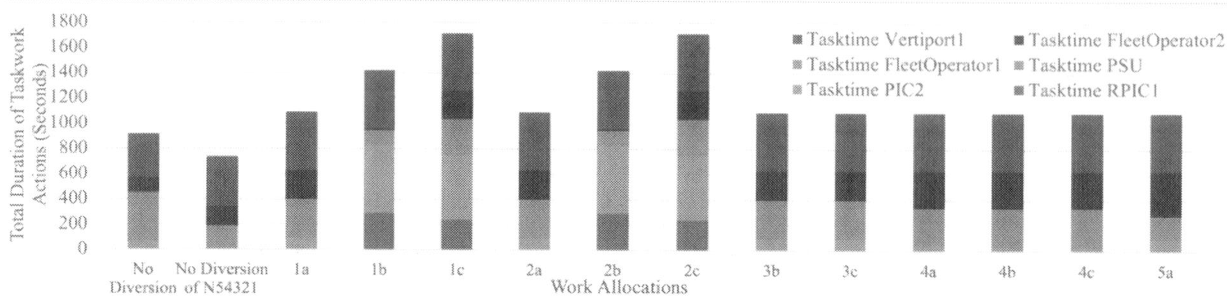

Figure 3. Total agent taskload duration with loss of C2 link at 240s

Discussion and Conclusion

The simulation helped uncover two trade-offs in task design for a lost link contingency procedure. First, there is a fundamental tradeoff between local responses (in this case having the lost C2 link aircraft continue to fly, creating potential for significant risk and uncertainty about the aircraft's state and intentions) and system-level responses, that create significant taskload for all parties involved. Note that the uncertainty associated with an aircraft without a C2 link operating in the airspace was not tested as part of the simulations in this study, but this remains a rich area of exploration to consider for future work. Second, there is a tradeoff between the number of agents monitoring, the interval of monitoring, the required information exchange, and the overall taskload on agents and the entire system. The specific risks, inefficiencies, and other tradeoffs among the procedures are likely to be highly context-dependent according to factors such as the airspace design, secondary traffic that might be impacted, the specific procedure for replanning, and the complexity of the replanning task.

The simulations also revealed how the success of contingency management is determined by the ability of the system actors to collectively stay in sync with the high tempo of operation. The dynamics of the UAM, UTM, and AAM systems are fast, resulting in the system's response to contingencies being highly time-pressured. Simulation of the aircraft dynamics relative to the envisioned procedures showed that when the system's response is slow and stale relative to the tempo of operations, system actors easily lose control. In this particular scenario, when reroutes were not approved or implemented quickly enough, separation was lost between the lost link vehicle and secondary vehicle. This points to potential performance requirements for supporting high tempo work during contingency response.

Even with fast responses, the lost C2 link procedures envisioned in this study have a period of time in which secondary vehicles continue to fly on their original (approved) flight plan while a lost C2 link aircraft (automatically) implements its preprogrammed contingency response. Thus, when lost C2 link aircraft automatically implement alternate flight paths, there is a risk of loss of separation while secondary aircraft are reconfiguring their own flight paths to accommodate the lost C2 link vehicle. Likewise, while generating new flight plans for airborne aircraft during a contingency, replanning needs to account for the time it will take to plan and negotiate reroutes (Atkins et al., 2018). During this time, aircraft continue to fly their original flight plan. Any reroutes need to account for the stretch flown on this original path before the aircraft will receive its amended operational intent. Thus, the simulation revealed the need for every operational intent amendment to account for a Minimum Trajectory Negotiation Duration (MTND) (Atkins et al., 2018), which is yet to be defined for AAM concepts.

The simulation results also showed that there is significant potential for cascading effects. For example, an aircraft with a lost link might require other vehicles to reroute, which can create secondary conflicts that need to be resolved, or changes in airspace demands that need to be accommodated. This implies that if one vehicle alters its operational intent, system-wide replanning may very well be necessary to maintain safety. Note that this may apply in more than just contingency scenarios.

Finally, this study demonstrates how the computational simulation of envisioned architectures can be a useful tool for analyzing and evaluating contingency management procedures. Modeling and simulation provides a way to assess the feasibility of envisioned task designs and provides quantitative data that can be a basis for making informed design decisions around distributed work. The process of implementing the lost C2 link scenario in the WMC simulation capability helped address the complexity of a distributed system responding to a dynamic event, which led to documentation of assumptions, models of the tasks under investigation and reasonable (as opposed to unreasonable) designs, as well as some of our functional and performance requirements. The process also helped uncover gaps in procedure designs and architecture candidates as areas of possible concern that warrant more detailed attention in future work, such as the complexity associated with system replanning.

Acknowledgements

This work is supported by a NASA Small Business Innovation Research (SBIR) grant with Jason Prince serving as NASA Technical Monitor under Grant No. 80NSSC22PB098. The views of the research reported does not reflect the views of the granting organization.

References

ASTM. (2021). *UTM DRAFT Standard v0.3.*

Atkins, S., Evans, M., Bell, A., Kilbourne, T., Kirk, J., & Jackson, M. (2018). *Concept of Operations for Management by Trajectory.*

Boeing. (2022). *Concept of Operations for Uncrewed Urban Air Mobility.*

IJtsma, M., Ma, L. M., Pritchett, A. R., & Feigh, K. M. (2019). Computational Methodology for the Allocation of Work and Interaction in Human-Robot Teams. *Journal of Cognitive Engineering and Decision Making, 13*(4), 221–241. https://doi.org/10.1177/1555343419869484

Mosaic ATM. (2021). *UTM Evaluation 4 Final Report*. Mosaic ATM, Inc.

National Aeronautics and Space Administration (NASA). (2019). *Unmanned Aircraft System (UAS) Traffic Management (UTM).*

National Aeronautics and Space Administration (NASA). (2020). UAM Vision Concept of Operations. In *NASA.gov* (Vol. 1, pp. 0–94). https://ntrs.nasa.gov/citations/20205011091

Pritchett, A. R., Bhattacharyya, R. P., & IJtsma, M. (2016). Computational Assessment of Authority and Responsibility in Air Traffic Concepts of Operation. *Journal of Air Transportation, 24*(3), 93–102. https://doi.org/10.2514/1.D0024

Pritchett, A. R., Feigh, K. M., Kim, S. Y., & Kannan, S. K. (2014). Work models that compute to describe multiagent concepts of operation: Part 1. *Journal of Aerospace Information Systems, 11*(10), 610–622. https://doi.org/10.2514/1.I010146

A COMPARISON OF VIRTUAL REALITY AND HIGH-FIDELITY FLIGHT TRAINING DEVICES FOR AB INITIO PILOT TRAINING

Naomi Paul
University of Waterloo
Waterloo, ON, Canada

Brad Moncion
University of Waterloo
Waterloo, ON, Canada

Shi Cao
University of Waterloo
Waterloo, ON, Canada

High-fidelity flight training devices (FTDs) have value for ab initio pilot training, but their high cost is a major limitation. Researchers want to know if low-cost simulators, including virtual reality (VR), may be effective for some aspects of ab initio pilot training, supplementing FTDs. This work used a between-subjects experiment with 20 participants – all student pilots at the University of Waterloo with under 20 hours of flight experience – to analyze performance changes after training using either a FTD or VR simulator for procedural and aircraft handling tasks over three training sessions. Performance was assessed using instructor evaluations for 18 criteria. Participants training on the VR simulator showed similar improvement in performance for some procedural tasks but less improvement for aircraft handling tasks compared to participants training on the FTD. The findings emphasized the need for future studies to identify tasks that can and cannot be trained using low-fidelity VR simulators.

Researchers globally have conducted studies suggesting that virtual reality (VR) may be an effective simulation method and hypothesize that it may be able to replace or be used in conjunction with traditional simulation methods for pilot training. These traditional methods include flight training devices (FTDs) which are high-fidelity simulators used for teaching and practicing flight. FTDs artificially re-create the experience of flight through a model of the cockpit including physical controls and instrument panels.

Fidelity of simulators refers to "how accurately a simulator represents the real-world experience," (Suzanne K. Kearns, 2021) and involves three aspects: physical, cognitive, and functional fidelity (Myers III et al., 2018).

VR is an interactive and immersive experience into a simulated environment (Mazuryk & Gervautz, 1996) created using a combination of hardware, a head-mounted display (HMD) and controllers, and computer software. In domains outside of aviation, VR has been found to offer an engaging learning experience, improve knowledge/skill retention, reduce the cost of training, and safely simulate potentially dangerous situations (Chittaro et al., 2018; Wallace et al., 2020)

Despite interest in the field and the plausible improvements to training, there are a lack of studies directly comparing training transfer, training effectiveness, and differences in performance between VR and existing simulation methods (Michelle P. Hight et al., 2022). The potential improvements have led some organizations such as the Embry-Riddle Aeronautical University (*New US University Flight Training Program Using Virtual Reality Cuts Time To Solo By 30%*, 2022) and Alaska Airlines (Kristin Goodwillie, 2022) to implement VR training in their pilot programs, successfully reducing the time it took students to complete their first solo flight by over 30 percent and aiding pilots in developing the muscle memory needed to quickly locate switches, saving hours of training in higher-fidelity simulators and actual aircraft which can better be used for training flight maneuvers.

However, without quantitative evidence of the benefits of VR training, demonstrating that VR is as effective as existing simulators at training pilots safely, civil aviation authorities (CAAs) such as Transport Canada (TC) and the Federal Aviation Administration (FAA) will not approve the use of VR for training. Thus, the overall research objective of this study is to address gaps in existing literature through an experimental comparison between FTDs and VR.

Methods

Participants

Twenty (n = 20) participants volunteered for the study, with 10 (n = 10) participants assigned to each of the FTD and VR simulator groups. Participants were student pilots at the University of Waterloo who have not yet completed their first solo flight, have a maximum of 20 hours of flight experience, and do not have any existing medical conditions that lead to increased risks from simulator use or susceptibility to simulator sickness.

No statistically significant differences were found in the demographics between groups including age, gender, hours of flight experience, self-assessed susceptibility to motion sickness, and self-assessed familiarity with the Region of Waterloo International Airport (CYKF).

Devices

ALSIM AL250. The ALSIM AL250 operated by the Waterloo Institute for Sustainable Aeronautics (WISA) was selected as the FTD for this study. This simulator was set up to model a standard single-engine aircraft, like the Cessna 172, a common aircraft used in ab initio training.

Virtual Reality Simulator. The VR simulator for this study comprised of the Oculus Quest 2 HMD from Meta with the included Oculus Touch controllers. The HMD was connected to a laptop using the Oculus Link functionality to enable the HMD to run the X-Plane 11 software. X-Plane 11 is developed by Laminar Research and compatible on both Mac OS and Windows, as well as VR (*X-Plane 11*, n.d.), and it has been a popular choice for many flight simulation studies (Rongbing Xu & Shi Cao, 2021). It is an FAA certified simulator when combined with approved hardware. For this study the default Cessna 172SP single-engine fixed-wing aircraft was used.

Methods of Evaluation

For student pilots working towards various licenses and ratings, their performance is evaluated through flight tests. During these flight tests, evaluators grade the students' performance using flight test guides developed by Transport Canada. A modified Transport Canada flight test guide (Transport Canada, 2021) was used by a flight instructor to provide evaluations of performance on Day 1 and Day 5 for 18 criteria divided into three tasks: before start checklist (4 criteria), takeoff (6 criteria), and steep turn (8 criteria). The modifications included division of the test guide criteria into smaller components, allowing performance to be analyzed for individual behaviors, rather than solely the overall performance of each maneuver.

Procedure

The study was reviewed and approved by a research ethics committee at the University of Waterloo. On Day 1, all participants were asked to complete a before start checklist, takeoff, and steep turn using the ALSIM AL250 to assess their baseline performance. On Day 2, participants were given a half hour of free time to practice these assigned flight tasks using their assigned simulator: FTD or VR. On Day 3, participants were asked to complete the flight tasks once using their assigned simulator. On Day 4, participants were again given a half hour of free time to practice the assigned tasks using their assigned simulator. On Day 5, all participants were again asked to complete the before start checklist, takeoff, and steep turn using the ALSIM AL250 to assess their final performance.

Results

Participants scores for the 18 instructor evaluation criteria were averaged to determine and overall score for each flight task: the before start checklist, takeoff, and steep turn. The improvement in performance was then defined as *Final performance (Day 5) – Initial performance (Day 1)*. A repeated measures ANOVA was used to analyze this improvement as a result of the two independent variables: a between-subject's variable, the type of simulator used for training (FTD or VR), and a within-subjects variables, the type of flight task (before start checklist, takeoff, or steep turn).

The results revealed a significant interaction between the type of flight task and the type of simulator used for training ($F(2,36) = 4.604$, $p = .017$, $\eta_p^2 = .204$). The within-subject effect was not significant ($F(2,36) = .225$, $p = .800$, $\eta_p^2 = .012$). The between-subject effect was significant ($F(1,18) = 159.018$, $p < .001$, $\eta_p^2 = .898$). More specifically, using one-way ANOVA, it was found that there was a significant impact of the type of simulator used for the before start checklist ($F(1,18) = 14.308$, $p = .001$, $\eta_p^2 = .443$) and steep turn ($F(1,18) = 13.769$, $p = .002$, $\eta_p^2 = .433$) with the FTD group improving more than the VR group, whereas there was not a significant difference for the takeoff ($F(1,18) = 1.101$, $p = .308$, $\eta_p^2 = .058$).

A further one-way ANOVA was performed to understand for which specific criteria the groups improved similarly versus differently. The results of this analysis, included in Table 1, show that for the majority of the criteria, the hypotheses were confirmed; the hypotheses being that VR would be as effective as FTDs for procedural tasks, but less effective for aircraft handling tasks. For criteria which are highlighted, these hypotheses were violated.

Discrepancies between the hypothesis and results for procedural tasks are expected to have been caused by the high-fidelity simulator's lack of resemblance to the actual Cessna cockpit, which was accurately represented by the VR simulation. As such, the VR group became familiar with the layout of an actual Cessna 172 cockpit, while their performance was evaluated upon returning to the less-accurate layout of the ALSIM AL250 simulator.

Discrepancies between the hypothesis and results for aircraft handling tasks occur for criteria "Maintain directional control", which requires less controller feedback than the majority of aircraft handling tasks and criteria "Maintain angle of bank", which was on the cusp of being a significant result.

Table 1. *Results for Improvement in Performance of Instructor Evaluations Between Groups*

Criteria	Type	F (1,18)	Sig.	η_p^2
Before Start Checklist				
Demonstrate an awareness of other persons and property before and during engine start	P	10.6	.004	.370
Use the appropriate checklist provided by the manufacturer or aeroplane owner	P	7.2	.015	.286
Accurately complete the engine and aeroplane systems check	P	3.2	.089	.153
Check flight controls for freedom of operation and correct movements	P	6.1	.024	.253
Takeoff				
Complete appropriate checklist	P	0.0	1.000	.000
Check for traffic	P	0.9	.355	.048
Advance throttle smoothly to takeoff power	P	0.4	.530	.022
Maintain directional control during the takeoff roll	H	1.0	.340	.051
Rotate at recommended airspeed (+10/-5 knots)	P	0.5	.511	.024
Accelerate to an maintain recommended climb speed (+10/-5 knots)	H	6.7	.019	.271
Steep Turn				
Perform and maintain an effective lookout before and during the turn	P	0.0	1.000	.000
Roll into and out of turns using smooth and coordinated pitch, bank, yaw, and power control	H	10.3	.005	.364
Roll into a coordinated turn with an angle of bank of 45°	H	6.9	.017	.278
Maintain coordinated flight	H	10.0	.005	.357
Maintain the selected altitude (+/- 100 ft)	H	9.2	.007	.339
Maintain airspeed (+/- 10 knots)	H	7.4	.014	.290
Maintain 45° angle of bank (+/- 10°)	H	3.9	.065	.176
Visually recover from the turn at the pre-selected recovery reference point (+/- 10°)	H	9.3	.007	.341

Note. Type P refers to procedural training tasks, Type H refers to aircraft handling tasks.

Discussion and Conclusion

One of the major concerns with implementing virtual reality training in place of existing training is the lack of natural tactile interaction, which previous research has implied is essential for successful training. However, in discussion with current pilots, it was identified that natural tactile interaction, that is the feel of and feedback from flight controls and instrument panels, is not necessary in all aspects of pilot training, specifically procedural tasks.

While the instructor evaluations revealed significant differences in improvement in performance for the before start checklist, it is hypothesized that this is due to discrepancies between the layout of the ALSIM AL250 and the actual Cessna 172. For procedural tasks during takeoff and the steep turn, there were no significant differences between training on VR and the FTD, providing some evidence that shows VR can be used in ab initio pilot training of procedural tasks.

Future studies may expand upon this work by conducting a follow-up study with a larger sample size that could provide insight into results with small effect sizes. Additionally, a similar study may be conducted using and FTD which more closely replicated the Cessna 172 cockpit layout to investigate the potential use of VR for procedural tasks without any effects caused by the dissimilarities in cockpit configurations of the simulators.

Considering differences in improvement in performance for ab initio aircraft handling tasks, the instructor evaluations showed, at least for the majority of handling tasks, that there are statistically significant differences between the VR and FTD groups. As predicted, due to the lack of natural tactile interaction in VR, this evidence shows that VR should not replace existing high-fidelity flight training devices for ab initio pilot training of aircraft handling tasks.

It is hoped that this data provides a foundation for further research which may allow flight training schools to conduct a cost benefit analysis comparing VR and FTDs more. This data may also allow training schools to balance the use of these simulators in a way which maximizes the effectiveness of training by utilizing the benefits of both simulators, while simultaneously minimizing the overall training cost. The findings from the current study emphasized the need for future work to identify tasks that can and cannot be trained using low-fidelity VR simulators.

Acknowledgements

This paper is based on a thesis presented to the University of Waterloo in fulfillment of the thesis requirement for a master's level degree. This work was partially supported by an NSERC Discovery Grant (RGPIN-2015-04134) to S.C.

References

Chittaro, L., Corbett, C. L., McLean, G. A., & Zangrando, N. (2018). Safety Knowledge Transfer Through Mobile Virtual Reality: A Study of Aviation Life Preserver Donning. *Safety Science*, *102*, 159–168. https://doi.org/10.1016/j.ssci.2017.10.012

Kristin Goodwillie. (2022, November 14). Alaska Airlines one of first US Airlines to use virtual reality in pilot training. *King 5 Media Group*. https://www.king5.com/article/tech/alaska-airlines-use-virtual-reality-pilot-training/281-3cb512f2-6539-41af-8d17-addbdaa9d31d?fbclid=IwAR01Dl1J0NIwu6Klg0TMQZCzOqVvhlDtJVVWx6FMbD0-rgkndnsNt3Ee1po

Mazuryk, T., & Gervautz, M. (1996). *Virtual Reality History, Application, Technology and Future*. 1–72.

Michelle P. Hight, Stephanie G. Fussell, Martin A. Kurkchubasche, & Ian J. Hummell. (2022). Effectiveness of Virtual Reality Simulations for Civilian, Ab Initio Pilot Training. *Journal of Aviation/Aerospace Education & Research*, *31*(1), 1–17. https://doi.org/10.15394/jaaer.2022.1903

Myers III, P. L., Starr, A. W., & Mullins, K. (2018). Flight Simulator Fidelity, Training Transfer, and the Role of Instructors in Optimizing Learning. *International Journal of Aviation, Aeronautics, and Aerospace*, *5*(1). https://doi.org/10.15394/ijaaa.2018.1203

New US University Flight Training Program Using Virtual Reality Cuts Time To Solo By 30%. (2022, November 25). AFM. https://afm.aero/new-us-university-flight-training-program-using-virtual-reality-cuts-time-to-solo-by-30/?fbclid=IwAR3gqX8WVMXDDeOJZgOCS_7esgDnvspL4ban6OFi9zTLhwbNSC_M61eM2co

Rongbing Xu & Shi Cao. (2021). Modeling pilot flight performance in a cognitive architecture: Model demonstration. *Human Factors and Ergonomics Society Annual Meeting*, *65*.

Suzanne K. Kearns. (2021). *Fundamentals of International Aviation* (2nd ed.). Routledge.

Transport Canada. (2021). *Flight Test Guide—Private Pilot Licence—Aeroplane—TP 13723 (Sixth Edition—Revised)*. Government of Canada. https://tc.canada.ca/en/aviation/publications/flight-test-guide-private-pilot-licence-aeroplane-tp-13723#g18

Wallace, J. W., Hu, Z., & Carroll, D. A. (2020, December). Augmented Reality for Immersive and Tactile Flight Simulation. *IEEE Aerospace and Electronic Systems Magazine*, *35*(12), 6–14.

X-Plane 11. (n.d.). X-Plane. Retrieved May 20, 2021, from https://www.x-plane.com/

FATIGUE AND RECOVERY IN PILOTS AND AIR TRAFFIC CONTROLLERS: A MILITARY CASE STUDY

Pedro Piedade
Portuguese Air Force Psychology Centre
Lisboa, Portugal

Fatigue is a known threat to individuals, especially in high demanding and risky tasks, because of its physical, mental, and emotional consequences. In high reliability organizations (HRO), fatigue is a hazard to safety management systems and requires effective mitigation strategies. At the individual level, recovery experiences enable the ability to manage stressful situations at work but can also be effective in dealing with fatigue. In this research, 137 military pilots and air traffic controllers from the Portuguese Air Force (PoAF) responded to an online survey. The results showed that both samples were especially impacted by mental fatigue, but air traffic controllers are more able to use recovery to deal with their job´s demands. Mitigation strategies and possible explanations for the differences between participant groups are analyzed and discussed.

Fatigue is a main concern in organizations nowadays, not only because of its impact on work efficiency, but especially because of the way it affects people, both physically and mentally, but also emotionally. This is a worrying reality when we talk about high reliability organizations (HRO), since these work in an inherently dangerous and threatening environment, which demands an effective risk management strategy regarding desirable safety outcomes (Salas et al., 2020).

This insidious and unforgiven phenomenon is usually caused by sleep loss, sleep debt, desynchronization of normal circadian rhythms and work stress and demands (Chang et.al., 2019). In the aviation context, the International Civil Aviation Organization (ICAO) defines fatigue as *the physiological state of reduced mental or physical performance capability resulting from sleep loss or extended wakefulness, circadian phase, or workload (mental and/or physical activity) that can impair a crew member's alertness and ability to safely operate an aircraft or perform safety-related duties* (ICAO, 2011, 2-1).

While fatigue can cause strain to individual psychological and physical resources, recovery experiences promote the reduction of professional demands, allowing for a progressive detachment from work stresses by engaging in different experiences during leisure time (Sonnentag & Fritz, 2015; Sonnentag et al., 2008).

Although different types of recovery are found in the literature, Sonnentag and Fritz (2007) identify 4 main opportunities to recover from stress: psychological detachment, relaxation, control and mastery, with the first two being considered, in the research on recovery, the most prominent in the promotion of well-being (Demsky et al., 2018). Psychological detachment, which relates to the ability to separate both mentally and physically from job demands, as well as the quality of sleep, are also pinpointed as effective means of fatigue mitigation (Hülsheger, 2016). Relaxation relates to a feeling of tranquility, and positive affect (Sonnentag et al., 2008), while control links to the perception of being in control and able to choose what to do during recovery, whereas mastery means the individual ability to engage in activities that allow him/her to challenge himself/herself (e.g., learning a foreign language) (Sonnentag & Fritz, 2007).

Fatigue and Recovery in Operational Settings

Since both commercial and military aviation are critical systems, the understanding and control of fatigue are paramount for safety to prevail. One of the concerns regarding fatigue is that it affects the performance of aviators (Keller et al., 2022a), but also reduces the *situational awareness* of air traffic controllers, while increasing mistakes and slowing down their response times (Bongo & Seva, 2021).

According to the *National Transportation Safety Board* (NTSB), fatigue is a contributing factor to accidents in commercial aviation and flight training (Keller et al., 2022b), while Caldwell et al. (2004) states that around 8% of incidents in the US Air Force are related to fatigue. Hu and Lodewijks (2019) distinguishes active fatigue (continuous task related efforts) from passive fatigue (concerning monitoring tasks, but with reduced response actions), and state that both cause reduced alertness and increased drowsiness, affecting performance of both pilots and drivers.

In the air traffic control (ATC) setting, Chang et al. (2019) asserts the importance of duty cycles and breaks during shifts as an important mechanism to prevent fatigue effects, since, for instance, ATC personnel tends, in high workload conditions, to focus on the information provided by radar, and less on the aircraft itself, but they also have the tendency to rely more on their internal memory, which can be a source of error (Bongo & Seva, 2021).

The pressures and demands imposed by the aviation industry to its professionals are unquestionable and relate to the high stakes involving both commercial and military settings. Although this is understandable, the operational and social working environment can build up additional pressure in the individual, leading to more fatigue, which in turn leads to a performance decrement. This is the focus of what Sonnentag (2018) described as a recovery paradox, since high job demands and stressors can prevent recovery strategies to be effective, leading to more fatigue and emotional exhaustion (Sonnentag et al., 2010) because the person is unable to effectively recover during his/her off duty time. This is also true in our everyday lives, since we are surrounded by digital devices and social media that prevent us from detaching from the information overload, while our phones, *whatsapp* and emails keep a permanent link to our professional life.

At the operational level, there are fatigue mitigation strategies that can be applied by organizations. Keller et al. (2022b) mentions training, just culture and the support and assessment made by flight instructors (regarding their students), while Keller et al. (2022a), also referring to training organizations, states that fatigue can be minimized with proper sleep, reduced workload and proper scheduling. Concerning air traffic controllers, Bongo and Seva (2021) refers to proper assignment of ATC personnel and shift rest time as ways of alleviating fatigue and its consequences.

In this paper, we'll explore the effects of fatigue, but also the use of recovery experiences in a sample of military pilots and ATC professionals from the Portuguese Air Force (PoAF).

Methodology

Participants. In the present study, 137 participants (88 pilots and 49 air traffic controllers) from different PoAF Units were invited to answer an online questionnaire. The majority of pilots were assigned to transport and search and rescue (S&R) missions (67%), with an experience between 11 and 15 years in the air force, married, aged between 27 and 32. Around 47% of the ATC sample were airspace vigilance and interception controllers from the same unit (Air Command), most of the controllers had between 11 and 15 years of tenure, married and less than 40 years of age.

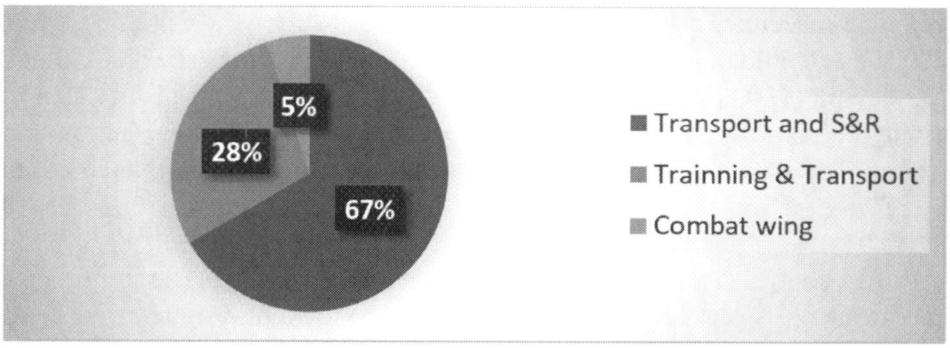

Figure 1.
Type of operational mission assigned to pilot participants.

Materials and Procedure. An online survey (using *surveymonkey*®) was sent to the participants professional email boxes with the previous agreement of their commanding officers. Fatigue was assessed using the *Three Dimension Work Fatigue Inventory (3D-WFI)* (Frone & Tidewell, 2015), measuring 3 types of fatigue: emotional, physical and mental (e.g. "during the past 12 months of often did you feel physically exhausted at the end of the workday?"); while for recovery experiences the *Recovery Experience Questionnaire* (Sonnentag & Fritz, 2007) was provided to respondents, measuring the 4 types of experiences: psychological detachment, relaxation, control and mastery (e.g. "during time after work, I do relaxing things"). Both instruments use a 5-point *Likert* scale. Sociodemographic data was also inquired in the survey. The data provided was then analyzed using *IBM SPSS® 28.*

Results. As shown in table 1, mean results indicate that pilots are generally more fatigued than ATC participants, and their scores for recovery strategies are lower than those for air traffic controllers. Mental fatigue had the highest score for both samples, but emotional fatigue was the only to show significant differences, with pilots having a higher score compared with ATC ($t(135)=2{,}053$, $p<.05$). For recovery experiences, the ATC sample had higher, and significant, results compared with pilots for psychological detachment (t(135)=-2,611, p<.05), relaxation (t(135)=-2,256, p<.05) and especially control (t(135)=-3,862, p<.05).

Table 1.
Descriptive statistics of the main variables for both Pilots and ATC.

		Mean	*S.d*	*Min.*	*Max.*
Pilots	Physical Fatigue	3,44	,901	1	5
	Emotional Fatigue	3,26	1,02	1	5
	Mental Fatigue	**3,70**	,864	1	5
	Psych. Detachm.	2,34	,854	1	5
	Relaxation	3,44	,710	1	5
	Control	3,13	,880	1	5
	Mastery	**3,51**	,886	1	5
ATC	Physical Fatigue	3,22	,992	1	5
	Emotional Fatigue	2,87	1,11	1	5
	Mental Fatigue	**3,52**	,818	1	5
	Psych. Detachm.	2,77	,988	1	5
	Relaxation	**3,71**	,601	1	5
	Control	3,66	.698	1	5
	Mastery	3,55	,771	1	5

An ANOVA procedure analysis showed that age and tenure didn´t influence air traffic controllers' fatigue or use of recovery strategies, although single controllers tended to use more relaxation experiences than married ATC professionals. Air traffic controllers posted in a shared Portuguese/American air base (Lages Field, in the Azores), seemed to display more mental and emotional fatigue, while Air Command personnel were more physically tired. There were no differences in pilots regarding age, tenure, marital status or military unit of origin.

A multiple regression analysis, focusing on pilots, showed that emotional and physical fatigue explain 75% of the dependent variable mental fatigue, with physical fatigue (β=.572) being the main predictor (t(83)=7,307, p<.001). For the military ATC professionals, the variance of mental fatigue was explained by physical and emotional fatigue (64%), with physical fatigue (β=.489) being the main predictor (t(44)=4,57, p<.001).The recovery strategies were not significant in the model for both samples.

Discussion and Concluding Remarks

The results found in the present study showed the importance of mitigating fatigue in operational settings, especially in high reliability organizations like the military. Mental fatigue had the highest scores in both samples, which was somehow expected, considering the high demands of the tasks inherent to pilots and air traffic controllers. Pilots were more impacted by emotional fatigue, when compared to ATC, and their mental tiredness related especially with physical exhaustion. This could be explained by the fact that most pilot participants were posted in flight squads that have demanding tasks, like search and rescue, but also because in these squads they are on a short notice alert and are regularly deployed away from home for consecutive periods of several weeks. For controllers, physical fatigue is also a main predictor of mental tiredness. Another interesting result emphasized the fact that controllers are more prone to use recovery experiences as strategies to cope with fatigue when compared to pilots. One hypothesis that could explain this is the recovery paradox, in other words, pilot´s fatigue, especially physical fatigue, was preventing them from effectively recovering. Considering these findings, the importance of having effective mitigation strategies is vital for the well-being of the individual, but it is also crucial for the safety of operations and missions. Resting opportunities, better perceived control over scheduling and mission assignments, but also training on how to cope with fatigue, including its consequences to self and others, could be helpful. Our future research in this area will try to complexify the present research model and look at the impact fatigue might have on safety attitudes, while testing the moderating effects of recovery but also organizational resilience.

Acknowledgements

The author would like to thank the Portuguese Air Force and the PoAF Command for the opportunity of developing this research. Also, a special acknowledgment to all the participants and their commanding officers for allowing this paper to exist.

References

Bongo, M. & Seva, R. (2021). Effect of Fatigue in Air Traffic Controllers' Workload, Situation Awareness, and Control Strategy. *The International Journal of Aerospace Psychology, 32,*1, 1-23.

Caldwell, J., Caldwell, J., Brown, D., & Smith, J. (2004). The Effects of 37 Hours of Continuous Wakefulness on the Physiological Arousal, Cognitive Performance, Self-Reported Mood, and Simulator Flight Performance of F-117A Pilots. *Military Psychology*, *16*(3), 162-181.

Chang, Y., Yang, H. & Hsu, W. (2019). Effects of work shifts on fatigue levels of air traffic controllers. *Journal of Air Transport Management, 76*, 1-9.

Demsky, C. A., Fritz, C., Hammer, L. B., & Black, A. E. (2018). Workplace Incivility and Employee Sleep: The Role of Rumination and Recovery Experiences. *Journal of Occupational Health Psychology*, 1-13.

Frone, M. & Tidwell, M. (2015). The Meaning and Measurement of Work Fatigue: Development and Evaluation of the Three-Dimensional Work Fatigue Inventory (3D-WFI). *Journal of Occupational Health Psychology, 20(3),* 273-288.

Hu, X. & Lodewijks, G. (2019). Detecting fatigue in car drivers and aircraft pilots by using non-invasive measures: The value of differentiation of sleepiness and mental fatigue. *Journal of Safety Research, 72,* 173-187.

Hülsheger, U. (2016). From Dawn Till Dusk: Shedding Light on the Recovery Process by Investigating Daily Change Patterns in Fatigue. *Journal of Applied Psychology, 101* (6), 905-914.

International Civil Aviation Organization (ICAO), (2011). *Fatigue Risk Management Systems – Implementation Guide for Operators*. ICAO.

Keller, J, Ziakkas, D & Mendonça, F. (2022a). Comprehensive Assessment of Fatigue and Stress Research on Collegiate Aviation Pilots in the United States. *Transportation Research Procedia, 66*, 40-48.

Keller, J., Mendonça, F. & Adjekum, D. (2022b). Understanding Factors Underlying Fatigue among Collegiate Aviation Pilots in the United States, *Safety, 8*, 1-21.

Salas, E., Bisbey, T., Traylor, A. & Rosen, M. (2020). Can Teamwork Promote Safety in Organizations? *Annual Review of Organizational Psychology and Organizational Behavior*, 7, 6.1-6.31.

Sonnentag, S. & Fritz, C. (2007). The Recovery Experience Questionnaire: Development and Validation of a Measure for Assessing Recuperation and Unwinding From Work. *Journal of Occupational Health Psychology, 12 (3),* 204-221.

Sonnentag, S. & Fritz, C. (2015). Recovery from job stress: The stressor-detachment model as an integrative framework. *Journal of Organizational Behavior, 36*, 72-103.

Sonnentag, S. (2018). The recovery paradox: Portraying the complex interplay between job stressors, lack of recovery, and poor well-being. *Research in Organizational Behavior, 38*, 169-185.

Sonnentag, S., Binnewies, C. & Mojza, E. (2008). "Did You Have A Nice Evening?" A Day-Level Study on Recovery Experiences, Sleep, and Affect. *Journal of Applied Psychology, 93 (3),* 674-684.

Sonnentag, S., Binnewies, C. & Mojza, E. (2010). Staying Well and Engaged When Demands Are High: The Role of Psychological Detachment. *Journal of Applied Psychology, 95 (5)*, 965-976.

Easy as ABC: A Mnemonic Procedure for Managing Startle and Surprise

Matteo Piras[1], Annemarie Landman[1,2], René van Paassen[1], Olaf Stroosma[1], Eric Groen[2], Max Mulder[1]
[1]Delft University Of Technology, Delft, The Netherlands, [2]Netherlands Organisation for Applied Scientific Research, Soesterberg, The Netherlands

Background. Mnemonic procedures are currently being taught to airline pilots to manage startle and surprise. We previously tested the effectiveness of a four-item mnemonic. Pilots generally rated it as useful but some remarked that it induced too much additional workload. Therefore, we tested whether a simpler mnemonic, Aviate-Breathe-Check, would be more useful. **Method.** The experiment took place in a hexapod simulator with a Piper Seneca aerodynamic model and a generic cockpit. Airline pilots ($n = 25$) were divided into an experimental ("ABC") and control group. All received ground training on startle and surprise, which included instructions on the ABC mnemonic for the ABC group. The mnemonic aims to support prioritization of flight-path management (Aviate), followed by physiological and mental stress management (Breathe), followed by troubleshooting (Check). All pilots performed four familiarization scenarios, during which the ABC group practiced the ABC mnemonic. Two test scenarios were then performed to evaluate performance, mental effort, stress, and pilot evaluations of the ABC mnemonic. **Results.** The pilots' evaluations of the ABC mnemonic were significantly higher than those were for the previously-tested mnemonic in the same scenarios. There were no significant differences between the ABC and control group in mental effort and stress, whereas there were trends towards higher mental effort and stress with the previous mnemonic. No significant effects on performance were found. **Conclusions.** The results suggest that the ABC mnemonic was more useful and easier to apply than a previously tested mnemonic. This is promising for the development of effective pilot training interventions for startle and surprise.

Startle and surprise have the potential to seriously impair pilots' abilities of troubleshooting and immediate procedural responses (Landman, Groen, Van Paassen, Bronkhorst, & Mulder, 2017b). "Startle" refers to a stress response to a sudden intense stimulus, whereas "surprise" is an emotional and cognitive response indicating a mismatch between expectation and reality (Landman, Groen, Van Paassen, Bronkhorst, & Mulder, 2017a). Dealing with an unexpected event may require "reframing" of the situation, meaning that the situation is analysed and the cognitive mismatch is resolved. This process is thought to be especially difficult to perform if the above-mentioned cognitive functions are impaired by stress (Landman et al., 2017a; Eysenck & Derakshan, 2011). Startle and surprise may thus have an interactive negative effect on performance, and can lead to pilots remaining "stuck" in this impaired state if the failures to reframe increase stress even further.

Startle and surprise management has been increasingly incorporated in (recurrent) pilot training (European Aviation Safety Authority, 2015; Federal Aviation Administration, 2015). However, empirical data on effective startle and surprise management training are still lacking. One type of training intervention that has been proposed is to teach pilots a short startle and surprise management procedure. This consists of one or more actions which could be useful for managing the effects of stress, facilitating the reframing process, or both. Examples of these actions are, in sequential order: 1) performing a stress reduction technique like taking a deep breath or releasing muscle tension (Field, Boland, Van Rooij, Mohrmann, & Smeltink, 2018; Landman et al., 2020; Martin, 2017), obtaining situation awareness with regard to available time (Gillen, 2016), primary flight parameters, (Landman et al., 2020; Gillen, 2016) and indications of the problem, (Martin, 2017; Landman et al., 2020; Field et al., 2018), or taking decisive action (Martin, 2017; Field et al., 2018; Landman et al., 2020). These actions are usually taught in the form of a mnemonic, which makes them easy to remember and apply in the appropriate order.

The effectiveness of such a mnemonic procedure was recently tested in a simulator experiment (Landman et al., 2020). The procedure was taught using the mnemonic *COOL*, with the steps *Calm down*: take a deep breath, sit up straight, release muscle tension in shoulders and arms, focus on exerted force on the controls *Observe*: check and call out primary flight parameters, *Outline*: focus on the problem and analyze it, and *Lead*: formulate a plan and act. Pilots in the experiment generally found the method useful. Results also indicated that the method led to better analysing of the problem as pilots were more likely to take actions to prevent exacerbation. However, there were also non-significant trends and anecdotal remarks by pilots indicating that the mnemonic procedure was too mentally demanding, and distracted from giving priority to crucial actions (e.g., recovering an upset).

Thus, the current study is aimed to test the effectiveness of a new mnemonic procedure that is shorter, more simple, and includes prioritization of restoring the flight path. This procedure was *ABC* with the steps: *Aviate*: ensure that the flight path is stabilized, *Breathe*: the same as *Calm down* in *COOL*, *Check*: the same as *Observe* in *COOL* with the difference of not having to call out instrument readings.

Method

Participants

The sample group consisted of 25 commercial airline pilots, who were assigned to an experimental group (ABC, n = 13) or control group (n = 12). Characteristics of the groups are listed in Table 1. Despite efforts to balance the groups, an independent-samples t test indicated that the ABC group scored significantly higher than control on trait anxiety as measured with the State-Trait anxiety index (Spielberger, Gonzalez-Reigosa, Martinez-Urrutia, Natalicio, & Natalicio, 1971), $p = 0.008$. There were no other significant or nearly significant differences between groups.

Table 1
Characteristics of the participants.

	ABC	Control
Age in years (mean, SD)	41.0, 9.9	44.5, 8.3
Flight hours large transport (mean, SD)	9,750 (6,899)	10,160 (5,732)
Working experience in years (mean, SD)	17.6 (11.2)	14.0 (9.5)
Trait anxiety score range 20-80 (mean, SD)	31.4 (5.3)	26.7 (3.3)
Captains / FOs / SOs*	6/6/1	7/4/1
Type rating instructors or examiners	2	1
Men / Women	11/2	12/0

Apparatus

The experiment was performed using the SIMONA research simulator located at the Delft University of Technology. This is a full motion flight simulator featuring six hydraulic actuators, and allowing pilots a 180 degrees field of view. One projector malfunctioned during the experiment, resulting in a field of view of 120 degrees instead. The cockpit mock-up and aerodynamic model were based on the Piper Seneca PA-34, a multi-engine piston aircraft. All participating pilots had flown a similar type during their initial training. Controls consisted of a column with electric pitch trim, rudder pedals, throttle, flaps, and gear. The left seat was used, see Fig. 1.

Figure 1. The experimental setup (left seat).

Procedure

The experimental procedure was very similar to (Landman et al., 2020). Both the ABC and control group received familiarization with the simulator and a briefing on startle and surprise to prevent differences in expectations between groups. The ABC group received an explanation of and instruction to use the "ABC" procedure (see Introduction). They were told that the goals of the procedure were to support proper prioritization of actions, and recognize and manage psychological and physiological effects of startle and surprise. Both groups then performed four training scenarios with non-normal events so that the ABC group could practice the *ABC* procedure. Finally, both groups performed two test scenarios: the Cargo Shift scenario in which cargo shifted towards the tail during take-off, and the Flap Asymmetry scenario which occurred at base leg (see Landman et al., 2020). Pilots had no checklists for these failures. Both failures required timely control responses and a quick analysis of the limited controllability. Both issues also allowed for making the decision to land with partial flaps or flaps up to prevent further exacerbation of controllability problems.

Dependent measures

Immediately after each test scenario, pilots rated perceived mental effort experienced during the scenario on the Rating Scale for Mental Effort (RSME) (Zijlstra, 1995), and perceived stress on a 1-10 point Likert scale (Houtman & Bakker, 1989). Baseline measures of stress and mental effort were also obtained in the last familiarization scenario. These were subtracted from the measures in the test scenarios to correct for individual differences. Perceived startle and surprise were both rated on custom scales similar to the one used for stress. This was done to check if the scenarios succeeded in startling and surprising the pilots. The ABC group rated perceived usefulness of the procedure after the test scenarios on a 1-10 point Likert scale. As a measure of performance, the decision to divert from the normal flaps LAND setting in each scenario was used as a binary measure. Using flaps LAND following the failure would severely reduce controllability in each scenario.

Statistical analysis

The baseline-corrected mental effort and stress scores were compared between the ABC and control group using Mann Whitney U tests, which is a non-parametric between-subjects test. Perceived usefulness ratings were compared using the same test between the ABC group and data of the COOL group obtained in the same scenarios from a previous study (Landman et al., 2020). Decisions to divert from normal flap settings were compared between groups using a Pearson Chi-squared test.

Results

Two participants of the ABC group were excluded from the Flap Asymmetry analysis due to either not noticing the failure or due to noticing the failure too late for a response.

No significant differences were found between ABC and control on perceived mental effort and stress (see Table 2. On average, the Flap Asymmetry scenario was rated 4.7 (SD = 2.1) on startle, 5.6 (SD = 2.2) on surprise, 61.3 (SD = 19.7) on mental effort, and 4.0 (SD = 2.0) on stress. The Cargo Shift scenario was rated on average 6.6 (SD = 2.1) on startle, 7.2 (SD = 1.7) on surprise, 75.0 (SD = 21.3) on mental effort and 5.6 (SD = 2.1) on stress.

In the Flap Asymmetry scenario, we observed 6/11 pilots in the ABC group and 5/12 pilots in the control group select flaps LAND, with 4/11 and 3/12 also landing with this

Table 2
Change from baseline to the post-test scenarios in perceived mental effort and stress.

	ABC Mean (*SD*)	Control Mean (*SD*)	*p*
Flap Asymmetry scenario			
Δ Mental effort (1-150)	8.3 (23.0)	10.6 (12.4)	0.688
Δ Stress (1-10)	1.3 (1.9)	1.1 (2.0)	0.640
Cargo Shift scenario			
Δ Mental effort (1-150)	21.4 (28.8)	25.6 (16.7)	0.479
Δ Stress (1-10)	3.5 (2.8)	2.6 (1.4)	0.614

setting, respectively. In the Cargo Shift scenario, we observed 2/13 pilots in the ABC group and 5/12 pilots in the control group selecting flaps LAND, with 2/13 and 3/13 also landing with this setting, respectively. There were no significant differences between groups.

Perceived usefulness of the method was significantly higher in ABC, mean = 7.0, SD = 0.8, than COOL, mean = 5.2, SD = 1.8, in the Cargo Shift scenario, $p = 0.004$, but not in the Flap Asymmetry scenario, $p = 0.814$.

Discussion

The ABC procedure did not have a significant effect on pilots' perceived mental effort, stress and performance in the scenarios. Whereas a previous experiment indicated a trend towards more mental workload when using the COOL procedure than control, no such trends were observed in the current study.

The perceived usefulness of the ABC procedure was scored significantly higher than the COOL procedure was scored by a different sample group (Landman et al., 2020). This was only the case in the Cargo Shift scenario, which requires an immediate response to recover a pitch up upset. This recovery was not easy, as the backwards shifting of the center of gravity reduced authority in the pitch axis, and in some cases required roll and throttle changes to prevent loss of control. After recovering, pilots were seen to test the effect of pitch control inputs to get themselves acquainted with the changed dynamics. Thus, the step *Aviate* of the ABC procedure could help pilots in this scenario to focus on regaining and ensuring stability and control in this scenario. A second reason why the procedure was possibly most effective in the Cargo Shift scenario is that this scenario was also rated as the most startling and stressful scenario. The Flap Asymmetry scenario was moderately successful in inducing startle and surprise in pilots, as subjective ratings of startle, surprise, stress and mental effort were around the midpoints of the scale. The Cargo Shift scenario was more successful, as scores were above the midpoint of the scales.

One limitation is that the ABC group scored significantly higher on trait anxiety than the control group, which may have caused the ABC group to respond with relatively more stress to the scenario events. A second limitation is that the experiment featured a simple aircraft model with scenarios that did not involve crew resource management or complex system failures. Whether the effects also translate to large transport aircraft operations is not certain.

Remarks by pilots suggested that parts of the procedure could be selected based on the situation at hand. Some preferred calling out the steps, whereas others preferred not to. Calling out either out loud or in one's mind of (one of) the steps, such as "Aviate", or a different calming phrase, could be an effective self-talk method for managing stress (Tod, Hardy, & Oliver, 2011). Future research could focus on the effectiveness of such self-talk, and on the usefulness of startle management procedures in varying types of startling situations. The results suggest that brevity and simplicity are important aspects of an effective startle and surprise management procedure.

References

European Aviation Safety Authority. (2015). Loss of control prevention and recovery training (notice of proposed amendment 2015-13).

Eysenck, M. W., & Derakshan, N. (2011). New perspectives in attentional control theory. *Personality and Individual Differences*, *50*(7), 955–960.

Federal Aviation Administration. (2015). Stall prevention and recovery training. *(Advisory Circular No. 120-109A)*.

Field, J. N., Boland, E. J., Van Rooij, J. M., Mohrmann, F. W., & Smeltink, J. W. (2018). *Startle effect management. (report nr. NLR-CR-2018-242)* (Tech. Rep.).

Gillen, M. W. (2016). *A study evaluating if targeted training for startle effect can improve pilot reactions in handling unexpected situations in a flight simulator.* The University of North Dakota.

Houtman, I., & Bakker, F. (1989). The anxiety thermometer: a validation study. *Journal of personality assessment*, *53*(3), 575–582.

Landman, A., Groen, E. L., Van Paassen, M. M., Bronkhorst, A. W., & Mulder, M. (2017a). Dealing with unexpected events on the flight deck: A conceptual model of startle and surprise. *Human factors*, *59*(8), 1161–1172.

Landman, A., Groen, E. L., Van Paassen, M. M., Bronkhorst, A. W., & Mulder, M. (2017b). The influence of surprise on upset recovery performance in airline pilots. *The International Journal of Aerospace Psychology*, *27*(1-2), 2–14.

Landman, A., van Middelaar, S. H., Groen, E. L., van Paassen, M. M., Bronkhorst, A. W., & Mulder, M. (2020). The effectiveness of a mnemonic-type startle and surprise management procedure for pilots. *The International Journal of Aerospace Psychology*, *30*(3-4), 104–118.

Martin, W. (2017). *Developing startle and surprise training interventions for airline training programs.* https://pacdeff.com/wp-content/uploads/2017/08/PACDEFF-FC-Forum-Presentation-on-Startle.pdf.

Spielberger, C. D., Gonzalez-Reigosa, F., Martinez-Urrutia, A., Natalicio, L. F., & Natalicio, D. S. (1971). The state-trait anxiety inventory. *Revista Interamericana de Psicologia/Interamerican journal of psychology*, *5*(3 & 4).

Tod, D., Hardy, J., & Oliver, E. (2011). Effects of self-talk: A systematic review. *Journal of Sport and Exercise Psychology*, *33*(5), 666–687.

Zijlstra, F. R. H. (1995). Efficiency in work behaviour: A design approach for modern tools.

LVC, WHAT IS IT GOOD FOR? TRADE-OFFS IN TRAINING VALUE OF LIVE VIRTUAL CONSTRUCTIVE AIR COMBAT TRAINING IN LARGE FORCE EXERCISES

Robert Ramberg[1,2], Henrik Artman[1,3], Rogier Woltjer[1], Sanna Aronsson[1], Mikael Mitchell[1,4]
[1] Swedish Air Force Combat Simulation Centre (FLSC),
Swedish Defence Research Agency (FOI), Stockholm, Sweden
[2] Department of Computer and Systems Sciences, Faculty of Social Sciences,
Stockholm University, Stockholm, Sweden
[3] Division of Media Technology and Interaction Design, School of Electrical Engineering and
Computer Science, KTH Royal Institute of Technology, Stockholm, Sweden
[4] 211 sqn, Swedish Air Force, Luleå, Sweden

This article reports the results of a workshop study with fighter pilots about the potential benefits and drawbacks of introducing Live Virtual Constructive (LVC) training (combining real aircraft, simulators, and computer-generated AI forces) into Large Force Exercises (LFEs). The study elaborates on a questionnaire study conducted during Arctic Challenge Exercise (ACE) 2021, to investigate pilots' attitudes towards including virtual and constructive entities in LFEs. In order to get a better understanding of training value of LVC, and explanations for questionnaire answers, two workshops with a total of eight fighter pilots were conducted. Results classify the statements made by the pilots into various themes, i.e., categories of training value trade-offs to consider when designing LVC scenarios and planning for future LVC LFEs. Results provide depth to the argumentation on training value of future LVC, properties of future AI Constructives in LVC, and training scenario design mixing Live, Virtual, and Constructive entities.

Live, Virtual and Constructive (LVC) is an air combat training concept where real aircraft (Live, L), manned simulators (Virtual, V) and computer-generated aircraft (Constructive, C) act in the same scenario. The concept holds promise for future fighter pilot training (Best & Rice, 2018). Some studies have surfaced potential risks with the introduction of LVC training, such as flight safety aspects of uncertainty about whether entities are Live, Virtual, or Constructive, and increased risk-taking, as well as the need for effective tools and methods for exercise management (Sherwood et al., 2020). Although technical development around LVC matured over the years, research on the training concept LVC and its training effect is not available to the same extent (Best & Rice, 2018), including which learning objectives LVC can contribute to and how to evaluate this potential improvement in training scenarios and tactics, techniques, and procedure development (Mansikka et al., 2021; Stacey & Freeman, 2016). Our research focuses on these aspects, and in particular on training value, defined here as the value (in terms of skills, experience, and/or knowledge) a participant acquires from participating in the training. As the LVC training concept is not currently implemented in Sweden, studying it requires innovative approaches, such as LVC scenario design and allocation workshops (Aronsson et al., 2022b), as well as evaluation of LVC scenario and allocation in a Virtual-Constructive (simulator) environment (Aronsson et al., 2022a), focusing on training value for both Live and Virtual pilots.

At the Large Force Exercise (LFE) Arctic Challenge Exercise 2021 (ACE 2021), a questionnaire study about LVC was conducted with a Live environment as the starting point (Woltjer et al., in press). The study was conducted to gain insight into the participating fighter pilots' attitudes and opinions towards LVC as a training concept, and more specifically as implemented in LFEs. The study also focused on whether participating pilots experienced that they received intended training value relative to defined learning objectives (DLOs) during the missions flown at ACE, and whether the introduction of Virtual- and Constructive entities in ACE and similar LFE training would contribute to training value. General results from statistical analyses of the answers to the questionnaire (Woltjer et al., in press) show that the fighter pilots received good training value during ACE 2021, especially regarding DLOs flying a complex mission and deconfliction. Blue Air experienced higher training value than the Red Air did, as expected. The pilots did not hold a strong position on whether V- or C-entities would improve training value. The pilots' responses were clear on that the training value would be maintained to a greater degree if a Live pilot within the four-ship were to be replaced by a Virtual pilot, compared to being replaced by an (unmanned) Constructive entity, implying that pilots prefer all members of a group to be human-in-the-loop actors in LFEs. The majority of pilots believe that V-entities should have the same rules and restrictions, but a minority disagrees. Similarly, there appear to be divided opinions about whether the displays should show whether an entity is an L-, V-, or C-entity (see also Aronsson et al., 2022a, 2022b). In order to probe the reasons for these and other differences in opinion, qualitative data was sought for. Hence, workshops were planned and carried out with pilots that had participated in ACE 2021. The purpose of the workshops was to seek possible reasons and explanations to differences in opinion. The research question formulated in this article is: Which expected values can be identified with regard to implementing Live Virtual Constructive (LVC) in Large Force Exercises (LFEs)?

Method

Two workshops with a total of eight fighter pilots (four per workshop) were conducted. Each workshop took three hours to complete. The workshops were divided into two separate defined parts with specific tasks. The first part of the workshop dealt with the strengths and weaknesses of V- and C-entities relative to the pre-defined learning objectives applicable to ACE 2021. Data was collected using statements written by the pilots on post-it notes categorized by the pilots into a matrix prepared by the research team. On the matrix, one axis represented the learning objectives which were structured into three main categories (DLOs related to Preparation and coordination, Tactics, and Cockpit), the other axis represented whether the statements were Weaknesses, Strengths or Issues (to be resolved) respectively of V- and C-entities relative to learning objectives and main category. Additional data consisted of notes taken by the research team.

The second part of the workshops dealt with selected excerpts of descriptive statistics based on responses to the ACE 2021 questionnaire, which was compiled and presented in a paper booklet. The pilots worked in pairs to formulate potential explanations for selected outlier responses. For example, for the question "If Virtuals are to participate they must have the same rules and restrictions as Live aircraft", the two extreme positions "Virtuals should have the same restrictions because…" and "Virtuals should NOT have the same restrictions because…" were presented as free-text answer questions, together with a histogram of actual answers and the opportunity to further contextualize the question with an "It depends…" field. Qualitative explanations behind the specific answers to the ACE 2021 questionnaire were thus elicited. After the pairwise discussion, all four pilots (in two pairs) and researchers discussed both the questionnaire answers (histograms) as well as their extreme position explanations and contextualization. The data that was collected consisted of pilots' statements in the booklets, and two researchers' discussion notes. The qualitative data from the different parts of the workshops were analyzed using thematic analysis, resulting in five distinct themes from the first workshop part, as well as five distinct questionnaire statements with Yes/No questions asked during the second part. For brevity, statements made by the pilots were abstracted into their essential and common meaning. The themes and questionnaire statements were subsequently combined to compactly present the results. Each statement was, besides the classification into the themes/statements from the (first/second) workshop part from which it originated, also assigned a classification in the statements/themes from the other (second/first) workshop part, resulting in Table 1.

Results

Identifying expected benefits of and trade-offs in LVC-training is a work in progress, and is meant to assist mission scenario designers and exercise planners responsible for LVC allocation (see Aronsson et al., 2022b) as well as LVC acquisition officers to consider and articulate a number of relevant LVC training value questions, given which DLOs and tasks are to be emphasized. Generally there are two notable overarching observations in the data. The first is whether to allocate human (L/V) versus constructive (C) entities, where there is a mistrust in C-entities to act realistically, the second whether V- and C-entities are regarded to be more suitable acting as adversaries on the red side. By using V- and C-entities, adversary performance can be simulated while simultaneously reducing the risk for mid-air collision between blue and red air, and hence facilitating deconfliction. There is a trade-off between whether the displays should identify, show, and distinguish aircraft LVC-entity types. If entity identity is not shown there is a risk of creating, in the fighter pilots' words, a "false SA [Situation Awareness]", and flight safety might be compromised. If entity identity is shown this could, however, result in pilots acting differently towards the different LVC-entity types. Similar trade-offs are applicable regarding whether or not L- and V-entities (as well as Constructives) should abide by the same rules and regulations to avoid tactical advantages and in relation to upholding flight safety.

Including Virtual entities in training scenarios would keep a human-in-the-loop, which in turn ascertains dynamic and realistic behavior, which would enable the extension of training goals and opportunities, in particular if the Virtual entities are acting as adversaries. As V-entities can simulate the performance of actual adversary aircraft, both technical as well as human aspects contribute to adversary realism. This also reverberates the issue of LVC-allocation, as there is a risk of misprioritization if not being careful in the design of the scenario and mission goals (Aronsson et al., 2022b) for example when considering low-altitude flying and weather conditions. However, the use of Virtual entities might also give rise to another very human behavior, that of gaming the game, that is, gaining tactical advantage because of the features of being in a simulator, e.g., not being subjected to g-forces.

Table 1.
Pilots' Statements from both Workshop Parts Combined in one Table, Illustrating Trade-Offs.

First part / Second part		Credibility & Trust	Safety	Resources	Task allocation	Collaboration & Coordination
If V or C are to participate they must have the same rules and restrictions (R&R) as Live A/C	Yes	Realistic adversary performance	Ensures flight safety	-	Having the same R&R will influence decision making w.r.t. L and V	More complex deconfliction-planning
	No	Gaming the game Unfair fight if not the same R&R	Risk for confusing the pilots' SA Same R&R will influence deconfliction planning and execution	Flexible use of V entities w.r.t. training goals	Not having the same R&R will influence decision making w.r.t. L and V	V cannot utilize their potential with the same R&R, within a mixed L/V-fourship Blue L-side can act on all altitudes with different R&R with Red as V
It is important that my displays show whether other A/C are Live Virtual or Constructive	Yes	-	Improves flight safety	-	-	Shared SA w.r.t. entity identity
	No	Knowing which is which may result in acting differently towards entities	May create false SA when entities look the same	-	-	May create false shared SA w.r.t. entity identity
The mission would be suitable for V or C	Yes	V- and C-entities allow for spatial and temporal flexibility	Use of V- and C-entities would reduce risk	Substitution of entities Extension of entities/crew	-	C-entities can take over simple tasks
	No	Too much flexibility and lack of trust in C-entities may induce stress	-	-	C-entity proficiency questionable	C-entities do not communicate, indicate intent
If any of my Live flight members would be flying as V or C I would still get good training value?	Yes	If V, i.e., a human in the loop	V-entity should act as adversary	-	V-entities can act dynamically	Enables tactics experimentation V-entity ensures human-in-the-loop coordination
	No	Influence of different prerequisites of L-, V- and C-entities	-	C-entities require resources for game lead to act realistically	C-entities are less realistic handling complex missions	C-entities cannot be lead (or lead) appropriately
Additional training value with LVC	Yes	Extend training value; supersonic flight (V/C), size of training range (L/V/C), altitude changes (V/C)	-	V-entity numbers and performance increases realism Less environmental restrictions (e.g., noise) for V/C enable more flexible resource use	V-and C-entities enable larger scenarios V entities are able to do things L-entities cannot	More accurate missile simulation facilitates implicit coordination of kill/no kill
	No	Low-altitude flying less realistic (V) Weather less realistic (V)	Prioritization of safety on behalf of realism	Reduced Live training experience	C-entities lack dynamic behavior	Use of C-entities risk mission adaptability

Note. Statements classified into both the themes from the first part of the workshop (horizontal/top headers), and questionnaire statements from the second part of the workshop (vertical/left headers), with positive and negative comments, respectively, illustrating trade-offs.

Pilots acting as Virtual entities will also be exerted to and experience less stress, which gives them an advantage in terms of air combat, while at the same time decreasing realism. In any case, the pilots express more trust in Virtual than they do in Constructive entities, having a human-in-the-loop is a recurring argument for Virtual over Constructive entities. Constructive entities are not believed to be able to communicate intent, or to take orders or understand commander's intent, to act dynamically, or handle complex missions in a realistic and adaptable (i.e., human-like) way. Either C-entities act unreasonably skilled, or the opposite. To current technological standards, C-entities are believed to be suitable for simpler, less dynamic, tasks.

V- and C-entities present positive aspects in terms of minimizing the risk of disturbing noise and other environmental impact. LVC at large, where everything is connected in one and the same scenario, will enable simulating missile ballistics in real-time and thus human judgements of hit or miss will be less debatable in debriefings. Furthermore, LVC is thought of as a resource-efficient complement to current training as bigger scenarios with more entities can be assembled even when fewer Live aircraft are available by adding V- and C-entities. Another benefit is that LVC may facilitate including virtual pilots from distant physical locations in a training scenario. In all, LVC will likely increase training value and extend training opportunities.

Discussion and conclusions

So, LVC, *what is it good for? Absolutely somethin'!* In our studies, fighter pilots prioritize Live aircraft and human decision making. Taking the Live Large Force Exercise of ACE as a starting point, the pilots in this study find Virtual (V) and Constructive (C) entities to be most suitable as adversaries, more than part of their own constellations. This is to be expected since the explicit focus of training value in the ACE exercise is on blue air, so that LVC would enable more pilots to train as blue air, while reducing Live aircraft assignments to tasks that give less training value, such as red air. Another argument is that V- and C-entities can simulate adversary aircraft more realistically than nationally-owned or coalition aircraft. The results present a picture that broadly coincides with previous studies (Aronsson et al., 2022a, 2022b; Sherwood et al., 2020). The pilots do not fully believe that C-entities can replace real pilots and real aircraft. This may be an effect of the status of C-entities existing today, i.e. they are far from being comparable to human decision-making, communication, and coordination. It appears that pilots do not want to exchange their own forces for either C- or V-entities, which cannot be explained in terms of a lack of human decision-making and communication in the case of V-entities. The pilots are, however, far from skeptical towards LVC as a training concept and expect that, properly applied, it will contribute to training value. Another aspect concerns whether entities should adhere to the same set of rules or not. If the same rules and restrictions are not applied to L- and V-entities, a tactical imbalance can be created between the actors and flight safety may be compromised. This however would also mean that V-entity simulated flight performance cannot be utilized to the same degree (e.g., low-altitude flight, supersonic speed).

For LVC to become a training practice there are still unresolved questions. Some of these are technological, some social and organizational. Some tasks, contexts or situations might not be appropriate for some entities for reasons of both factual conditions (noise, weather) but also trust, safety and resource efficiency. The role that V- and C-entities can have in large force exercises must be placed in relation to the learning objectives that are defined for the exercise, as well as the roles and tasks to which the entities are assigned. It is hence pivotal that pilots in both L- and V-entities are assigned roles and tasks where the intended training values can be met. There is therefore a need in the planning and design of LVC training scenarios for an assigned dedicated function that focuses on and seeks to ensure this (Aronsson et al., 2022b).

Acknowledgements

We wish to thank all participating pilots for their cooperation. This research has been funded by the Swedish Armed Forces' Research and Technology development program.

References

Aronsson, S., Artman, H., Mitchell, M., Ramberg, R., & Woltjer, R. (2022a). A live mindset in Live Virtual Constructive simulations – a spin-up for future LVC-air combat training. *The Journal of Defense Modeling and Simulation: Applications, Methodology, Technology*, online. doi: 10.1177/15485129221106204

Aronsson, S., Artman, H., Mitchell, M., Ramberg, R., & Woltjer, R. (2022b). LVC Allocator: Aligning training value with scenario design for envisioned LVC training of fast-jet pilots. *The Journal of Defense Modeling and Simulation: Applications, Methodology, Technology, 19*(3), 287-298. doi: 10.1177/1548512920958079

Best, C., & Rice, B. (2018). Science and technology enablers of live virtual constructive training in the air domain. *Air & Space Power Journal, 32*(4), 59-73.

Mansikka, H., Virtanen, K., Harris, D., & Salomäki, J. (2021). Live–virtual–constructive simulation for testing and evaluation of air combat tactics, techniques, and procedures, Part 1: assessment framework. *The Journal of Defense Modeling and Simulation, 18*(4), 285-293.

Sherwood, S. M., Neville, K. J., McLean III, A. L., Walwanis, M. M., & Bolton, A. E. (2020). Integrating New Technology into the Complex System of Air Combat Training. In H. A. H. Handley, & A. Tolk (Eds.), *A Framework of Human Systems Engineering: Applications and Case Studies* (pp. 185-204). Wiley Online Library.

Stacy, W., & Freeman, J. (2016). Training objective packages: enhancing the effectiveness of experiential training. *Theoretical Issues in Ergonomics Science, 17*(2), 149-168.

Woltjer, R., Ramberg, R., Artman, H., Aronsson, S., Mitchell, M., & Oskarsson, P.-A. (in press). The future of fighter pilot training? Live Virtual Constructive in Large Force Exercises: Perceived and expected training value. *International Journal of Aerospace Psychology*.

UTILIZING FRAM AND DIGITAL MATERIEL MANAGEMENT TO EVALUATE SYSTEM RESILIENCE AND INFORM DESIGN DECISIONS

Hannah Rennich[1,2], Michael E. Miller[1], John M. McGuirl[1]
[1]Air Force Institute of Technology, Wright Patterson AFB, OH, USA
[2]Air Force Life Cycle Management Center, Wright Patterson AFB, OH, USA

As technology upgrades become more complex and introduce sources of performance variability into the system, human factors engineers must identify and mitigate the risks involved. As opposed to more traditional methods like Human Factors Failure Mode and Effects (HF-FMEA), this research explored the use of the Functional Resonance Analysis Method (FRAM) and Model-Based Systems Engineering (MBSE) activity diagrams to better understand how variability of human behavior in complex socio-technical systems affect overall performance and how redesign may address performance shortfalls. FRAM analysis was conducted to detect potential failures and deviations. MBSE activity diagrams were then developed to decompose the actions of the aircrew and analyzed to determine specific areas that human factors engineers should address. This paper will discuss the observations that each individual technique supports while also providing a discussion of how both methods can be used together to create a more resilient design.

Failure analysis has been used by engineers to identify what went wrong during an accident and to develop methods to avoid similar future failures. Resilience engineering, on the other hand, focuses on what goes right the majority of the time in a system and evaluates how to propagate this behavior throughout a design to improve system resilience. These methods allow engineers to concentrate on the times that systems work well to pinpoint concepts that could be used on future designs to reduce additional failures. By evaluating a system's variability, resilience engineering identifies ways to exploit the system variability and thereby minimize failures. In this way, resilience engineering focuses on human variability as "the ability to make performance adjustments is an essential human contribution to work, without which only the most trivial activity would be possible (Hollnagel & Leonhardt, 2013)."

The Functional Resonance Analysis Method (FRAM) is used to examine cognitive work, understand how it is performed, or how this work could be performed to systematically and reliably understand and represent this work using a well-defined format (Hollnagel, 2018). Several examples have been published implementing the theory of FRAM in multiple fields. For example, Karikawa and colleagues applied FRAM to investigate air traffic control operations, illustrating how the controllers must adapt to situational changes to maintain aircraft separation (Karikawa et al., 2019). While this analysis provides insight into resilient behaviors that must be emphasized and maintained during operation, it does not illustrate how this knowledge might affect future system design concepts. In another example, FRAM is applied to flood protection and is used to make a subjective assessment of some potential system alternatives (Anvarifar et al., 2017). However, neither approach provides a systematic and repeatable approach to identifying and selecting design alternatives.

Digital Materiel Management (DMM), an updated term for Digital Engineering, is defined as an "initiative that shapes the culture and workforce to collaborate and work more efficiently with an authoritative source of truth (Baldwin, 2018)." The DMM initiative extends beyond simply converting the design of a system to a digital format, and instead focuses efforts on digitizing and integrating all aspects of the system's design, evaluation, and manufacturing into a single, connected model. Model-Based Systems Engineering (MBSE) creates a digital representation that enables designers to design, evaluate, and document a system prior to any physical components being manufactured and then to update and modify this model throughout the remainder of the product's lifecycle (Team, n.d.).

As more design and subsequent evaluations use DMM methods that often include MBSE, human factors engineers must identify ways to utilize DMM to test and verify engineering designs which overcome any potential improvement opportunities identified using FRAM. This work explores utilizing a FRAM analysis to understand sources of variability during engineering design and then applies a MBSE artifacts to model and analyze design alternatives with the intent to identify specific design recommendations.

Aircrew responses to electronic warfare attacks on an aircraft with multiple crew members was chosen as the case study for this research. This environment involves coordinated interaction between a complex system and multiple crew members in an environment where errors or time delays can have catastrophic consequences. The overall goal of this research was to attempt to define a method for combining FRAM and traditional MBSE methods using the Systems Modeling Language (SysML) to identify and select design alternatives.

Method Overview

The case study involved reviewing existing system documentation to form an initial FRAM model. Data was then collected from a combination of interviews with an experienced flight crew, in-flight observations, and crew briefs before and after flight. The FRAM model was developed. Sources of variability and potential resonance were identified. Methods to address this variability were then developed and SysML-based MBSE models were then developed and evaluated to form design recommendations.

FRAM Model Development

Using the method defined by Erik Hollnagel (Hollnagel, 2012), a FRAM model was created to evaluate the process that crew members perform to scan, detect, and defeat Infrared (IR) and Radio Frequency (RF) threats launched at the aircraft. The FRAM Model Visualizer (FMV) was used to develop the initial steps of the FRAM analysis and can be found below in Figure 1.

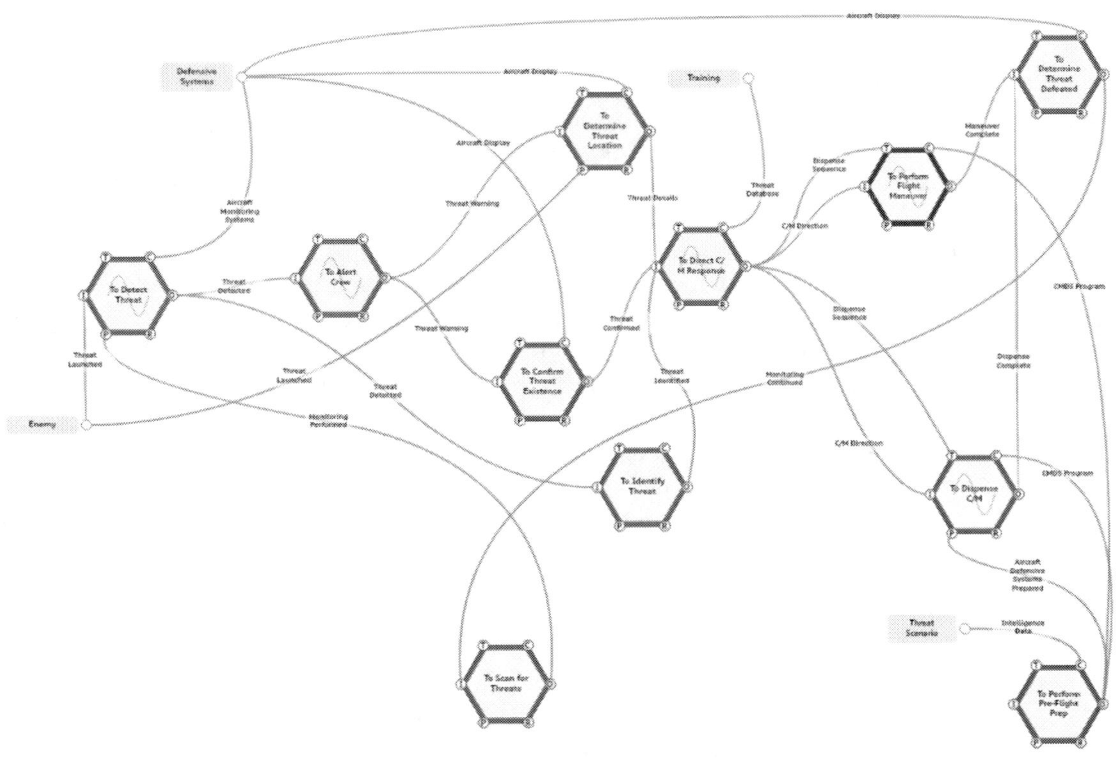

Figure 1. FRAM modeling of crew responses to threat while airborne.

A functional analysis of the aircraft defensive systems was created with several foreground functions. Since FRAM is an evaluation of functions rather than a series of actions, the associated model looks different from how a typical chain of events would be mapped out. As an example, engineers would typically map out a sequence of events that includes deploying a Countermeasure (C/M) and then instigating a maneuver based off the C/M. However, as shown it is the identification of the threat and associated C/M that determines the maneuver and there is not a specific requirement that the C/M be deployed before the maneuver is instigated. Additionally, to perform flight-prep is shown towards the right-hand side of the diagram, which might be associated with the end of the process in a process diagram. However, this step has multiple influences on the dispensing of C/M's, as shown in Figure 1. Instead of focusing the analysis on a set of actions that created a cause-and-effect loop, the FRAM analysis focuses on how each of the main functions are connected using the six main aspects of input, output, time, control, resources, and preconditions.

The case study system of defending against EW threats was decomposed into eleven foreground functions. Several background functions were included in the model to help feed different aspects of the model's main functions. The background functions are considered stable during the activities being evaluated, so they were not part of the model variability analysis. Assumptions made when building the model were that the aircraft was loaded with the necessary C/M's, the crew members were trained in their specific responsibilities, and all pre-flight preparations were performed correctly.

Once the model was developed, the next step was to identify which functions exhibited variability. Four functions in this model were labeled as having either technological or human variability. These variabilities included precision, timing, and wrong action variability. These functions are denoted in Figure 1 with the sinusoidal symbol behind the function name.

The third step in the FRAM analysis is to determine the possibility of functional resonance based on the potential for variability in the other functions. For example, if a threat is detected too late, the crew members will have less time to react and may not have sufficient time to precisely identify the type or location of a threat. If the threat is identified incorrectly, the crew will have to perform shortcuts to undo the effects of initial actions and then respond appropriately in a timely manner to defeat the threat. The information gathered in this part of the FRAM analysis provides the link between identifying the functional variability and determining methods for mitigating controllable variability.

The final step in the FRAM analysis process is to propose methods to mitigate the variability and resonances that are exhibited in the system's functional model. Reducing the variability within a system's function will enable the operators to respond more appropriately and reduce failure modes. Several mitigation recommendations were made based on the case study FRAM analysis that simulated a reduction in variability and improvement in resiliency of the crew members.

The FMV analysis provided a graphical representation of the decomposition of the functional interactions between the aircraft, crew members, and a launched threat. The analysis also aided the understanding of the sources of the variability that increased response variability. This allowed the authors to propose potential recommendations for reducing variability within the system. While these recommendations could be presented to engineers as improvements to the overall system performance and failure reduction, it was not clear which recommendations might have the most benefit. Therefore, these results did not provide adequate information to aid the discussion of cost, schedule, and benefit tradeoffs which are necessary during alternative selection. DMM's were used to evaluate both the as-is system as well as two of the potential design recommendations that were created based upon knowledge of the variability sources that were identified during the FRAM process.

MBSE Model Development

All DMM modeling was accomplished using Cameo Systems Modeler (version 19.0) and basic SysML constructs and diagrams. An MBSE model was created to define the structure and critical process steps performed by the aircraft defensive systems and crew members. As a proof of concept, the MBSE model focused on evaluating the actions of the crew and aircraft systems after a threat was detected and identified rather than modeling the entire FRAM system model.

As-Is Model Creation

The as-is design was employed a Block Definition Diagram (BDD) to define the structure of the system. It included the aircraft defensive systems, specifically the threat monitoring and dispensing systems as well as the crew members. The process conducted by the system and crew was then modeled as an activity diagram in Cameo. This activity diagram for the "as-is" system set a baseline for the model evaluation process. The activity diagram addressed the variability associated with the actions identified during the FRAM process through decision nodes and probabilities for each path estimated from the information gathered from interviews and flight observations. Duration constraints were assigned to each activity to provide ranges of time that each action might take, depending on the scenario encountered, including the initial settings of the systems and actions taken by crew members. A discrete-event simulation was then defined to execute the activity diagram. Opaque actions were placed at different points within the activity diagram to interrogate the times that certain milestones were achieved during the simulation and to store these times as value properties for later evaluation. Requirements were set for each value property to check compliance with design standards. For this research, Total Time and Response Time were selected as the value properties to evaluate. Total Time relates to the entire time required for the crew to perform all actions related to defeating a launched EW threat. On the other hand, Response Time looked specifically at the time required for the crew member responsible for directing the C/M responses to perform their duties. This later time was important as the design of their interface to the system was believed to contribute to an increase in response time variability.

Monte Carlo simulation relies on repeated random sampling and statistical analysis to understand the steady state response of a system that includes variability (Raychaudhuri, 2008). By using a Monte Carlo simulation with a large enough sample size, histogram plots of the Total and Response Times were created. The simulation was run such that the sample size for the least probable path in the activity diagram would be taken at least fifty times. A simulation with more runs demonstrated similar results, thus the sample size was deemed large enough to meet the intent of the Central Limit Theorem. The Monte Carlo simulation of the as-is design produced results that put 16% of the total time scenarios and 8% of the response times out of a predefined specification. Figure 2 provides an example of the histogram plot created by Cameo for the Total Time. This figure provides the number of runs completed, mean, standard deviation, and out of spec percentage in the upper right-hand corner.

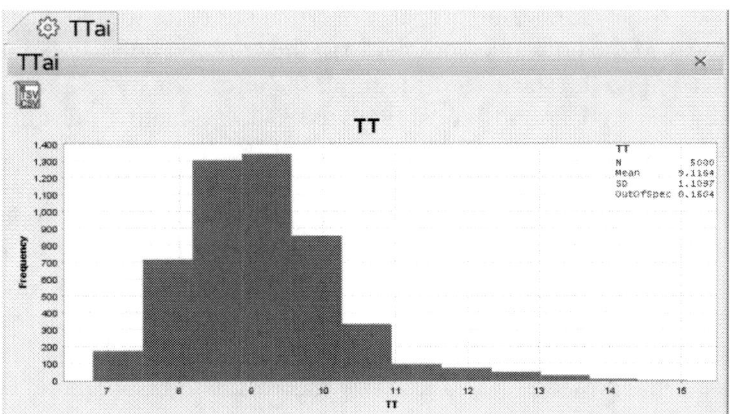

Figure 2. Monte Carlo results of total time required to run the "as-is" activity diagram.

Design Proposals based off FRAM Analysis

Activity diagrams were created for two of the recommendations using the same process as for the as-is model above. The Monte Carlo simulations were run matching the number of runs performed on the as-is design and results were then compared.

Design Recommendation #1 decreased the out of spec Total Time percentage to 9.6% while the out of spec Response Time percentages increased to 9.5%. Design Recommendation #2 on the other hand, increased the out of spec Total Time percentage to 19% and decreased the out of spec Response Time percentage to 1%. Figure 3 below shows how the time distribution changed for Design Recommendation #1 as compared to the as-is model.

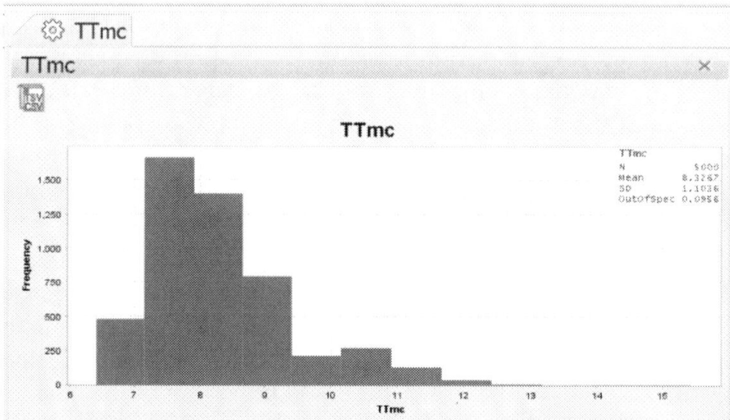

Figure 3. Monte Carlo results of total time required to run the design #1 activity diagram.

Discussion and Conclusion

The FRAM analysis provided an understanding of several sources of variability in response accuracy or time and permitted one to understand how this variability affected total system response. This analysis permitted the human factors practitioner to focus on attributes of the current design that led to the sources of variability that had the potential to significantly influence system performance. As a result, the designer could focus on identifying specific design changes that deserved further analysis.

Each variability mitigation recommendation correlated to a component or actor within the system. Once the system components were tied to the FRAM recommendations, a more detailed analysis of them could be performed. The Total Time value property is critical to determining whether the mission is successful against a threat. If the crew does not perform the required defensive maneuvers and deploy the correct countermeasures within the specified time limit, the aircraft is more likely to succumb to the threat and the mission and all lives on board will most likely be lost. Evaluating the simulated results from the Total Time Monte Carlo provides a basis for determining if the design recommendations will hinder or benefit the crew members in their threat response. Reduction of the Total Time allows the crew more time to perform additional maneuvers if necessary to defeat a threat and save the mission and crew.

The Response Time value property determines how long the rest of the crew members have to react to the threat and defeat it with defensive maneuvers after the countermeasure has been called out. The longer the crew member directing the threat response takes to communicate with the system and the crew, the less time the rest of the crew will have to defeat the threat with the defined evasive maneuver. Reducing the Response Time provides the entire crew more time to respond and ensure the threat is defeated and the mission is safe.

One might argue that a simple way to reduce the time values entirely would be to completely automate the entire defensive system and completely remove the need for human reasoning. While this recommendation might initially seem like a situation that would both reduce response time and crew workload, due to the variability of threats and the potential for false positive identification of threats, the crew should not be removed entirely from the functional process.

The results of the Cameo simulations exhibit promise as a method to evaluate FRAM variability mitigation recommendations. By utilizing a quantitative method like DMM combined with a qualitative method like FRAM, human factors engineers can bring combination of well-reasoned qualitative and quantitative results to other engineers to aid the team's understanding how design changes will affect the system and the lives of those within the system.

Acknowledgements

The views expressed in this article are those of the authors and do not necessarily reflect the official policy or position of the Department of the U.S. Air Force, U.S. Department of Defense, nor the U.S. Government. The authors gratefully acknowledge the study participants for their support of this work.

References

Anvarifar, F., Voorendt, M. Z., Zevenbergen, C., & Thissen, W. (2017). An application of the Functional Resonance Analysis Method (FRAM) to risk analysis of multifunctional flood defences in the Netherlands. *Reliability Engineering and System Safety*, *158*(September 2016), 130–141. https://doi.org/10.1016/j.ress.2016.10.004

Baldwin, K. (2018). Journal of Defense Modeling and Simulation (JDMS) special issue: Transforming the engineering enterprise—applications of Digital Engineering and Modular Open Systems Approach: *Https://Doi.Org/10.1177/1548512917751964*, *16*(4), 323–324. https://doi.org/10.1177/1548512917751964

Hollnagel, E. (2012). *FRAM: the Functional Resonance Analysis Method*. CRC Press.

Hollnagel, E. (2018). *The Functional Resonance Analysis Method*. http://functionalresonance.com/the fram model visualiser.html

Hollnagel, E., & Leonhardt, J. (2013). From Safety-I to Safety-II: A White Paper EXECUTIVE SUMMARY. *European Organisation for the Safety of Air Navigation*, 1–32.

Karikawa, D., Aoyama, H., Ohashi, T., Takashi, M., & Kitamura, M. (2019). Resilience of Air Traffic Controllers in Control Tower. *REA Symposium Embracing Resilience*, 7. https://doi.org/https://doi.org/10.15626/rea8.16

Raychaudhuri, S. (2008). Introduction to Monte Carlo Simulation. *Winter Simulation Conference Proceedings*, 91–100. https://doi.org/10.1063/1.3295638

Team, S. A. F. (n.d.). *Model-Based Systems Engineering - Scaled Agile Framework*. https://www.scaledagileframework.com/model-based-systems-engineering/

CONCEPT OF A COGNITIVE AGENT SUPPORTING COLLABORATION IN HUMAN TEAMS

Wolfgang Sachsenhauser & Axel Schulte
University of the Bundeswehr Munich
Munich, Germany

This contribution aims at the support of human teamwork between crew members of next generation combat aircraft by means of a distributed and adaptive assistant system. In future combined air operations several aircraft, manned and unmanned, operate together to achieve a common mission objective. That requires a high degree of coordination amongst the pilots, each of them being highly charged with e.g., managing unmanned vehicles from their cockpits. Our approach is to develop a distributed assistant system that observes each pilot in their cockpits. By use of a task model, it shall create and update a shared representation of the team members' activities, pending tasks, and available mental resources. From that, adaptive teamwork supporting interventions shall be generated. Currently, we are developing a laboratory prototype that shall be integrated and evaluated in pilot-in-the-loop experimentation in our fighter aircraft cockpit simulator in Manned-Unmanned Teaming (MUM-T) missions.

MUM-T describes the interoperability between manned and unmanned mobile military assets to pursue a common mission objective. Both, the manned and the unmanned assets, need to be employed in the same confined spatial, temporal, and mission-related context. In MUM-T, the unmanned platform(s), as well as its/their mission payloads will be commanded by the manned asset(s). MUM-T requires to master the high work demands posed on the human user(s) arising from the multi-platform mission management and execution tasks. Fig. 1 shows two MUM-T cells, in each of which a pilot managing an unmanned team consisting of a small number of UAVs (Unmanned Aerial Vehicles) from aboard their command fighters. To tackle the human mental capacity related challenges, we take a cognitive automation approach inspired from both, cognitive ergonomics, and AI (Artificial Intelligence) methods. Meitinger (2008) developed a decentralized multi-agent system to coordinate a team of UAVs under human command. Uhrmann (2012) and later Dudek (2020) investigated the design of that delegation relationship. Heilemann (2019) as well as Meier (2022) took a centralized planning and scheduling approach to coordinate the UAV-team under human supervision. These works solely focused on one MUM-T cell incorporating only one command vehicle with a single pilot cockpit. However, in larger scale future Combined Air Operations (COMAOs) more than one MUM-T cell will work together, creating a hybrid Manned-Unmanned Team (hMUM-T) (cf. Fig. 1).

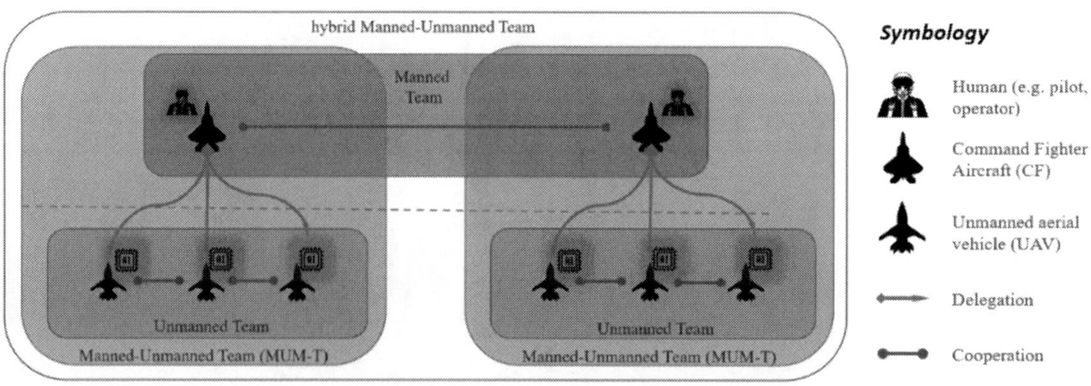

Figure 1: Forms of Collaboration in MUM-T.

In current, purely manned COMAOs, pilots are well trained to deal with the challenges of manned-manned teaming, taking benefit from established hierarchical responsibilities, i.e., mission commander, flight group leaders, and wingmen. However, in hMUM-T we expect manned hierarchies becoming flatter, because each participating fighter will be a command vehicle. These command-and-control activities will harder to be observed by the other human teammates in their cockpits. Furthermore, the high work demands arising from the UAV mission management will take mental capacity away from the manned-manned teaming task. To address these challenges, Brand and Schulte (2021) developed an assistance system in a helicopter application that used a team task model to predict the tasks of a two-person cockpit crew. Although, team-oriented model structures were already provided, the assistance mainly concentrated on individual crew-member support. In this contribution we present a concept to provide team-oriented assistance.

Concept of a Team Moderation Assistant System on Behavioral Level

To address these challenges, we introduce an assistant system, referred to as the Team Moderation Agent (TMA), to support the coordination of a team of pilots. To describe the work relation of such a TMA with the human pilots, we propose a cognitive work system design (see Fig. 2). Schulte, et al. (2016) suggested a graphical description language to describe complex Human Autonomy Teaming (HAT) systems. Here, two distinct modalities of cognitive agents can be introduced: Tool Agents ($_T$) and Worker Agents ($_W$). A Tool Agent receives tasks from a human user and performs them on a high level of automation. In our work system Tool Agents are used to control the UAVs in a MUM-T cell. A Worker Agent acts on its own initiative and assist the human user, e.g., pilot ( ) in achieving the mission objective. In our configuration, a Worker Agent shall be developed as TMA to support the coordination of the human pilot team, as indicated by the heterarchical, i.e., cooperative work relationship between the TMA and the human team, as illustrated in Figure 2.

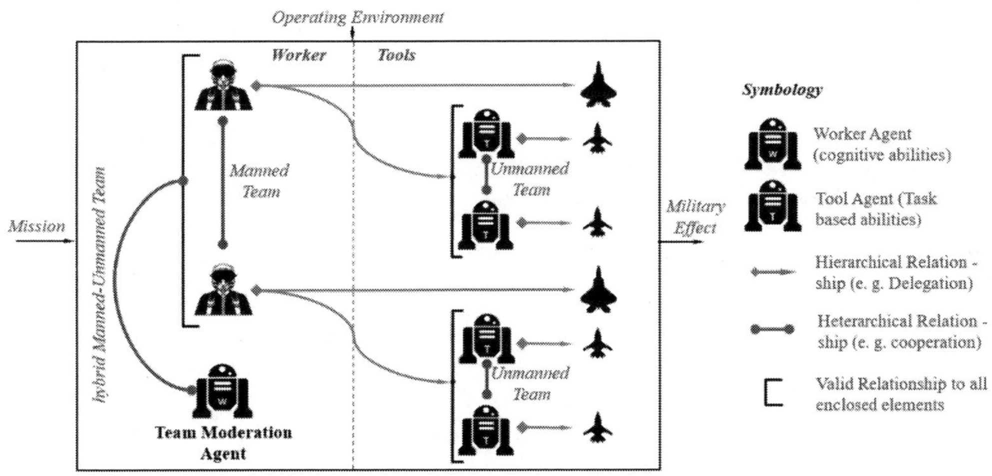

Figure 2: Work system of the presented approach.

To establish functional requirements for the TMA, collaboration rules based on literature on effective human collaboration must be established. The literature summary that follows provides guidance for interpreting the agent's behavior toward this objective. To attain a shared work objective, all team members must carry out task and teamwork processes (Dyer, J.L. 1984). According to Fisher (2013), taskwork processes are directly related to the team's tasks and objectives, while teamwork processes refer to extra efforts required to facilitate coherent teamwork. The relative importance of these two process types cannot be generally stated, as it depends on the domain, situation, and mission role. For both process types, there are known factors that improve performance. Adequate resources and sufficient mental capacity (Wickens 2008), timely access to all relevant information, the opportunity to work independently (Feyerherm 2002), and the opportunity to work without interruption (Chism 2011) are necessary for effective performance in taskwork processes. On the other hand, teamwork processes' performance depends on several factors, including clear shared goals, effective communication, clear roles and responsibilities, equal participation (Chism 2011), trust and respect for others, and strong leadership support (Feyerherm 2002).

Based on these findings, **Rules of Collaboration** can be established to facilitate effective co-operation and error prevention. Taskwork processes and teamwork processes require distinct sets of rules. In Taskwork processes Walsdorf (2001) highlights the importance of commitment to own tasks, timely task completion, avoidance of unnecessary obligations, and regular mission goal review. Pilots should report task execution problems promptly, offer support to teammates in resource-constrained situations, and maintain an overview of pending tasks to achieve the mission goal. In teamwork processes, even workload distribution in terms of time and quantity, efficient allocation of shared resources, avoidance of redundant task assignments and unnecessary dialogues, early and frequent task coordination, and consideration of user expertise and skills during the coordination process are essential (Meitinger 2008).

The following step includes deriving **Rule Violations** from the rules of collaboration. In the case of taskwork processes, violations consist of missed individual or team deadlines, disregard of team-wide task coordination, resource bottlenecks, and uncoordinated usage of shared resources. In teamwork processes, violations consist of uneven task coordination in quantity or time, assigning tasks to unqualified members, coordinating parallel tasks for the same member, failure to prioritize urgent tasks, and unguided pursuit of mission objectives.

Based on this, we derived functional requirements for the TMA to recognize rule violations, to develop solutions, and to initiate interventions. To accomplish this, the TMA needs knowledge about the current mission progress and the mental state of the team members. This includes:

- all mission tasks necessary to achieve the mission goal, including their sequence and temporal or logical dependencies (for a representation of the mission plan, see **Planner**),
- all past, present, and (after a successful coordination) future pilot activities of the entire team (see **Activity Determination**), and
- a domain-specific representation of how mission tasks can be divided into pilot tasks and pilot sub-tasks, supplemented by time and resource requirements for the pilot tasks (also a hierarchical task model, see **Static Task Model**).

This knowledge must be continuously updated and made available to the agent in the form of a **Dynamic Team Task Model**. Through this model, the agent is capable of identifying both unprocessed pilot tasks and results of pilots' task coordination. This allows for the creation of two schedules. The first one contains a chronological arrangement of pending tasks in terms of pilot coordination, known as the **Team Time Plan** (TTP). The second one, on the other hand, includes only pending tasks necessary for mission fulfillment and arranges them in terms of optimal, evenly distributed utilization of all human resources, known as the **Ideal Time Plan** (ITP). The TTP serves as a means for the agent to support pilots in their solution approach and bring improvements. The ITP, on the other hand, serves as an evaluation measure of targeted mission tracking and even task distribution and can be used to make major coordination adjustments if pilot coordination deviates too far from the optimum. To initiate interventions, it is necessary for the TMA to actively follow the rules of collaboration when interacting or communicating with the affected pilot or pilot group.

Concept of the Team Moderation Agent on Functional Level

We have adopted a modular structure for the operationalization of the TMA, as depicted in Figure 3. On the left, there is the supervisory control of the pilots controlling their command fighters and teams of UAVs, either by conventional aircraft control interfaces, or through different modalities such as gaze, touch, voice, and gestures to task the UAVs. The cockpit displays provide the pilots with up-to-date

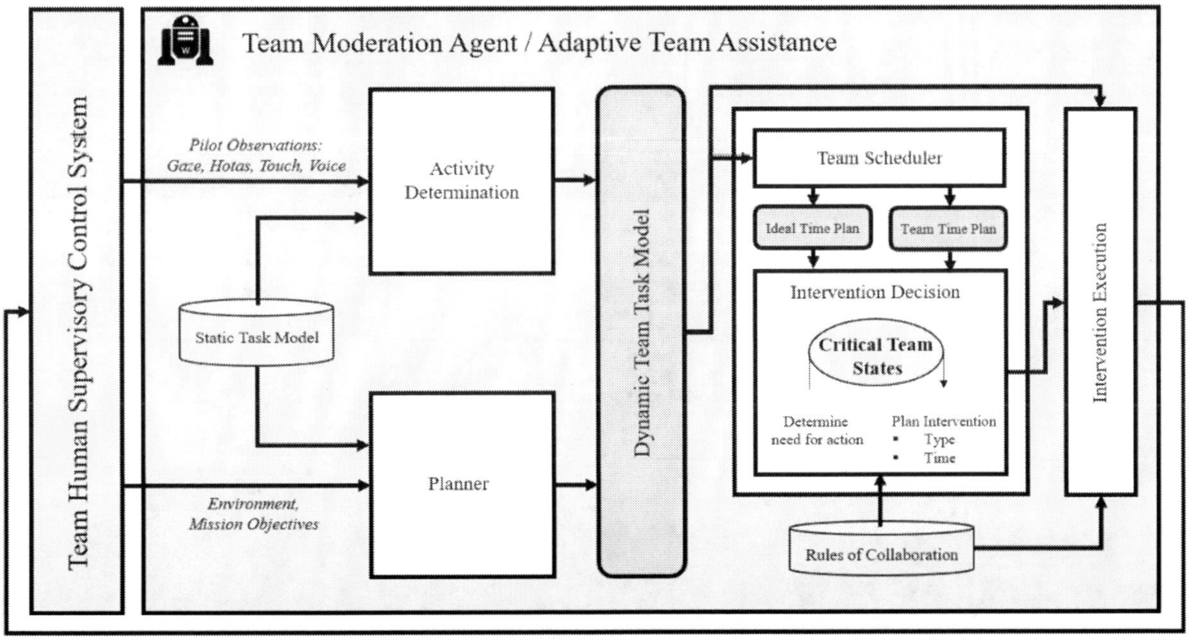

Figure 3: Functional architecture of the Team Moderation Agent.

information about the status of their controlled vehicles as well as the environment and other mission data. The entirety of this information serves as input to the **Activity Determination** module, developed by Tschurtschenthaler and Schulte (2023), that determines the current, past, and future activities of the pilots. The **Planner** module, developed by Maier and Schulte (2022), creates a mission plan based on the information entered. For the further processing, both modules require a domain-specific hierarchical task model (Static Task Model), from that pilot tasks, sub-tasks and actions are derived.

Static Task Model

As per Tschurtschenthaler and Schulte (2023), the Static Task Model (STM) is a representation of task knowledge within a particular domain. This hierarchical model provides a definition of the relationships between mission tasks, pilot tasks, pilot sub-tasks, and actions. Additionally, the model stores an estimated duration for each pilot task. Since humans are able to effectively handle complex situations by breaking them down into tasks, possessing task knowledge is crucial for an agent that aims to collaborate with a human team.

Planner

A logical planning module analyzes the current mission order and determines the necessary mission tasks to achieve the mission objective (Maier and Schulte 2022). This is done while considering temporal and logical constraints from the mission briefing and ensuring that the result is human-readable and meets human expectations. The planning module also determines the latest possible execution time for each mission task while considering available resources. Afterward, the module interprets these mission tasks using the static task knowledge from the STM and continually updates a **Dynamic Team Task Model** (DTTM) with the derived pilot tasks. The structure of the DTTM is derived from the STM, but it only contains tasks that need to be performed in the current situation. The last possible time for the execution of a mission task is used to determine the last possible time for the execution of the associated pilot tasks while considering their temporal sequence. Finally, the DTTM represents all pilot tasks, including their deadlines, which the team must complete to achieve the mission goal after the planner processes have been completed.

Activity Determination

The Activity Determination module performs the analysis of the pilot-system interactions to determine the pilot's current activity (Tschurtschenthaler and Schulte 2023). Observations of these interactions are stored in the DTTM, linking them to specific pilot actions in the action layer (as depicted in Figure 4 on the left). The hierarchical structure of the DTTM enables the identification of the current pilot task and its associated mission task based on the pilot's actions. The module populates the previously created DTTM with the pilot's past (black connection), current, and future tasks (yellow/green connection), providing a comprehensive overview of the mission status and team mental state for both the human and the agent.

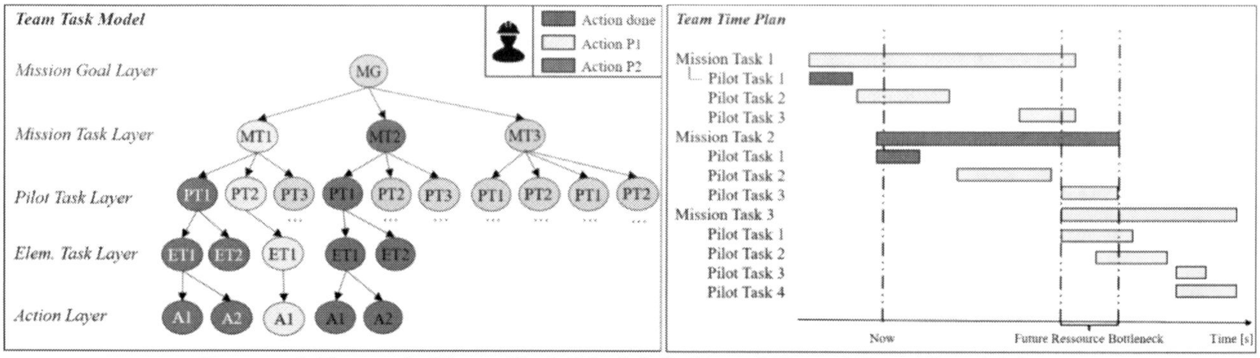

Figure 4: Dynamic Team Task Model (left) and Team Time Plan (right).

Team Scheduler

The Team Scheduler module utilizes the DTTM as an input, where each unprocessed pilot task contains a last-possible processing start based on planning results and constraints. Human coordination results are also stored in the DTTM. The module then generates two schedules: the Team Time Plan (TTP) based on human coordination processes and the Ideal Time Plan (ITP) generated by an optimizer that distributes open tasks among team members in an evenly distributed way taking into account their capabilities, without considering human coordination. The TTP is displayed in Figure 4 on the right for a two-person team of fighter pilots. The TMA can detect and react to future resource bottlenecks based on an evaluation scheme. In this scenario, the TMA could on basis of the ITP suggest how to distribute tasks among the team to avoid a resource bottleneck by properly coordinating tasks.

Intervention Decision

Based on the DTTM, the TTP and the ITP, the TMA's cognitive abilities allow it to detect violations of the rules of collaboration, which may require intervention. In the event of a rule violation, the TMA should intervene in an adaptive and incremental manner:

- utilization of the resources of the affected pilot or group,
- utilization of the resources of the entire human team,
- generation of solution proposals by TMA's machine resources and
- takeover of tasks by the TMA's machine resources.

However, the TMA should prefer to use human resources to prevent pilots from being excluded from the decision-making process. The primary goal is to maintain human responsibility and prevent automation-induced errors.

Intervention Execution

The TMA initiates an automated communication with the affected pilot or group of pilots to inform them of the violation of the rules of collaboration. The timing and method of communication are crucial and should follow communication rules derived from the collaboration rules. These rules include avoiding task changes or interruptions during the work process, informing team members of changes in task coordination in a timely manner, and providing support only when necessary. The TTP provides the agent with an opportunity to identify suitable intervention times and assess whether the level of automation needs to be adjusted based on the pilot's workload.

Next Steps

The subsequent step involves the TMA module's completion and integration into the multi-fighter aircraft cockpit simulator, where up to four pilots can collaborate on a prospective MUM-T mission. Following this, the scenario will undergo testing with German Air Force pilots in human-in-the-loop experiments. The assessment will address the following research inquiries: In what way can a cognitive agent enhance human team collaboration? How do humans appraise the task coordination of a cognitive agent in diverse hierarchical relationships to humans? What rules should a cognitive agent adopt to develop a dialogue with individual or a group of human users?

Reference

Brand, Y., & Schulte, A. (2021): Workload-adaptive and task-specific support for cockpit crews: design and evaluation of an adaptive associate system. In *Hum.-Intell. Syst. Integr.* 3 (2), p. 187–199.

Chism, G. (2011): Book Review: Group Genius: The Creative Power of Collaboration by R. Keith Sawyer. *Journal of Food Science Education* 10 (3), p. 26.

Dudek, M., Lindner, S. & Schulte, A. (2020). Implementation of Teaming Behavior in Unmanned Aerial Vehicles. In *Intelligent Human System Integration 2020.*

Dyer, J. L. (1985). Team research and team training: A state-of-the-art review. *Applied Ergonomics* 16 (4). p. 306.

Feyerherm, A. E., & Rice, C. L. (2002): Emotional Intelligence And Team Performance: The Good, The Bad And The Ugly. In: *The International Journal of Organizational Analysis* 10 (4), S. 343–362.

Fisher, D. M. (2013). Distinguishing Between Taskwork and Teamwork Planning in Teams: Relations With Coordination and Interpersonal Processes. In *Journal of Applied Psychology*.

Heilemann, F., Schmitt, F., & Schulte. A. (2019). Mixed-Initiative Mission Planning of Multiple UCAVs from Aboard a Single Seat Fighter Aircraft. In *AIAA Session Human-Automation Interaction.*

Meitinger, C., & Schulte, A. (2008). Human-UAV Co-operation Based on Artificial Cognition. In *D. Harris (Ed.). Engineering Psychology and Cognitive Ergonomics.*

Maier, S., & Schulte, A. (2022). A Cloud-based approach for synchronous multi-pilot multi-UAV mission plan generation in a MUM-T environment. In *AIAA Session Human-Automation Interaction.*

Schulte, A., Donath, D., & Lange, D. S. (2016). Design Patterns for Human-Cognitive Agent Teaming. In *Don Harris. Engineering Psychology and Cognitive Ergonomics, Bd. 9736. Cham: Springer International Publishing (Lecture Notes in Computer Science)*. S. 231–243.

Tschurtschenthaler, K., & Schulte, A. (2023). Concept for an Automated Activity Determination in the Temporal Domain for Adaptive Pilot Assistance. In *International Symposium on Aviation Psychology 2023.* Rochester.

Uhrmann, J., & Schulte, A. (2011). Task-based Guidance of Multiple UAV Using Cognitive Automation. In *The Third International Conference on Advanced Cognitive Technologies and Applications.*

Walsdorf, A., & Onken, R. (2001). Assistant Systems for aircraft guidance: cognitive man-machine cooperation. In *Aerospace Science and Technology*. p. 501- 520.

Wickens, C. D. (2008). Multiple resources and mental workload. In *Human factors* 50 (3), S. 449–455.

PERSONALITY FACTORS AND EDUCATION OUTCOME IN SWEDISH MILITARY PILOT EDUCATION

Malcolm Sehlström
Department of Psychology, Umeå University
Umeå, Sweden
Markus Nyström,
Psychology, Department of Health, Education and Technology, Luleå University of Technology
Luleå, Sweden
Jessica Körning Ljungberg,
Psychology, Department of Health, Education and Technology, Luleå University of Technology
Anna-Sara Claeson,
Department of Psychology, Umeå University
Umeå, Sweden

Profiling pilot personality is a common effort within aviation. We examined whether there are personality-related differences in who passes or fails the Swedish military pilot education. Assessment records of 182 applicants, accepted to the education between the years of 2004 and 2020 were studied (Mean age 24, SD 4.2. 96% male, 4% female). Descriptive discriminant analysis (DDA) was used to explore which personality traits and suitability ratings might be related to education outcome. Analysis included suitability assessments by senior pilots and by a psychologist, a number of traits assessed by the same psychologist, as well as the Commander Trait Inventory (CTI). The resulting discriminant function was significant (Wilk's Lambda = .808, (20) = 32.817, $p = .035$) with a canonical correlation of .44. The modeling suggests that senior pilot assessment and psychologist assessment contribute to the structure. Also contributing were the traits energy, professional motivation, study forecast and leader potential.

There has been long-standing interest in understanding what makes someone a good pilot, and how to identify them. While pilot aptitude tests or general cognitive ability scores might be informative of being able to learn to be a pilot, it has been argued since the earliest aviation researchers that personality might be more indicative of professional success in the long run (Sells, 1956). When it comes to military aviation, the pilot needs not only be suited for piloting but also the military life in general (Retzlaff and Gibertini, 1987). In recent years, more and more studies of military pilot personality have been conducted and published, with many looking into who makes it through the education to become a pilot. A meta-analysis by Campbell, Castaneda and Pulos from 2009 included 26 American studies that had examined personality traits in relation to aviation training outcome. Grounded in the five-factor model (Costa & McCrae, 1992), it was shown that higher levels of the personality trait *neuroticism* were related to failing training, while higher levels of *extroversion* were related to succeeding. King et al. (2012) also noted that individuals who are goal-oriented and confident tend to succeed whereas those with low levels of aggression, impulsivity and risk-taking tended to drop out of training voluntarily. However, there is still more mapping to be done.

This paper covers the preliminary report on data from the Swedish military pilot education. In Sweden, the Special Selection Department at the Swedish Armed Forces Human Resources Centre collects test data during the selection process for military pilot education. The Swedish model relies on their own assessment procedures for pilot applicants and this study contributes to furthered understanding of how personality factors might relate to military pilot outcomes by studying this data. With the previous research being mainly concerned with the American system and populations, differences between our populations could carry factors hitherto unaccounted for. We have thus examined personality traits related to officer suitability, psychologist assessed personality traits related to military aviatior suitability, as well as general military aviatior suitability ratings based on interviews with psychologists and senior pilots, in relation to education outcome.

Method

Participants

A representative study sample was provided by the Special Selection Department at the Swedish Armed Forces Human Resources Centre and contains pilot aptitude test results for eligible military pilot applicants from the year 2004 until the year of 2021. For applicants that were accepted to begin education, it also contains a recorded education status. Included in our study were those who had been accepted into education and had a noted education status. After controlling for these inclusion criteria, the sample size was 182 individuals. Ninetysix % were men and 4 % women, with a mean age of 24 years (SD= 4.2).

Measures

Education outcome. Education outcome is noted in the obtained dataset and a binary variable was created which indicates for each cadet whether they completed their education or had it terminated. A noted termination outcome can reflect either a voluntary dropping out or a separation based on abnormal training progression or compatibility issues.

Commander Trait Inventory. The Commander Trait Inventory (CTI) is an 11-scale personality inventory that assess the respondents cognitive style and officer relevant personality aspects. (Carlstedt & Widén, 1998). The 6 cognitive style scales are *abstract thinking, concrete thinking, ideological value orienting, superficial value orienting, sensation orienting,* and *intuitive decision making.* The 5 officer relevant personality aspects are *empathy, leader motivation, egocentricity, impulsivity* and *ethnocentrism.* All 11 scales are stanine-transformed.

Psychologist assessment.
Employed by the Swedish Armed forces, licensed and extensively experienced psychologists conduct semi-structured interviews to discern pilot suitability. Beyond giving a general recommendation score, psychologists rate candidates on the traits of *social ability, energy, emotional stability, maturity, leadership potential, professional motivation* and give a *study forecast*. Ratings are given on a 1-9 scale.

Senior pilot assessment.
Coming from the background of being experienced senior pilots, these assessors rate cadet suitability for the pilot profession. Ratings are given on a 1-9 scale.

Statistical analysis
We took an exploratory approach to examining the relationship between the assessment variables and education outcome using descriptive discriminant analysis (DDA), conducted in IBM SPSS Statistics software. Using DDA, we attempt to assess variable importance by creating discriminant functions to explain group membership; discriminant analysis provides a structure of the weight or correlation from each variable to the discriminant function. In predictive efforts, one might rerun modeling using custom sets of independent variables in order to find the model with the greatest classification success, but for this initial exploratory endeavor we included every assessment measure to note their relation to the model function. In this manner, the analysis was conducted with the binary education outcome variable

(completion/termination) as grouping variable and the 11 CTI subscales, 7 psychologist assessed traits, general psychologist assessment and senior pilot assesment as independents.

Results

Of the 182 pilot students in the register sample, 36 had their education terminated (19.7%). In conducting analysis, there were some cases of missing data. The sample in carrying out the discriminant analysis thus included 31 failing students and 135 passing.

Because of the binary outcome, completion or termination, the discriminant analysis produced a single discriminant function. The function produced was statistically significant: Wilk's Lambda = .808, (20) = 32.817, p = .035, with a canoncial correlation of .44, indicating a performance of 44% prediction of the variances by the relationship of factors and group membership. The structure matrix for the discriminant function is displayed in Table 1.

Table 1
Discriminant Analysis Structure Matrix

	Function 1
Senior pilot assessment	.598
Energy	.551
Professional motivation	.428
Study forecast	.388
Leader potential	.376
Psychologist assessment	.372
Emotional stability	.306
Empathy	.306
Social ability	.228
Impulsivity	-.192
Superficial value orienting	.183
Intuitive decisionmaking	.179
Maturity	.169
Egocentricity	.137
Leader Motivation	.118
Etnocentricism	.113
Abstract thinking	-.071
Sensation orienting	.037
Concrete thinking	.025
Ideological value orienting	-.001

Note. N = 166.

As can be viewed in Table 1, the variable that most heavily loads into the discriminant function is senior pilot assessment, which is followed by the traits of energy, professional motivaiton, study forecast, leader potential and the general psychologist assessment. Emotional stability and empathy might be considered on the edge of contributing significantly to the function if considering simply function weights. Testing does however not support there being significant group differences based on emotional stability (Wilk's Lambda = .978, (1) = 3.652, p = .058,) nor empathy (Wilk's Lambda = .978, (1) = 3.644, p = .058,). None of the other variables beyond the primary six in the structure matrix display significant group differences either.

Discussion

In this study we examined if personality traits and suitability ratings from the selection process are related to education outcome in the Swedish military pilot education. Our results do suggest that there are a number of factors from assessment that indeed appear related to whether a cadet completes training or not, even after they have been accepted into the education. The analysis shows that senior pilot ratings of cadet suitability are most strongly related to education outcome, which is followed by the traits; energy, professional motivation, study forecast and leader potential, as well as the general psychologist suitability assessment.

Assessment of cadets by observers has been researched in some studies, and personality and performance have been shown to be related to these assessments (Barron et al., 2016, Skoglund et al., 2021). To our knowledge, professional assessment has not previously been found more strongly related to pilot success than other assessment. In this way, our results highlights the proficiency of active pilots in understanding the demands on and expectations of pilot students.

The trait scores, which are set by the same psychologists that give the general suitability rating, carry more specificity and it can in this way be expected that they would appear more strongly related to the outcome than the general score, which was observed. As for these traits: the trait of energy encompasses initiative, perserverance and stress tolerance. The relation of this variable to outcome could be related to previous studies that have relied on the five-factor model traits (Campbell et al., 2009, King et al., 2012). Indeed, low stress tolerance in the form of higher neuroticism scores has been shown as related to not finishing aviation training, and higher conscientiousness and extroversion as connected to success. Our results are in this way in line with much of previous research.

Previous job analysis research has military pilots pointing out the importance of motivation for the profession (Damos, 2011) and general motivational research support its value for e.g. finishing higher level education (Liu, Bridgeman & Adler, 2012). Our results regarding professional motivation add to this notion of motivational importance in military aviation education as it remains predictive of outcome in our study population. Leadership potential encompasses some motivational aspects also, but more so diplomacy. This diplomatic aspect can be related to *openness* from the five-factor model, which has been noted higher than normal in some pilot populations (King et al., 2012).

We found no relationship between any of the CTI subscales and education outcome. It is noteworthy that neuroticism, which has been noted in previous research as tied to failed military pilot education (Campbell et al., 2009), has a similar subscale to the impulsivity subscale of the CTI. The lack of relation between our impulsivity scale and education outcome could reflect the scales not measuring the same construct, a difference in range within the populations, or a sample size issue. Indeed, while theorizing around nonsignificant effects should be minimal, higher impulsivity does have a negative relation, while small, to education completion via its function weight, as visible in Table 1. With empathy as a CTI subscale on the brink of significance as well as social ability, it is possible that there might be effects hidden by the sample size.

In general, there might be conceptual overlap between our results and much of that in previous research, but since both the traits assessed in the psychologist interviews and the scales of the CTI are not based on the five-factor model, the level of overlap cannot be specified without further research.

Summarily, these preliminary results suggest that specific individual traits as well as general suitability ratings, from applicant assessment data, are related to education outcome in the Swedish military pilot education. This applies even after these measures have been used in selecting for applicants.

Acknowledgements

The authors would like to thank the SAAB Support and Services division and the Industrial Doctoral School at Umeå University for financial support of the project. Jessica K. Ljungberg is funded by a grant from Vinnova (Grant number 202100-2841). Much thanks to Gerhard Wolgers, licensed psychologist at the Special Selection Department at the Swedish Armed Forces Human Resources Centre for his work in providing the data and sharing his knowledge. Thanks to the Special Selection Department in general.

References

Barron, L. G., Carretta, T. R., & Bonto-Kane, M. V. A. (2016). Relations of personality traits to military aviator performance: It depends on the criterion. *Aviation Psychology and Applied Human Factors, 6*(2), 57.

Campbell, J. S., Castaneda, M., & Pulos, S. (200). Meta-analysis of personality assessments as predictors of military aviation training success. *The International Journal of Aviation Psychology, 20*(1), 92-109.

Carlstedt, L. & Widén, H. (1998). *CTI – Commander Trait Inventory*. Ledarskapsinstitutionen Serie T:5. Karlstad: Förvarshögskolan

Costa, P. T., & McCrae, R. R. (1992). *The Revised NEO Personality Inventory Manual*. Odessa,FL: Psychological Assessment Resources.

Damos, D. 2011. *KSAOs for Military Pilot Selection: A Review of the Literature*. Report number AFCAPS-FR-2011-0003. Randolf AFB, TX: Air Force Personnel Center Strategic Research and Assessment.

King R. E., Retzlaff, P., Barto, E., Ree, M. J., & Teachout, M. S. (2012). *Pilot personality and training outcomes.* School of Aerospace Medicine Wright Patterson AFB OH.

Liu, O. L., Bridgeman, B., & Adler, R. M. (2012). Measuring learning outcomes in higher education: Motivation matters. *Education Researcher, 41*(9), 352-362.

Retzlaff, P. D., & Gibertini, M. (1987). Air force pilot personality: Hard data on the" right stuff". *Multivariate Behavioral Research, 22*(4), 383-399.

Sells, S. B. (1956). Further developments on adaptability screening of flying personell. *The Journal of Aviation Medicine, 27*(5), 440-451.

Skoglund, T. H., Fosse, T. H., Lang-Ree, O. C., Martinsen, Ø. L., & Martinussen, M. (2021). Candidate personality traits associated with ratings in a military officer selection setting. *Skoglund, TH (2022). A short-form personality measure for military personnel selection: Psychometric investigation and perspectives on usage. (Doctoral thesis). https://hdl.handle.net/10037/24840.*

EFFECTIVE INTEGRATION OF HUMAN FACTORS ENGINEERING INTO FAA SYSTEM DEVELOPMENT ACQUISITION PROGRAMS

Philip J. Smith
The Ohio State University, Columbus OH

This was a research effort focused on developing recommendations to improve the understanding and application of human factors (HF) by Federal Aviation Administration (FAA) acquisition program personnel. Broadly speaking, the goal was to investigate methods to help ensure the successful integration of human factors over the lifecycle of the FAA acquisition process. Structured interviews were conducted with 24 individuals with relevant program management and HF experience from the FAA and industry. Relevant FAA resources (documents and websites) and past examples of products developed by individual acquisition programs were reviewed. Based on this research, 22 recommendations were made. Key recommendations along with insights from the interviews are presented in this paper.

Introduction

This was a research effort to support improved understanding and application of human factors by FAA acquisition program personnel to improve compliance with FAA Acquisition Management System (AMS) policy and guidance. Broadly speaking, the goal was to investigate methods to help ensure the successful integration of Human Factors (HF) over the lifecycle of the FAA acquisition process and to provide recommendations for refinement of the FAA HF Acquisition Job Aid (which provides guidance for human factors specialists responsible for managing the HF aspects of FAA acquisitions) responsible for contributing to and managing the integration of HF into the acquisition process. Following Research for Service Analysis to better document needs and potential solutions, the AMS process includes the following stages: Strategic Analysis and Strategic Planning, Concept and Requirements Definition, Investment Analysis, Solution Implementation and In-Service Management.

More specifically, this effort was framed as addressing the following more detailed questions:

- How can the FAA influence Program Managers and HF Coordinators for FAA development projects to more effectively integrate HF in the acquisition process?
- How can the FAA improve the transfer of knowledge gained from the initial Research for Service Analysis to the Solution Implementation stage?
- What resources are available to support the more effective integration of HF within an FAA acquisition program?
- How can the FAA HF Acquisition Job Aid (2013) be enhanced based on these findings?

Methods

The findings presented below are based on:

- Interviews with FAA staff responsible for program management and HF support for the acquisition of new hardware and software tools to support air traffic control, air traffic flow management and technical operations.
- Interviews with HF staff providing management and HF support for a range of different companies involved with the development of new hardware and software tools.
- Interviews with FAA Air Traffic Organization (ATO) staff who have participated as members of user teams within acquisition programs and who have experienced and observed the introduction of new tools and procedures into FAA facilities.

- Samples of documents produced by FAA acquisition programs that provide insights into the methods and results associated with the integration of HF considerations in the development of new software tools.

Procedures

Relevant experts were interviewed individually for 1-2 hours. The questions addressed included:

- (For FAA staff and contractors) Do you make use of the HF Acquisition Job Aid? If so, how?
- (For FAA staff and contractors) How would you improve the HF Acquisition Job Aid?
- What are the steps in the acquisition/development process used by your FAA program or company?
- How is continuity regarding HF insights communicated across successive steps in terms of documentation and personnel?
- What roles and responsibilities do HF specialists play in these steps?
- What strategies do you find most effective to help assure the successful integration of HF in the process? (including support from program managers)
- What do you do (or what could be done) to make the integration of HF in the process more cost-effective/efficient?
- What are barriers to the integration of HF in the process? What strategies could be employed to overcome these barriers?
- What human factors methods are used?

In addition, FAA documents and the relevant HF literature were reviewed.

Interview Participants

- 12 industry participants were interviewed. All of them work as human factors specialists who collectively have had experience in the application of human factors to software development in the following industries: Agriculture; air transportation; communication; consumer goods; ground transportation; healthcare systems and medical devices; large-scale computing; personal computing; smart cities. Their years of experience ranged from 5-35 years.
- 8 interviews were conducted with individuals representing experience in either program management and/or HF engineering as part of FAA acquisition programs. This included FAA staff and contractors hired by the FAA to provide HF expertise in support of the management of acquisition programs. These individuals represented 10-38 years of experience associated with the eight FAA acquisition programs.
- 3 STMCs from three different Air Route Traffic Control Centers (ARTCCs) were interviewed regarding their experiences as members of user teams associated with the acquisition of FAA software tools and their experience with the introduction and use of such tools.
- 1 FAA staff member was interviewed who has HF responsibilities within the FAA Air Traffic Organization (ATO).

Findings and Discussion

For the purposes of this discussion, we will characterize the FAA lifecycle management process for hardware and software development projects as typically including several stages: Research for Service Analysis preceding the formal AMS process which includes the following stages - Strategic Analysis and Strategic Planning, Concept and Requirements Definition, Investment Analysis, Solution Implementation and In-Service Management.

he transition from the concept exploration, development and evaluation stage and the Solution Implmentation stage can be characterized as adhering to a waterfall model, where the concept exploration, development and evaluation stage produces a CONOPS (Concept of Operatoins) as well as a Statement of Work for Solution Implementation, with associated requirements.

Potential Users of the Job Aid

The HF Job Aid is primarily designed to support HF specialists who are part of the FAA *systems engineering project management team* that oversees and supports HF integration during the research, development and implementation conducted by contractors hired for an acquisition. As a secondary audience, the Job Aid is useful to the HF specialists who are part of the contractor team responsible for actually conducting the HF research, development and implementation for an acquisition, as it lets them understand the role and expectations of the FAA management team.

> *Job Aid Recommendation 1.* Frame the contents of the Job Aid assuming the primary target audience is the FAA practitioner supporting and monitoring HF research activities during Research for Service Analysis and the FAA HF Coordinator supporting and monitoring HF design and evaluation activities during Strategic Analysis and Strategic Planning, Concept and Requirements Definition, Investment Analysis, Solution Implementation and In-Service Management.

Findings from Interviews and Document Reviews

The interviews indicated a wide variation in actual practice regarding HF involvement as part of these FAA and contractor teams. At one extreme (the lowest end in terms of participation), there sometimes has been no participation by a HF specialist as part of the FAA management team or the contractor team. At the other extreme, there have been HF specialists who have been an integral part of the FAA management team and of the contractor team conducting the Research for Service Analysis or Solution Implementation. There also have been FAA projects where an HF specialist has played a more limited reactive role, only providing input when questions have been raised by engineers or program managers on the project.

The same range of involvement by HF specialists was indicated in the interviews with industry staff. At the low end of HF involvement in industry projects:

- "The involvement of HF is hit and miss. Some project managers and development staff think that everyone knows human factors: 'I'm human so I know what people need'. Of course they don't really have the understanding that can be provided by a HF expert. But they may just cherry pick to decide when to ask a HF expert for input, or they may not involve the HF expert at all."
- "There are some projects that don't include a human factors specialist. They just show their result to the design group [which may include HF experts] later in the development to get input. That is too late."

In terms of industry examples with more effective HF integration:

- "Technically we're using an agile development model. If you have a HF specialist dedicated to the project like the other engineers, you can develop good relationships and have effective input. But you need the core competencies, including HF, on the agile team. Even then, though, you have to have good management. If I miss a meeting and they make a decision that I don't agree with, that can be a problem. Decision making needs to be managed."
- "The waterfall approach tends to put HF way too late. It's almost a checkbox item. Everything at our company is now agile with HF involvement very early. End users are also involved early. We had one major project where the engineering lead reached out to the HF professional for feedback too late and then decided to leave the recommended changes for future revisions. The project failed because there were too many usability challenges."

- "With the agile approach you're able to add/extend requirements pretty easily. It allows you to learn quickly. But they can get caught up in playing with a design and may use this as an excuse to skip important HF steps. … They can fixate on a particular design too quickly."

This emphasis on the value of integrating HF starting early in the design process is consistent with FAA (2013) which notes: "The funding necessary to conduct a comprehensive human factors engineering program for a solution has been estimated to be between 0.5% and 6% of developmental costs (depending upon the sensitivity of the solution to human factors issues). The benefit from conducting a comprehensive human factors program has been estimated at between 20% to 30% of total acquisition costs."

> *Job Aid Recommendation 2.* Emphasize the need for the FAA and contractor HF Coordinators to be integral parts of their respective systems engineering teams rather than relying on a model where systems engineering requests HF input only when they recognize that a particular design or evaluation issue has arisen that they feel requires such expertise.

> *Job Aid Recommendation 3.* Communicate that, whether an agile systems engineering model or a more traditional systems engineering model is used, the critical issue is whether the HF expert is involved as an integral member of the team and whether the lessons learned are adequately documented so that they can be communicated to later stages in the acquisition process.

Links to Examples of Best Practices for HF Content in Acquisition Documents. To "tune" the expectations of the FAA HF Coordinator regarding the contents of documents required as part of the acquisition process, there was unanimous agreement in the interviews that it would be valuable to provide links in the Job Aid to sample documents illustrating the needed HF content:

- "You need good examples that are not just shared among practitioners."
- "A lot of stuff is boilerplate. You could use model documents. You sometimes write CDRLs [Contract Data Requirements Lists] from scratch when you could have saved 2 months of work by modifying a boilerplate document. Provide links to sample documents in a website."
- "It would be absolutely useful to have links to sample documents."
- "It [the Job Aid] should be helping people to understand how to tailor."

> *Job Aid Recommendation* 4. One suggestion, supported by all of the FAA staff interviewed, is the inclusion of links to "model documents" in the Job Aid.

As an illustration, below is a slightly edited section from an FAA HF Plan outlining the activities associated with the responsibilities of the FAA HF Coordinator during Solution Implementation.

Activity Schedule

- Develop FAA Integrated Human Factors Plan.
- Evaluate contractor Integrated Human Factors Plan.
- Monitor and support contractor HF activities during Solution Implementation.
  - Coordinate with Project Management and Systems Engineer on the HF Coordinator role on the project with respect to the Responsible, Accountable, Support, Consult, Inform (RASCI) matrix
  - Establish Human Factors Working Group and CHI Team Sub-group
  - Review and comment on HF Contract Data Requirements Lists (CDRLs)
  - Review and comment on CDRLs for the Screening Information Request (SIR), including the Human Engineering Design Approach Document, Operator (HEDAD-O) and Human Engineering Design Approach Document, Maintainer (HEDAD-M) documents
  - Participate in TIMs, working groups, and relevant engineering, development, and management meetings
  - Participate in and observe contractor HF-related activities

- Evaluate proposed data collection instruments for Early User Involvement Events (EUIEs) to evaluate Graphical User Interface/Computer Human Interaction (GUI/CHI) prototypes before Critical Design Review (CDR)
- Evaluate plans and observe/support EUIEs as necessary
- Develop HF reports and update documentation
- Provide HF input to the completion of the In Service Review (ISR) checklist.

Knowledge Transfer to the Solution Implementation Stage. If contractors are employed to conduct the Research for Service Analysis, Strategic Analysis and Strategic Planning, Concept and Requirements Definition and Investment Analysis stages, the SOWs should explicitly address how HF findings will be communicated to downstream phases. As noted earlier, the transition from the Research for Service Analysis to the Solution Implementation stage is a major step in the development "waterfall". The interviews indicated that, in many but not all cases, there is a considerable loss of HF insights across these two stages.

The impact is a loss of efficiency in the development of the final implementation, as well as a potential loss of effectiveness of some design decisions embedded in the final implementation, as HF insights that have been gleaned during the Research for Service Analysis, Strategic Analysis and Strategic Planning, Concept and Requirements Definition, and Investment Analysis stages may not be communicated to the Solution Implementation team. As indicated in the interviews:

- "There can be important HF insights uncovered during concept development that are not adequately communicated in a requirements document."
- "It's hard to take a prototype and turn it into requirements that capture everything that is important and then to take those requirements and turn them back into a system."

The interviews indicated that there is a wide variation in the extent to which the documents and artifacts produced during the Research for Service Analysis, Strategic Analysis and Strategic Planning, Concept and Requirements Definition, and Investment Analysis stages are included in a transfer package provided to the contractor responsible for the Solution Implementation stage. There was a strong consensus that: "There needs to be a mechanism for HF documents from the initial concept development to follow the concept through implementation. Otherwise, you lose many of the HF insights found during concept exploration." In addition to the transfer of such documents produced during the Research for Service Analysis, Strategic Analysis and Strategic Planning, Concept and Requirements Definition and Investment Analysis stages, there are other artifacts that can be fruitfully transferred to the Solution Implementation stage. This potentially includes interface designs, storyboards (with or without specific interface designs) and prototypes.

The potential value of allowing the contractor who is responsible for Solution Implementation to view such artifacts was emphasized by one Program Management Officer for an FAA acquisition program: "We found that the contractor [who was implementing the solution] was revisiting things we solved 2 years ago, so we decided to let the contractor see the prototype developed during concept development. We'd spent years doing the original concept exploration to develop the CONOPS and requirements. Why throw all that expertise away? There's a lot that they have learned that isn't captured in the CONOPS and requirements. Let's give the contractor a head start. They will have new ideas too. I wish I had done this earlier. It would have saved a lot of money."

Job Aid Recommendation 5. As part of the transfer process, support the transfer of HF insights from the the Research for Service Analysis, Strategic Analysis and Strategic Planning, Concept and Requirements Definition and Investment Analysis stages to the Solution Implementation team by providing the latter team with access to relevant documents, prototypes, storyboards and/or screen displays. Require this as part of the Statement of Work for the Contractor completing the Research for Service Analysis.

Human Factors Methods to Consider. One important consideration in terms of planning for HF activities in a plan focuses on the different types of HF methods that can be considered. Below is a list of HF methods that the interviews indicated have been used in various FAA and industry projects. This list is not intended to imply that all such methods should be used for every project as it is necessary to tailor the selection of methods to the needs and constraints of specific projects.

- Ethnographic studies at work sites, including consideration of the "larger ecosystem" (observations; interviews; studies of current tools and studies of other artifacts such as training materials).
- Review of failure reports.
- Structured interviews (individually or in focus groups) and surveys.
- Task, workflow and shortfall analyses, including cognitive and critical task analyses (FAA, 2009).
- User analyses (defining the range of important defining characteristics for the user populations).
- Development of use cases and scenarios. ("While they are useful, use cases alone are too narrow. The engineers will find a way to say their design supports the individual use cases if they are specified. You need to identify broader scenarios and consider how users will perform on them.")
- Design of mockups/wireframes/storyboards and prototypes.
- Heuristic analyses, including use of the Human Factors Design Standard (FAA, 2016), cognitive walkthroughs, design reviews within development teams, workload assessments, think-aloud studies and part-task or full mission laboratory studies (with evaluations at the individual, group and systems levels).
- Shadowing, demonstration or Human-in-the-Loop (HITL) studies and/or field studies.

It should be further noted that human factors practitioners frequently use the analytical methods in an informal, back-of-the-envelope sense to structure their thinking. As one person interviewed said: "They are routinely applying heuristics in the Human Factors Design Standard in their heads as they propose or evaluate design concepts or implementations and they are thinking through how particular scenarios could play out as a user interacts with a particular design."

> *Recommendation 6.* Develop resources to help program managers, HF Coordinators and practitioners to tailor the HF activities for particular programs for integration in a cohesive HF plan.

Conclusion

This project was designed to produce insights into current and recommended practices for more effective integration of HF into the FAA acquisition process. Additional details and the full set of 22 recommendations are available upon request.

Acknowledgements

The authors thank Dan Herschler, Bill Kaliardos and Ben Willems for the oversight and guidance of this project and the FAA NextGen Human Factors Division (ANG-C1) for sponsorship. Our gratitude also goes out to the FAA and industry staff who participated in this project.

References

FAA (2009). FAA-HF-004, Version: A Critical Task Analysis Report. Retrieved from http://everyspec.com/FAA/FAA-General/FAA_HF_004A_31241/

FAA. (2016). FAA HF-STD-001B: Human Factors Design Standard. Retrieved from https://hf.tc.faa.gov/publications/2016-12-human-factors-design-standard/full_text.pdf

FAA (2013). Human Factors Acquisition Job Aid. Retrieved from HFAcqJobAid.doc (live.com)

THE WEAK SIGNALS OF CYBER
DISCERNING AND LEARNING
THAT WHICH IS MEANT TO BE IMPERCEPTIBLE, ILLUSORY, AND TO INVEIGLE

Elena St Amour, Phat Ngo, Tameah Young, James Ness
Federal Aviation Administration
William J. Hughes Technical Center
Atlantic City International Airport, NJ 08405

Cyber threats are often weak signals designed to exploit targeted systems. These signals manipulate cyber, psyber, and risk communication components of the signal to diminish signal-to-noise ratio. Cyber components are the physical aspects of the signal that can range from viral code to the use of aberrant signals from the electromagnetic spectrum to confound operations such as global positioning systems. The psyber component includes the behavioral propensities of the individual operator and level of experience detecting and managing threats. Risk communication is the tension set by the organizational culture priming individual operator propensities. The psyber components affect the ability to perceive contingencies. The risk communication sets the signal detection threshold for distinguishing true threats from false alarms. This paper describes current simulation efforts to afford the application of evidence-based methods to discern weak signals and to accelerate the experience of operators in discriminating weak signals via immersive training simulations.

In compliance with the Aircraft Certification, Safety, and Accountability Act (2020), the National Academy of Sciences, Engineering, and Medicine initiated a 10-year program to identify, categorize, and analyze emerging safety trends in air transportation. In the report, an identified critical need is to discern anomalous patterns in the aviation system visible only as "weak signals" (National Research Council, 2022). Cyber threats are often in the form of weak signals with the signal-to-noise ratio typically manipulated along cyber, psyber, and risk communication parameters. Cyber refers to physical aspects such as hardware, software, and the electromagnetic spectrum used in information technology. Psyber is the influence of cyber on that which is apprehended by the targeted system's operators. Risk communication is the level of tension set in the organizational culture influencing the degree of operator attention from complacency to overreaction.

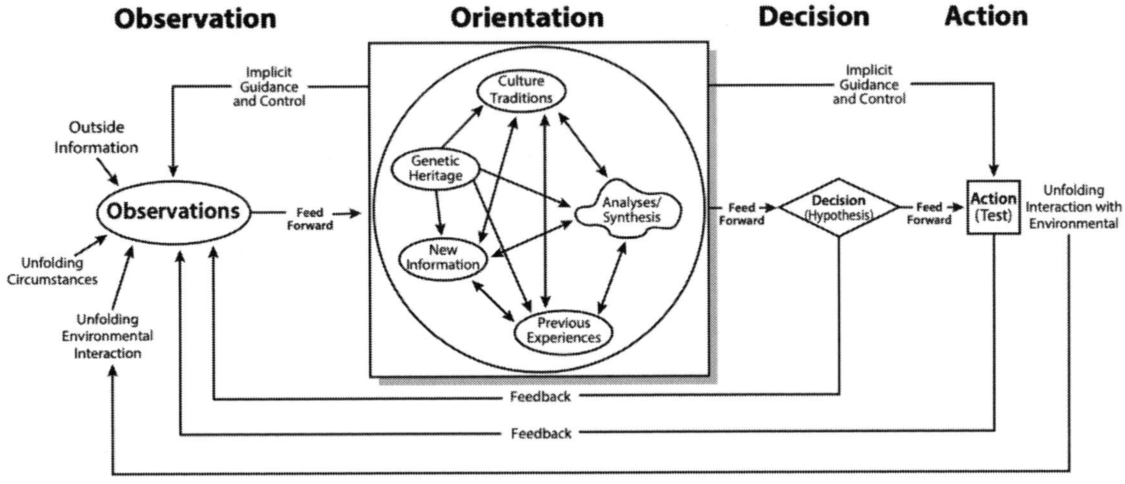

Figure 1. Boyd's OODA Loop

The causal attribution of cyber is perceived in that the operator must detect, either directly or by way of automation, the antecedent/consequent events associated with the signal pursuant to attributing a cause. The causal attribution is the perception of the cyber threat as being either present or absent. In either case the attribution can be correct or mistaken, with the intent of the cyber threat to promote a mistaken perception. The decision process is influenced by the signal's cyber, psyber, and risk communication aspects. Given these aspects, the decision process assigning cause is best described in the OODA loop (MacCuish 2012). Figure 1 shows the OODA loop decision process, which is a feedback loop integrating the steps of Observe, Orient, Decide and Action (Boyd, 2018). Orientation is central in the feedback loop process; previous experience and organizational culture shape Orientation to the signal influencing Decisions and Actions. By nature, cyber signals are novel as the threats evolve. Given potential consequences of the evolving threat, organizations tend toward strict information technology cyber safety protocols. This action, in some ways prudent, does affect the organizational culture in its ability to use information technology to achieve organizational mission, which in turn influences its shared idea of cyber security as legitimate action to an imminent threat or overreaction.

Since all permutations of experiences in the future of air and space transportation cannot be known *a priori*, proscribed intentional controls, memorization of facts, or scripted sequences are likely to be of limited value. A more human centric approach to meeting the future is to note that the quintessential human means for diffusing lessons and experiences is through a tradition of passing on stories (Campbell, 1973). Representations of experiences, as in cave paintings and storytelling are the oldest traditions of recounting events, imparting lessons, and projecting affect (Lord, 1971). These formats structure information in part-whole relations affording the experiencer schematic frameworks to interpret past, present, or future analogous events (Mandler & Johnson, 1977). The diffusion of lessons through stories, using technology-mediated means diffuses lessons in a rapid and salient manner affording exploration of the art of the possible (Aldrich, 2005).

Within big data there exists the foundations of stories in that within big data is an extensive time series of information that cuts across contexts. This information can be compressed and presented in models and simulations to accelerate the experiences of the principals. This process is leveraged in the development of air traffic control simulations which are based on data from the Performance Data Analysis and Reporting System (PDARS). PDARS is the repository for key flight events such as flight transitions, facility handoffs, air space crossing, etc. Leveraging the PDARS, models and simulations can be developed to accelerate experience in the art of the possible in cyber threats, the mitigation of those threats, and in refining the organization's risk communication of cyber threats.

In shaping risk communication, the leadership must recognize that certain terms and actions have a psychological saliency that focuses collective attention on a concept (e.g., cyber) in a manner that can overshadow alternatives and exceptions to the collective idea (Ness, 2006). The replicated idea shared across individuals in the organization becomes the organizational culture's meme. A meme is a concept first introduced by Dawkins (1976) arguing that all life evolves by the differential survival of replicating entities. Extending the idea of the biological replicating entities, genes, the meme is a unit of cultural transmission. As a replicating entity a meme exhibits the properties of longevity, fecundity, and copying-fidelity, which make an established meme hard to undo. Thus, in conveying its meme of cyber, the organization should apply due diligence in forming and communicating its unit of cultural transmission through its actions and words, balancing along the continuum of complacency to overreaction.

This paper presents an ongoing effort to develop models and simulations to meet the challenge of detecting and acting appropriately on weak signals often associated with cyber threats. The purpose of these models and simulations is to optimize operator decision making as described in the OODA loop. Within this broader purpose, the methods presented are a framework for models, simulations, and digital twins of future potential strains on the National Airspace System such as challenges of remote piloted aircraft and commercial space transportation.

Method

In collaboration with other Federal Agencies, The Federal Aviation Administration's William J. Hughes Technical Center contributes to and leads efforts to defend the Nation's infrastructure from cyber threats. One such effort is the Cyber Rodeo Lab Intrusion Detection Event. For the 2022 event, a remotely accessible Standard Terminal Automation Replacement System (STARS) simulation (Stasiowski, Kaelin, & Prata 2021) was employed. Figure 2 depicts the system image of the remote simulation. The remote simulation differs from the test facility set up in that the remote simulation renders the trackball and keypad hardware as interactive virtual input devices.

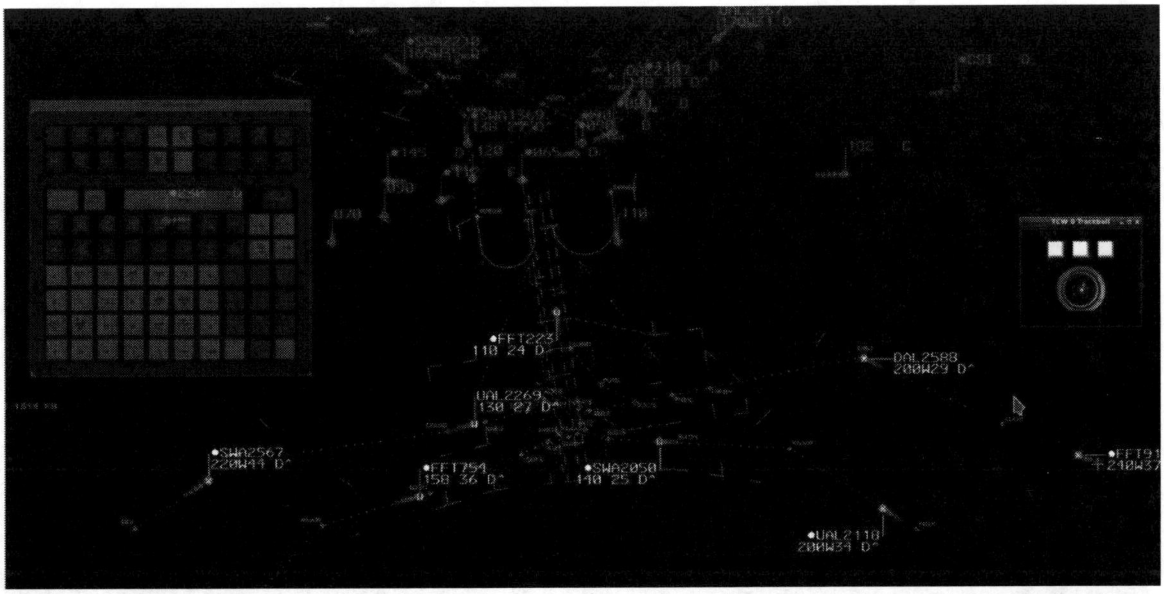

Figure 2. STARS interface showing west sector arrivals in white. Note that the trackball and keypad are virtual.

The Standard Terminal Automation Replacement System (STARS) is the fielded system used by Air Traffic Controllers to ensure the safe separation of military and civilian aircraft within the terminal airspace of the United States. STARS is a real-time digital processing and display system that replaced legacy air traffic control automation equipment at over 200 FAA and Department of Defense (DoD) Terminal Radar Approach Control (TRACON) facilities, over 600 FAA and DoD Air Traffic Control Tower facilities, and more than 100 systems installed and maintained at STARS support sites including Operational Support Facilities (OSFs) and the FAA Academy airspace (*FAA,* 2022).

Procedure

Air traffic scenarios were derived from Denver traffic flow archived in the PDARS. Figure 2 shows the virtual user interface with which the volunteer air traffic controller interacted. The controller was assigned the west sector for incoming traffic, which are the white airline track identifications. For a trial, the controller was briefed on their sector, within which they controlled the traffic for several minutes to establish baseline performance. Subsequent the baseline period anomalous targets were introduced into the traffic flow. Figure 3 shows a Google Earth Pro rendition of the Denver scenario depicting the anomalous target labeled "Spoof2" in conflict with UAL282. The insertion of anomalous targets was to test the effect on controller actions upon presentation of the anomalous target. The anomalous target simulated a drone signaling its position using an Automatic Dependent Surveillance-Broadcast (ADS-B). Thus, the anomalous target's flight characteristics were not typical of commercial aircraft, but its broadcasted information mimicked that of commercial aircraft. To verify that the target was anomalous the controller had to switch from a fused sensor mode to a single sensor radar mode turning off sensors registering the ADS-B information.

Single Sensor Mode is a mode that displays data from only one sensor/radar on the STARS Terminal Controller Workstation (TCW) display. Fused Mode is a mode that combines all data from all sensors/radars normally used by the site along with ASD-B data and displays the combined data on the TCW display. ADS-B is an advanced surveillance technology that combines an aircraft's positioning source, aircraft avionics, and a ground infrastructure to create an accurate surveillance interface between aircraft and air traffic control. ADS-B is a performance-based surveillance technology that is more precise than radar and consists of two different services: ADS-B Out and ADS-B In. ADS-B Out works by broadcasting information about an aircraft's GPS location, altitude, ground speed, and other data to ground stations and other aircraft, once per second. ADS-B In provides operators of properly equipped aircraft with weather and traffic position information delivered directly to the cockpit (*FAA, n.d.*).

Figure 3. Google Earth rendition of flight path showing the Spoof2 and UAL282 conflict.

Results and Discussion

The results of the simulation proved a successful test of the remote access STARS simulations. There was mention in the post-trial debriefing that using the virtual trackball presented some difficulties and that the hardware version of the trackball interface would improve immersion and realism. A hardware version for remote access simulations is being worked. Notwithstanding, the success of a remotely accessible system means greater access to principals involved in air traffic control toward greater representation of elements of the National Airspace System (NAS) informing models and simulations designed to discern the weak signal of the cyber threat.

During the post-trial debriefings, the controllers mentioned that "spoof" was not currently in the lexicon of Air Traffic Controllers. A discussion of communicating the risk of anomalous targets resulted in maintaining the current risk communication to the term "anomalous target" vice the promulgation of the term "spoof" or other terms that would bias the controller's orientation in the OODA loop process.

Figure 4 shows the Air Traffic Controller's action resolving the "Spoof2" and UAL282 conflict. The ADS-B signal displayed on the TCW from "Spoof2", which had no other associated identification, was efficiently identified as anomalous and tagged in yellow as "WATCH". This indicated that the air traffic controller was Observing and Orienting on information to discern the nature of the seeming conflict. Air traffic control was affected only in that some attention was resourced to the "WATCH" anomaly. Upon further OODA loop processing, the controller Decided that the anomaly did not pose a threat and renamed it "whodat", which was followed by the Action of moving the icon from the approach sector.

Figure 2. Anomalous target colored yellow and labeled "WATCH".

In conclusion, the simulation confirmed the centrality of Orientation in the OODA loop process. Moreover, the simulation informs future presentation of simulation generated system images. System images are the operator's conceptual models made manifest by the signals presented in the simulation (Norman, 2013). For example, signal qualities of "Spoof2" Oriented the controller to its track. Inferences concerning effects of controller experience and threshold differences between behaviors of commercial aircraft and anomalous target are plausible explanations of operator behavior but remain empirical questions. Future work will begin with storyboarding scenarios for simulations designed to titrate signal detection thresholds for art of the possible cyber threats. Simulations which best inform signal detection thresholds (Stanislaw & Todorov, 1999), will be candidates for development as immersive training simulations and for the development of digital twins to accelerate modeling of "what ifs". These simulations will provide evidence-based methods to discern "weak signals" and to accelerate the experience of operators in discriminating "weak signals" pursuant to mitigating safety threats, particularly those which evince from accumulated faults along the complex decision stream.

Acknowledgements

Stephanie Bell & Joe Pagano for their leadership ensuring a meaningful product.
Snezana Gatto for her brilliant orchestration of the simulations.
Joseph Stasiowski for his system engineering that afforded us a remotely accessible simulation.
Steve Jacobs & Wes Stoops for their ATC wisdom that guided scenario development.

References

Aircraft Certification, Safety, and Accountability Act (2020). Retrieved from https://www.congress.gov/bill/116th-congress/house-bill/8408

Aldrich, C. (2005). *Learning by Doing: A Comprehensive Guide to Simulations, Computer Games, and the Pedagogy in e-Learning and Other Educational Experiences*. Pfeiffer, New York.

Campbell, J. (1973). *Myths to Live By*. Bantam Books, New York.

Dawkins, R. (1976). *The Selfish Gene*. Oxford University Press, New York.

Boyd, J.R. (2018). *A Discourse on Winning and Losing*. Air University Press, Maxwell AFB, AL.

FAA. (2022, February 25). *Standard Terminal Automation Replacement System (STARS)*. Federal Aviation Administration. Retrieved from https://www.faa.gov/air_traffic/technology/tamr/arts/

FAA. (n.d.). *Automatic Dependent Surveillance - Broadcast (ADS-B)*. Federal Aviation Administration. Retrieved from https://www.faa.gov/about/office_org/headquarters_offices/avs/offices/afx/afs/afs400/afs410/ads-b

Lord, A. B. (1971). *The Singer of Tales*. Atheneum, New York.

MacCuish, D. (2012). Orientation: Key to the OODA Loop – Cultural Factor. *Journal of Defense Resource Management, 3*(2), 67-74. Retrieved from http://www.jodrm.eu/issues/volume3_issue2/05_maccuish_vol3_issue2.pdf

Mandler, J. M. & Johnson, N. S. (1977). Remembrance of things parsed: Story structure and recall. *Cognitive Psychology, 9*, 111-151.

National Research Council (2022). Emerging Hazards in Commercial Aviation Report 1: Initial Assessment of Safety Data and Analysis Process. The National Academies Press, Washington, D.C. Retrieved from http://doi.org/10.17226/26673

Ness, J. (2006). Communicating the Risk of Weapons of Mass Casualty. *U.S. Military Academy Combating Terrorism Center Biodefense Report, 1*(1), 10-12. Retrieved from https://www.files.ethz.ch/isn/26330/ctc_biodefense_rep_june_06.pdf

Norman, D. (2013). *The Design of Everyday Things*. Basic Books, New York.

Stanislaw, H. & Todorov, N. (1999). Calculation of signal detection theory measures. *Behavioral Research Methods, Instruments, & Computers, 31*(1), 137-149.

Stasiowski, J., Kaelin, C. & Prata, W. (2021, May 13). Making T&E Pandemic-Proof: Transforming T&E. *Paper presented at 24th ITEA Test and Training Instrumentation Workshop: Innovating for Tomorrow's Challenges*.

PSYCHOPHYSIOLOGICAL RESEARCH METHODS TO ASSESS AIRLINE FLIGHT CREW RESILIENT PERFORMANCE IN HIGH-FIDELITY FLIGHT SIMULATION SCENARIOS

Chad L. Stephens, Tyler D. Fettrow, Lawrence J. Prinzel III, Jon B. Holbrook, & Kathryn M. Ballard
NASA Langley Research Center
Hampton, VA

Daniel J. Kiggins
San Jose State Research Foundation
San Jose, California

New concepts in aviation system safety thinking have emerged to consider not only what may go wrong, but also what can be learned when things go right. This approach forms a more comprehensive approach to system safety thinking. A need exists for methods to enable a better understanding of human contributions to aviation safety and how they may inform Safety Management Systems (SMS). A high-fidelity 737-800 simulation study was conducted to study how current type-rated commercial airline flight crews anticipate, monitor, respond to, and learn from expected and unexpected disturbances during line operations. A number of dependent measures were collected that included traditional SMS data types, but also non-traditional safety data to include multiple psychophysiological metrics. This paper describes the psychophysiological measures results that evinced the capability of measures to help identify resilient flight crews. Implications for future research and design of future In-time Aviation Safety Management Systems are discussed.

The NASA System-Wide Safety (SWS) Project is focused on developing new technologies and operational concepts for the aviation industry to meet the increasing global demand while maintaining the current ultra-safe level of system safety. To achieve this, the project is studying safety producing behaviors (e.g., Hollnagel, 2016) and developing research priorities, including In-time System-wide Safety Assurance (ISSA) and In-time Aviation Safety Management System (IASMS; Ellis et al., 2019). Challenges currently being addressed include identifying data sources, analyzing data to detect and prioritize risks, and optimizing safety awareness and decision support. The project is focused on developing domain-specific safety monitoring and alerting tools, integrated predictive technologies, and adaptive in-time safety threat management to expand the knowledge base of resilience engineering and inform ISSA and IASMS for traditional and emerging operational concepts. One test case for this effort concerns non-adherence of area navigation standard terminal arrival route (RNAV STAR) procedures used at major airports.

Stewart, Matthews, Janakiraman, and Avrekh (2018) conducted a study on aircraft flight track data for over 10 million flights into 32 domestic airports and revealed that only 12.4% of flights fully complied with the published arrivals' vertical and lateral profiles. Based on that study, Holbrook et al. (2020) collected data from pilots, air traffic controllers, and airlines to examine safety behaviors during RNAV STAR arrivals at Charlotte Douglas International

Airport (KCLT). The takeaway was that the majority of non-adherences were to sustain operations under dynamic real-world conditions. These findings suggest that traditional approaches to risk and safety management may not be sufficient to address the misalignment between published procedures and routine safe operations, and a complementary approach that includes ensuring that "things go right" is necessary. The study by Holbrook et al. highlights that to maintain safety, humans will likely need to continuously adjust their work to match their operating conditions (Hollnagel, 2014).

Historically, resilience engineering research has centered on the theoretical aspects of productive safety. To address the gap in guidance on measuring resilient performance, we designed and conducted a human-in-the-loop (HITL) flight simulation study to gather empirical data to be used to understand productive safety (Stephens et al. 2021). Neuroergonomics research examining human operators in the context of safety-critical behavior has incorporated traditional human factors methods, including psychophysiological methods, to study human error (Dehais et al., 2020). We are extending this research by developing psychophysiological measures of resilient performance of pilots in simulated flight scenarios. Additionally, exploration of the data generated will determine how to analyze this data to prioritize risks and optimize decision-making support for safety awareness.

The main research objective for this study was to create a data testbed our team and the research community could explore to determine how commercial airline pilots manage routine contingencies and safety during RNAV arrivals. Studying actual operational events in airline operations is challenging because there is a limited amount of data that can be collected and analyzed for productive safety research due to pragmatic, logistical, procedural, or regulatory constraints. This research study involved gathering a comprehensive dataset of candidate measures to facilitate future data science efforts and to gain a better understanding of the phenomena of productive safety. To this end, traditional human factors data collection methods were employed including operator-generated data (e.g., self-report measures of workload, situation awareness, and resilient performance), observer-generated data (e.g., psychophysiological measures: electroencephalography, electrocardiography, galvanic skin response, and eye tracking) and system-generated data (e.g., simulated flight track data) were captured during the flight simulation. However, for the current analysis, we are focused specifically on the eye tracking data.

Methods

Data presented herein were collected during the SWS Operations and Technologies for Enabling Resilient In-Time Assurance (SOTERIA) flight simulation study conducted at NASA Langley Research Center in Hampton, VA USA during May-June 2022. Details of the full data collection plan and flight simulation scenarios are described in Stephens et al. (2021). Twenty-four (24) healthy airline transport pilots (9 women, M = 49.2 years) from a major US airline volunteered for the study. Subjects provided informed verbal and written consent to participate. The experiment was conducted under approval from NASA's Institutional Review Board.

After explaining the experiment and obtaining consent from each crew, each pilot was outfitted with a combined electroencephalography (EEG) and electrocardiography device (ABM X10, CA, USA), and a smart watch that measures galvanic skin response, skin temperature, and heart rate (Empatica, MA, USA). The impedance of each EEG electrode was verified to be less than 10 megaohms. Following the checkout of the outfitted systems, each pilot proceeded to the simulator flight deck and performed an eye tracking (Smarteye, MA, USA) calibration procedure.

All psychophysiological devices were time synced and triggered for recording through cyc3DX Multi-modal Analysis of Psychophysiological and Performance Signals (MAPPs; IA, USA). The data were exported from MAPPs for processing with custom python (Python3) scripts. At this time, eye tracking data analysis is ongoing; therefore only data processing details are discussed. Several metrics of interest were derived from the eye tracking data. These metrics were derived from different raw data generated by the eye tracking system, and had different methods of filtering, calculation, etc. For each variable, we averaged over time epochs of 10 seconds. We use the following definitions for each eye tracking metric:

- Head Heading Velocity: The rate (degrees/second) of the head turning left or right. We only retained indices where the reported % quality was greater than 60%.
- Pupil Diameter: The diameter of the pupils (mm). Because this variable is the most difficult to acquire, in order to keep sufficient indices, we retained indices where the reported percent quality was greater than 40%.
- Gaze Velocity: The velocity of the gaze vector (degrees/second). We retained indices where % quality was greater than 60%, and the gaze velocity of a particular frame did not exceed 700 degrees/second (Wilson et al. 1992).
- Gaze Variance: The variance (spread) score of the gaze vector. We converted the unit vector to a plane using standard stereographic mapping (Marcus, 1966). We retained indices where the % quality was greater than 60%, and the velocity of the raw gaze vector of respective indices did not exceed 700 degrees/second.

In addition to the psychophysiological sensors, we administered an array of traditional human factors measures including self-reported workload and situation awareness. We also created a custom resilience questionnaire, "Resilient Performance Self-Assessment" (RPSA). The RPSA consists of 16 questions that were modeled on American Airlines Learning Improvement Team (LIT) Proficiencies (American Airlines, 2020). The participants were required to specify whether they made use of a particular behavior, and if so, rate their perceived success of implementing that behavior. The choices consisted of a discrete scale from 1 (very unsuccessful) to 5 (very successful). Here, we are only focused on the RPSA scores, and not the other questionnaire data.

We investigated whether pilots exhibit behaviors that can be captured via eye tracking sensors (Smarteye system) that have a relation to their perceived resilience scores. We ran statistics for two questions. 1) Do resilience scores differ by crew? 2) Do the same crews that exhibit different resilience scores, exhibit differing psychophysiological behaviors, specifically in eye tracking measures?

To test our hypotheses, we used *lme4* (Bates, 2015) within R (version 4.1.2; R2021) to perform linear mixed effects analyses. We fit multiple linear mixed models and ran a single model for each variable of interest, including RPSA, Head Heading Rate, Pupil Diameter, Gaze Velocity, and Gaze Variance. For RPSA, we treated each of the 16 questions as repeated measures, assuming equal weighting, used fixed effects of Crew and Seat (left vs right), and subject as a random effect. The psychophysiological data consisted of varying total repeated measures per crew and scenario since we used the average across the 10 second epochs for each dependent variable. The models contained the same factors as the model for RPSA. We performed post hoc pairwise analyses for each model by calculating the least squares means and estimating the 95% confidence intervals, using a Kenward-Roger approximation implemented in the R-package *emmeans* (Lenth, 2016).

Results

All participants volunteered for all aspects of the experimental protocol. In general, all participants completed every scenario successfully, without any mishaps. Figure 1 shows the results by crew for the reported resilience scores (combined across questions). Crews 6, 8, and 10 showed the lowest RPSA scores, and were significantly different from 1, 2, 11, and 13 (95% confidence intervals did not overlap). Our primary goal here, is to identify psychophysiological measures that exhibit similar crew differences, and therefore indicate resilient or non-resilient behavior.

Here we are interested in identifying whether the same crews that had statistically significant RPSA scores, also showed differences in metrics we derived from the eye tracking data. Figure 2 shows the statistical results of the metrics derived from the eye tracking data. Crew 11 had a statistically significant difference in Gaze Variance. Crew 11 showed significantly higher variance scores compared to all other crews, which suggests that this crew was looking at more of the cockpit than the rest of the crews throughout the scenarios. The significant findings for Crew 11's Variance score did not transfer to any other metric. Crew 8 exhibited the lowest Gaze Velocity out of all crews. Low gaze velocity indicates less shifting of attention over time. In addition, Crew 8 exhibited the largest Pupil Diameter out of all crews. Crew 8 was one of the crews that showed relatively lower resilient scores, therefore Gaze Velocity and Pupil Diameter appear to be likely candidates for predicting resilient behavior (or lack thereof).

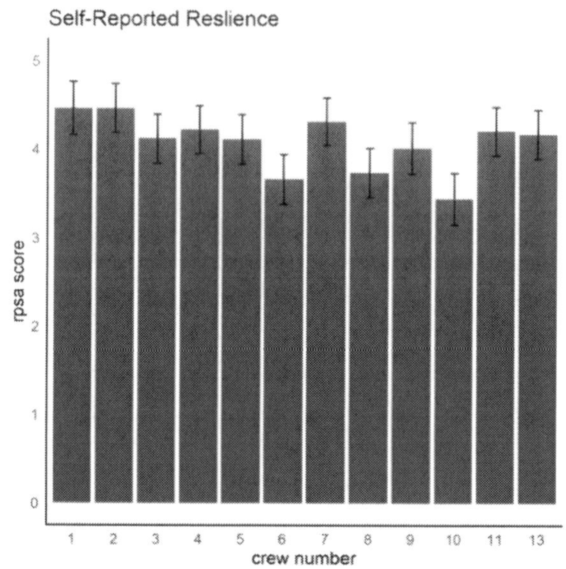

Figure 1: Mixed model results for RPSA scores. The bar graphs depict the estimated marginal mean (bar), and 95% confidence interval (error bar) for each crew's self-reported resilience scores. Lack of overlap between any crews' confidence interval indicates statistical significance between those crews.

Discussion

In the current preliminary analysis of a subset of the psychophysiological data captured during the study, we were interested in identifying metrics that can predict resilient (safe) behavior. In general, we showed significant differences between some crews in self-reported resilience scores and the psychophysiological measures.

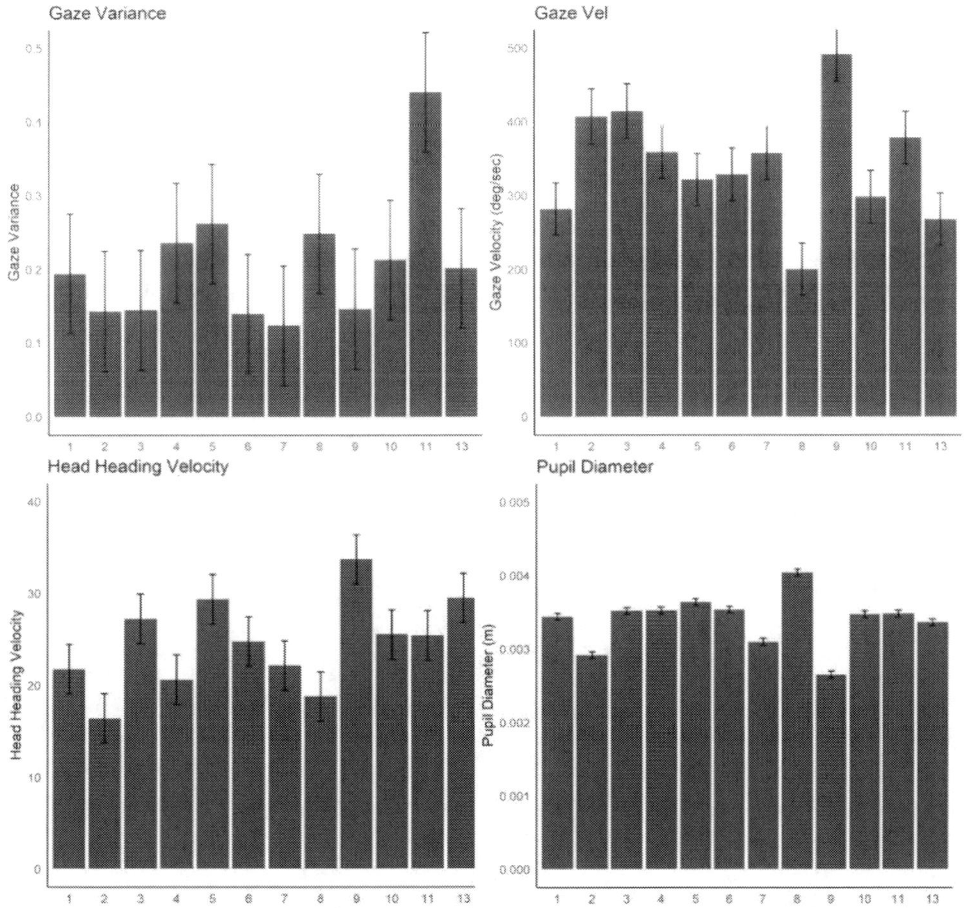

Figure 2: Mixed model results for eye tracking metrics. The bar graphs indicate the estimated marginal mean (bar), and 95% confidence interval (error bar) for each crew's self-reported resilience scores. Lack of overlap between any crews' confidence interval indicates statistical significance between those crews.

The self-reported resilience scores showed significant differences between crews, where Crews 6, 8, and 10 had the lowest scores. Despite these crews being significantly lower than the other crews, the average reported resilience score was well over 3, indicating a self-reported resilience of more than successful.

Psychophysiological measures also showed significant differences between crews, however, the same crews did not exhibit the same differences across all the psychophysiological measures. For example, Crew 11 showed the highest Gaze Variance, but was medial for all other metrics. Crew 8, which was one of the crews that reported lower resilient scores, showed the highest Pupil Diameter and the lowest Gaze Velocity. This finding might suggest that these two metrics could be used to predict resilient behavior. Future work will include direct analysis between resilience scores and the psychophysiological values.

There are several considerations that should be noted while interpreting this work. First, the psychophysiological analyses were performed without consideration of whether the data fell within a certain window or when an "event" occurred. Specifically, the reported results include data from the entirety of the scenarios, which may actually hide more significant effects if we focus the analyses on specific event timings. Second, we intentionally did not want to perform a direct analysis between the RPSA and psychophysiological measures. The RPSA was created for use in this study, but it is not a psychometrically validated measure. We assumed equal weighting of the individual questions towards the overall "resilience score", but it is possible that some participants showed resilience in one category (i.e., adapt) and not another (i.e., learn). There were also several missing responses which is reasonable if the participant was not able to exhibit a specific resilient quality, they were not able to rate themselves on the scale. Furthermore, we are still experimenting with ways to analyze both the RPSA scores and the psychophysiological scores. A direct comparison did not seem fair given all these considerations.

Future work will address the issues discussed in the Considerations section, but also expand on the current work. There are several other psychophysiological sensors that were used to collect data including electroencephalography and electrocardiography that we plan to analyze in similar format. Furthermore, we also plan to extract more detailed resilience scores for each crew. Each scenario had video and audio recording that we plan to have observations completed by The LOSA Collaborative and American Airlines LIT that will provide resilience metrics for each scenario and crew. This will improve our resolution and expand the types of analyses we could perform with the dataset.

Acknowledgements

This work was funded by NASA's System-Wide Safety Project, part of the Aeronautics Research Mission Directorate's Aviation Operations and Safety Program.

References

American Airlines' Department of Flight Safety. (2020). Trailblazers into Safety-II: American Airlines' Learning and Improvement Team, A White Paper Outlining AA's Beginnings of a Safety-II Journey.

Bates, D., Machler, M., Bolker, B., & Walker, S., 2009. Fitting linear mixed-effects models using lme4. Science 325, 883–885.

Dehais, F., Lafont, A., Roy, R., & Fairclough, S (2020) A Neuroergonomics Approach to Mental Workload, Engagement and Human Performance. Frontiers in Neuroscience 14, 268, 1-17.

Gray, D.E. (2013). Ethnography and participant observation (Chapter 17). Doing research in the real world. London: Sage.

Holbrook, J. Prinzel, L., Stewart M., & Kiggins, D. (2020). How do pilots and controllers manage routine contingencies during RNAV arrivals? In Proceedings of the 11th International Conference on Applied Human Factors & Ergonomics.

Hollnagel, E. (2014). Safety-I and Saftey-II: The Past and Future of Saftey Management. Farnham, UK: Ashagte.

Lenth, R.V. (2016). Least-Squares Means: The R Package lsmeans. Journal of Statistical Software 69.

Marcus, C.F. (1966). The stereographic projection in vector notation. Mathematics Magazine, 39, 2, 100-102.

Stephens, C., Prinzel, L., Kiggins, D., Ballard, K., & Holbrook, J. (2021). Evaluating the use of high-fidelity simulator research methods to study airline flight crew resilience. 21st International Symposium on Aviation Psychology, 140-145.

Stewart, M., Mathews, B., Janakiraman, V., & Avrekh, I., (2018). Variables influencing RNAV STAR adherence. IEEE/AIAA. Proceedings of the 37th Digital Avionics Systems Conference (DASC). London, UK.

Wilson, S.J., Glue, P., Ball, D., Nutt, D.J. (1992). Saccadic eye movement parameters in normal subjects. Electroencephalograph Clinical Neurophysiology. 86 (69-74).

RESEARCH AND TECHNOLOGY CHALLENGES FOR HUMAN DATA ANALYSTS IN FUTURE SAFETY MANAGEMENT SYSTEMS

Chad L. Stephens Lawrence J. Prinzel III, Ph.D. Kyle K. Ellis, Ph.D.
Michael J. Vincent Samantha I. Infeld Ph.D.
NASA Langley Research Center
Hampton, VA

Nikunj C. Oza, Ph.D. Misty D. Davies, Ph.D. Robert W. Mah, Ph.D.
NASA Ames Research Center
Moffett Field, CA

Paul A. Krois, Ph.D.
Crown Consulting, Inc.
Aurora, CO

James Ackerson
Flight Research Aerospace
Louisville, KY

Enabling new and novel concepts of operations for Advanced Air Mobility poses an important need to evolve current safety management systems (SMS) and is posited to be realized through advances in Machine Learning (ML) Data Sciences and Artificial Intelligence. The "In-time Aviation Safety Management System" (IASMS) concept of operations supports the need to evolve today's SMS to become more tailorable, scalable, and interoperable in response to forecasted changes expected for the future airspace system. Key to IASMS is integration of proactive and predictive ML algorithms trained to provide "in time" detection and mitigation of hazards and emergent risks through new methods and novel data types. IASMS research and technology development includes human factors design considerations for these systems to include human-system teaming, innovations in human interfaces and management of complex digital data information, human-system interaction/model-based system engineering, and verification and validation for data assurance and trust.

Expanding future sustainable operations for today's commercial air carriers, combined with envisioned transformations of the National Airspace System (NAS) integrating Urban Air Mobility (UAM) and other innovations that lead to Advanced Air Mobility (AAM), pose significant opportunities for advancing today's Safety Management Systems (SMS) (Prinzel et al., 2021; Ellis et al., 2022). Today's data aggregation, risk assessment, and decision making typically involve data assessed in silos with limited cross-silo comparisons. These and other constraints limit scalability and rapid identification of known and emergent risks. This time scale subsequently results in more time taken before mitigations can be determined and implemented. The augmentation of today's SMS forms part of the foundation of the In-time Aviation Safety Management System (IASMS).

Advanced data analytics technologies, as part of the SMS toolbox, can significantly enhance the human's capability to evaluate the growing volume of available safety data . The National Academies of Science (2018) argued that envisioned changes to the NAS, within the next decade or two, will vastly outpace the ability of the current system to identify precursors, anomalies, and indicators of early hazard emergence and risk; the organization argued that safety analysis will need to become more "in time" and integrated than exists today. To address the recommendation and to help achieve the Federal Aviation Administration (FAA) (circa 2035) and National Aeronautics and Space Administration (NASA) (circa 2045) visions for a transformed NAS, significant investments will be needed in Machine Learning (ML) Data Sciences and Artificial Intelligence (AI) methods that can provide needed in-time safety data analysis capabilities. However, the change also affects the role and responsibilities of the human analyst that currently is responsible for a large majority of monitoring, assessing, and decision-making (for mitigation or actions, if any, that need to be taken based on assessment of the risk or hazard). The future of integrated safety management and assurance will rely more and more heavily on big data analytics that will challenge, if not be impossible, for the human analyst to fully understand the underlying methods/process for the ML-based data outcomes. It is envisioned that, for the foreseeable future, the human will retain responsibility for what decisions are made on actionable data, but also of necessity may have to accept more reliance and trust in the system. The present paper discusses IASMS research and technology development needs for ML and data visualization with specific focus on human factors "use, misuse, disuse, and abuse" design considerations and concerns potentially involved with advanced data analytic systems (Parasuraman & Riley, 1997). Existing guidance may serve to mitigate some of these known historical human-system interaction issues. However, progress in innovations on how to achieve human-autonomy teaming in design and practice and new modeling and system engineering/integration will be critical to IASMS success (Ellis et al., 2022; Holbrook et al, 2020; NASA, 2022).

The FAA Concept of Operations (ConOps) for an information-centric NAS (ICN) takes an expansive, layered Integrated Safety Management approach to safety controls, data, and risks resulting from the integration of distributed and diverse systems (2022). As part of an ICN (circa 2035), timely analysis will correct issues at system boundaries and adapt the system to changes in risks or the operational environment. Safety management with new entrants will scale enabled by interoperability across air vehicle operations with their diverse performance and mission requirements.

Further, NASA foresees a "Sky for All" NAS (circa 2045) having a cornerstone for in-time Integrated Safety Management Systems and Safety Assurance (through verification and validation and new certification approaches) that will integrate monitoring and assessing known and emerging hazards, mitigating risk, and assuring the safe performance of the future aviation system (FAA, 2023; NASA, 2022). This includes establishing standards and metrics for the safety data architecture and management, integrating hazards monitoring for situation awareness and developing simulation tests for validating systems and performance-based airspace functional requirements and guidance for new operations, leveraging predictive modeling and cooperative in-time crowd-sourced information using ML-based automated systems identification of risks and mitigations. The maturation of in-time integrated safety management provides Safety for All with seamless, integrated and highly autonomous safety mitigation.

NASA has been maturing the IASMS concept to augment current SMS relative to forecasted changes of the NAS as recommended by the National Academies (2018). These changes include fusing existing, new, and underutilized data sources and adopting increasingly sophisticated data analytics and ML. With this architecture, IASMS will more quickly monitor and assess large data sets to identify known and emergent risks in-time for implementing risk mitigations. This vision for IASMS, viewed through the lens of the FAA ICN ConOps, highlights the tailored safety for different sized operators flying diverse aircraft and missions and the in-time safety assurance of automated systems with their safety-critical technologies.

SMS for Commercial Air Carriers

Enabling future visions of the NAS poses an important need to augment the current SMS to take advantage of integrating advances in ML and Data Science and meet the needs of in-time safety management. The concept of SMS was established by the International Civil Aviation Organization (ICAO) establishing policies and procedures requisite for managing safety (2018).

The FAA provides guidance and methods for developing and implementing an SMS through Advisory Circular (AC) 120-92B, titled "Safety Management Systems for Aviation Service Providers." (2015). The AC shows how a commercial air carrier can show means of compliance for meeting federal SMS regulations although there could be other means to meet requirements. Keys to a successful implementation of SMS include how data and information are analyzed and interpreted, how informed decisions are made, and how it leads to new operational and business methods. Another key is scalability of SMS relative to the size and complexity of the air carrier. FAA-sanctioned SMS programs include Flight Operations Quality Assurance (FOQA), Aviation Safety Action Program (ASAP), Aviation Safety Reporting System (ASRS), Line Operations Safety Audit (LOSA), and Continuing Analysis and Surveillance System (CASS). FAA AC 120-103A addresses fatigue risk management systems.

Aviation Safety Data Analysis and Sharing Systems

Commercial air carriers and other stakeholders share confidential and anonymous SMS data through the Aviation Safety Information Analysis and Sharing (ASIAS) system to improve NAS-wide safety. ASIAS aggregates data from carriers with data analysts manually fusing data sets together to undertake targeted and prioritized studies of safety issues. With today's SMS, an air carrier typically collects and analyzes data within the data silos built for different SMS methods. Analysts review their data and compare trends with analysts working with data from other methods. Data boards, management boards, safety executives, and others lay eyes on trends, make comparisons, ask questions, discuss operational conditions and risk controls, and decide on possible risk mitigations.

FAA planned evolution of ASIAS from today's system, called ASIAS 1.0, to future visions called ASIAS 2.0 and 3.0, is foreseen to replace today's data silos with integrated fusion of disparate data sources including using new and underutilized data sources (Office of Inspector General, 2021). Manual data analysis will be replaced with faster analysis using higher volumes and more varied data. ASIAS 3.0 will transform collaboration with more agile, innovative interactions with new communities and use increasingly sophisticated predictive analytics and advanced tools to identify emerging risks. ASIAS efficacy will increase through decision fusion leading to predictive safety and prescriptive risk mitigation.

Data Challenges

Commercial air carriers face numerous challenges with data coming from the range of SMS methods. Challenges include using new and underused sources of data, fusing data using novel techniques, and developing and verifying predictive analytic models. These challenges are compounded by the large volume of data associated with radar tracks of flight trajectories and weather data. The velocity of data involves the fast flow of these data types to be logged and stored each month. The veracity of the data relates to its variable fidelity, including missing data, duplicate tracks, tracks ending in midair, and reused or duplicate flight identifiers. Lastly, the variety of data spans numerical types (e.g., radar or Global Positioning System), air traffic control or aircraft voice recordings, textual voluntary safety reports, radar and airport meta data, and actual and forecast weather data.

Data Analytics

Data analytics involve different ML algorithms and statistical methods applied to the previously identified common performance data sources (e.g., FOQA data). Analysis may focus on known adverse events and identifying their precursors and associated trends, or be exploratory to identify hidden anomalies or emergent trends using predictive analytics. While there is a need for increasingly autonomous fusion of aviation safety data and advanced data analytics, these processes would be managed by human decision-makers to ensure acceptability and practicality of any findings. In other words, findings may be statistically significant but of limited operational value, or have high operational value even though there may be limited operational data to achieve statistical significance.

Unsupervised learning can identify existing topics and emerging trends. The methodology involves automatically parsing a report's narrative and partitioning narratives based on operational relevance and not according to some a priori taxonomy. Findings showed that maintenance reports; flight attendance reports; and cabin smoke, fire, fumes, or odor incidents were most consistently separable possibly due to the different vocabulary used compared to other reports.

Voluntary safety reports can also be analyzed using natural language processing (NLP). One air carrier successfully applied NLP to aircraft maintenance to improve safety and efficiency involving coding voice transcriptions of mechanics' findings and actions (Carvalho, 2022). An NLP study of safety reports on losses of separationcoded free-text narratives based on the Toolkit for ATM Occurrence Investigation taxonomy and found that unsupervised topic modeling successfully detected unknown recurrent behaviors or conditions (Buselli et al., 2022). Results reframed human behavior from being a sequence of errors leading to an undesired outcome and instead showed safety events to be emergent from complex interactions of the system.

Anomaly detection involves methods allowing FOQA or FOQA-like data to "speak for themselves" in revealing trends leading to degraded system safety and performance. The conundrum is that a complex engineering system involves multiple interdependently functioning components so the variety of ways in which problems can arise can be complex. Consequences from degraded or failed performance can result in damaged equipment, human injury, or other unacceptable outcomes. Oza et al. (2021), for example, used data containing nominal and anomalous states to identify statistical anomalies and the precursors that could be disrupted to mitigate the anomalies. Findings showed that anomaly detection with domain expert validation of the operational significance of identified anomalies can effectively detect and explain operationally significant anomalies during operations. These methods automatically identify precursors and allow domain experts to effectively explain their undesirable effects. This study demonstrated effective teaming of human domain experts trusting ML algorithms to identify sequences of events that lead to anomalous operations.

LOSA observations are coded using an extensive taxonomy comprised of threats, errors, and undesired aircraft states. Coded data are analyzed for prevalence as the percentage of flights involving particular threats, errors, and undesired aircraft states, and mismanagement as the percentage of taxonomy codes leading to a flight crew error. Important trends can be identified such as based on higher prevalence, mismanagement, or demographic factors (e.g., city pairs, fleet). For example, the LOSA archive shows for the threat of "aircraft malfunctions unexpected by the crew" that 13% were mismanaged and further analysis showed many flight crews flying one of several fleets failed to properly reference the Quick Reference Handbook.

Data Analyst Must Integrate Aviation Knowledge with Data Analytics Methods

Two vital pragmatic considerations necessary for data analysts to adopt ML methods are developing aviation domain knowledge to understand the origins and limitations of the data to be analyzed and establishing a working knowledge of data analytics methods. Previous SWS research efforts have involved collaboration with ML data scientists, but with limited or no experience with specific types of aviation safety data (e.g., human performance data, Napoli et al., 2022; flight operational data, Garcia et al., 2022). During these research efforts, the time required to enable sufficient working knowledge of the data to permit appropriate application of ML methods was approximately 6-12 months. A cross-functional and cross-discipline collaboration between data scientists and aviation domain experts is needed to mature these methods for future SMS/IASMS.

Overcoming Human Analyst Limitations with Big Data

The forecasted increase in volume and complexity of big data sets will necessitate integration of ML to overcome human limitations in analyzing and understanding these risks. Given the scale of changes with flight safety data between current and future data analysis needs, data analysts will increasingly address risks which cannot be adequately solved without ML. ML solutions are effective when data analysis rules for identifying safety issues involve too many factors and these rules overlap or need fine tuning. Furthermore, ML approaches are well suited to scale to ever increasing volumes of data which analysts cannot effectively analyze with traditional methods.

Human-System Interaction Issues

The complexity of human-system interaction (HSI) is reflected by critical design features needed to enable the human analyst to detect, identify, and understand data anomalies. Collaboration with operational experts can lead to informed decisions about operational significance of these anomalies. The design of algorithms and the processes for their use provide a context for potential issues with HSI involving "use, misuse, disuse, and abuse." Issues shaping the use of automation include trust, over-reliance on automation to detect problems, reduced attentiveness to deal with false alarms, and degradation of skills. For example, a data analyst might miss something in reviewing and coding an ASAP report when focused on something else. When automation and processes are used rigidly, such brittleness makes it difficult to handle gaps in the models used by automation and unanticipated emergent behaviors resulting from mismatching between multiple automated systems.

Data analysts will use visualization techniques to review and better understand safety risks and elevated risk states. Techniques for data visualization can involve statistical tables and figures represented by histograms and frequency distributions representing simple to complex statistics. Integrated results can be shown as a dashboard laying out key statistics such as events displayed by locations (e.g., airports), organization divisions (e.g., flight operations, ground handling, Mx), and types of data (e.g., FOQA, ASAP). Relationships could be shown between data types over time or binning data into a risk matrix (e.g., counts of events classified as red, yellow, or green based on frequency and consequence).

The human remains the decision maker using data analytics to inform effective safety assurance actions. Data visualization for the data analyst may be different for safety executives. These data may also provide understanding and insight into the air carrier's efficiency and environmental considerations. The IASMS will provide faster identification of precursors, anomalies, and trends and emergent risks that represent hidden, masked, or previously unknown risks. Architectures can be integrated and interoperable between operators for in-time safety management through design of automated Services, Functions, and Capabilities (SFCs). The size and complexity of these architectures and SFCs will be scaled based on complexity of operations and vehicles. SFCs represent what data will be aggregated and how data will be fused and analyzed using ML. SFCs may provide initial risk analysis although the human analyst will retain final decision making for risk analysis. SFCs may provide alerting when a risk threshold is reached and automated mitigation, as appropriate.

Discussion

The research and technology challenges we presented (i.e., aviation safety data analysis/data sharing systems, data analysts' roles and responsibilites, HSI issues, etc.) must be addressed to permit a key enabler of IASMS: safety intelligence (SI). ICAO considers SI to be an outcome of the process of analyzing safety data and safety information to support decision-making. SI is a prerequisite for in-time safety management being able to rapidly evaluate existing data patterns and discover new patterns that can lead to the next safety event before such an event might occur. The benefit from looking at safety management through the lens of SI is improved speed and characterization of system-wide risk using ML. SI integrates the knowledge gained from reactive, proactive, and predictive safety systems . The IASMS concept progresses today's SMS to be responsive to expanded use of new and underutilized data sources, advancements in use of ML and possibly other areas of AI for safety management, and future evolution of the NAS with AAM. Challenges surface with use of novel sources of data and innovations in predictive modeling. IASMS provides a framework for improving SI facilitated through integration of proactive and predictive algorithms trained to detect known hazards and identify emergent risks more quickly and effectively. Importantly, understanding and addressing the research and technology challenges for the human data analyst in future safety management systems is a paramount need to ensure a future "Sky for All" system that is assured to be "Safe for All" (NASA, 2022).

Acknowledgements

The authors extend their appreciation to Mrs. Laura Bass (Analytical Mechanics Associates, Inc.) for her valuable contributions in the development of this paper.

References

Buselli, I., Oneto, L., Damnbra, C., Gallego, C., Martinez, M., Smoker, A., Ike, N., Pejovic, T., and Martino, P. (2022). "Natural language processing for aviation safety: extracting knowledge from publicly available loss of separation reports," Open Research Europe.

Carvalho, T. (2022). "Natural Language Processing (NLP) in Airline Maintenance Operations," Aviation Week Aerospace IT.

Ellis, K., Prinzel, L., Krois, P., Davies, M., Oza, N., Stephens, C., Mah, R., Infeld, S., & Koelling, J. (2022). A Future In-time Aviation Safety Management System (IASMS) Perspective for Commercial Air Carriers. AIAA 2022 Conference. Chicago, IL: AIAA.

Federal Aviation Administration (2022). Charting Aviation's Future: Operations in an Info-Centric National Airspace System. Washington, DC: FAA.

Federal Aviation Administration. (2015). "Safety Management Systems for Aviation Service Providers," AC No. 120-92B.

Garcia, E.J., Stephens, C.L., & Napoli N.J. (2022). Detecting Risk and Anomalies in Airplane Dynamics through Entropic Analysis of Time Series Data. AIAA Aviation Forum 2022.

Holbrook, J., Prinzel, L., Chancey, E., Shively, R., Feary, M., Dao, Q., Ballin, M., & Teubert, C. (2020). Human-Autonomy Teaming Research Challenges and Recommendations. AIAA Aviation Forum 2022-3250. doi.org/10.2514/6.2020-3250.

International Civil Aviation Organization (2018). "Safety Management Manual," ICAO Doc 9859, 4th Edition.

Napoli, N., Stephens, C., Kennedy, K. Barnes, L., Garcia, E., Harrivel, A. (2022). NAPS Fusion: A framework to overcome experimental data limitations to predict human performance and cognitive task outcomes. Information Fusion, 91, 15-30.

National Aeronautics and Space Administration (2022). Sky for All Portal. https://nari.arc.nasa.gov/skyforall/

National Academies of Sciences, Engineering, and Medicine. (2018). In-time Aviation Safety Management: Challenges and Research for an Evolving Aviation System. doi.org/10.17226/24962.

Office of Inspector General, Department of Transportation. (2021). "FAA Has Made Progress in Implementing ASIAS, but Work Remains to Better Predict, Prioritize, and Communicate Safety Risks," FAA Report No. AV2021022.

Oza, N., Bradner, K., Iverson, D., Sahasrabhojanee, A., and Wolfe, S. (2021). "Anomaly Detection, Active Learning, Precursor Identification, and Human Knowledge for Autonomous System Safety," AIAA Sci Tech Forum.

Paradis, C., Kazman, R., Davies, M., and Hooey, B. (2021). "Augmenting topic finding in the NASA Aviation Safety Reporting System using topic modeling," AIAA Sci Tech Forum.

Parasuraman, R., & Riley, V. (1997). "Humans and automation: Use, misuse, disuse, abuse," Human Factors, 39, 230-253.

Prinzel, L., Ellis, K., Krois, P., Koelling, J., Davies, M., and Mah, R. (2021). "Examining the changing roles and responsibilities of humans in envisioned future in-time aviation safety management systems," International Symposium of Aviation Psychology.

INTERFACE DESIGN FOR COLLABORATION WITH SEMI-AUTONOMOUS AGENTS FROM AN AIRBORNE AIRCRAFT

Maj John A Stevenson, John McGuirl, and Michael E. Miller
Air Force Institute of Technology, Wright Patterson AFB, OH, USA

The development of intelligent standoff weapons presents several challenges to support effective human-machine teaming. A key issue involves leveraging the weapon's ability to sense and react to new threats and engage targets of opportunity in flight while balancing human fiduciary control. This research sought to develop a methodology to determine information requirements of weapons operators to support high levels of teaming during a mission. Baselining a population of operators and conducting cognitive task analyses of system experts from among Air Force Weapons School graduates helped to understand the cognitive requirements and teaming practices of currently fielded systems and elicit goals and interactions with future semi-autonomous systems. Model Based Systems Engineering activity diagrams captured the team's goals and visualization needs. Wireframe user interface designs presented notional displays of required data. This generalizable method proved beneficial to prototype use-centered, decision-driven user interfaces for novel systems in support of follow-on research.

In recent years, the divestiture of aging military materiel has been met head on with a sixth revolution in military affairs (RMA), and the ushering in of a seventh. The first five RMAs led modern society from the creation of Westphalian society and nation-states to the concept of nuclear cold war. Currently, the Information Revolution is opening the door for the Autonomous Revolution (Hoffman, 2017; Knox & Murray, 2001). This technology growth, or rather the cost associated with it, has aggravated policy in the United States (US), leading to "building down to build up," resulting in a net decrease in the number of weapon systems in the inventory (Gunzinger, 2021). While this is at odds with von Clausewitz's (1874) principle of mass, the inability to assume strategic dominance in each domain has incentivized the US, along with the rest of the world, to embrace the new RMAs and work for technological advantage (Morgan & Cohen). As Brig. Gen Y.S. (2021) noted "China and Russia are accelerating their AI capabilities, and the U.S. has no choice but to forge ahead and lead the field" (p. 110).

Accordingly, the US has begun to research and develop autonomous weapons systems (AWS) that can select and engage targets after being deployed, without the need of direct intervention from humans (Dombrovski, 2021). The defense news cycle has been inundated with stories of programs like the Air Force Research Laboratory's (AFRL) Golden Horde and Skyborg—which have conducted experiments with concepts of semi-autonomous, collaborative small-diameter bombs (SDB) and loyal autonomous wingmen respectively. This has been complimented with Defense Advanced Research Project Agency's (DARPA) AlphaDogFight trials, in which an AI piloted simulator scored aerial combat victories against a human F-16 pilot (AFRL, 2021a; AFRL, 2021b; Drubin, 2020; Hitchens, 2021). Golden Horde currently uses mission planners in a way they compare to play calling in football: the coach (mission planners) pre-load a mission, but, given the right conditions, swarm capabilities within established sets of rules or audibles to the play can be called by the swarm itself to other authorized missions in the playbook if a series of conditions are met (AFRL, 2021a). This playbook includes the rulesets for interaction and changes in authority among agents, both human and artificial. The SDB has a relatively short time of flight, and the proof of concept raises the question of post launch collaboration with munitions having lager standoff distances as a force-multiplying capability in a future conflict.

The purpose of the current study was to develop and assess a method for prototyping interfaces between a user and an autonomous agent that was informed by cognitive task analysis and a Function-

Behavior-Structure framework with the overall goal of eliciting requirements in support of a decision-driven design. One thing becomes increasingly clear as the focus shifts to the Autonomous Revolution; in the words of Kelly (2016), "This is not a race against the machines. If we race against them, we lose. This is a race with the machines" (p. 61). This research summary will first define autonomy and relationships among human-agent teams (HATs). Additionally, it will describe the specific approach developed to design HAT interfaces. Finally, it will provide discussion on outcomes and opportunities for continued learning and research.

Redefining Autonomy

Autonomy has been defined as an application of machine learning (ML) in which a system is able to independently learn from experience and take action to accomplish specific goals based on that experience (Haddal, Hayden, & Frazar, 2018). While this is an accepted definition in some of the computer science literature, it proves to be deficient. Autonomy in machines, like autonomy in humans, stems not only from the ability to perform a task, but from being allowed to perform it, or having the correct authority to do so (Miller, McGuirl, Schneider, & Ford, 2020). Expanding on this definition, one can glean from the literature and practice the conceptual model shown in Figure 1. Autonomy is situational and can be afforded and removed depending on the context and need. For an agent to have its own sovereignty, it must also have the capability to process information to make and execute appropriate decisions. The agent must be granted authority to make the decision. Finally, information must be available and the cognitive processing unit must have sufficient capacity to process the information.

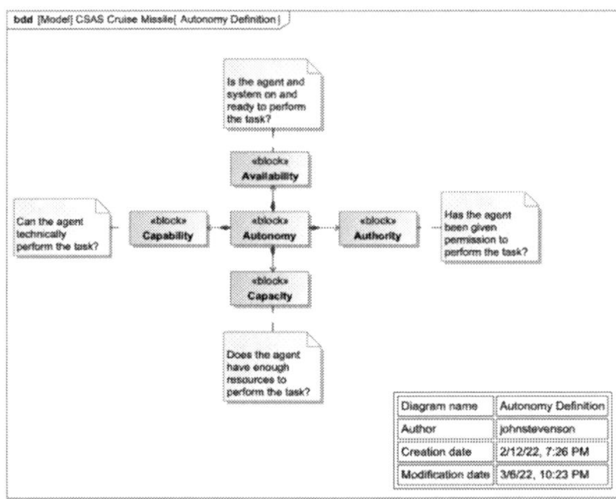

Figure 1. **Autonomy Redefined in Block Definition Diagram**

Relationships of Human-Agent Teams

Agents within the HAT must be constructed to share information and work with joint intention, i.e., work towards shared goals, in a symbiotic manner (Drubin, 2020). High-performance teams "set ambitious goals, shape motivations, and break external and internal obstacles" (Smolska, 2021, p. 65). Therefore, the basic objectives of crew resource management (CRM) can be applied to HATs as well. These include:

- Minimize human error,
- Maximize human performance,
- Enhance communication,
- Improve situation awareness,
- Strengthen decision making, and
- Improve teamwork. (Brunacini, 2015, p. 32).

While acknowledging that AI does not "think" like humans do, the overarching goal in HATs is to mirror the traits of highly effective human teams, only now the optimal solution minimizes both human and agent error while maximizing their joint performance. By challenging and checking each other's performance, a HAT can develop shared goals and trust in decisions they each make (Konaev & Chalal, 2021; Moz & Kleiner, 2021; Walliser, Mead, & Shaw, 2017).

Contrary to popular opinion that offloading tasks to autonomous agents will replace humans entirely, Murphy & Burke (2008) explain that as the machine eases the cognitive task load of the operator in menial tasks, the human is freed to take on a new role within the joint cognitive system, the role of knowledge worker. This worker focuses on big-picture goals and decisions for the HAT, rather than basic functions, leaving the human to ingest new information to aid or train the agent, even if geographically separated.

A Method for Developing Human-Agent Team Interfaces

Historically, the Department of Defense (DoD) and much of industry begin system design with well-defined boundaries. This presents a major difficulty for the design of HATs by limiting the interaction of agents viewed internal to the system with humans who are viewed as external to the system (Sterling & Taveter, 2009). To allow the design to be resilient to emergent interactions, McCaffrey & Spector (2018) recommend alternating "between top-down *problem framing* and bottom-up *problem solving*" iteratively. This process leads to a "use-centered" design, where rather than focusing on the perceived desires of the projected system, the design focuses on the use of the system in the real-world, and the actions and information, as well as the interaction of the two, ultimately specifying a design that better supports both the user and the agent (Flach & Dominguez, 1995).

This study utilized a Function-Behavior-Structure (FBS) framework to guide this outside-in design philosophy, by asking what the system is for, what it does, and what it is, respective to each of the three variables in its name, so that design becomes a "goal oriented, constrained, decision-making activity" (Gero 1990, p.28; Gero & Kannengieser, 2004). McCaffrey & Spector (2018) define these three variables with respect to human-agent teams as goals, interactions, and entities. This definition informed the design of human-agent relationships for this study.

User Baselines

One of the difficulties of developing AWS is that, with no true AWS fielded, the goals of the HAT need to be derived from as-is system users with an eye toward future systems. In this study, US Air Force aircrew were used as knowledge workers to help define the system. As a population baseline, a group of ten operators ranging from 1-11 years of experience were surveyed to determine their tendencies to anthropomorphize objects, trust other operators, and to be complacent or offload tasks to automation (Ross, 2008). They responded on a 1-5 Likert scale with 5 most favoring automation. Specific prompts included items such as:

- I sometimes berate or curse at objects when they annoy me.
- In dealing with strangers, one is better off to be cautious until they have provided evidence that they are trustworthy.
- Automation in aircraft, such as automatic landing systems, make air journeys safer.

A Student's t test on the results of each question from the survey elucidated information on the notional average user for the to-be system. The statistics found that the user was not significantly different from neutral in Anthropomorphic Tendencies ($\bar{x}=3.07$, $s^2=2.05$, $p=0.61$) and Interpersonal Trust ($\bar{x}=2.95$, $s^2=1.16$, $p=0.63$). Complacency Potential, on the other hand returned a significant result ($\bar{x}=4.02$, $s^2=0.99$, p= 1.62e-20). These results revealed a significant desire to offload tasks to an autonomous agent, but no preference to anthropomorphizing objects or working with other humans.

Eliciting Goals - Knowledge Worker Interviews

Next, the study conducted a cognitive task analysis (CTA) with systems experts. Two 90-minute interviews were conducted with a pair of USAF Weapons School graduates. The participants were introduced to the project and provided with an organizational diagram, a domain model, and a framing concept map. The interviews were conducted using the four sweeps model of the Critical Decision Method (CDM): incident identification, timeline verification/decision point identification, progressive deepening, and expert-novice differentiation (Crandall, Klein, & Hoffman, 2006, p. 81). Since CDM focuses primarily on events where the decisions made saved or ruined the goals of the team, adjustments were made because in a combat mission in which their decision making directly impacted success, it is likely the operator would be so cognitively tasked that memory would be poor after the fact.

The first sweep asked the user to recall a relevant event or set of vignettes from multiple events that involved interactions with their systems under challenging circumstances. The second sweep established the timeline of the selected event and defined how the goal was accomplished. The third sweep probed deeper and elicited information about how they used their expertise (as well as the tactics, techniques, and procedures [TTPs] and standard operating procedures [SOPs]) to identify the important factors to attend to as the scenario played out. Sweep 4 changed the focus to how things could be done differently, whether doing something else would change the outcome, and importantly whether a lesser-experienced operator would have done things differently. It asked about how an autonomous agent would differ from an inexperienced human operator, and whether steps are better accomplished by humans or computers.

The Knowledge Worker Interviews were recorded to allow the interviewer to review them and build and annotate concept maps to directly identify key steps in the task and decision points that affect mission success in an elegant document, while preserving the conceptual data of the lengthy interviews (Cañas, 2006). Both experts' concept maps were compared to define interface design requirements.

From CTA to Interface Design

The two interviews brought to light information needed to construct a goal diagram with roles, decisions, and information elements. Model-based Systems Engineering (MBSE) applications within Magic System of Systems Architect (MSOSA) were used to model the overarching goals and their subgoals, associated at the proper levels to roles and decisions. These goals were used to determine capabilities of the system, the first step in the FBS framework (Sterling & Taveter, 2009).

Next, a Display Task Description (DTD) was created in MSOSA. This analysis provided a multi-agent method for developing a functional abstraction hierarchy for a user interface (UI). First, joint decision- making requirements within the HAT were identified. These decision requirements were traced to goals identified in a goal diagram. The decision requirements (DRs) demonstrated emergent dependencies on information being provided between the operator and the agent, documented as information requirements. The information requirements (IRs) were substantiated by visualization needs (VNs), graphical/auditory implementations within a prototype user interface. The visualization needs were defined to satisfy the structure in the FBS framework. (Potter, Elm, Roth, Gualtieri, & Easer, 2001)

After developing the DTD, the behaviors, or interactions of the system needed to be modeled from the goals and entities. This was accomplished by producing an architectural control pattern with an internal block diagram (IBD) in MSOSA to define the flows of information between the agents (human and machine) of the system. It was anticipated that there may be multiple autonomous agents in the design based on the length of flight and complexity of missions, and that there would be a negotiation interaction with a central node. The interactions were defined by the need of the goals and the decisions.

Finally, wire frame diagrams were developed in Balsamiq to demonstrate the feasibility of a notional implementation of the proposed user interface within a tactical display system (TDS) on current bomber systems. A briefing showing the user interface screen wireframes with the traceability to the requirements from their own interviews was prepared and provided to the knowledge workers to solicit feedback and elicit any new visualization needs or updates necessary to provide the proper information. The complete process this study proposed for HAT interface design can be seen in Figure 2.

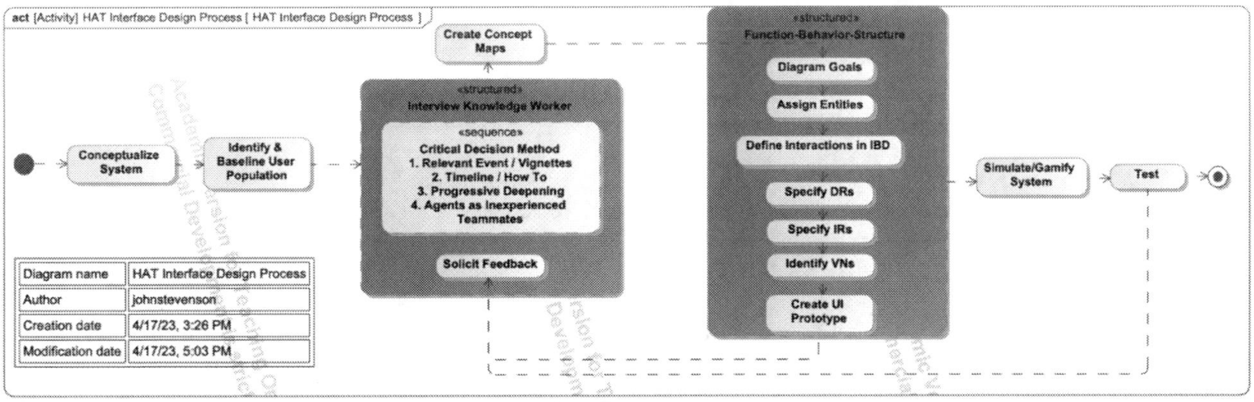

Figure 2. **Human-Agent Team Interface Design Process**

Discussion

It would benefit industry to adopt the definition of autonomy presented here. The comprehension that authority is always derived from a human element and that is provided situationally is a strong counterpoint to the "killer-robot" fallacy. The knowledge that capacity is situational helps in function allocation—when a task can be performed satisfactorily by any agent, it should be a shared task, and the agent with the lowest current cognitive task load (or highest capacity) is primed to perform the task. Viewing autonomy as a composition of availability, capability, capacity, and authority can guide a socio-technically mature design process.

Crew resource management (CRM) is a tool to support high performing human teams to build a shared mental model. It may be possible that CRM techniques can be abstracted and made useful within human- agent teams (HATs). This study developed an effective technique for using CTA to reveal the functional flows and operational sequences that build intention within the HAT: the Knowledge Worker Interview.

These techniques extend beyond the domain explored in this study. This method of assessing the domain, interviewing the experts, and designing based on cognitive requirements is universally applicable, and the techniques presented should apply to multi-agent systems regardless of the domain. When the design seeks to exploit the unique competencies of the human and the agent in the team, the capability of the system is greater than the capability of each individual agent.

Acknowledgements

The content of this paper is derived from a study conducted as part of a student thesis in partial fulfillment of the requirements for the degree of Master of Science in Systems Engineering.

The views expressed are those of the authors and do not reflect the official policy or position of the United States Air Force, Department of Defense, or the United States Government.

References

Air Force Research Laboratory [AFRL]. (2021a). Golden Horde Coliseum: Department of the Air Force Vanguard Program. Retrieved from https://cdn.afresearchlab.com/wp- content/uploads/2020/02/20154350/AFRL_Golden-Horde_FS_0921.pdf

Air Force Research Laboratory [AFRL]. (2021b). Skyborg: Open...Resilient...Autonomous. Retrieved from https://cdn.afresearchlab.com/wp- content/uploads/2020/02/20155247/AFRL_Skyborg_FS_0921.pdf

Brunacini, A. (2015). Positive Power. Fire Engineering, 168(7), 32-36.

Crandall, B., Klein, G., & Hoffman, R. (2006). Working Minds: A Practitioner's Guide to Cognitive Task Analysis. Cambridge: MIT Press.

Dombrovski, A. (2021). The Unfounded Bias Against Autonomous Weapons Systems. Információs Társadalom, 21(2), pp. 13-28.

Drubin, C. (2020). AlphaDogFight Trials Foreshadow Future of Human-Machine Symbiosis. Microwave Journal, 63(10).

Flach, J. M., & Dominguez, C. O. (1995). Use Centered Design. Ergonomics in Design, 19-24.

Gunzinger, M. (2021). Why America Could Lose Its Next War. Retrieved from Defense News: https://www.defensenews.com/opinion/commentary/2021/07/09/why-america-could-lose-its-next-war/

Haddal, R., Hayden, N., & Frazar, S. (2018). Autonomous Systems, Artificial Intelligence and Safeguards. Sandia National Laboratories.

Hitchens, T. (2021, Sep). AFRL's Golden Horde 'Gladiator' drones to compete virtually. Retrieved from Breaking Defense: https://breakingdefense.com/2021/09/afrls- golden-horde-gladiator-drones-to-compete-virtually/

Hoffman, F. (2017). Will War's Nature Change in the Seventh Military Revolution? Parameters, 47(4).

Kelly, K. (2016). The Inevitable: Understanding the 12 Technologies That WIll Shape Our Future. New York: Penguin Books.

Knox, M., & Murray, W. (2001). The Dynamics of Military Revolution. Cambridge: Cambridge University Press.

Konaev, M., & Chahal, H. (2021). Building Trust in Human-Machine Teams. Tech Stream.

McCaffrey, T., & Spector, L. (2018). An Approach to Human-Machine Collaboration in Innovation. AI EDAM, 32(1), pp. 1-15.

Miller, M. E., McGuirl, J. M., Schneider, M. F., & Ford, T. C. (2020). Systems Modeling Language Extension to Support Modeling of Human-Agent Teams. Systems Engineering, 23

Morgan, F. E., & Cohen, R. S. (2020). Military Trends and the Future of Warfare. RAND Corporation. Retrieved from https://www.rand.org/pubs/research_reports/RR2849z3.html

Moz, E., & Kleiner, B. H. (2021). Motivating High Performance Teams. Industrial Management, 63(2), 25-30.

Novak, J. D., & Cañas, A. J. (2006). The Origins of the Concept Mapping Tool and the Continuing Evolution of the Tool. Information Visualization, 5, 175-184

Potter, S. S., Elm, W. C., Roth, E. M., Gualtieri, J., & Easter, J. (2001). Bridging the Gap Between Cognitive Analysis and Effective Decision Aiding. In M. D. McNeese, & M. A. Vidulich, State of the Art Report (SOAR): Cognitive Systems Engineering in Military Aviation Environments: Avoiding Cogminutia

Ross, J. M. (2008). Moderators of Trust and Reliance Across Multiple Decision Aids. Orlando: University of Central Florida.

Smolska, M. (2021). A Team Development Process Based on the High Performance Team Coaching Model: A Case Study of Team Maturity Management. Scientifc Journals of the Maritime University of Szczecin, 67, 65-72.

Sterling, L., & Taveter, K. (2009). The Art of Agent-Oriented Modeling. Cambridge: MIT Press.

von Clausewitz, C. (1874). On War. Retrieved from Project Gutenberg: https://www.gutenberg.org/files/1946/1946-h/1946-h.htm

Walliser, J. C., Mead, P. r., & Shaw, T. H. (2017). The perception of teamwork with an autonomous agent enhances affect and performance outcomes. Proceedings of the Human Factors and Ergonomics Society Annual Meeting, 231-235.

Y.S, B. G. (2021). The Human-Machine Team. eBookPro.

EXAMINING THE RELATIONSHIP BETWEEN WORKLOAD AND RESILIENT PERFORMANCE IN AIRLINE FLIGHT CREWS

Michael Stewart[1], Kathryn M. Ballard[2], Chad L. Stephens[2], Lawrence J. Prinzel III[2], Jon B. Holbrook[2], Tyler D. Fettrow[2], & Daniel J. Kiggins[3]

Datum Aero LLC[1]
NASA Langley Research Center[2]
San Jose State University Research Foundation[3]

Workload has long been associated with human performance in aviation. High workload is typically viewed as a system design problem. The aim of this work was to find observable behaviors that guard cognitive resource margins; thus, improve handling of perturbations that arise. Twelve commercial airline flight crews participated in an experiment at NASA Langley Research Center. We explored self-reported workload data (NASA Task Load Index) after completing flight simulation scenarios to see how different individuals, using the same technologies, cope with challenging situations. A trained observer (from the same airline as the study participants) assessed performance. We differentiated a lower workload crew from a higher workload crew and analyzed their workload and performance. Results indicate that the workload may be a component of pilots' resilient performance. Initial strategies that reduced workload were classified and these types of behaviors, if implemented correctly, might assist pilots in maintaining functional workload resource margins.

Imagine a situation where two people faced the same challenging scenario. If both handled it successfully, how would you tell whether one had skills that made it easier to be successful? Specifically, skills that afford additional performance margin for tasks, making them more flexible and reliable. In commercial aviation, task management can be critically important in order to maintain the ability to tend to essential flight tasks. In this paper, we attempted to identify and categorize skills in airline pilots that afford additional margin for handling the complexity and variability of flight operations. Our goals were to determine if the skills were observable behaviors and if they could be identified by observing a simulated series of scripted flight scenarios.

Our interest in this has several motivations. First, our understanding of what humans do that contributes to success in aviation is limited. Commonly, it is thought that humans are the weakest link; thus, a movement towards autonomy is the answer. But if this is the future, we should know what we are replacing to test that assumption. Second, we believe that these skills might be trainable and would be an example of human resilience for performance enhancements rather than risk reduction. Last, we wanted to compare an observational technique that was derived from Hollnagel's (2011) Resilience Analysis Grid (RAG) with resilient performance potentials that were modified and adopted by American Airlines' Learning and Improvement Team (LIT), with the NASA Task Load Index (TLX) workload self-assessment. The comparison was intended to give us a method of differentiating low and high workload individuals (TLX scores) and determine if there was a relationship between them and the observed resilient performance (LIT potentials) of the individual.

Background

Resilience and human contributions to safety. Safety in aviation has traditionally been defined in terms of what went wrong through a forensic analysis of accidents and incidents. This view is limited in both definition and measurement of safety. Such approaches have proven utility, but they are ill-equipped to enable learning how humans contribute to safety in events which do not result in an incident due to human performance. Research efforts under the NASA System-Wide Safety (SWS)

Project has focused on the "human contribution to safety" intended to capture and describe what human flight crews routinely do right. This work builds upon a growing literature on resilience engineering and new approaches to safety (Hollnagel, 2014).

Operational Data Capture. American Airlines has been pioneering efforts to capture operational data about safety producing behaviors flight crews exhibit in operations through its LIT observations and interviews (American Airlines, 2020). LIT has created its own language and model for gathering data to guide its program development. This language is based on resilience potentials identified by Hollnagel (2017), and each potential is broken down into specific proficiencies. The language is firmly rooted in Safety-II principles (Hollnagel, 2014) and considers factors that are unique to American Airlines and modern airline flight decks. The proficiencies are observable indicators of performance that can be recorded during flight deck observations. By collecting this data during routine flights, LIT enables pilots to learn from their everyday experiences.

What is workload? Resilience is maintaining functions during times of disruption (Hollnagel, 2019). In order to maintain functions, resources must be allocated to match the need of the situation. However, resources are typically finite. One such resource in operations involving human operators is the capacity to continue to perform tasks effectively. This capacity to perform is multivariate and can be related to physical and cognitive abilities that can be affected by temporal demands. The term *workload* is used to describe the scale between no effort and max effort or task failure.

There are several theories about optimal workload, but questions remain regarding their practical application in aviation. Notably, Yerkes and Dodson (1908) looked at how arousal affected time to learn a simple habit in mice and found that pressure in the form of electrical shocks improved performance to a point, but then decreased. This "inverted U" chart of pressure and performance has become widely cited and applied to areas from self-improvement to human performance. Subsequent research has found that the curve depends on pressures, task type and the individuality of people. Pilots must be adept at many different types of tasks from simple actions to complex reasoning. Thus, it is unclear how task variability affects flight operations, but for the purpose of this study we are concerned with task performance when self-reported workload is near the upper extremes or the failure point.

Internal vs external modulation of workload. It is known that context and situational circumstances in aviation play a large role in task loading at any given moment. As previously discussed, the ability for a person to perform at varying task loads is dependent on individual differences as well as the nature of the task. Our interest in workload was specifically to see if self-modulation was occurring when the individual thought that workload needed to be reduced. This contrasts with other human performance work that is specifically aimed at reducing the workload of the operator. We care about operational strategies and not necessarily the overall workload.

Types and limitations of workload measures. Workload measurements currently fall into three categories: subjective, performance based, and physiological. The NASA TLX is a subjective test and is designed to elicit the following categories of workload: mental, physical, frustration, and temporal. Since workload is mainly a construct that depends on the individual, the TLX was used instead of other methods.

Hypotheses

1. We hypothesized that observed resilient strategies can be identified and cataloged.
2. We hypothesize that pilots that display more resilient behaviors will show lower workload self-assessment scores.
3. We hypothesize that self-reported performance will increase with increased resilient behaviors.

Method

Participants

Twenty-four (24) healthy airline transport pilots (9 women, M = 49.2 years) from a major US airline volunteered for the study. Subjects provided informed verbal and written consent to participate.

The experiment was conducted under approval from NASA's Institutional Review Board. The analysis in this paper only involved data from four of the participants in the study.

Procedure

Sim Scenarios Data presented herein were collected during the SWS Operations and Technologies for Enabling Resilient In-Time Assurance (SOTERIA) flight simulation study conducted at NASA Langley Research Center in Hampton, VA USA during May-June 2022. Details of the full data collection plan and flight simulation scenarios are described in Stephens et al. (2021).

Differentiating workload groups. To streamline the video viewing process, only two crews were chosen for the observational scoring process. The runs from the two most different crews in terms of workload ratings were selected using hierarchical clustering. Clustering is a set of methods used to group similar items together. Hierarchical clustering, specifically Ward's method (1963), is a recursive algorithm that groups pairs of items iteratively until all items are grouped. This clustering was applied to the reported NASA-TLX scores from all the crews. First, the six subscale scores (mental demand, physical demand, temporal demand, effort, frustration, and performance) were averaged between each crew's pilots for each scenario. With five scenarios and six subscale items, each crew had 30 workload measurements to compare among the crews. JMP Pro v16.0.0 was used to generate clusters and calculate distances between crews. Two were found to have highest distance measures and labeled as high or low workload.

1	Applied prior knowledge (that I brought with me to the simulation) to solve a problem	Learn	9	Intervened to address or resolve an unwanted condition.	Adapt
2	Verbally shared prior knowledge/experience with the other crewmember.	Learn	10	Changed automation or system to respond to an unanticipated new pressure.	Adapt
3	Heightened my own awareness/focus to prepare for an impending challenging task.	Adapt	11	Cross-checked other pilot's actions, over and above SOP-required cross-checking.	Coordinate
4	Monitored aircraft status in response to a detected anomaly, surprise, or inconsistency.	Coordinate	12	Initiated an action to decrease my own or other crewmember's workload/task load.	Adapt
5	Discussed expected actions before an anticipated challenging task.	Plan	13	Asked external source (e.g., ATC) for input/assistance.	Coordinate
6	Developed "what if" scenarios to support a possible change in plan.	Plan	14	Asked other pilot for input/assistance.	Coordinate
7	Gathered information to support a possible change in plan.	Plan	15	Managed time effectively to complete tasks.	Adapt
8	Identified countermeasures to an identified problem/pressure.	Adapt	16	Debriefed what happened after a problem occurred.	Learn

Figure 1. List of behaviors used to score videos

Observational scoring. With regard to our first hypothesis, we scored the videos using an abbreviated version of the LIT code book (see Figure 1) and added some additional categories of information: a goal description and whether it was explicitly stated by the pilot, inferred by the observer, or unknown. Outcome was captured on a scale of one to five; one was a failed goal and five was a flawless success. Goals were also coded as "no longer relevant" if the situation warranted or "unknown" if it wasn't possible to determine. A dichotomous Success score was also coded when an outcome was greater than a 3 and failure if less than 3. Unknown or No Longer Relevant were not coded as success or failures. Although we used this scoring system to capture broad categories, there was a review of descriptions of the specific behaviors to determine if we could validate the LIT codes as well as identify additional behaviors not being captured. If the behavior or the goal included explicit (i.e., 12, 13, 15) or implicit reference to workload then the behavior was coded as a workload behavior.

Results

Our second hypothesis that 'self-reported workload would be reduced when resilient behaviors were increased' was partially confirmed. We found that pilots in the *low workload* group exhibited significantly greater resilient behaviors than the *high workload* group. However, this effect was limited to behaviors with outcomes that were *successful* and pertained specifically to altering workload $\chi^2(1, N = 22) = 4.545, p = .033$. Although greater *total* and *successful* resilient behaviors were seen in the *low workload* group, they were both non-significant (see Table 1).

Table 1. Self-Reported Workload Scores by Success and Relevance

Behavior Type	*Count for Low Workload Crew*	*Count for High workload Crew*
All Behaviors	49	42
Successful Behaviors	28	19
All Workload Behaviors	26	18
Successful Workload Behaviors *	16	6

* Statistically significant
Note: proportions of success vs. totals are not indicative of percent failed due to coding methods.

Our third hypothesis, that 'self-reported performance would be greater with higher occurrences of resilient behaviors' was not confirmed. Mean differences between the two performance groups were non-significant. Surprisingly, the *low workload* group that also showed more resilient behaviors, reported a lower mean performance score, which is the opposite of our hypothesis and will be addressed in the limitations section of discussion.

Table 2. *Observed Resilient Behaviors Related to Workload*

Observed Behavior	*Inferred Goal*	*Possible Strategy*
Pilot manipulating autopilot controls for other pilot	Assuming other pilot's additional tasks for maintaining teammate's workload at acceptable level	[L]Pick-up other's tasks (offer your additional resources)
Pilot asked ATC to square the vectored pattern	Lengthening the mileage to approach and creating more time to prepare by changing the rate of externally driven events	Change tempo
Asked other pilot whether they were in the "green"	Explicitly inquire about teammate status	Monitor teammate
Pilot asked ATC for turbulence ahead	Seek information not easily or accurately obtainable within the aircraft	[L]Ask external resource
Asked other pilot what the ATC instructions were	Ensure correct flight path by using other pilot's memory/attention	[L]Delegate own tasks (conserve own resources)

[L] LIT codebook behaviors that were observed as successful workload behaviors.

Observed strategies. We observed five unique strategies within the behaviors we coded (see Table 2). Of the five, two were not included in the LIT codebook. The remaining three fit generally into the LIT categories and were validated as successful in the simulated operational environment. We provided an example behavior and listed its inferred goal and possible overarching strategy. In addition, proportions of the observed strategies are displayed (see Figure 2).

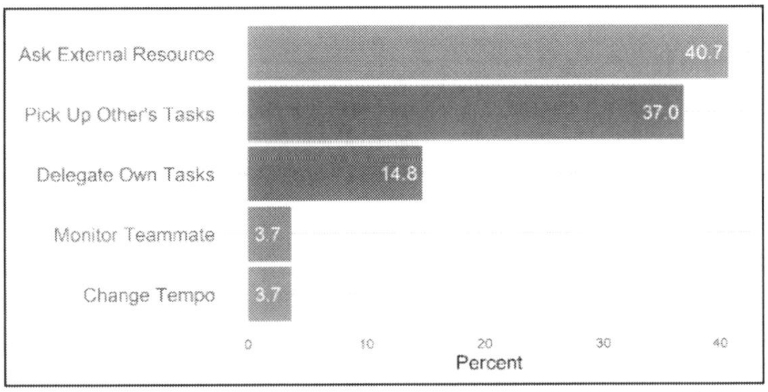

Figure 2. Proportions of Observed Workload Strategies

Discussion

Potentials for resilient performance. The framework Hollnagel described characterizes that the potential for resilient performance exists within the high-level categories: respond, monitor, anticipate, and learn. A critically important point to consider is that **behaviors which fit the description of resilient performance only have a potential to produce resilient performance, they do not guarantee it**. Our initial findings support this by showing that behaviors with unsuccessful outcomes did not have the same impact on workload as did the successful behaviors. This may seem obvious but is currently not how resilient data collection is being accomplished in aviation. Therefore, we suggest that this methodology be incorporated for further understanding of how potentials may be operationally relevant.

This logic applies to most behaviors and is not embodied in current methods of managing safety. Sometimes good behaviors do not work; sometimes bad behaviors do, and sometimes we get things right. Understanding more of the success rates and other factors that influence them would be beneficial to enhance safety. A need remains to examine resilience through multivariate analysis of observed behaviors and self-reported experiences of flight crews to understand possible advancements in aviation safety.

Outcomes. During observations, when able, we captured the inferred or stated goals and outcomes. Initially, we did not know how identifiable the outcomes were, so we did not include this information in our hypothesis. Further, we did not have any occurrence rate data for successful outcomes. Thus, we had low confidence in our understanding of outcome impacts. Our updated information suggests that outcomes are critically important to understanding how resilient behavior impacts airline crew performance.

Performance over workload. We feel that criteria for positive performance should be developed in addition to workload measures. Currently, negative performance indicators known as "safety performance indicators" (SPIs) identify things like unstable approaches and hard landings, but we lack a common analogous list for the characteristics we would like to see in operations. We should look at SPIs and develop a list of antonyms for them as a starting point.

Granularity of the TLX. During the five simulated flights there were multiple opportunities for resilient performance and multiple examples of resilient performance. Over the ten observations we recorded 81 behaviors. The simulations were between 15 and 20 minutes and the TLX questionnaire was given after the end of each scenario. Given the amount of time between when the TLX was given and the many behaviors and events within the flight, it is likely that the TLX is not sensitive enough to capture the true effect of a single behavior or possibly even multiple behaviors that might have impacted workload for a brief time. This is an issue for looking at workload in future studies and should be considered.

Limitations

Self-reported performance issue. The NASA TLX scale captures the participant's impression of how they performed. The scale used is similar to workload in directionality (i.e., lower is "better"). However, this may be counterintuitive and the researcher administering the TLX was catching abnormal responses that did not match the performance of the crews. He asked participants to review and reminded them about the reversed scaling. Thus, it is possible that the design of the TLX form potentially inverted the answers of some participants. This could explain the opposite result in the self-assessed performance data and will be reviewed and possibly administered separately from the TLX in future work.

Statistical power. Although this study saw significant differences in workload groups, we acknowledge that we lacked statistical power due to the small sample size; we plan to repeat this process with the remaining SOTERIA data to confirm the results.

Scenario comparability. Although scenarios were scripted and external events identical between crews, each participant's response differed in certain ways that impacted the flight and other pilot's work. These differences likely hid or made effects. In addition, it highlighted the fact that we cannot see the impact on workload from preventative behavior unless the same person were to perform the same scenario twice and choose differently each time – with no memory of the first action. Thus, we can only see results between subjects and not within subjects, which obscures effects.

Conclusions

We compared self-reported workload and performance data with resilient behaviors and found that they do impact workload when performed successfully. This information needs to be validated with additional observations using the same methods. However, we are optimistic in this approach to identify resilient performance. Possible applications of this work could be to implement TLX questionnaires after standardized airline training events to capture and train how experienced and skilled pilots manage crew workload for enhancing flexibility, performance, and ultimately safety.

Acknowledgements

This work was funded by NASA's System-Wide Safety Project, part of the Aeronautics Research Mission Directorate's Aviation Operations and Safety Program.

References

American Airlines' Department of Flight Safety. (2020). Trailblazers into Safety-II: American Airlines' Learning and Improvement Team, A White Paper Outlining AA's Beginnings of a Safety-II Journey.

Hollnagel, E. (2011). RAG – the resilience analysis grid. In: Hollnagel, E., Pariès, J., Woods, D.D., Wreathall, J. (eds.) Resilience Engineering in Practice. A Guidebook. Ashgate, Farnham.

Hollnagel, E. (2014). Safety-I and Safety-II: The Past and Future of Safety Management. Farnham, UK: Ashgate.

Hollnagel E. (2019, November 9). *What is Resilience Engineering?* Resilience Engineering Association. https://www.resilience-engineering-association.org/blog/2019/11/09/what-is-resilience-engineering/

Hart, S.G., & Staveland, L.E. (1988). Development of NASA-TLX (Task Load Index): Results of empirical and theoretical research. In P.A. Hancock & N. Meshkati (Eds.), Human Mental Workload (pp. 139–183). Amsterdam: North-Holland. JMP Pro v16.0.0, SAS Institute Inc., Cary, NC, (2022); https://jmp.com

Ward, J.H., & Hook, M.E. (1963). Application of an Hierarchical Grouping Procedure to a Problem of Grouping Profiles. *Educational and Psychological Measurement, 23*, 69 - 81.

Stephens, C., Prinzel, L., Kiggins, D. Ballard, K., & Holbrook, J. (2021). Evaluating the Use of High-Fidelity Simulator Research Methods to Study Airline Flight Crew Resilience. 21st International Symposium on Aviation Psychology, 140-145.

Yerkes RM, Dodson JD (1908). "The relation of strength of stimulus to rapidity of habit-formation". *Journal of Comparative Neurology and Psychology*. 18 (5): 459–482.

RESILIENT STRATEGIES IN COMMERCIAL AVIATION

Michael Stewart
Datum Aero LLC
La Palma, CA

Bryan Matthews
KBR Inc.
NASA Ames Research Center, Moffett Field, CA

When we fly and nothing scary happens, is the system's design affording this success? Not always — sometimes humans are the cause of success. This resilient performance is often overlooked. We explore two types of resilient performance *strategies*: *countermeasures* and *modifications*. countermeasures are behaviors triggered by variables anticipated to be challenging or problematic (i.e., pressures). To capture this, we look at examples of how a problem was avoided. For example, a country road may have a hairpin turn where accidents more frequently occur. With this pressure identified, we look at successful drivers for insights. Modifications are changes that are created to fill a gap between work-as-imagined and work-as-done. This strategy is from the design of systems. In aviation, work-as-imagined is often explicit, so it can be compared to behaviors using data. These two resilient potentials aim to better understand how systems function, as well as how people contribute to unrecognized successes.

Our understanding of how humans contribute to successes in aviation organizations is limited, because we do not systematically investigate this area. One assumption is that when safety performance indicators do not exceed unacceptable thresholds, things are going as planned. However, this is sometimes not the case. Notably this is due to the capacity for humans to adapt and achieve goals despite being given poor tools. Hollnagel's (2011) "work-as-imagined" versus "work-as-done" concept provides us with the language to illustrate the gap where compensatory behavioral strategies exist that create the appearance of normality and mask contextual variables (*pressures*) that render the imagined work unfeasible.

The term "pressures" describes operational, environmental, or other forces that may be challenging and that may stress the resources of the individual (Blajev & Holbrook, 2022). We are using this terminology to help describe what is triggering the resilient performance of interest.

Although many behaviors exist that can enable resilient performance, two behavioral *strategies* that we posit help provide the appearance of normalcy in the face of *pressures,* and that may indicate a need for organizational intervention are: *countermeasures* and *modifications.* A countermeasure is an action that sets a barrier or mitigation against an anticipated pressure; thus, increasing the likelihood of goal success (American Airlines' Department of Flight Safety, 2020). A modification describes the augmentation or change, specifically to a procedure or policy that also increases the likelihood of goal success. Although similar, the distinguishing factor between countermeasures and modifications is that countermeasures are heuristics deployed in a variety of situations. These may become modifications if a systemic issue is present, and the countermeasure has been adopted unofficially by users.

From an organizational perspective, these adaptive strategies and the pressures (i.e., context-dependent variables triggering them) are the targets of this methodology. Identification of pressures can help with redesigning systems aimed at expanding the range of work-as-imagined to include more of the total distribution. The goal is to enhance predictability by learning from one's own workforce. Our approach to this opportunity is to leverage existing concepts and data collection methods but alter the indicators of interest.

We acknowledge that many strategies that are preventative could be classified as countermeasures. Modifications are also essentially the same behavior as countermeasures, but related to a policy or procedure. Thus, modifications are specifically relevant to organizations and should not be used to classify the strategies themselves initially, as they are a sub-group. We suggest investigating when the goal of the strategy is similar to the basic goals of the organization. That is, when people are trying to ensure critical organizational functions are successful. If so, domain experts are necessary to make that determination.

This provides us with an opportunity for new learning. These issues are especially critical now since there is a push toward increasingly autonomous systems in aviation where these strategies may need to be factored into autonomous operations. We are proposing an approach to capture these strategies by utilizing a variety of data sources that are currently in-use.

Human vs. organizational resilience. Humans have evolved the abilities that are necessary to adapt and handle challenges; organizations however, are groups of people, systems, and are entities of their own. Even with resilient performers within the organization, the organization must deliberately design-in resilient potentials.

To begin, one method that organizations can use is developing the potential to *learn* from their naturally resilient human performers. We use the term *learn* as a potential for organizational resilient performance as described in Hollnagel's (2011) Resilience Assessment Grid (RAG). The organization must be able to introspect and understand how its systems and policies perform – at least to a level that is meaningful for their success.

Positive deviance. The concept of investigating what works is not new. Positive Deviance (PD) is the review and understanding of high performers in situations where challenges exist and has been around since the 1970s (Positive Deviance Collaborative, 2023). Identifying and understanding success cases from high performers follows a general process: 1) Differentiate high/low performers; 2) study what makes them perform differently; 3) test hypotheses. This methodology has been successful in environmental health and hospital care domains (Bradley, et al., 2009).

Resilient performance indicators. Safety has generally been defined in terms of its absence. This is noted by the generally negative theme of safety performance indicators (SPIs). For example, loss of separation, ground proximity warning, and bird strike are all examples of current SPIs (International Civil Aviation Organization, 2023). These events are important to measure, but are a small minority of the overall occurrences in the system (PARC/CAST, 2013). Therefore, we intend to start an analogous catalog of resilient performance indicators (RPIs). That is, a list of events that are deemed to be desired performance and not merely under the threshold of what is unacceptable.

To search for RPIs we can leverage the massive amounts of data generated by the aviation system. A variety of sources exist, which include: 1) Aircraft centric data such as Flight Operational Quality Assurance (FOQA) data that can be leveraged to determine how the aircraft was flown; 2) Surveillance data such as ADS-B or radar track data that reveals how multiple aircraft interact within air traffic patterns; and 3) text reports and narratives from NASA's Aviation Safety Reporting System (ASRS) or airline Aviation Safety Action Program (ASAP) that captures the context of the operations and why safety events mishaps happen. Other rich text narratives from Learning and Improvement Team (LIT) or Line Operations Safety Audit (LOSA) observations offer additional insights into behaviors that capture the context from a different perspective. Indicators from these various data sets can be informative in determining what resilient behavior humans are performing to make the system run safely.

Resilient performance may not be positive for everyone. Although resilient performance may be a positive indicator that people are essential to success, it can also highlight issues that need to be improved within an organization. If there are cases where users of a system feel compelled to alter or augment it, there is likely a need for change. Organizations should embrace this as continual improvement for all stakeholders and not criticism.

Case Study 1. Wake Turbulence Countermeasures

Event-Report Initiated Analysis

Countermeasures can potentially be more generalizable than modifications and not tied to a particular procedure or policy. Thus, searching for these strategies can be initiated around observing operator actions as well as event reports. LOSA, ASRS, and ASAP may trigger an investigation into the objective data such as FOQA to quantify the occurrence rate. This is achieved by running a targeted search within the numerical data to detect points in the flight that match a Subject Matter Expert's (SME's) query parameters. When undesirable events are identified, mitigating strategies can then be crafted and implemented. Subsequently, the numerical data can be monitored to measure whether the mitigations are working. With this well-established methodology already in practice, it can be leveraged to capture successful operations as well.

Step 1. Identify the strategy occurrence in operations

Example: We used flight deck observation data collected during a simulated series of flights at NASA Langley Research Center (Stephens et al., 2021). The observations were a subset of two crews' data (Stewart et al, 2023). When pilots were managing wake turbulence events on arrival, some requested speed relief to increase distance from the previous aircraft. Another strategy was a request for lateral offset on the arrival to avoid the turbulence altogether (See Table 1.).

Proficiency	Pressure	Description	Goal	Outcome	Description
countermeasures	ATC/ Traffic	Asked ATC for 1 mile offset to avoid wake	Avoid wake turb	Success	No wake was observed
countermeasures	ATC/ Traffic	Asked ATC to slow for additional spacing for A330	Avoid wake turb	Failure	Hit wake

Table 1. Observation examples of countermeasures used to avoid a wake turbulence event.

Step 2. Identify contextual pressure variables

Example: In this example we searched the real ASRS database for reports that are related to our procedure of interest (BOOVE arrival). Pressures that may trigger a countermeasure response could be due to high traffic flow which results in reduced spacing when following a heavy aircraft on arrival. Recommended spacing behind a heavy is 7 NM for large and 8 NM for small aircraft. Thus, ATC and traffic were both coded as pressures.

ASRS Report 1. "*SOCAL Approach Control cleared our flight for the ILS 24R via the CRCUS transition. We were following a B787-9. To help increase the space between our airplanes the Los Angeles Center Controller instructed us to slow to 250 KIAS while on the ANJLL4 arrival which we complied with. Looking at our TCAS display, I estimated the 787 was approximately 5 miles ahead of us. SOCAL approach appropriately cautioned us for wake turbulence since we were following the heavy 787. Our flight was normal until we reached CRCUS waypoint where we encountered the 787's wake*".

In this scenario, the reduced traffic spacing was anticipated as a *pressure* that would result in wake turbulence. A countermeasure to reduce speed was applied; however the desired spacing was not achieved and wake turbulence was encountered.

Step 3. Compare outcomes with and without strategy

Example: This step is key to having all data sources available to properly assess the outcome. FOQA data can objectively determine how the wake turbulence event is managed, while radar track data can provide the distance and aircraft type of the proceeding aircraft. Being able to fuse these data sources together would facilitate an assessment of whether the *strategy* was successful or not and what *pressures* were involved either internally or external to the aircraft.

Outcome. This example does not have a real-world outcome as it used simulated observation data. However, this methodology could be employed if enough observational data are collected, and a consensus is reached on the efficacy of the countermeasure.

Descriptions of countermeasures and modifications can be found that address and resolve the safety issue being reported. Evidence of these actions may be present in the numerical data during these adverse situations. It is also possible to determine if these countermeasures are being implemented in consistent geographical locations, which may indicate a hot spot where positive deviations are necessary. This approach can provide insight into commonly-used strategies to handle adverse situations. Furthermore, the intervening actions that are implemented can be examined to determine if they are safe strategies or if a systemic change is needed to address the problem in the system that is requiring positive deviations by the operators in the first place.

Case Study 2. DFW Arrival Modification

Numerically Initiated Analysis

To identify modifications, we examine work-as-imagined (WAI) versus work-as-done (WAD). Procedures are examples of WAI, which are used in many aspects of aviation. This case study is a standard terminal arrival route (STAR) serving Dallas Fort Worth International Airport (DFW). By comparing the lateral and vertical confines of the procedure with radar tracks of aircraft that flew the arrival, we can see when adherence to the criteria of the procedure is, or is not, occurring. This observation was accomplished using a system called RADI (Stewart & Matthews, 2017). In some cases, achieving the high-level goal of safe and expeditious movement of traffic to the airport may require a positive deviation from WAI. While the work imagined in the procedures is to automate the arrival to facilitate lower workloads for air traffic controllers (ATC) and provide optimal profile descents to save on fuel, this is not always achievable due to compounding factors such as weather or traffic loads Knowing what the procedure restrictions are, we can look for systemic areas where adherence is low or if flights are missing restrictions by a consistent margin. This can point to possible modification techniques that ATC uses to re-route traffic to meet the higher-level objective of flights reaching their destination safely. Once a systemic non-adherence is identified, the location or waypoint fix can be searched for in ASRS to help ascertain why a restriction was not met.

Step 1. Identify systematic difference between work-as-imagined and work-as-done

Example: Altitudes not being adhered to on BOOVE arrival procedure into KDFW: Crossing DELMO waypoint at 12,000ft and 10,000ft instead of the published 11,000ft.

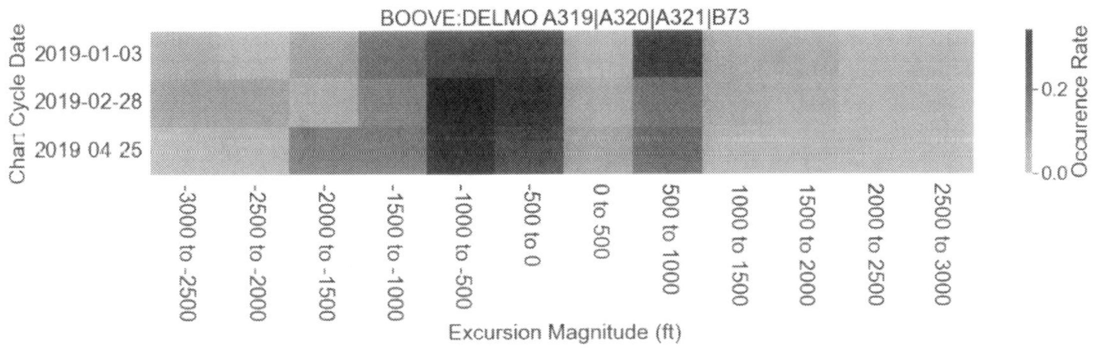

Figure 1. Proportions of altitude crossings relative to the restriction altitude at DELMO over time.

Step 2. Identify contextual pressure variables

Example: Look at subjective event reports (ASRS, ASAP, and company specific) for context clues and search based on commonalities or fusion points. This case would be the arrival (BOOVE) and the waypoint (DELMO).

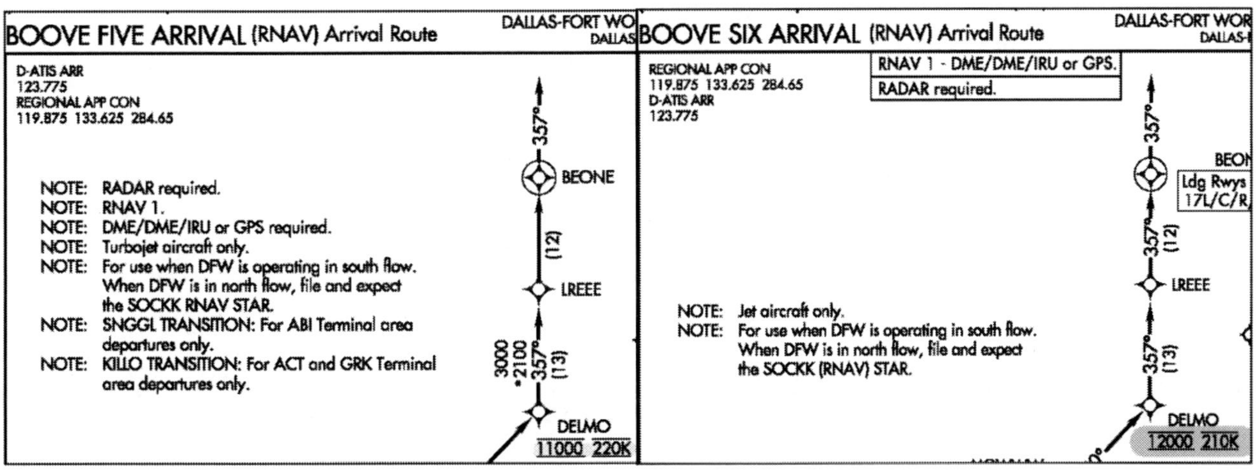

Figure 2. Navigation chart depicting the change to the procedure altitude.

ASRS Report 1. *"During the BOOVE4 arrival into DFW. We were descending out of 11400 just prior to DELMO for 11000. Approach advised us of traffic at our 1 o'clock climbing. Seconds after, we had a traffic advisory from the TCAS that immediately changed to an RA with a climb advisory.* ***Traffic alerts from ATC and TCAS into DFW occur on almost every arrival and departure."***

After identifying a candidate pressure – traffic in this case, we could determine that there is likely a pressure that a modification is being used to manage.

Step 3. Compare outcomes with and without strategy

Example: For this portion of the example, we would need to have access to the airline's internal data sources. In this case, FOQA data for TCAS Resolution Advisories would be the target variable.

Outcome. In this example, the waypoint DELMO was changed in the procedure from 11,000ft to 12,000ft. This structural change to the procedure illustrates that the modification may have been necessary and was included in the subsequent BOOVE6 iteration of the procedure (see Figure 2.). Figure 2 is a real chart that is used daily at DFW; we highlighted the altitude change to illustrate the change.

Conclusion

We described a general process using currently available safety data that can be used to capture two different resilient performance strategies: countermeasures and modifications. Investigating the effectiveness, and how these strategies are used to counter pressures may help to identify systems that are not functioning as intended, while simultaneously offering possible solutions. This approach should be tested and further developed to maximize its operational value. Our next steps are to provide empirically validated results using real-world data. When these solutions are captured, understood, and built into an organization, it has an increased potential to learn and adapt to changing conditions.

References

American Airlines' Department of Flight Safety (2020). Trailblazers into Safety-II: American Airlines' Learning and Improvement Team, A White Paper Outlining AA's Beginnings of a Safety-II Journey

Blajev, T., & Holbrook, J. (2022, June 1). *Learning From All Operations Concept Note 6: Mechanism of Operational Resilience.* Flight Safety Foundation. Retrieved April 18, 2023, from https://flightsafety.org/wp-content/uploads/2022/06/LAO-Concept-Note-6_rev2.pdf

Bradley, E.H., Curry, L.A., Ramanadhan, S., et al. (2009) Research in action: using positive deviance to improve quality of healthcare. *Implementation Sci* 4, 25. https://doi.org/10.1186/1748-5908-4-25

International Civil Aviation Organization. (2023). *Indicator Catalogue.* ICAO. Retrieved April 18, 2023, from https://www.icao.int/safety/Pages/Indicator-Catalogue.aspx

PARC/CAST Flight Deck Automation Working Group. (2013). Operational use of flight path management systems. Final Report of the Performance-based operations Aviation Rulemaking Committee/Commercial Aviation Safety Team Flight Deck Automation Working Group. Washington, DC: Federal Aviation Administration.

Holbrook, J.B., Stewart, M.J., Smith, B.E., Prinzel, L.J., Matthews, B.L., Avrekh, I., Cardoza, C.T., Ammann, O.C., Adduru, V., Null, C.H. (2019) Human Performance Contributions to Safety in commercial Aviation. NASA-TM-2019-220417.

Hollnagel, E. (2011). RAG – the resilience analysis grid. In: Hollnagel, E., Pariès, J., Woods, D.D., & Wreathall, J. (eds.) Resilience Engineering in Practice. A Guidebook. Ashgate, Farnham.

Positive Deviance Collaborative. (2023). *THE BEGINNINGS.* Positive Deviance Collaborative. Retrieved April 18, 2023, from https://positivedeviance.org/

Stephens, C., Prinzel, L., Kiggins, D. Ballard, K., & Holbrook, J. (2021). Evaluating the Use of High-Fidelity Simulator Research Methods to Study Airline Flight Crew Resilience. Proceedings of the 21st International Symposium on Aviation Psychology, 140-145.

Stewart, M. and Matthews, B. Objective assessment method for RNAV STAR adherence, *2017 IEEE/AIAA 36th Digital Avionics Systems Conference (DASC)*, St. Petersburg, FL, USA, 2017, pp. 1-9, doi: 10.1109/DASC.2017.8102034.

Stewart, M., Ballard, K., Stephens, C., Prinzel, L., Holbrook, J., Fettrow, T., & Kiggins, D. (2023). Examining the Relationship between Workload and Resilient Performance in Airline Flight Crews. Proceedings of the 22nd International Symposium on Aviation Psychology.

MEASURING THE LOOKOUT BEHAVIOR OF STUDENT PILOTS IN A VIRTUAL REALITY FLIGHT SIMULATOR

Ivo V. Stuldreher
Erik Van der Burg
Wietse D. Ledegang
Mark M.J. Houben
Eric L. Groen
TNO Human Performance, Soesterberg, The Netherlands
Danny van der Horst
Erik A.M. Starmans
Guido Almekinders
Royal Netherlands Air Force, Soesterberg, The Netherlands

Learning adequate gaze behavior is essential in flight training. In this exploratory study we investigated the development of gaze behavior in flight training in a virtual reality (VR) flight simulator. Following standardized study material, fifteen participants without flying experience repeatedly practiced three basic flight maneuvers in a VR simulator of a small aircraft. During some runs, participants performed an additional N-back task to measure cognitive spare capacity. From the recorded gaze data we computed the percentage of time during which the gaze was directed outside the cockpit, i.e., the "Lookout". This outside dwell ratio differed between flight maneuvers. A higher outside dwell ratio was associated with better flight performance. Remarkably, the outside dwell ratio increased with the additional N-back task. A heatmap indicated staring behavior during the N-back. In a follow-up study we will extend the analysis of gaze behavior with more dynamic measures than only the dwell ratio.

Aspiring military pilots within the Royal Netherlands Air Force (RNLAF) undergo Elementary Military Pilot Training (EMVO) in a turbo-prop trainer aircraft (the Pilatus PC-7). During their flight training, the student pilots also learn how to perform adequate visual scanning during flight, as this is leads to improved flight performance (Ziv, 2016). During the EMVO, emphasis is made on directing a large proportion of gaze to the outside environment. Student pilots are instructed to perform a structured lookout procedure during which they systematically scan the horizon, alternated by brief cross-checks of the relevant flight instruments inside the cockpit.

The RNLAF is interested in the possibility to incorporate Virtual Reality (VR) as training means within the EMVO. VR means have already been implemented by the Royal Air Force, United States Air Force and Royal Australian Air Force (Pope, 2019; Air Education and Training Command, 2020; Lewis & Livingston, 2018; Pennington et al., 2019). Nowadays VR systems have a built-in eye tracker, which allows for the monitoring of gaze behavior of the student pilots. There are indications that pilots whose gaze behavior better corresponds to that of expert pilots show better flight performance (Wickens et al., 2008). Besides learning how to control the aiplane it is thus also important to learn how to direct your gaze during flight.

In this study, we investigated the development of gaze behavior of student pilots in a VR flight simulator during a mini-training of three sessions in which they practiced three basic flight maneuvers. In a related paper, we discuss the flight performance measures (Ledegang et al., in press). Here, we examine 1) how gaze behavior develops over sessions in which student pilots learn to fly, 2) how their gaze bahavior relates to flight performance, and 3) how their gaze behavior is affected by additional cognitive load.

Method

Participants

Fifteen military cadets (12 males and 3 females) of the Royal Military Academy participated in this study. The participants had a mean age of 23.7 years (± standard deviation of 2.4 years), an average of 3.6±7.8 hours of flight experience on powered- and glider aircraft and 2.4±7.7 hours on flight simulators. Prior to the experiment, all participants signed an informed consent, stating that the details of the experiment had been sufficiently explained, and that they participated voluntarily. The experiment was conducted with approval of the institutional ethics committee and was in accordance with the revised Helsinki Declaration.

Materials

The simulator environment (see Figure 1), developed by the company multiSIM BV, consisted of a fixed-base cockpit (front-seat) of the Pilatus PC-7 turboprop trainer aircraft, includingcontrol devices with control loading. A VARJO-Aero VR device with built-in eye-tracker (200Hz) was used to present the cockpit and virtual environment near Woensdrecht Air Force Base, the Netherlands, rendering at 90Hz. The flight model characteristics were comparable to the PC-7 aircraft and were validated by EMVO flight instructors. During the experiment, audio instructions and an auditory secondary task were presented through a headphone.

Procedure

The participants repeatedly practiced three basic flight maneuvers: Straight-and-Level flight (SAL), Speed Change (SC) and Level Turn (LT). Each manoeuvre was performed three times in runs of 210 seconds each, followed by a test run in which the same manoeuvre was performed while simultaneously executing an additional N-back memory task as a measure of cognitive spare capacity. Each session of four consecutive runs was repeated three times, divided over two days, cumulating to twelve runs per manoeuvre.

The primary task consisted of manual control of the aircraft, including the instrument scan. This instrument scan was part of the lookout procedure, during which the particpant scanned the horizon and performed an instrument crosscheck each time when gaze passed the airplane nose. As secondary task, an auditory N-back memory task (Kirchner, 1958) was used, which required the participant to continuously update their working memory. The applied 2-back task required the participant to remember the last two letters of an auditory sequence of continuously changing letters at a fixed 3-seconds interval with a 25% repetition probability. The participant was instructed to make a self-paced response by pressing a dedicated button on the throttle if the letter heard was identical to the letter two trials back and to withhold a response if the letter was different.

Prior to each block, the participant was asked to study the instruction material (including a video from a flight instructor explaining each maneuvre), so that the instructions were identical for each participant. The participant received no feedback during the experiment.

Figure 1. (A) Setup of the VR simulator during the experiment, with the experimental test leader behind the instructor station and the participant inside the cockpit mock-up. (B) Pilot view with gaze direction overlaid in green. Note: the gaze direction was not shown during the experiment.

Measurements

From the recordings we analyzed the gaze direction in pitch, roll and yaw directions. We processed data to obtain the episodes where the gaze was directed inside the cockpit, and episodes where the gaze was directed to the outside environment. We calculated the outside dwell ratio, i.e., the percentage of time that the gaze was directed to the outside environment, during each run.

Objective measures of flight performance were also extracted, normalized and combined into one normalized performance measure on scale from zero to one per maneuver. This procedure is described in the accompanying paper by Ledegang et al. (in press).

Statistical analysis

We conducted a repeated-measures ANOVA on the outside dwell reatio with run and flight maneuver as within-subject variables to examine how the outside dwell ratio varied over runs across the three flight maneuvers. Note that runs 4, 8 and 12 were excluded from these analyses, as for these runs participants also perfomed the additional N-back task.

We then used Pearson correlations to investigate whether the outside dwell ratio related to flight performance, both averaged across flight maneuvers.

Finally, we examined whether the outside dwell ratio differed between runs with the N-back task (runs 4, 8, and 12) and the preceding runs without the N-back task (3, 7, and 11, respectively) using three separate two-tailed t-tests. For all analyses, alpha was set to .05.

Results

Figure 2(A) shows the group mean outside dwell ratio over the twelve runs for the SAL, SC and LT flight maneuvers, separately. A repeated measures ANOVA revealed that the outside dwell ratio varied between flight maneuvers, $F(2, 8) = 25.659$, $p < .001$ and varied over runs, $F(2, 8) = 2.069$, $p = .043$. Post-hoc paired t-tests revealed that the outside dwell ratio was significantly lower during the SC, $t(16) = -7.35$, $p < .001$, and LT $t(16) = -6.35$, $p < .001$, maneuvers compared to the SAL maneuver, but did not vary between SC and LT maneuvers, $t(16) = 0.99$, $p = .335$. A second set of post-hoc paired t-tests showed that for SAL, SC and LT maneuvers in none of the runs the outside dwell ratio significantly differed from the outside dwell ratio of run 1. This indicates that there was no systematic learning effect, even though the outside dwell ratio varied over runs.

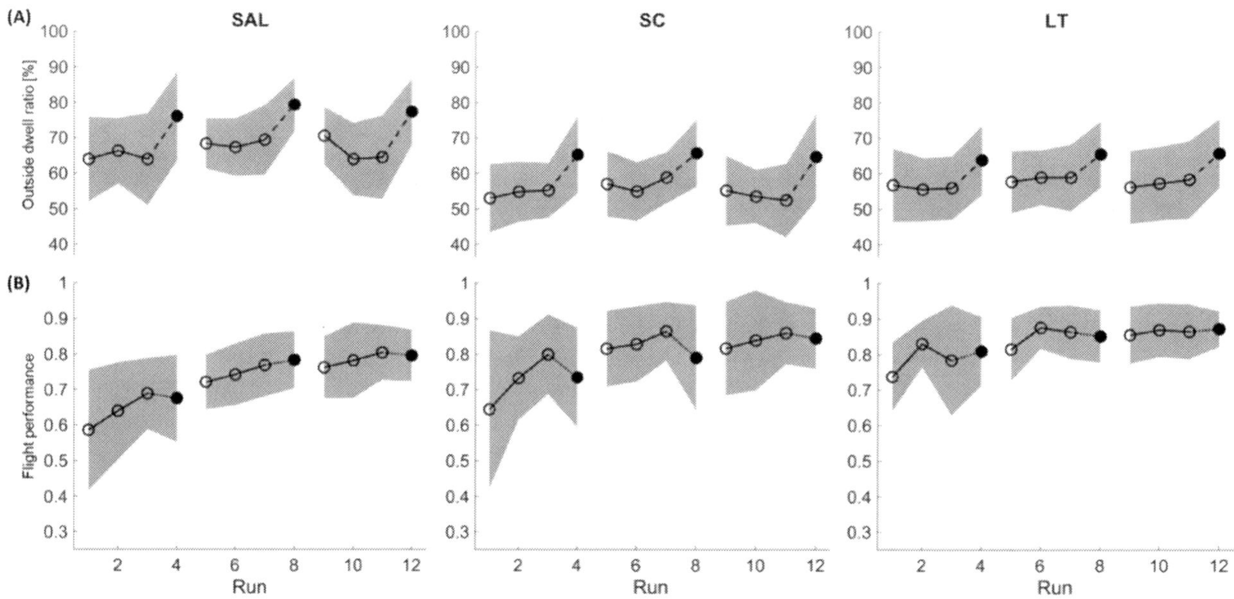

Figure 2. (A) Mean outside dwell ratio and (B) Flight performance as a function of the twelve runs for SAL, SC and LT flight maneuvers. Shading reflects standard deviation across participants. Filled markers correspond to runs with N-back task..

Figure 2(B) illustrates the group mean normalized flight performance over the twelve runs for SAL, SC and LT flight maneuvers, seperately. We found a significant positive correlation ($r = 0.18$, $p = .045$) between the normalized flight performance and the outside dwell ratio averaged across flight maneuvers, which means that participants with higher outside dwell ratios also showed better flight performance.

With respect to the effects of the additional N-back task, the black dots in Figure 2(A) illustrate that the outside dwell ratio was higher during runs with N-back task compared to runs without this task (open dots). Separate paired t-tests revealed that this was the case for each flight maneuver and each session, $p < .007$. To further investigate this effect, we examined differences in gaze direction between runs with and without N-back task in a post-hoc analysis. The results are shown in Figure 3(A), depicting a heatmap of the difference in gaze proportion during runs with versus runs without the N-back task. Figure 3(B) depicts the difference in gaze proportion in four areas-of-interest for the three flight maneuvers. It shows that with the additional N-back task a larger proportion of the gaze is being directed outside, and mostly straight ahead. This seems to indicate staring behavior.

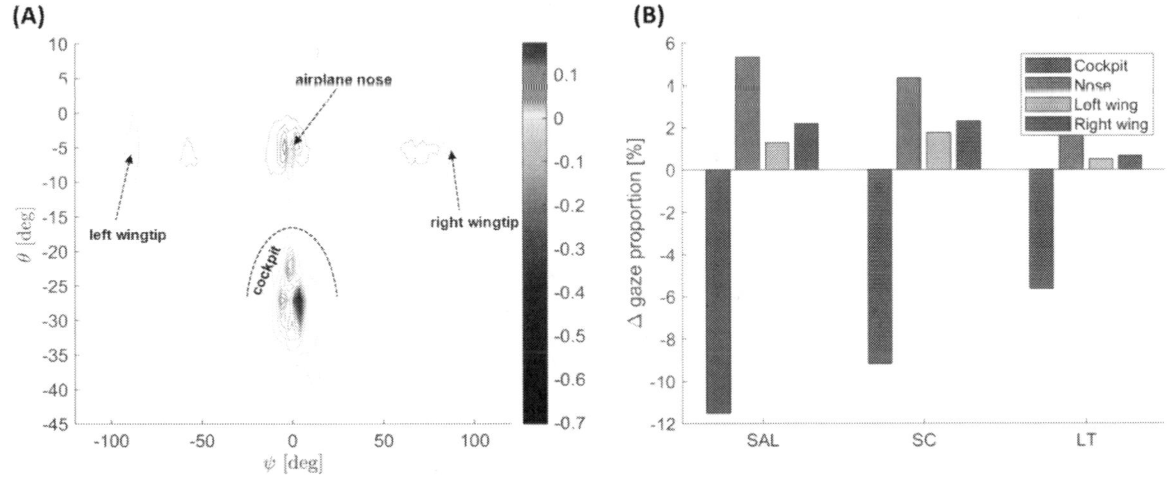

Figure 3. (A) Heatmap of difference in gaze proportion between runs with and without N-back task, averaged over all maneuvers. A positive number means a larger proportion of gaze is directed in that area during runs with N-back. (B) Difference in gaze proportion between runs with and without N-back for four main areas of interest: the cockpit, the outside view around the airplane nose, the outside view over the left wing and the outside view over the right wing.

Discussion

We examined how the Lookout behavior of student pilots changes during flight training in a VR flight simulator, how this Lookout behavior relates to flight performance and how extra cognitive load affects gaze behavior. Regarding the first research question, the outside dwell ratio varied significantly between flight maneuvers. In the SC and LT conditions, the outside dwell ratio was significantly lower than in the SAL condition. We explain this difference by the observation that compared to SAL, during the SC and LT maneuvers pilots have to monitor their instruments more closely to check the progression of their speed change or turn, respectively. Although the outside dwell ratio varied significantly over runs, we did not find evidence that the outside dwell ratio systematically increased over runs. Regarding the second research question, we found that the outside dwell ratio was positively correlated with flight performance. Thus, it seems that the participants with better flight performance were also directing a larger proportion of their gaze towards the outside environment.

Regarding the third research question we observed that with an additional N-back task the outside dwell ratio increased, indicating that the participants were staring. This suggests that with extra cognitive load the participants did not have the cognitive capacity to process the information of the flight instruments inside the cockpit.

Conclusions

Our findings indicate that the dwell ratio did not show any progression in Lookout performance in student pilots during a mini-flight training course without feedback from an instructor. This suggests that the dwell ratio is too rudimentory to measure progression. Therefore, future work should also consider measures capturing the scan pattern dynamics when assessing the lookout and instrument cross-check while learning to fly.

Acknowledgements

This research was supported by the Defence Research and Development Programmes V1917 and V1903, and RNLAF AIR. The authors acknowledge all RNLAF subject matter experts from EMVO, FCL and CMA for their guidance and advise in the preparation of the experiment and interpretation of the results. Furthermore, we thank multiSIM BV for their VR simulator developments and all participants for their enthusiastic contribution.

References

Air Education and Training Command. (2020). About PTN

Kirchner, W. K. (1958). Age differences in short-term retention of rapidly changing information. *Journal of Experimental Psychology, 55*(4), 352.

Ledegang, W. D., Van der Burg, E., Stuldreher, I. V., Houben, M.J., Groen, E. L. Van der Horst, D., Starmans, E. A. M, Almekinders, G. (in press). Acquiring manual flying skills in a virtual reality flight simulator. *International Symposium on Aviation Psychology*

Lewis, J., & Livingston, J. (2018). Pilot Training Next: Breaking institutional paradigms using studentcentered multimodal learning. *Interservice/Industry Training, Simulation, and Education Conference (I/ITSEC)*, Orlando, FL.

Pennington, E., Hafer, R., Nistler, E., Seech, T., & Tossell, C. (2019, 26/04). Integration of advanced technology in initial flight training. *Systems and Information Engineering Design Symposium (SIEDS)*, University of Virginia, USA.

Pope, T. (2019). A cost-benefit analysis of Pilot Training Next Air University. Wright-Patterson Air Force Base, OH, USA.

Sheets, T. H., & Elmore, M. P. (2018). Abstract to action: Targeted learning system theory applied to adaptive flight training.

Wickens, C. D., McCarley, J. S., Alexander, A. L., Thomas, L. C., Ambinder, M., & Zheng, S. (2008). Attention-situation awareness (A-SA) model of pilot error. *Human Performance Modeling in Aviation*, 213-239.

Ziv, G. (2016). Gaze behavior and visual attention: A review of eye tracking studies in aviation. *The International Journal of Aviation Psychology, 26*(3-4), 75-104.

PILOTS' EYE BEHAVIOR AND ELECTRODERMAL ACTIVITY DURING FLIGHT

Gajapriya Tamilselvan
Aviation Science, Saint Louis University
Saint Louis, MO
Stephen M. Belt
Aviation Science, Saint Louis University
Saint Louis, MO

The purpose of this study was to investigate how a given flight task manifested in pilots' eye behavior and electrodermal activity and how they changed over the course of an instrument flight session. This study implemented a quasi-experimental one-group time series design to identify attentional and emotional states of pilots. Pilots were emotionally aroused during flight segments like takeoff and climb, approach and landing-1, missed approach, and approach and landing-2. Pilots felt visually demanding during takeoff and climb, approach and landing-1, missed approach, en route-2, and approach and landing-2. Missed approach was the most visually taxing flight segment, where pilots rapidly shifted their attention from one fixation to another on their primary flight display (PFD). The research findings from this human-in-the-loop experiment inform the aviation community about the precise evaluation of pilots' mental workload during flight.
Keywords: AATD, gaze, visual attention, skin conductance, workload

Many aviation researchers favored the use of subjective scales of workload like the NASA Task Load Index (NASA-TLX) scale, Modified Cooper-Harper (MCH) scale, Bedford Workload Scale, and Subjective Workload Assessment Technique (SWAT) (Belt et al., 2021). In aviation settings, there was a need to estimate mental workload in a more objective fashion without any pilot intervention. The psychophysiological evaluation of pilots' performance and workload became prominent in the literature (Wilson, 2002; Dahlstrom & Nahlinder, 2009; Wang et al., 2020). Combined measures like heart rate, eye movement, blinks, and subjective workload scales quantified mental workload in basic civil aviation training (Dahlstrom & Nahlinder, 2009). Three flight segments like takeoff brake release, engine failure, and cruise were statistically significant in the study. Pilots experienced higher mental workload during engine failure, whereas their workload was low during rejected takeoff, climb, and descent segments (Dahlstrom & Nahlinder, 2009).

Another study attempted to understand pilots' cognitive demands by measuring psychological measures like heart rate, heart rate variability, eye blinks, electrodermal activity, electrical brain activity, and subjective mental workload data during both Visual Flight Rules (VFR) and Instrument Flight Rules (IFR) conditions (Wilson, 2002). Pilots exhibited significant electroencephalography (EEG) signals during various flight segments like landings, VFR touch and go, IFR missed approach, IFR climb-out, Instrument Landing System (ILS) tracking, and the final landing (Wilson, 2002). Allsop and Gray (2014) conducted a study to understand whether anxiety impacted visual attention and flight performance during the landing phase in a simulated setting. Participants spent more time outside the visual stimulus presented to them when the anxiety condition was introduced in the experiment. Anxiety influenced both attentional control and flight performance (Allsop and Gray, 2014).

Understanding the synergy between task management, attention, and mental workload is vital to perceive how pilots maintain situation awareness in a multitasking environment (Wickens, 2002). The current study aimed to study this synergy by employing a galvanic skin response (GSR) device and eye-tracking glasses. The purpose of this study was to investigate how a given task manifested in pilots' eye behavior and electrodermal activity and how they changed over the course of instrument flight session. In particular, this study sought to understand attentional and emotional states of pilots while they met task demands throughout the instrument flight session.

Methods

Participants

Our research sample comprised 17 pilots (2 females) with varying experience and certifications. The mean age was 47.29 ($SD = 13.3$) years. The mean of pilots' experience in flight hours was 5523.24 ($SD = 6142.21$) hours. All participants were instrument rated in our study.

Materials and Stimuli

Our experimental stimuli involved an instrument flight session in an Advanced Aviation Training Device (AATD) equipped with Prepar3D software. Weather conditions were set to ceilings overcast at 1000 feet, 3 miles visibility, and calm wind conditions. The circuits were divided into seven segments: takeoff and climb, en route, arrival, approach and landing-1, missed approach, en route-2, and approach and landing-2. Participants were outfitted with a wrist-based galvanic skin response (GSR) device and eye-tracking glasses. The sampling frequency of GSR and eye-tracking devices were 128 Hz and 51 Hz.

This research utilized a high-fidelity computer installed with iMotions - human insights software to collect, synchronize, and time-stamp biometric signals for statistical analyses (iMotions, 2023). We conducted gaze mapping analysis and examined the scan paths of each participant to understand their eye behavior. We aggregated gaze mapping data on certain areas of interest (AOI) and generated heat maps to gain deeper insights into how pilots maintain visual attention during the session. We analyzed the phasic component of skin conductance response (SCR) and detected GSR peaks to understand the electrodermal activity of participants. Descriptive statistics were used to understand how pilots' eye behavior and electrodermal activity change over the course of instrument flight session.

Design and Ethics

This study employed a quasi-experimental one-group time series design to explore pilots' eye behavior and electrodermal activity during flight. This research was approved by the Institutional Review Board (IRB# 32091) to satisfy legal and ethical requirements.

Results & Discussion

Electrodermal Activity

As the sympathetic nervous system governs sweat glands on the skin, this study used a GSR device to monitor the electrodermal activity of pilots during flight. Sweat gland activity is directly proportional to skin conductance response (SCR). Analyzing the phasic component of SCR revealed GSR peaks during flight segments. This study utilized GSR peak count and peak rate as the primary metrics to discover emotional states and demanding tasks during the instrument flight session. The GSR peak count revealed the degree of electrodermal activity in total, while the peak rate indicated the extent of emotional arousal and task demands that occurred every 60 seconds on average. The findings presented in Table 1 suggested flight segments that were emotionally arousing and demanding for pilots. Participants were emotionally aroused, and their task demands were intensive during takeoff and climb, approach and landing-1, missed approach, and approach and landing-2.

Table 1.

Summary of Galvanic Skin Response (GSR) Metrics

Flight Segment	GSR Peak Count	GSR Peak Rate
Take Off & Climb	27.71	5.04
En Route	11.82	5.32
Arrival	9.12	3.88
Approach & Landing-1	26.29	4.69
Missed Approach	21.65	4.94
En Route-2	13.41	3.41
Approach & Landing-2	22	3.16

Eye Behavior

Blinking is the physiological process recognized by rapidly closing the eyelids. As the semi-autonomic nervous system regulates blinking, this process offers psychological insights into how pilots maintain visual attention during the instrument flight session. This study utilized blink count and blink rate as the primary metrics to discover demanding flight tasks. Blink count demonstrated the degree of task demands during flight, while the blink rate showed the extent of task demands that occurred every 60 seconds on average. Based on the findings presented in Table 2, participants experienced takeoff and climb, approach and landing-1, missed approach, en route-2, and approach and landing-2 segments as having more demanding flight tasks during the instrument flight session.

Table 2.

Summary of Eye Blink Metrics

Flight Segment	Blink Count	Blink Rate
Take Off & Climb	92.82	17.86

En Route	39.71	20.51
Arrival	32.82	19.99
Approach & Landing-1	140.71	20.98
Missed Approach	104.12	20.93
En Route-2	90.59	21.85
Approach & Landing-2	126.88	20.45

The data analyses for this study involved gaze mapping process, where the eye-tracking data was mapped onto a dynamic experimental stimulus presented to the participants. This allowed aggregating the gaze mapping data on certain areas of interest (AOI) like primary flight display (PFD), multi-function display (MFD), interface (control yoke and parking brake), and stimulus (CCTV monitors of AATD). This algorithm enabled us to generate heat maps, which identified what areas captured the most visual attention of pilots during a particular flight segment. This study utilized time to first fixation (TTFF), time spent on a given AOI, percentage to notice, and saccade velocity as primary metrics to gain insights into how pilots maintain visual attention during the instrument flight session.

Time to first fixation (TTFF) indicated the average time lapsed until the first fixation was detected inside the AOI, which was measured in seconds. Time spent indicated the average of how long the pilots fixated at the AOI, which was measured in seconds. Percentage to notice represented the average of how long the pilots fixated at the AOI with respect to the time during the active AOI, which was expressed in percentage (%). Saccade represented the rapid transition of eyeball movements from one fixation to the next. Saccade velocity specified the average peak velocity of all saccades detected inside the AOI, which was measured in degrees per second (deg/s).

Table 3.
Visual Attention During Flight Segments

Flight Segment	AOI	TTFF *(s)*	Time Spent *(s)*	% to Notice	Saccade Velocity *(deg/s)*
Take Off & Climb	Interface	194.24	1.31	0.39	79.64
	MFD	80.68	1.29	0.4	105.42
	PFD	41.78	13.11	4.26	81.36
	Stimulus	133.39	188.05	58.57	138.9
En Route	Interface	74.47	0.83	1.52	81.8
	MFD	64.43	0.49	0.35	102.42
	PFD	29.69	7.27	6.84	77.52
	Stimulus	1.05	77.67	59.49	145.57
Arrival	Interface	11.19	0.71	0.52	48.26
	MFD	60.44	0.72	0.74	101.94
	PFD	43.09	4.6	4.85	93.99
	Stimulus	3.93	79.97	56.89	134.66

Approach & Landing-1	Interface	89.67	0.9	0.33	68.95
	MFD	66.77	1.27	0.31	94.05
	PFD	127.4	7.95	2.84	112.88
	Stimulus	7.83	194.14	58.32	135.5
Missed Approach	Interface	154.49	0.93	0.32	94.02
	MFD	62.78	0.9	0.34	99.76
	PFD	64.49	108.49	3.75	144.59
	Stimulus	0.60	166.11	59.5	136.35
En Route-2	Interface	156.09	1.33	1.57	92.81
	MFD	84.95	0.46	0.27	96.44
	PFD	30.49	8.38	4.65	88.73
	Stimulus	4.41	111.4	60.34	129.33
Approach & Landing-2	Interface	131.38	3.96	1.05	53.56
	MFD	103.92	0.45	0.17	101.04
	PFD	36.7	9.67	5.15	80.2
	Stimulus	26.79	119.53	54.32	135.26

Based on the data presented in Table 3, participants were mostly fixated on the primary flight display (PFD) and experimental stimulus (CCTV monitors of AATD) throughout the flight session. Pilots spent 108.49 seconds on average observing critical data presented on the primary flight display during the missed approach. This finding revealed the most visually taxing flight segment was the missed approach during the instrument flight session. Saccade velocity indicated how pilots shifted their attention from one fixation to another. The average saccade velocity during the missed approach was 144.59 deg/s, which indicated the extent of how pilots shifted their attention between different visual cues and met their task demands during the transition to missed approach.

Figure 1. Heat Map of Approach & Landing-2 Segment

Heat maps represent clusters of eye-tracking data and are a visualization tool to understand how pilots maintain visual attention during the instrument flight session. Figure 1 specified which areas of experimental stimulus captured visual attention during the approach and landing-2 flight segment.

Conclusion

The present study sought to understand how attentive and emotional the pilots were while executing tasks during the instrument flight session. The intensity of pilots' emotional arousal was higher when they met task demands during takeoff and climb, approach and landing-1, missed approach, and approach and landing-2. Pilots felt visually demanding during flight segments like takeoff and climb, approach and landing-1, missed approach, en route-2, and approach and landing-2. Missed approach was the most visually taxing flight segment, where pilots rapidly shifted their attention from one fixation to another on their primary flight display (PFD). These findings were limited in terms of generalizability to a broader population, convenient sampling strategy, and sampling error. The results of this study provide insights into the relationships between task management, attention, and mental workload. Future research will expand to include Electroencephalography (EEG) into the current mix of psychophysiological measures and precisely evaluate mental workload during flight.

References

Allsop, J., & Gray, R. (2014). Flying under pressure: Effects of anxiety on attention and gaze behavior in aviation. *Journal of Applied Research in Memory and Cognition, 3*(2), 63-71.

Belt, S., Gai, Y., Gururajan, S., Tamilselvan, G., & Bollock, N. K. (2021, July). Exploring Pilot Workload During Professional Pilot Primary Training and Development: A Feasibility Study. In *International Conference on Applied Human Factors and Ergonomics* (pp. 193-201). Springer, Cham.

Dahlstrom, N., & Nahlinder, S. (2009). Mental workload in aircraft and simulator during basic civil aviation training. *The International journal of aviation psychology, 19*(4), 309-325.

iMotions (2023). iMotions A/S (Version 9.3) [Computer software], Denmark.

Wang, X., Gong, G., Li, N., & Ding, L. (2020). Use of multimodal physiological signals to explore pilots' cognitive behaviour during flight strike task performance. *Medicine in Novel Technology and Devices, 5*, 100030.

Wickens, C. D. (2002). Situation awareness and workload in aviation. *Current directions in psychological science, 11*(4), 128-133.

Wilson, G. F. (2002). An analysis of mental workload in pilots during flight using multiple psychophysiological measures. *The International Journal of Aviation Psychology, 12*(1), 3-18.

CONCEPT OF AN AUTOMATED ACTIVITY DETERMINATION IN THE TEMPORAL DOMAIN FOR ADAPTIVE PILOT ASSISTANCE

Karl Tschurtschenthaler & Axel Schulte
Institute of Flight Systems, University of the Bundeswehr
Munich, Bavaria, Germany

In this contribution, we describe our initial conceptual thoughts on how to determine pilot activity in real-time within a multitasking environment.
The presented concept extends our previous research on determining the pilot activity which analyses manual and gaze interactions by evidential reasoning. This approach resulted in a fragmented pattern of activities over time due to the high frequency of gaze shifts. Our concept suggests concatenating the activities and using the resulting sequence as a feature set for classification. We hypothesize that this representations of activities reflects the complex nature of concurrent and serial multitasking more appropriately. For probabilistic inference, we aim to use Conditional Random Fields.
We are currently developing an activity recognition prototype supported by pilot interaction datasets from experiments. Our application is the management of a team of unmanned vehicles guided from the cockpit of a fast jet. We aim to use activity determination for adaptive assistance purposes.

Manned-Unmanned Teaming (MUM-T) describes the cooperation of manned and unmanned aircrafts, known as Unmanned Aerial Vehicles (UAVs), in a joint air operation. In these missions, the pilot of the manned air vehicle delegates the unmanned platforms from the aircraft cockpit to pursue mission-relevant tasks. Hence, monitoring unmanned systems is an important task in MUM-T missions. As a result, the pilots' mission performance is heavily dependent on multitasking (Trent & Barron, 2021).

Human performance issues also play a significant role when teaming with intelligent artificial agents. In this context, loss of situational awareness (Chen & Barnes, 2014a) or increased workload situations (Gaydos & Curry, 2014) are frequently reported issues. Our research tackles such human performance issues through adaptive assistant functions. To achieve this, we aim to develop automation that works cooperatively with humans. Adaptive assistant systems shall be able to adapt to the user in mentally demanding multitasking situations. De Visser & Parasuraman (2011) showed that such assistant systems can achieve a reduction of workload at high task loads. Here, adaptation is based on the estimation of certain users' mental states.

Mental state estimation forms the basis for adaptive assistance for many published approaches. However, it is known that these systems also must be task-sensitive (Fuchs et al., 2006). Therefore, incorporating the task context is required to trigger appropriate assistance in the form of intervention. For example, Dorneich et al. (2012) show that user-adaptive interruptions can only be effectively designed if they rely on the operator's task context. Such systems are also referred to as *Intelligent Adaptive Assistance Systems* (Besginow et al., 2018). Adaptive assistant systems that support teams of human operators in task allocation (e.g., cross-team task allocation) must consider the task context when making assistance decisions (Feigh & Pritchett, 2014). Thus, determining the task context is an essential subcomponent of assistant systems with adequate intervention strategies. Furthermore, Schulte et al. (2016) propose some requirements for an activity determination which serves adaptive assistance purposes: An activity determination must be *continuous*, *non-intrusive*, and *context-rich*.

The Task Context in MUM-T Scenarios

Human supervisory control (HSC) in MUM-T missions is strongly characterized by multitasking. However, the majority of monitoring tasks are performed purely visually, and demand only gaze input. According to the multitasking theory, this cannot be considered concurrent multitasking (CM) (MacPherson, 2018). Thus, HSC task execution is primarily defined by serial multitasking (SM). Similarly, SM in this context can be explained by the limited availability of resources for the parallel execution of tasks (Fischer & Plessow, 2015) or the dual-task paradigm (Reissland & Manzey, 2016). The latter refers to dual-task interference due to the similarity of information processing resources. However, Trent & Barron (2021) argue for their model of remotely piloted aircraft that the working memory of the pilots is decisive for the multitasking context. They claim that HSC tasks in their area occur in parallel.

Regardless, HSC tasks in the use of unmanned systems are always described by fast task switches. These are characterized by the user's attention shifts from one task to another. We have observed such behavior, particularly in fighter pilots in MUM-T missions (Schwerd & Schulte, n.d.). Their behavioral patterns involve routines to maintain situational awareness continuously. Task switching causes the user to interrupt their current task and continue to the next. It is commonly known that this attention shift leads to task-switching costs. These costs result in lower situation awareness, slower response times, and lower user performance (Chen & Barnes, 2014b).

Activity Determination to assess the multitasking context

Previous work on Activity Determination

The presented work is motivated by the research of Honecker and Schulte (2017) on activity determination of pilots. In their approach, observations (gaze and touch interactions in the cockpit) are interpreted as conditional evidence by use of probabilistic reasoning and Dempster-Shafer Theory. Continuous evidential reasoning (to infer the performed pilot tasks) is applied to the observed pilot interaction data. For the modeling of tasks, they use a fine-granular task model (Honecker et al., 2016). (Honecker & Schulte, 2019) evaluated this modeling approach in full-mission human-in-the-loop experiments. Their results show a highly fragmented pattern of recognized tasks over time. To overcome this, they experimented with a-posteriori low-pass filtering. They concluded that this reflected the task context more appropriately but ignores the richness of the contextual information it contains.

Related work on probabilistic activity recognition can be found in pattern recognition. Here, recognition is accomplished by structured prediction. Kim et al.

(2010) report challenges (e.g., ambiguity of observations or interleaved activities) that are similar to our considerations for observing activities in a multitasking environment. For this they propose methods like Conditional Random Fields (CRFs). CRFs have gained popularity in the field of activity recognition because they offer advantages over methods like Hidden Markov Models (HMM). They can be used to model dependencies between features (Vail et al., 2007). An activity recognition benefits from this, since observed pilot actions are executed sequentially. Furthermore, CRFs show higher performance in classification (Liao et al., 2007). This is primarily achieved through inference of the underlying discriminative classification model. Here, the joint probability is determined directly using the *argmax* $P(Y|X)$.

Concept of an Activity Recognition to assess the multitasking context

Our concept aims to provide a more efficient basis for the decision-making of an adaptive assistance agent. Since activity recognition provides a user's task context to an adaptive assistance system, it is imperative that our recognition must deal with uncertainty. Therefore, it must be based on probabilistic reasoning. As a starting point, we want to extend the work of Honecker and Schulte (2019) and directly address the fragmentation of the recognition results.

Fragmentation is most likely caused by neglecting dependencies between recognized tasks over time. In our previous approach, the tasks are semantically independent. However, this assumption is not consistent with the multi-tasking context: tasks are always performed in the context of a high-level task. In addition, recurring tasks are *re-visited* repeatedly. This conclusion means for an activity determination:

1. For an external observer, observable pilot actions have semantic relations to each other. Capturing the structure/pattern of the recognized activities must be part of an activity determination.
2. Pilots perform tasks in the context of higher-level tasks. This makes the probabilistic activity recognition a classification problem. Hereby each task can be interpreted as a state. A classifier estimates in which state a pilot is currently active in. Regarding a probabilistic reasoning during recognition, we see the maximum probability as the current attention of the pilot.

To meet these requirements, we want to deduct the task relations from a *hierarchical task model* and transform them into a *hierarchical state model*. Thus, a real-time activity determination automatically labels observed pilot actions as states. The states assigned with the highest probabilities represent the activities with the most attention of the pilot. Thus, a hierarchical task model serves as the central data structure of our model for activity recognition. We think that following aspects can be incorporated with this approach:

1. We consider this approach as a context-rich activity recognition since tasks are determined in context to higher-level tasks. This information is crucial for adaptive assistance that works adaptively to pilots' activity. Furthermore, we hypothesize that it is a more accurate representation of activities that occur in the pilot's working memory.

2. We also see the consideration of task relations as a solution for dealing with ambiguous observations (e.g., looking at a moving unmanned vehicle). These are common observations taken from semantic gaze tracking. They can only be interpreted in the context of other related tasks determined over time.
3. By estimating the higher-level activity states, we want to capture task switching. We see this as a key aspect of properly describing monitoring tasks in HSC that are executed in parallel. By detecting attention shifts on higher-level tasks (change in the maximum states probability over time), we aim to detect task switching.

For inference of the activity states, we intent to use a Linear-Chain-CRF. As already mentioned, it is a reported classification method which addresses the sequential nature of observed data.

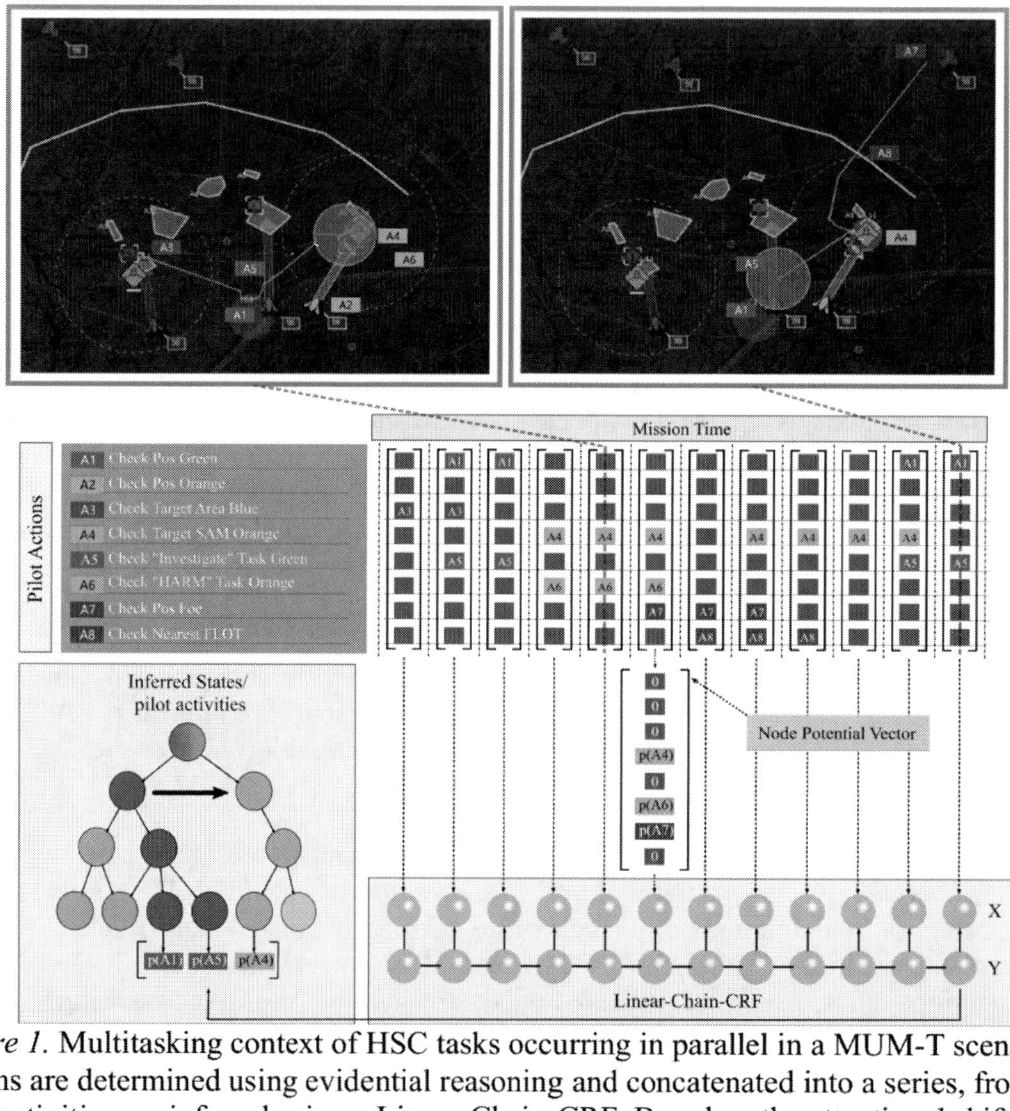

Figure 1. Multitasking context of HSC tasks occurring in parallel in a MUM-T scenario. The actions are determined using evidential reasoning and concatenated into a series, from which pilot activities are inferred using a Linear-Chain-CRF. Based on the attentional shifts in the inferred hierarchical state model, task switches can be detected over time.

Application of our Conceptual Model on a MUM-T Multi-Tasking Situation

Figure 1 illustrates our approach in a MUM-T multitasking context. It shows a task switch from one monitoring task to another (visual scan-path in blue). The pilot tries to monitor the progress of the delegated UAV tasks (arrow indicating the task to be processed, top left of Figure 1). Shortly after, an enemy appears. Then the pilot visually checks the threat (top right of Figure 1). Figure 1 (below) shows how task states can be inferred probabilistically from the observed actions of the pilots. These are determined by evidential reasoning. We want to concatenate the actions and use the resulting series as observations for a linear chain CRF. To do this, we need to convert the actions into a set of feature vectors and pass them to the CRF nodes. After inference, we extract the final activity state distribution and transfer it to the task model graph. By detecting a shift of attention from one branch to another in the activity state model, we aim to capture task switches.

Current and Future Objectives

We are currently working on a prototype that classifies activities based on CRFs and a hierarchical state model. Further development will use interaction datasets from experiments. We expect to gain further understanding of some modeling aspects for our application, such as the length of the linear-chain CRF or the determination of the feature vectors for classification. Furthermore, we intend to collect pilot interaction datasets from experiments for training.

Our goal is to integrate real-time activity recognition into a fast jet research cockpit simulator for MUM-T flight missions. The recognition will be evaluated in human-in-the-loop experimentation with fighter pilots for adaptive assistance purposes.

References

Besginow, A., Büttner, S., & Röcker, C. (2018). Intelligent Adaptive Assistance Systems in an Industrial Context - Overview of Use Cases and Features. *Lecture Notes in Informatics (LNI), Proceedings - Series of the Gesellschaft Fur Informatik (GI), September 2018*, 853–862. https://doi.org/10.18420/muc2018-ws18-0533

Chen, J. Y. C., & Barnes, M. J. (2014a). Human - Agent teaming for multirobot control: A review of human factors issues. *IEEE Transactions on Human-Machine Systems*, *44*(1), 13–29. https://doi.org/10.1109/THMS.2013.2293535

Chen, J. Y. C., & Barnes, M. J. (2014b). Human - Agent teaming for multirobot control: A review of human factors issues. *IEEE Transactions on Human-Machine Systems*, *44*(1), 13–29. https://doi.org/10.1109/THMS.2013.2293535

De Visser, E., & Parasuraman, R. (2011). Adaptive Aiding of Human-Robot Teaming:Effects of Imperfect Automation on Performance, Trust, and Workload. *Journal of Cognitive Engineering and Decision Making*, *5*(2), 209–231. https://doi.org/10.1177/1555343411410160

Dorneich, M. C., Ververs, P. M., Mathan, S., Whitlow, S., & Hayes, C. C. (2012). Considering Etiquette in the Design of an Adaptive System. *Journal of Cognitive Engineering and Decision Making*, *6*(2), 243–265. https://doi.org/10.1177/1555343412441001/ASSET/IMAGES/LARGE/10.1177_15553434 12441001-FIG2.JPEG

Feigh, K. M., & Pritchett, A. R. (2014). Requirements for effective function allocation: A critical review. *Journal of Cognitive Engineering and Decision Making*, *8*(1), 23–32. https://doi.org/10.1177/1555343413490945

Fischer, R., & Plessow, F. (2015). Efficient multitasking: Parallel versus serial processing of multiple tasks. *Frontiers in Psychology*, *6*(September), 1–11. https://doi.org/10.3389/fpsyg.2015.01366

Fuchs, S., Hale, K. S., Stanney, K. M., Berka, C., Levendowski, D., & Juhnke, J. (2006). Physiological Sensors Cannot Effectively Drive System Mitigation Alone. *Foundations of Augmented Cognition (2nd Ed.), January*, 193–200.

Gaydos, S. J., & Curry, I. P. (2014). Manned-unmanned teaming: Expanding the envelope of UAS operational employment. *Aviation Space and Environmental Medicine*, *85*(12), 1231–1232. https://doi.org/10.3357/ASEM.4164.2014

Honecker, F., Brand, Y., & Schulte, A. (2016). A Task-centered Approach for Workload-adaptive Pilot Associate Systems. *The 32nd Conference of the European Association for Aviation Psychology (EAAP)*.

Honecker, F., & Schulte, A. (2017). Automated online determination of pilot activity under uncertainty by using evidential reasoning. *Lecture Notes in Computer Science (Including Subseries Lecture Notes in Artificial Intelligence and Lecture Notes in Bioinformatics)*, *10276 LNAI*, 231–250. https://doi.org/10.1007/978-3-319-58475-1_18

Honecker, F., & Schulte, A. (2019). Full-mission human-in-the-loop experiments to evaluate an automatic activity determination system for adaptive automation. *Advances in Intelligent Systems and Computing*, *903*, 731–737. https://doi.org/10.1007/978-3-030-11051-2_111

Kim, E., Helal, S., & Cook, D. (2010). Human activity recognition and pattern discovery. *IEEE Pervasive Computing*, *9*(1), 48–53. https://doi.org/10.1109/MPRV.2010.7

Liao, L., Fox, D., & Kautz, H. (2007). Hierarchical conditional random fields for GPS-based activity recognition. *Springer Tracts in Advanced Robotics*, *28*. https://doi.org/10.1007/978-3-540-48113-3_41

MacPherson, S. E. (2018). Definition: Dual-tasking and multitasking. *Cortex*, *106*, 313–314. https://doi.org/10.1016/j.cortex.2018.06.009

Reissland, J., & Manzey, D. (2016). Serial or overlapping processing in multitasking as individual preference: Effects of stimulus preview on task switching and concurrent dual-task performance. *Acta Psychologica*, *168*, 27–40. https://doi.org/10.1016/j.actpsy.2016.04.010

Schulte, A., Donath, D., & Honecker, F. (2016). Human-System Interaction Analysis for Military Pilot Activity and Mental Workload Determination. *Proceedings - 2015 IEEE International Conference on Systems, Man, and Cybernetics, SMC 2015*, 1375–1380. https://doi.org/10.1109/SMC.2015.244

Schwerd, S., & Schulte, A. (n.d.). *Experimental Assessment of Fixation-Based Attention Measurement in an Aircraft Cockpit*.

Trent, J. D., & Barron, L. G. (2021). Multitasking as a predictor of simulated unmanned aircraft mission performance: Incremental validity beyond cognitive ability. *Military Psychology*, *33*(3), 128–135. https://doi.org/10.1080/08995605.2021.1897450

Vail, D. L., Veloso, M. M., & Lafferty, J. D. (2007). Conditional random fields for activity recognition. *Proceedings of the International Conference on Autonomous Agents*, 1331–1338. https://doi.org/10.1145/1329125.1329409

RECONSTRUCTION OF CREW'S BEHAVIOURS USING COCKPIT IMAGES AND THE SUGGESTION OF THE DEVIATION FROM STANDARD PROCEDURE

USAMI Kenji, TSUDA Hiroka and FUNABIKI Kohei
Japan Aerospace Exploration Agency
Mitaka, Tokyo, JAPAN

The images from the cameras installed in the cockpit are useful for Flight Operational Quality Assurance (FOQA), the investigation of an inccident and the validation the design of the cockpit and a pilot's procedure through the certification process of the aircraft. JAXA has developed a new tool for the reconstruction of crew's behaviour using the images recorded by the cameras in the cockpit under the machine learning. This paper reports that the accuracy of the estimation of the behaviour is improved with a novel function.

The video images recorded in the cockpit can display the pilots' behaviour including the status of the instruments of the aircraft. Hence the images are very useful for the investigation of the aircraft incident, FOQA, and the validation of the design of the cockpit and a pilot's procedure through the certification process. On the other hand, it is inefficient to take much time to review those images and to write out the all of time history of the pilots' behaviour and the aircraft status. Our motivation is to develop a new effective tool to estimate and reconstruct the sequence of pilots' actions in chronological order automatically. JAXA has tried to develop the new tool by applying the machine learning algorithms to estimate which procedure is conducted.

Concept Design and Objective

The functional structure of the proposed system is described in Figure 1. Pilots' body motion is captured by a camera and recorded in the cockpit. The authors selected to use a 2D camera and to extract the skeleton by using OpenPose (Zhe et al., 2017) before the process of behavioral analysis (Tsuda et al, 2020). The 2D position data of joints of the subject's body skeleton with the label is input to machine learning and the estimator is generated. Here the label is the kind of the subject's actions. When the new position data of joints are input to this estimator, the estimator outputs the subject's actions as the results of its estimation.

Although the tool with real-time analysis will be needed in the future, the one with post-flight analysis is considered appropriate at this stage of this study. The nowadays objective is determined to develop a function to identify whether the pilots' procedure is conducted or not.

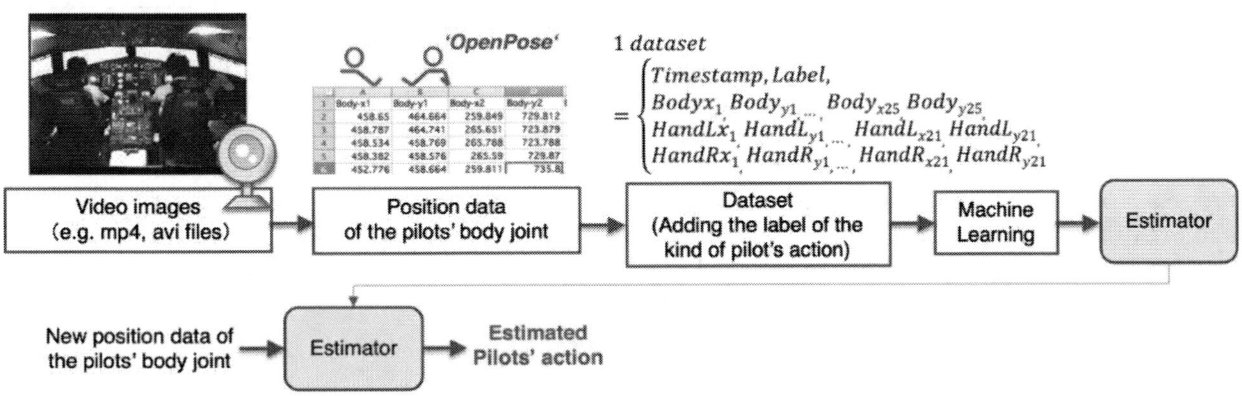

Figure 1.
Functional structure of the proposed system.

Procedure

Generating Datasets for Machine Learning

Simulating and recording pilots' behaviours. Some movies are taken in the cockpit of the simulator of Mitsubishi Aircraft Corporation (MITAC[1]). Two subjects sat on the seats as a pilot and a copilot, and conducted the following actions as part of Take Off (T/O) procedure:

1) Set the thrust lever to the advance position,
2) Push the Take Off and Go Around (TOGA) button and set the thrust lever to TOGA position,
3) Set the Landing Gear (LDGR) control lever to up position, and
4) Set Flap Slat control lever to the stowed position.

Addition to the normal T/O procedure, the simulated flights were conducted by intentionally omitting one of the above steps. One camera was utilized in this analysis, positioned to take from the right rear of the subject. The resolution of the camera was 3840 x 2160 pixels and its frequency was 60 Hz.

Extracting skeleton and labeling the procedures. The movies were analyzed by OpenPose to extract a series of datasets, the 2D positions of joints of the body. The appropriate labels of pilot actions were attached to the datasets manually. The labels used in this report are listed in Table 2. The labels were attached to the frames which showed the actual pilot action. For example, the frame which contained the action of the pilot to push the thrust lever from the idle position to the advance position was labeled as Thrust_advance_pos. On the other hand, the frame which contained the action to move pilot's hand close to the thrust lever or the action just to put pilot's hand on the lever was labeled as No_Status.

Table 2.
Type of Actions and Labels.

Type of Action	Label
Change Flap/Slat Control Lever position from 25 to 10.	FSControlLever_Up
Change Flap/Slat Control Lever position from 1 to 0.	FSControlLever_Up_Flap0
Change Flap/Slat Control Lever position from 10 to 1.	FSControlLever_Up_Flap1
Change Landing Gear Control Lever from down to up.	LDGR_up
Other than those.	No_Status
Change Thrust Lever position from idle to advance.	Thrust_advance_pos
Change Thrust Lever position from advance to TOGA.	TOGA_Push

Machine Learning Process

Datasets constucted from frames labeled by pilot actions were used as input to machine learning process. For analysis, scikit-learn (Pedregosa et al., 2011) was utilized as library for random forest classifier in this report. The analysis was performed in two methods. One method was to split one dataset into test data and training data in a ratio of 8:2. Another way was that the different datasets were used for training data and test data.

---

[1] MITAC changed its company name to MSJ Asset Management Company on April 25, 2023.

Experiments and Results

We conducted three types of experiments. The first is an attempt to check the difference of estimation's accuracies caused by the difference of training data. The second is to verify whether the pilots' action can be classified using training data prepared separately from test data. The third and final, the extent of detection of skipped actions by analyzing data was investigated.

Estimated Accuracy

Test data and training data. Simulated flights in the simulator were conducted 20 times, including normal T/O procedures and T/O procedures with some pilot actions skipped intentionally. Table 3 lists the recorded contents and the number of frames in movies as the results of simulated flights. The objects were two pilots (called as A and B symbolically). The pilot and copilot were switched according to "Datasets No." described in Table 3. The skelton data was extracted from the all datasets listed in Table 3 by OpenPose.

Table 3.

The Contents and the Numbers of Data Frames of Each Datasets Obtained by Simulated Flights.

Datasets No.	Contents	Number of data frames	Pilot	Copilot
1	Normal take off procedure	10029	A	B
2	Take off procedure without Flap/Slat Control Lever position	11936	A	B
3	Take off procedure without Thrust Lever position change from idle to advance	10336	A	B
4	Take off procedure without Thrust Lever position change from advance to TOGA	10817	A	B
5	Take off procedure without Landing Gear Control Lever from down to up	10563	A	B
6	Normal take off procedure	10291	A	B
7	Same as above	10094	A	B
8	Same as above	10727	A	B
9	Same as above	10052	A	B
10	Same as above	9856	A	B
11	Normal take off procedure	11183	B	A
12	Take off procedure without Flap/Slat Control Lever position	11674	B	A
13	Take off procedure without Thrust Lever position change from idle to advance	10513	B	A
14	Take off procedure without Thrust Lever position change from advance to TOGA	10292	B	A
15	Take off procedure without Landing Gear Control Lever from down to up	11835	B	A
16	Normal take off procedure	10223	B	A
17	Same as above	10186	B	A

18	Same as above	10171	B	A
19	Same as above	10217	B	A
20	Same as above	10338	B	A

Accuracy estimated by random forest classifier. A total of 20 datasets were utilized for input to the random forest classifier process. The three combinations of test data and training data were analyzed by random forest classifier. The combinations and estimated accuracies were shown in Table 4. In this report, the labels listed in Table 2, excluding the No_Status, were tried to be estimated and reconstructed as the pilot and copilot actions. Therefore, in addition to the normal accuracies, the accuracies estimated excluding the No_Status label (refer to Table 2) was estimated.

Table 4 showed that the accuracy was less than half of other cases when test data and training data were different datasets and accuracy was estimated excluding the No_Status label. There was not much difference in accuracy between the accuracy estimated when 9 training data was used and that estimated when 19 data was used. It means that it might not improve the accuracy if additional training data is prepared.

Table 4.

Accuracy Estimated by Random Forest Classifier.

No.	Combination of Datasets for Test Data and Training Data (For dataset, refer to Table 3)	Average Accuracy (%)	Accuracy Estimated Excluding the No_Status Label (Refer to Table 2) (%)
1	Test data: 20 % of dataset No. 1 Training data: Remaining 80 % of dataset No. 1.	99.40	89.13
2	Test data: Dataset No. 1. Training data: Datasets No. from 2 to 10.	96.30	42.45
3	Test data: Dataset No. 1. Training data: Datasets No. from 2 to 20.	95.47	42.01

Detail investigation of predicted probabilities of each label. Although the estimated accuracy was 42.01 %, authors tried to detect the pilot and copilot action by investigating the values of predicted probabilities obtained from the result of random forest classifier process using combination of datasets No. 3 listed in Table 4.

Figure 2 showed that the pilot/copilot actions and the predicted probabilities of the six labels, excluding No_Status. The horizontal axis showed the time (sec.). In this figure, authors confirmed that the peak timing of plots constructed from the probabilities of the six estimated labels matched with the timing of operations corresponding to each label. This indicates that even with low accuracy, we can estimate the pilots' behaviour by investigating the predicted probabilities of each label.

In Figure 2, in addition to these operation timing, the peak of Thrust_advance_pos was found around 166 seconds. This peak was corresponding to the hand position shown in Figure 3. Around 166 seconds, the pilot's right hand passed near the thrust lever. Other plots of FSControlLever_Up, FSControlLever_Up_Flap1 and FSControlLever_Up_Flap0 were also found around from 78 to 138 seconds. We confirmed that the copilot's left hand had been always on the Flap/Slat Control Lever around from 78 to 138 seconds. In normal T/O procedure, we can consider that the thrust lever position change to Thrust_advance_pos and Flap/Slat Control Lever position changes from 25 to 10, from 10 to 1, and from 1 to 0 were performed only once respectively. Based on this assumption, the timing of the the actual operation can be known from the timing at which the maximum values in the graphs of the predicted probabilities.

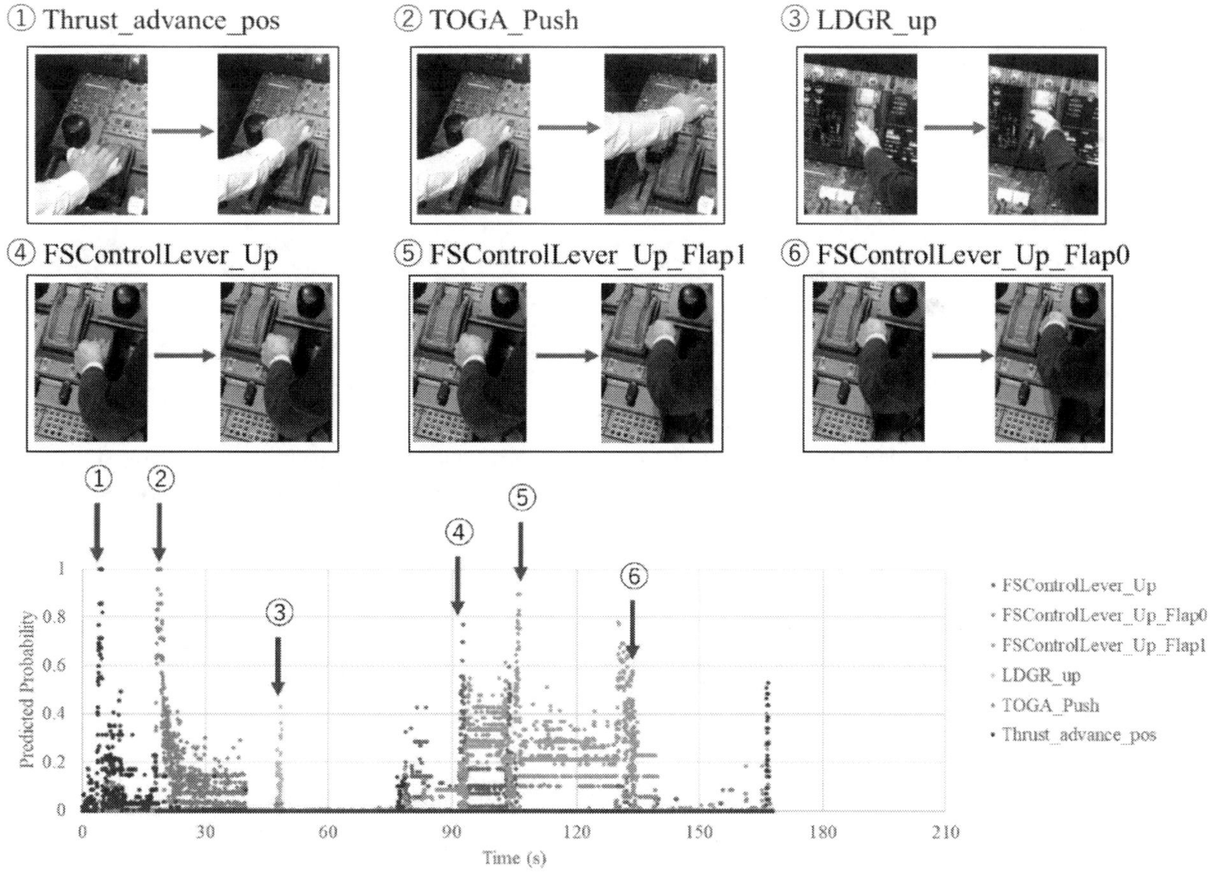

Figure 2.
The plots of the results by machine learning.

Figure 3.
The pilot's hand position at around 166 seconds.

Detection of skipped action by probabilities investigation. Additionaly, random forest classifier process were performed to confirm whether skipped pilot/copilot action can be detected. The input data was datasets No. 2 listed in Table 3 as test data and the all other 19 datasets as training data. The results of random forest classifier process was shown in Figure 4 and there were no peaks of FSControlLever_Up, FSControlLever_Up_Flap1 and FSControlLever_Up_Flap0. This result indicates that the skipped action of pilot and copilot can be found by investigation of predicted probabilities.

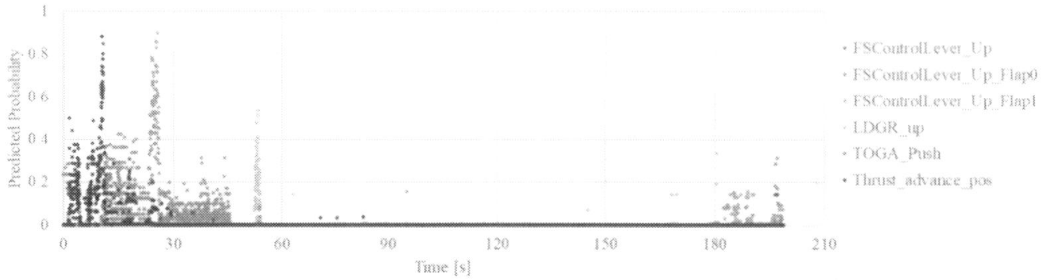

Figure 4.
The plots of the results by machine learning for skipped actions.

Conclusion

The accuracy estimated by random forest classifier process was less than half, when test data and training data were different datasets and accuracy was estimated excluding the No_Status label.

To detect actions of pilot and copilot from these test data, predicted probabilities were analyzed. We found that the timing at which the maximum values in the plots of predicted probabilities of each labeled action showed the timing of pilots' behaviour in normal procedure. We also found that the deviated action from normal procedure can be detected by investigation of predicted probabilitiy.

References

Pedregosa, F., Varoquaux, G., Michel, V., Thirion, B., Grisel, O., Prettenhofer, P., Weiss, R., Dubourg, V., Vanderplas, J., Passos, A., Cournapeau, D., Brucher, M., Perrot, M., and Duchesnay, E. (2011, Octorber 11), Scikit-learn: Machine Learning in *Python, Journal of Machine Learning Research*, Vol. 12, 2011, pp. 2825–2830.

Tsuda, H., Stroosma, O., & Mulder, M. (2020). Reconstruction of Pilot Behaviour from Cockpit Image Recorder. In *Proceeding of the AIAA SciTech 2020 Forum*. Orland, FL: doi: 10.2514/6.2020-1873.

Zhe, C., Tomas, S., Shih-En, W., & Yaser, S. (2017). Realtime Multi-Person 2D Pose Estimation using Part Affinity Fields. In *Proceeding of the IEEE Conference on Computer Vision and Pattern Recognition (CVPR)*, Honolulu, HI: pp.7291-7299.

ON THE APPROPRIATE PARTICIPANT EXPERTISE FOR DISPLAY EVALUATION STUDIES

M. M. (René) van Paassen, Clark Borst, Max Mulder and Gijs de Rooij
Aerospace Engineering – Delft University of Technology
Delft, The Netherlands
Ferdinand Dijkstra - LVNL Netherlands, Schiphol, The Netherlands
Adam Balint Tisza - EUROCONTROL, Maastricht, The Netherlands

Expert participants may not always be available for evaluation of new displays or support systems, and in some cases, it might be better to use novice participants, particularly when the display or support significantly changes existing work practices. To provide tools and arguments for selecting the expertise level of participants, we propose the use of Rasmussen's decision ladder to analyze where and how a new visualization or a support tool changes the task, and identify steps where a novice participant may learn to perform the task to an acceptable level. A comparison to the support with the current operational interfaces then shows where an expert might have difficulty in stepping away from learned practice. This analysis is applied to the domain of air traffic control, and a selected set of relevant past research with both expert and novice participants is reviewed, revisiting the decision for a participant level in the study.

Introduction

New designs for interfaces, or new support tools, are commonly tested in controlled experiments with human participants. Ideally, the existing version of the interface or support system is tested against the new development in a study that replicates daily operations to such an extent that differences in performance, and expert opinion, indicate whether the new implementation is more efficient and safe. A properly performed evaluation should be indicative of effects in practice, or in other words have external validity (Libby et al., 2002).

Expert participants may not always be available for these interface evaluation studies, and, often out of need, non-expert participants are invited. This may affect the validity of an experiment, primarily by changing the capability of an experiment to correctly assess the effect of manipulations in its experimental conditions, i.e., its internal validity (Libby et al., 2002). But given that there is sometimes no opportunity to use expert participants, and particularly in testing early prototypes, one would prefer to not use scarce opportunities for access to experts, there is often a need to invite and train novices for evaluation.

Expertise by itself is difficult to define; Gobet (2015) defines it in terms of performance with respect to others. Chase and Simon (1973) argue that expert chess players have a vast memory for structures in chess, and can therefore better code and remember chess positions, resulting in improved chunking, indicating how long-term memory and trained perception play a large role in expertise. Given that training for many expert level jobs takes several years, and developing senior expertise in most positions requires at least a decade, it can be argued that any training of a novice or relatively inexperienced participant before experimental evaluation sessions can never approximate true expertise. On the other hand, commonly the number of different conditions presented in an experiment are limited, and only a fraction of an expert's vast store of knowledge will be needed to perform the task. Given due care, it should in many cases be possible to use results obtained with novice participants, properly trained to perform the experiment tasks, to provide a realistic evaluation of a new display, or provide insight into how a (partial) task is approached and performed.

To provide a handle on judging the effect of using participants with an expertise level that differs from the intended end-users, we will apply cognitive task analysis with the "decision ladder" (Rasmussen, 1983; Rasmussen, 1986). The decision ladder model describes processes and knowledge stages in human information processing. It will be used here to assess the potential effect of a difference in expertise level on these different processing stages, and from there on to infer the effect on performance in an experiment. Using the distinction between the processes for different cognitive stages may support a systematic review of the effect of expertise; alternative cognitive models, such as IDA or ACT-R (Anderson et al., 1997; Ritter et al., 2019; Smidts et al., 1997) would offer the means to model and implement these same information stages and knowledge states, but the lack of distinction between these processes does not offer a systematic checklist for evaluating the effect of expertise.

In addition to discussing the decision ladder as a tool to evaluate the effect of using participants with different expertise levels, the paper gives a small overview of some past experiments or evaluations, and reviews these with the new approach.

Cognitive Task Analysis as guidance

Air Traffic Control (ATC) is responsible for the safety and efficiency of air traffic. Commonly, air traffic is monitored and directed through plan views reflecting radar screens, in which fused data from radar systems and on-board navigation systems is used to position aircraft symbols. When providing support with radar vectors, the Air Traffic Controllers (ATCos) provide speed, heading and altitude instructions to the aircraft under their control, both to solve possible conflicts and ensure an efficient and regular traffic flow in the sector. To provide support in these tasks, several interfaces and support systems are in use, and new ones are being developed and evaluated. The continuing shortage of ATCos, the considerable investments needed for selection and training and requirements on safety provide a need for innovation and development of support systems, at the same time there is a lack of available experts to properly test and evaluate these systems.

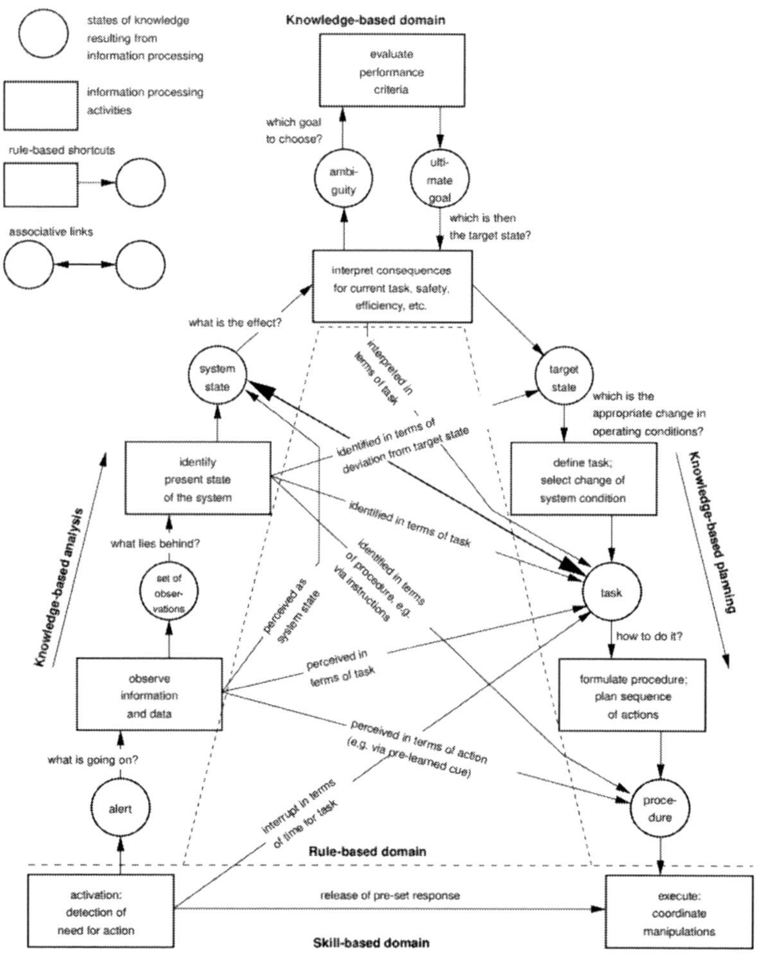

Figure 1: Decision ladder, showing perceptual and cognitive processes, and information stages. After (Rasmussen, 1983)

For the argument in this paper, we assume that a full or partial simulation is set up for the task (ATC tasks in this case), and any interfaces or support systems are tested in representative task scenarios. In the following, we will discuss which factors might affect experiment outcome when participants with an expertise level differing from the target users are invited.

As an example, the research by Somers et al. (2019) used an experiment with both expert ATCo and less experienced participants to generate the data for evaluating different ATC complexity metrics. This project used a simplified, hypothetical sector shape, with a simplified traffic sample and control of the traffic through a mouse-operated interface.

The "decision ladder" model by Rasmussen (Rasmussen, 1986) is used here as a template to consider the effect of participant expertise level on the outcome of experiments. This model is, as explained in (Vicente, 1999), fashioned after models for stages in cognitive processes, augmented with options for rule-based shortcuts that represent the repertoire of standard inferences and responses available to an operator or controller. The decision ladder model distinguishes a number of cognitive processes and their resulting knowledge states; using these distinctions, the potential effect of the level of expertise of experiment participants is discussed.

Consider the graphical representation of the decision ladder in Figure 1. The experience level of a participant might affect these processes in the following ways:

activation Activation provides the initiation of an activity. This starts at a skill-based level, with perception and pattern recognition, in our example with the identification of a need for action, e.g., with an aircraft entering the sector, an aircraft leaving, and thus requiring a hand-over, or the detection that aircraft might become involved in a conflict. Participants with less experience can be expected to have less efficient activation, leading to missing events, or inefficient detection, initializing activities when none are needed. Since currently in ATC the presentation of information is steady and clear, there will likely not be a large effect of experience, in contrast to situations where raw radar images need to be interpreted, or where unlabeled objects would have to be monitored, e.g., in the case of a radar operator responsible for detecting incoming attacks (Klein, 1999).

observation A next step, after a cognitive task has started, is commonly to assemble necessary information. We can assume that if an experiment closely resembles an expert's working environment, the expert can more efficiently gather information, spending less time and effort in this state. On the other hand, if an experimental interface is used, or even a simple re-arrangement of the information has taken place between the working environment and the experiment, an expert's "advantage" quickly disappears.

identification This is the process of understanding and classifying the state of the system to be controlled. If the fidelity of the experimental environment is good enough, the routine and the repertoire or experience of the expert will greatly support this step. Experts will be able to see more nuances, and know more conditions in which the system can be. Applied to the case of ATC, an expert ATCo will likely better understand the flow patterns in the sector, and will be able to group flights, rather than work on a case-by-case basis. This implies that for experiments that require a complete, high-fidelity task, there will be a larger difference between expert and novice participants. On the other hand, interfaces that facilitate the identification of the process (e.g., by presenting the information at a higher level in an integrated fashion), might support novices better, because these will more readily accept and use the new support.

interpretation In this step, the state of the controlled system is checked against the desirable or goal state. Experiment participants with a higher level of expertise most likely have a better definition of their goal state; in a further analysis of Somers' experiment, de Jong and Borst, 2022, found that the expert participants created a regular structure for merging the traffic, where the less experienced participants used more direct-to instructions to the exit waypoint. Thus, the participant's interpretation of the goal state shapes the experimental results. If an experimental interface allows or even promotes a different approach to the work, this might also be the point where experts might raise most objections, most likely ignoring the support and persisting with learned approaches to the work, where novices accept the structure proposed by a support system. This might also indicate a case where more familiarization with the interface, and efforts to explain new support, might help "win over" experts.

evaluation and criteria In the diagram, this is presented as an optional pathway. It is a meta-cognitive phase, in which performance and goals are evaluated, and goals are adjusted if that is deemed necessary. In ATC, the decision to start using an approach stack, or the decision to divert flights when runways need to be closed, e.g., due to weather conditions, fall in this step. Such conditions are seldom investigated in evaluation experiments, and when considered, require the participation of experts, as the performance of non-experts on these tasks is likely to be significantly different.

task definition In this step, the activities needed to achieve a desired state are planned and/or formulated. For experts, these steps may be immediately clear, being associated and known solutions to recognized deviations. Participants with less expertise will require more effort in this stage. Providing support in task definition, for example by offering a menu of resolutions, or a menu of actions, might improve novice performance, but any mismatch between the implementation preferred by experts and offered by the interface might result in rejection of the support.

procedure formulation This will largely rely on experience by the operator. In most experiments, there is a focused, relatively simple task to be performed, and the procedure formulation has limited variation, and can be quickly trained. Again, this is an aspect of the work that might be seen as disruptive to an expert if it is different from what is used in practice.

execution In many cases, the execution step in an experiment differs from the one in actual practice, e.g., entering vectors with a command interface versus radio communication with pilots. Since, contrary to perception skills, execution skills are in many cases not critical in computer supported work (Vicente, 1999), differences between simulation environments used for experimental evaluation and practice will seldom affect the outcome.

In addition to effects of expertise on different cognitive steps, one can expect differences in rule-based behavior depending on the level of expertise. An expert might have better recognition of routine situations, and know more routine responses to handle these, leading to shunts (shortcuts to knowledge states), and leaps, known associations between recognized knowledge states (Vicente, 1999). When offering practice runs to novices, enough training should be given so that at least common situations in the experiment task can be handled in a rule-based, recognition-driven manner.

From reasoning with the decision ladder model, one can see that the effects of providing additional display support or decision support may in a number of cases be amplified when using novice participants. One should particularly expect larger differences when novices have difficulty with task aspects, and the interface or support system can provide clear distinctions (perception and interpretation stages) or structure (definition and execution stages) needed in the task. In the following, we will review a number of experimental evaluations and try to assess the effect of the expertise level of the participants on the outcomes of the comparison. Table 1 gives a summary of the assessed effects of expertise.

Application to past experiments

The experiments described in Somers et al. (2019) and De Jong and Borst (2022) were intended to provide subjective rating data for comparison against different candidate workload metrics calculated from the traffic state. In the experiments, traffic scenarios in simplified sectors were controlled with a menu-based interface. The modifications to the control input remove the need for radio phraseology, and remove any uncertainty in the execution. This simplifies task planning, and removes any differences due to expertise in execution. The absence of wind in the scenario made them more predictable, and thus reduced the requirements on interpretation of consequences, and removed the need to consider wind in observation. By normalizing the rating data, differences in expertise are largely eliminated. The experiment also consisted of regular traffic, with the exit waypoint shown for each of the aircraft in the sector, reducing planning and structuring effort. Data from both expert and non-expert participants could thus be used; the only difference in behavior was that experts produced a more regular traffic pattern, while non-experts would accept a slightly less regular pattern, with more aircraft on direct headings towards their exit points.

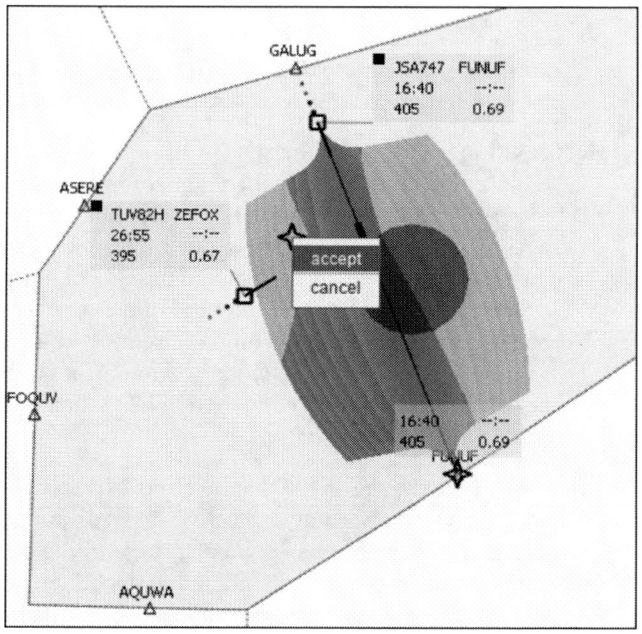

Figure 2: Screenshot of a 4D trajectory manipulation interface (R. E. Klomp et al., 2013).

In a study to investigate strategies of conflict evaluation in ATC, (Rantanen & Nunes, 2005), invited both expert and non-expert participants. The task focused on conflict classification only, with the presentation of only a conflict pair, essentially replacing activation by a fixed cue (the presentation of the next experiment condition) and replacing all execution steps by a press of a key on the keyboard. The effect of expertise on the further development of traffic pattern in the experiment is thus not an issue. The study showed a consistent effect of altitude differences on the required time to analyze conflicts for both the expert and non-expert groups, with the non-experts requiring somewhat more time, and displaying a larger variation when presented with "difficult" conflict geometry of aircraft at the same flight level.

Table 1: *Summary of consideration for the effect of expertise level on efficiency and behavior at different cognitive processing stages*

Stage	Non-expert	Expert
Detect/alert	Relies on basic features, less efficient, may miss or exhibit spurious false alert	Detect complex patterns, efficient
Observe	More laborious, possibly inefficient, have misses or serendipitous hits	Efficient, look only for needed information, serendipitous misses in low probability scenarios
Identify	More effort needed, coarse identification	Often recognized as pattern, sparse, little superfluous data needed
Interpret	Missing threats, or make mountains out of molehills	Easily evaluate goal state relation
Evaluate criteria	Likely invalid, not enough expertise	Valid only in high-fidelity scenario and simulation, difficult to elicit and interpret in an experiment
Define task	Produces single, simple tasks	May produce more complex tasks
Formulate procedures	Need to provide enough training, may require more time, may limit task formulation	Requires little effort if matching work situation
Execute	Need to provide enough training	Automatic if matching work situation
Shortcuts	Provide enough training to enable rule based shortcuts	Check that learned rules are applicable in the experiment set-up

While Somers' experiment used a largely conventional interface, the research by Klomp, and further developments in that field (R. Klomp et al., 2016; R. E. Klomp et al., 2013; ten Brink et al., 2019), focused on new 4D ATC concepts, see Fig. 2. The controller in these situations effectively produces or modifies four-dimensional planned trajectories, by adjusting a speed, height and lateral profile defined by waypoints and speed and altitude targets. Many facets of a current ATCo's expertise become less relevant; activation is supported by the automation, through highlighting of parts of the trajectory with a future loss of separation, and selection of one of the conflicting aircraft gives an overview of both the currently planned 4D trajectory and the options to modify this trajectory into a conflict-free one, largely supporting observation, identification, interpretation and task definition stages. Most of the work will be new, both to experts in the current system and non-experts. A probable advantage of experts will be the evaluation of the 4D trajectories against aircraft performance limits. Since so much of the task, interface and work instructions is new for these evaluations, initial evaluations can well be performed with non-experts in ATC, and any further evaluations with participants trained in current ATC practice will need ample introduction into the new practice and tools.

An interesting middle ground is found in the evaluation by Mercado Velasco et al. (2021), where a "solution space" display is added to support present-day tactical ATC. This display shows combined speed and heading solutions that are clear from surrounding traffic. Participants at all three expertise levels; novices, intermediate and expert air traffic controllers showed improvement when using the tool. Experts did use it in a different manner, formulating their own solutions, and then using the tool for confirmation, effectively ignoring the support of the tool for most of the cognitive processing stages. Novices and intermediate level participants relied more heavily on the tool, using it for guidance, and they would also select solutions indicated by the tool with smaller separation margins, solutions that the experts, using a more conservative approach to generating solutions, would not consider.

Conclusions

In many types of experimental evaluation or fundamental research, one may be forced to select non-expert participants. The use of the decision ladder model as a template for the cognitive steps in a task, provides a list of cognitive processes, each of which may be influenced by the expertise level of participants in a task. A systematic check should be used to analyze what the effect is of the participants' expertise level in performing each cognitive process, and whether the experiment environment matches or differs from the targeted operational environment at this level. In addition, the repertoire of trained rule-based behavior for both non-expert participants has to enable routine-based responses to a reasonable degree.

When comparing two or more interface variants or support systems, the participation of non-experts may lead to increased variation in performance, and also to an increased contrast, when participants strongly rely on support given in certain experimental conditions, where expert participants might be able to perform the task without support. Experimental set-ups may also be simplified in comparison to work situations, and care should be taken that the simplification does not unnecessarily constrain the response option of expert participants.

References

Anderson, J. R., Matessa, M., & Lebiere, C. (1997). ACT-r: A theory of higher level cognition and its relation to visual attention. *Human–Computer Interaction, 12*(4), 439–462.

Chase, W. G., & Simon, H. A. (1973). Perception in chess. *Cognitive Psychology, 4*(1), 55–81.

de Jong, T., & Borst, C. (2022). Determining air traffic controller proficiency: Identifying objective measures using clustering. *IFAC-PapersOnLine, 55*(29), 7–12.

Gobet, F. (2015). *Understanding expertise: A multi-disciplinary approach.* Palgrave Macmillan.

Klein, G. A. (1999). *Sources of power how people make decisions.* MIT Press.

Klomp, R., Borst, C., van Paassen, M. M., & Mulder, M. (2016). Expertise level, control strategies, and robustness in future air traffic control decision aiding. *IEEE Transactions on Human-Machine Systems, 46*(2), 255–266.

Klomp, R. E., Borst, C., Mulder, M., Praetorius, G., Mooij, M., & Nieuwenhuisen, D. (2013). Experimental evaluation of a joint cognitive system for 4d trajectory management.

Libby, R., Bloomfield, R., & Nelson, M. W. (2002). Experimental research in financial accounting. *Accounting, Organizations and Society, 27*(8), 775–810.

Mercado Velasco, G., Borst, C., van Paassen, M., & Mulder, M. (2021). Solution space decision support for reducing controller workload in route merging task. *Journal of Aircraft, 58*(1), 125–137.

Rantanen, E. M., & Nunes, A. (2005). Hierarchical conflict detection in air traffic control. *The International Journal of Aviation Psychology, 15*(4), 339–362.

Rasmussen, J. (1983). Skills, rules, and knowledge; signals, signs, and symbols, and other distinctions in human performance models. *IEEE Transactions on Systems, Man and Cybernetics, SMC-13*(3), 257–266.

Rasmussen, J. (1986). *Information processing and human-machine interaction : An approach to cognitive engineering.* North-Holland.

Ritter, F. E., Tehranchi, F., & Oury, J. D. (2019). ACT-r: A cognitive architecture for modeling cognition. *WIREs Cognitive Science, 10*(3).

Smidts, C., Shen, S., & Mosleh, A. (1997). The IDA cognitive model for the analysis of nuclear power plant operator response under accident conditions. part i: Problem solving and decision making model. *Reliability Engineering & System Safety, 55*(1), 51–71.

Somers, V., Borst, C., Mulder, M., & van Paassen, M. (2019). Evaluation of a {3d} solution space-based ATC workload metric. *IFAC-PapersOnLine, 52*(19), 151–156.

ten Brink, D. S. A., Klomp, R. E., Borst, C., van Paassen, M. M., & Mulder, M. (2019). Flow-based air traffic control: Human-machine interface for steering a path-planning algorithm. *2019 IEEE International Conference on Systems, Man and Cybernetics (SMC)*, 3186–3191.

Vicente, K. J. (1999). *Cognitive work analysis: Toward safe, productive, and healthy computer-based work.* Lawrence Erlbaum Associates.

CHALLENGES AND OPPORTUNITIES OF LEARNING AVIATION ENGLISH BY CHINESE PILOTS

Chi Hang Wong, Ion Juvina
Department of Psychology, Wright State University
Dayton, Ohio
Meredith Pitts
Department of English, Literature, and Modern Languages, Cedarville University
Cedarville, Ohio
Steve Chung
Guangzhou Civil Aviation College
Guangzhou, China
Jennifer Roberts, Andrew H. Schneider
College of Aviation, Embry-Riddle Aeronautical University
Daytona Beach, Florida

We explored the perspectives of Chinese pilots, a rapidly expanding sector in commercial aviation, on how they learn and use Aviation English (AE). A focus group with ten Chinese aviation professionals and in-depth semi-structured interviews with three Chinese commercial pilots were conducted to investigate their views of AE's pertinence from flight training to line operations. The findings indicate the reliance on rote learning in AE training and the importance of experiential learning for interacting with different AE varieties. The pilots highlighted increased workload when interacting with unfamiliar accents and country-specific phraseology; understanding written communication was also mentioned as a challenge. This research enriches the account of non-native-English-speaking pilots on how they achieve communicative competence and how current standards aid and affect their flight operations. We discuss the relevance of an immersive, peer-assisted, and technology-aided training strategy that would take advantage of the inherent knowledge and workload asymmetries in AE speakers.

In recent years, China's aviation industry has grown to become one of, if not the, largest aviation sectors in the world. In China, which has witnessed growth in commercial traffic possibly surpassing the volume thereof in North America, the need for training and licensing pilots *ab initio* and at home (Bieswanger et al., 2020) has expanded to meet the demands of China's traffic volume. While such demand is likely to rise due to overall sector growth, an imminent linguistic challenge may result from the declining rates of Chinese pilots receiving training and licensure in native English environments or simply in situations warranting cross-cultural learning.

Estival and Farris (2016) defined Aviation English (AE) as "a lingua franca and a variety of English" (p. 1) used in international aviation radiotelephonic communications, which rely on extensive prescribed exchange formats, vocabulary, syntax, and pronunciation (Tosqui-Lucks & Silva, 2020). Meanwhile, plain language remains inseparable from AE where standard phraseologies are not available, demanding high oral competence from AE users (Estival & Farris, 2016). To prevent communication errors and ensure the consistency and safety of radio communication, ICAO has worked for the past few decades toward a set of English Language Proficiency (aka. AE) standards, which were implemented by most Member States in 2008 (ICAO, 2010).

Aviation English in China

As a Member State, China adapted ICAO's language proficiency requirements (LPRs) into a test called Pilot English Proficiency Examination of the Civil Aviation Administration of China, commonly known as PEPEC. All Chinese pilots must pass the PEPEC to fly commercially.

Summarizing the Chinese literature on AE, Deng and Xiao (2013) found that researchers had focused mostly on communication pedagogy and test designs, with very little emphasis on needs analysis as well as cultural comparisons and communication contexts. Wu et al. (2012) studied the washback effect of PEPEC testing on AE instruction and observed that, although PEPEC promoted AE learning and improved pedagogical foci, both AE instructors and students agreed that oral (plain) English should be given greater emphasis inside and outside the classroom. Also, the lack of proficient (fluent) teachers may have hindered student pilots from learning English beyond the classroom setting. Similarly, Xia and Huang (2012) recommended greater emphasis on developent student pilots' vocabulary and communicative skills. A later study by Guo (2018) highlighted that despite receiving training in both general English and AE, using plain English in non-routine and unscripted operations "challenges Chinese student pilots to the utmost difficulties." In addition to language prodcution difficulties, the student pilots' comprehension skills are frequently put to the test as native-English-speaking (NES) controllers might speak at a rate (150+ words per minute) much higher than what the pilots are used to (100-120 wpm) (Guo, 2018). Meanwhile, challenges to effective AE communication may also come from NES operators who underutilize standard phraseology and use longer sentence structures, which are more likely to be confusing for non-native-English speakers (NNES) (Kim & Billington, 2018). Furthermore, the AE discourse context often occurs in a noisy audio-only environment, in which NNES are especially disadvantaged. Thus, Kim and Billington (2018) recommended comprehension training on a wider range of AE accents by both NES and NNES operators.

The Current Study

Although there is some research on the communication needs and challenges of NES pilots (Kay et al., 2021), little of this topic is known from the perspectives of Chinese (non-student) pilots (to authors' best knowledge). Given the frequency and criticality of AE in the international airspace (ICAO, 2010), this study aimed to improve our understanding of the experiences of Chinese operators to inform AE pedagogy and support operators conducting cross-linguistic interactions by qualitatively exploring the research questions of (1) whether AE training fits the needs of Chinese pilots, (2) the pilots' experiences in learning AE, and (3) their experiences of interacting with foreign operators (controllers).

Method

Participants

We conducted an online focus group ($n = 9$), which was recruited through an online invitation sent to a group of experts in aviation communication inside the Air League, a non-profit organization of aviation professionals in China. As such, the participants in the focus group were already certified professionals in the field by their membership in the Air League. Although the original focus group included a variety of aviation professionals (i.e., airline pilots, AE instructors), only participants who were commercial pilots contributed substantially, reducing the effective group size to six.

After the focus group discussions, three of the focus group participants with commercial flight experience were contacted to conduct in-depth, semi-structured interviews to complement the discussions. The three pilots varied in age, seniority, English background, and international flight experience, thus providing a snapshot of how these variables could affect AE communication among Chinese pilots.

All participants provided informed consent before participating in the focus group and interviews. This study was approved by the Institutional Review Board of Cedarville University. The three interviewees were male, with age ranged from 39-53 ($M = 44$). Their commercial flight experience ranged from 2 to 24 years ($M = 13$). One pilot was a captain, and the other two had been first officers. All pilots had some interantional flying experience, although two mainly operated domestic routes. All pilots had passed PEPEC with the certification of ICAO Level 4 AE proficiency (ICAO, 2010).

Data Collection and Analysis

The focus group and semi-structured interviews took place from January to March 2022. The focus group was conducted over a WeChat group chat to enable easy access for the Chinese participants, and the interviews were conducted online using video calls. Both procedures were in Chinese. The researcher was unable to collect demographic information beyond the occupation and experience of the focus group members due to confidentiality concerns in the online format.

In the focus group, questions were posted in the group chat, and the participants were asked to respond individually to the questions while commenting on others' answers. Due to the lack of response from the focus group, three of the more vocal participants (who were also airline pilots) were invited to the interviews, all of which were around an hour in length. A question pool structured around the three research questions was developed. The interviewees were first asked the three research questions, and based on their responses, they were asked follow-up questions from the pool under each research question. Throughout the interviews, the researcher remained neutral and abstained from providing physical or verbal (dis)approvals to the responses.

The coding of the focus group and interview data followed an inductive method (Maxwell, 2012) such that constructs were the products inferred through careful, iterative identification of common codes. The interviews were transcribed in Chinese, but the codes were written in English.

Findings

Aviation English Training in China

The strength of Chinese AE training is in its comprehensiveness. The participants consistently agreed on the overall sufficiency of the AE training they received and indicated that the training covered most of the routine and unusual communication scenarios. Still, they noted that a strong English background could often supply the need to communicate effectively in highly unusual cases.

"The 900 sentences [that we had to memorize for PEPEC] include the vast majority of use cases, such as bird strikes, lightning strikes" (Pilot B).

A drawback of Chinese AE training is rote learning. The participants critiqued the extent of rote learning demanded by the nature of the test. For example, memorizing the "fearful PEPEC 900" was a huge hurdle for test-takers because their score depended on how accurately they repeated the sentences.

"The [PEPEC 900] tests more for memory than for comprehension of the sentences" (Pilot C).
Yet, rote learning may be necessary in the China because not all pilots possess high basic English proficiency. PEPEC needs to render testing compatible with those with lower proficiency (Pilot C).

English proficiency among Chinese pilots has grown. Younger pilots might demonstrate greater versatility and adaptability toward AE communication. Also, airlines have started to require higher English proficiency standards for promotions.

"[My generation]...had little interaction with foreigners, so our English background was quite weak...Kids with a college education nowadays can speak English fluently" (Pilot A).

Aviation English in Use

Interacting with foreign operators requires familiarity with other AE styles. The pilots identified southeast Asian controllers as somewhat challenging to communicate with as their English contained strong accents and suggested that Chinese crews might benefit from additional training to improve the recognition of certain speech styles. For Pilot A, American personnel not only "had an accent" but also "often added stuff to standard AE," and his airline required him to practice with specialized speech tapes to enhance his recognition and adaptation. Likewise, Pilot C, who used to fly international routes in Asia, became acquainted with how Korean controllers substituted the [f] sound with [p] and how the Thai often stretched vowel sounds—not through AE training but exposure.

Issues with local phraseology. While AE instruction addressed most use cases, the pilots reported that the standard phraseology they learned to speak and comprehend must expand to include international variations if they often flew internationally. For example, Taipei Taoyuan and Singapore airports named their taxiways "WP" as "West Path" instead of using the standard radiotelephony alphabets "Whiskey Papa," and "WC" as "West Cross" instead of "Whiskey Charlie" (Pilot C).

Notoriety of Chinese AE. The pilots were aware that Chinese pilots' English skills were "among the worst."

When foreign controllers encounter a Chinese crew, they know that your English is not very good, so they will reduce their rate of speech and articulate more clearly. This is done routinely.... They are quite intentional towards a Chinese crew" (Pilot A).

PEPEC's exacting pronunciation examination seems to reflect "Chinese authorities' lack of confidence in Chinese pilots' English" (Pilot C).

AE is not commonly used domestically. Although PEPEC is required for commercial certification, the main usage of AE in the Chinese airspace involves weather broadcasts (ATIS), flight plans, aeronautical data, and radio communication with other flight crews (Pilot B).

Issues with acronyms. Two pilots cited the challenge of interpreting acronyms in weather reports and flight plans. Memorizing acronyms does not aid understanding.

"A crew from our airline encountered a rare meteorological event for which there was an acronym, but they just couldn't guess its meaning. It was only after calling [the airline] that they identified the acronym's meaning" (Pilot B).

Discussion

The current research provides some of Chinese pilots' perspectives on AE, its training, utility, and their practical experiences therewith. Overall, the qualitative portrayal largely conforms to the extant literature on the test-oriented AE pedagogy in China (Wu et al., 2012), the challenge of communicating in unusual or unscripted circumstances (Estival & Farris, 2016), the deficit in Chinese pilots' oral English versatility (Xia & Huang, 2012), and the need for greater basic English proficiency (Guo, 2018).

However, as the participants revealed, these insufficiencies result partially from the NNES educational system in which most Chinese pilots have participated, such that PEPEC cannot demand AE proficiency levels greatly beyond the average English proficiency of Chinese pilots other than aim for a set of working, or minimum, proficiency requirements. Moreover, reports of the limited utility of AE within the Chinese airspace and of the rarity of scenarios that exist outside of the pilots' AE curriculum might also help explain the lack of emphasis on flexible oral proficiency. Nevertheless, the concern that memorization should not replace proficient comprehension skills underscores likekly the greatest peril (ICAO, 2010, para. 7.4.3) of the current standards, as pilots may not be taught to respond appropriately to situations beyond the memorized schemes.

When AE is used to interact with foreign personnel, Chinese pilots may more frequently encounter NNES, rather than NES, operators. Thus, the pilots emphasized the benefits of receiving additional training to familiarize themselves with the unique linguistic features they might encounter on international flights (i.e., first-language phonological influences, see Kim & Billington, 2018), as well as local variations on standard AE phraseology. Compared to the abundance of emphases on oral proficiency, literature on the Aviation English in its written format (ICAO, 2010, para. 5.2.1.5) is scant; however, in China where AE is reportedly the minority language, pilots' main exposure to AE may come through written means. Specifically, interpreting aviation acronyms/abbreviations and foreign airport regulations can be a challenge to Chinese and NNES pilots as some of these materials may not be comprehensively covered in basic AE training. Although ICAO (2010) instructed that "approved ICAO abbreviations…be converted into unabbreviated words or phrases…except for those [in] common practice" (para. 5.2.1.5.5) before a message is transmitted, it is possible that NNESs may still find common abbreviations more challenging than the full terms.

The sample size of the present study not only limits the breadth and depth of the information collected about Chinese pilots' language experiences but also necessitates further measures to validate the findings obtained. Additionally, while the participants represented a diversity of learning experiences concerning languages and cultures, diversity within a small group also meant that certain constructs, though explained in-depth, were only exposited by one or two individuals. To address these limitations, we used some of the findings to formulate a list of multiple-choice survey, which we plan to administer to a larger sample of Chinese pilots from Air League. Table 1 presents five of the twenty items generated.

Table 1.
Potential Survey Items for Validation of Findings

1. My AE training in China is based on memory-based learning.
2. My AE training in China has not prepared me to communicate well in unusual scenarios.
3. I can comprehend foreign controllers' messages despite their different speech styles.
4. I find it challenging to interpret/use Aviation English acronyms/abbreviations.
5. I can recognize and respond to English phraseologies that are not native to my training. (For example, items such as wind shear (e.g., minus/plus vs. loss/gain), taxiway designations (e.g., West Cross vs. Whiskey Papa) are reported differently in different countries.)

Note. A potential five-point response scale for these items is [Strongly Disagree – Strongly Agree].

These preliminary findings, notwithstanding their limitations, suggest a few necessary adjustments to the current approach to developing AE proficiency. First, the instructional materials should represent the actual work domain with higher fidelity: the language used for instruction should include terms and expressions that may not belong to the standard AE phraseology but may be likely to be encountered by pilots in uncommon contexts of use. To collect these uncommon terms and expressions, we recommend a large, crosscultural task/work analysis study (Jonassen et al., 1999). To be efficient, such a study can target uncommon work situations in which the participants have noticed usage of unusual language. For example, the critical incident technique (Butterfield et al., 2005) could be used to elicit uncommon situations. Second, we recommend an immersive, peer-assisted, and technology-aided training strategy that would take advantage of the inherent knowledge and workload asymmetries in AE speakers. Current instructional technologies afford bringing together learners from various geographical locations and diverse cultural backgrounds or linguistic skills. Specifically, flight simulators and multi-player gaming apps can be employed to develop highly interactive and immersive training environments tailored to the goal of developing AE proficiency.

Acknowledgements

This work was supported by Air League Non-Profit Cultural Organization, Hong Kong. The views of the research reported does not reflect the views of the Air League organization. The research was conducted as part of an undergraduate thesis.

References

Bieswanger, M., Prado, M., & Roberts, J. (2020). Pilot training and English as a lingua franca: Some implications for the design of Aviation English for ab initio flight training courses. *The Especialist*, *41*(4). https://doi.org/10.23925/2318-7115.2020v41i4a7

Butterfield, L. D., Borgen, W. A., Amundson, N. E., & Maglio, A.-S. T. (2005). Fifty years of the critical incident technique: 1954-2004 and beyond. *Qualitative Research, 5*(4), 475–497. https://doi.org/10.1177/1468794105056924

Deng, X., & Xiao, L. (2013). 近二十年国内民航英语研究述评 [Review of aviation English study in the last two decades]. *Journal of Civil Aviation Flight University of China, 24*(2), 9-13. https://doi.org/10.3969/j.issn.1009-4288.2013.02.002

Estival, D., & Farris, C. (2016). Aviation English as a lingua franca. In Estival, D., Farris, C., & Molesworth, B. (Eds.), *Aviation English: A lingua franca for pilots and air traffic controllers* (pp. 1-21). Routledge. https://doi.org/10.4324/9781315661179

Guo, X. (2018). *Aviation English training in China: Current trends, challenges, and future directions* [Presentation]. International Civil Aviation English Association, ICAEA Workshop 2018, Building on the ICAO LPRs – Communication as a Human Factor. https://commons.erau.edu/icaea-workshop/2018/thursday/12/

International Civil Aviation Organization (ICAO). (2010). *Doc. 9835 Manual on the implementation of ICAO language proficiency requirements* (2nd ed.).

Jonassen, D. H., Tessmer, M., & Hannum, W. H. (1999). *Task analysis methods for instructional design.*

Kay, M., Bullock, N., Charaslertrangsi, R., Coconnier, C., Johnson, D., Koyama, S., Pichon, L., Pulido, J., Silva, T., Sugimoto, K., & Tepnadze, I. (2021, July 12). *The effectiveness of ATC-PILOT radio communication around the world: Pilot and ATCO panel discussion* [Webinar]. International Civil Aviation English Association (ICAEA). https://www.icaea.aero/webinars/webinars-2021/

Kim, H., & Billington, R. (2018). Pronunciation and comprehension in English as a lingua franca communication: Effect of L1 influence in international aviation communication. *Applied Linguistics*, *39*(2), 135-158. https://doi.org/10.1093/applin/amv075

Maxwell, J. (2012). *Qualitative research design* (3rd ed.). Sage.

Tosqui-Lucks, P., & Silva, A. L. B. de C. e. (2020). Aeronautical English: Investigating the nature of this specific language in search of new heights. *The ESPecialist*, *41*(3), 1-27. https://doi.org/10.23925/2318-7115.2020v41i3a2

Wang, A. (2007). Teaching aviation English in the Chinese context: Developing ESP theory in a non-English speaking country. *English for Specific Purposes*, *26*(1), 121-128. https://doi.org/10.1016/j.esp.2005.09.003

Wu, T., Xin, X., & Zhang, Y. (2012). PEPEC 对院校飞行英语教学的反拨效应 [Study on washback effect of PEPEC to flight English teaching and learning]. *Journal of Civil Aviation Flight University of China*, *23*(2), 51-55. https://doi.org/10.3969/j.issn.1009-4288.2012.02.015

Xia, L., & Huang, D. (2012). 民航飞行学员 ICAO 英语学习中的需求分析 [Needs analysis for civil aviation students' ICAO English learning]. *Time Education*, (7), 39-40. https://doi.org/10.3969/j.issn.1672-8181.2012.07.026

DEVELOPMENT AND PSYCHOMETRIC EVALUATION OF US AIR FORCE APTITUDE COMPOSITES FOR AIRCREW TRAINING

Montana R. Woolley
Infoscitex, Inc.
Dayton, OH

Thomas R. Carretta
Air Force Research Laboratory
Wright-Patterson AFB, OH

The Air Force Officer Qualifying Test (AFOQT) is used to qualify applicants for officer commissioning and for aircrew training. Although the current aircrew aptitude composites have shown predictive validity against initial aircrew training outcomes for many years, they also have demonstrated moderate to large mean score subgroup differences (SGDs) for females and racial/ethnic minorities. Historically, AFOQT aptitude composites have been computed from a combination of the cognitive subtests. The current study examined the utility of Predictive Success Models (PSMs) which added personality facets from the Self-Description Inventory for Officers to the existing cognitive composites. Three statistical methods were utilized to create new PSMs: Nonlinear Multiple Regression, Corrected Linear Multiple Regression, and Corrected Pareto Optimization. The best performing models created from each method were tested against each other, and against the current cognitive composites. The new models were found to be successful in increasing criterion-related validity and maintaining or decreasing SGDs.

The AFOQT has been an important component of the Air Force Personnel Testing Program since 1953. It is used for officer selection and aircrew training qualification and is widely accepted by personnel specialists as a useful, cost-effective instrument. It has been the primary selection test for the Air Force Reserve Officer Training Corps, Officer Training School, and the Airman Education and Commissioning Program. It is also used in the selection process for Specialized Undergraduate Pilot Training (UPT), Undergraduate Remotely Piloted Aircraft Training (URT), Combat Systems Officer (CSO) training, and Air Battle Manager (ABM) training. Since its inception, the AFOQT has undergone several revisions to improve both its performance prediction and officer classification (see Drasgow et al., 2010).

The AFOQT has 9 cognitive subtests that measure verbal, math, spatial, perceptual speed, and aircrew knowledge. They are combined to form aptitude composites (Kantrowitz et al., in press) that have shown a strong track record of criterion-related validity (Carretta, 2010; Carretta & Ree, 2003).

With the implementation of AFOQT Form S in 2005, an experimental personality test, the Self-Description Inventory + (SDI+), was added. Following a thorough psychometric evaluation of the SDI+ (Manley & Weissmuller, 2017), it was revised and renamed the SDI for Officers (SDI-O) when Form T was implemented in 2015. The SDI-O includes additional facets and has higher reliability than did the SDI+ (Woolley et al., 2022).

The goals of including personality assessment were to (1) broaden the assessment of critical officer and aircrew attributes and (2) examine the utility combining the personality and cognitive scores to improve predictive validity and diversity/inclusion for women and racial/ethnic minorities. This decision was influenced by recent personnel selection literature suggestions that compensatory models

Case number AFRL-2023-1540 was cleared for public release on 3 April 2023.

(e.g., higher personality scores being used to offset low cognitive scores) can help minimize the diversity-validity dilemma (Rupp et al., 2020).

Method

Participants

Archival AFOQT Form T data collected between 2016 and 2020 were used. Most examinees attended manned aircraft pilot training ($N = 1{,}187$), followed by RPA pilots ($N = 719$), CSOs ($N = 658$), and ABMs ($N = 267$). Most of the examinees were male (73%-92%). White (69%-78%), and non-Hispanic (75-81%).

Measures

Predictors. The AFOQT composite scores-of-record for pilots, CSOs, and ABMs were used as a baseline. The personality scores included the SDI-O facet scores from examinees' first testing attempt. The SDI-O has 30 facets but only 26 were used due to an administrative error which affected the scores on four facets. The data ($N = 60{,}066$) were cleaned for missingness and carelessness (see Arias et al., 2020; Bowling et al., 2021; DeSimone et al., 2018; Huang et al., 2012). Carelessness was identified using several post-hoc statistical procedures: (1) long-string analysis (identified 187 cases); (2) intra-individual response variability (identified 61 cases); and (3) odd-even consistency (identified 381 cases).

Training criteria. The main criteria were the Merit Assignment Selection System (MASS) scores. MASS scores are composites that indicate the overall assessment of a trainee's airmanship based on academic grades, check flight scores, daily flight scores, and flight commander ratings. MASS scores range from 0 to 100. The primary criterion for manned aircraft pilots was for SUPT Primary training. The Introductory Flight Training (IFT) and from SUPT Advanced training MASS scores were used to cross-validate the new pilot composite. For CSOs, the primary criterion was the MASS score for Primary training. The IFT and Advanced training MASS scores were used to cross-validate the new CSO composite. For ABMs and RPA pilots, we had access to one set of MASS scores, so no cross-validation was possible for either ABM or RPA training.

Technical Approach

The AFOQT composites were used as a baseline to examine the utility of adding the SDI-O facets. We narrowed down the number of facets for each career field by examining their inter-correlations, theoretical linkages, and inclusion in stepwise regression models. Next, three approaches to create new composites were applied: (1) non-linear multiple regression (NLMR), (2) linear multiple regression (LMR) with range restriction correction, and (3) Pareto optimization (PO) with range restriction correction.

Non-linear multiple regression. NLMR was used due to interest in the potential non-linear relationship between personality and performance (e.g., Benson & Campbell, 2007). These analyses were limited to the facets displaying significant quadratic relationships to the training criteria. First, the facet scores were transformed into *z*-scores (for mean centering). Next, the linear terms were entered into regression Model 1. Then, the quadratic terms were entered into regression Model 2. If Model 2 outperformed Model 1 ($p < .10$), the quadratic term was considered for inclusion in the new composite. If the quadratic term was used in a composite, it was used in conjunction with its linear counterpart. One limitation of NLMR is that the predictors were not corrected for range restriction, due to violations of the linearity assumption underlying range restriction corrections (Lawley, 1943). Although failing to account for range restriction can result in biased validity coefficients and underestimated explained variance, we

Case number AFRL-2023-1540 was cleared for public release on 3 April 2023.

believe that the range restriction corrections will have only small effects on the personality facets because they were affected only by indirect range restriction. Additionally, the correlations between the cognitive composites and the personality facets were weak, meaning that the indirect range restriction should have had minimal impact. This speculation was supported by examining the changes in correlations after correcting for range restriction between predictors and criteria (Woolley et al., 2023).

LMR with range restriction corrections. Lawley's (1943) multivariate correction for range restriction was applied and we ran LMR models to find the optimal composites. All the predictors were simultaneously entered in the model. The statistical significance of each predictor was examined to determine which to retain. A limitation of this method is the inability to assess non-linear relationships.

Corrected PO with range restriction corrections. PO is a statistical technique that can help mitigate the diversity-validity dilemma (De Corte et al., 2007, 2022). The goals were to maintain validity while reducing mean score SGDs. Using PO allows us to generate regression weights that optimize achievement of both objectives. Multivariate correction for range restriction was performed prior to running these analyses. Limitations of PO include the inability to assess non-linear relationships and all predictors need to positively predict the outcome. To circumvent the latter limitation, we reversed the signs of any negative correlations so that they might be used.

Results

NLMR Results

As previously discussed, if quadratic components were statistically significant, both the quadratic and linear components were kept in the model. If the quadratic component was not statistically significant, it was not included, but the linear component was kept in the model. Initial models included all possible SDI-O facets (26), as well as the current aircrew composite for each career field (pilot, RPA, CSO, or ABM). Each successive model dropped any predictors that were not statistically significant (with the exception of linear components when the quadratic term was significant). This process continued until only statistically significant predictors remained. Since no range restriction corrections were performed on these models, the amount of variance explained by the cognitive composite is likely underestimated, and the increase in variance associated with the addition of personality facets is likely overestimated.

For manned aircraft pilots, the cognitive composite explained 2.2% ($p < .001$) of the variance in the SUPT Primary MASS score. The final model with the Pilot composite *and* SDI-O facets explained 5.8% of the variance in the criterion ($p < .001$). No quadratic terms were included in the final model. For CSOs, the cognitive composite explained 4.3% ($p < .001$) of the variance in the Primary MASS score. The final model including the CSO cognitive composite *and* SDI-O facets explained 7.4% of the variance in this criterion ($p < .001$). In the final model, two quadratic components were included. For ABMs, the cognitive composite explained 8.3% ($p < .001$) of the variance in the MASS score. The final model the ABM cognitive composite *and* SDI-O facets explained 13.3% of the variance in this criterion ($p < .001$). In the final ABM model, two quadratic components were included. For RPA pilots, the Pilot composite alone explained 9.67% ($p < .001$) of the variance in the RPA IFT MASS score. When testing for non-linear relationships, we found that the cognitive composite had a positive quadratic relationship with the RPA IFT criterion, such that those with an above average cognitive score performed better, but for those scoring below average, lower cognitive scores did not influence performance. Therefore, we included the quadratic components of the AFOQT Pilot composite in these models. Inclusion of the quadratic component increased explained variance to 10.9% ($p < .001$). The final RPA model which included the pilot cognitive composite *and* personality facets explained 16.4% of the variance ($p < .001$). The SDI-O personality facets had no significant non-linear components.

LMR Results

Initial models included all SDI-O facets and current AFOQT composite. Successive model dropped any non-significant predictors. We used data which had been corrected for Multivariate range restriction (Lawley, 1943). Therefore, variance estimates will be more accurate than in the non-linear NLMR analyses.

For pilots, the Pilot composite explained 5.2% ($p < .001$) of the variance in the SUPT Primary MASS score. The final model with the Pilot composite *and* SDI-O facets explained 8.2% of the variance in this criterion ($p < .001$). For CSOs, the cognitive composite explained 7% ($p < .001$) of the variance in the Primary MASS score. The final model with the CSO composite *and* SDI-O facets explained 8.2% of the variance in this criterion ($p < .001$). For ABMs, the cognitive composite explained 12.6% ($p < .001$) of the variance in the MASS score. The final model explained 12.8% of the variance in the criterion ($p < .001$). Finally, for RPA pilots, the AFOQT cognitive composite explained 19.7% of the variance in RPA IFT MASS ($p < .001$). The final RPA model explained 24.1% of the variance in this criterion ($p < .001$).

PO Results

Using PO, we attempted to maximize validity while minimizing mean score SGDs. We calculated separate PO models to examined gender mean score SGDs and racial majority v/minority mean score SGDs. No gender model was run for ABMs because the applicant gender ratio was lower than the ABM gender ratio. We examined all potential models and selected the ones that did not reduce validity provided by the cognitive composite alone and provided the highest adverse impact ratio (i.e., smallest mean score SGDs).

Model Testing

Next, we compared the highest-performing models identified by each of the three methods against one another. Models were tested several ways. First, validity coefficients were produced. These coefficients were produced for the main criterion for each sample, as well as for any alternative criteria previously described. Next, the effect sizes for the mean score SGDs were computed. The effect sizes, expressed as Cohen's *d*, were produced for both gender and race. Further, these SGDs were calculated in both the incumbent samples and in the applicant sample across all models.

The best models were not always clear for the career fields. For manned aircraft pilots, the LMR model was the best performer. This is because of the interest in predicting the SUPT Primary MASS criterion (as opposed to the alternative pilot training criteria), as well as mostly lower SGDs.

For CSOs and ABMs, the best performing models were clearly the NLMR models. These had the strongest validity coefficients and lower or lowest SGDs in comparison to all other models. For RPA pilots, the NLMR model showed the highest validity coefficient and lowest SGDs for gender. However, the NLMR model had slightly higher racial mean score SGDs ($d = .56$) compared to the cognitive composite alone ($d = .53$) and PO ($d = .53$) models. Notwithstanding, we observed improved SGDs across the new models and higher validity coefficients compared to the existing cognitive composites. See Table 1 for a summary.

For the new pilot Predictive Success Model (PSM) vs. the AFOQT composite, there was a 6.49% *increase* in criterion-related validity for the SUPT Primary MASS score, a 0.76% *decrease* for the SUPT Advanced MASS score, and an 8.78% *decrease* for IFT the MASS score. When examining the qualification rates (QR) in the applicant sample ($N = 46{,}440$; using the current minimum qualifying score

(25th percentile), there were no changes in QRs for any subgroups. The new Pilot PSM provided improvements only for predicting the SUPT Primary MASS score.

Table 1.
Comparisons between the Current Composites and Best Predictive Success Model

Sample	Differences in Validity for Main Criterion	Differences in SGD Effects Size (*d*)	
	ΔR^2	Racial/Ethnic Minorities	Gender Minority
Pilots	+6.49%	+.01	+.01
CSOs	+3.76%	-.02	+.16
ABMs	+6.99%	+.22	+.33
RPA Pilots	+7.53%	-.03	+.20

Note. Pilot N = 1,187, CSO N = 658, ABM N = 267, RPA Pilot N = 719, Applicant N = 46,440. CSO's and RPA Pilots had trivial increases in SGDs for racial/ethnic minorities.

For the new CSO PSM vs. the current AFOQT composite, there was a 3.76% *increase* in criterion-related validity for the Primary MASS score, a 1.31% *increase* for the Advanced MASS score, and a 3.62% *increase* for the IFT MASS score. When examining the QRs in the applicant sample (N = 46,440; using the current minimum qualifying score (25th percentile), 4% more women qualified for CSO training, and 5% more women of a racial/ethnic minority obtained passing scores.

For the new RPA pilot PSM vs, the AFOQT Pilot composite, there was a 7.53% *increase* in criterion-related validity for the RPA MASS score. When examining the training QRs in the applicant sample (N = 46,440; using the current minimum qualifying score (25th percentile), 5% more racial/ethnic minorities, 8% more women, and 14% more women of a racial/ethnic minority obtained passing scores.

For the new ABM PSM vs. the AFOQT composite, there was a 7.0% *increase* in criterion-related validity for the ABM MASS score. When examining the training QRs in the applicant sample (N = 46,440; using the current minimum qualifying score (25th percentile), 6.7% more women overall and 5.7% more women of a racial/ethnic minority obtained passing scores.

Discussion

The purpose of this study was to examine new PSMs for rated career fields with the dual objectives of maintaining/improving predictive validity and improving QRs for women and minorities. The new PSMs reflect a compensatory approach to selection and classification. Some individuals may be a good personality fit for a specific career field, but their cognitive scores are not high enough to qualify. The new PSMs provide an increase in criterion-related validity when compared to the cognitive-only counterparts, and either maintain or decrease the SGDs for women and racial/ethnic minorities. The Air Force plans to implement the PSMs as an alternate means to qualify for these career fields. These new PSMs will be revalidated as additional criterion data become available.

References

Arias, V. B., Garrido, L. E., Jenaro, C., Martínez-Molina, A., & Arias, B. (2020). A little garbage in, lots of garbage out: Assessing the impact of careless responding in personality survey data. *Behavior Research Methods, 52*(6), 2489-2505.

Case number AFRL-2023-1540 was cleared for public release on 3 April 2023.

Bowling, N. A., Gibson, A. M., Houpt, J. W., & Brower, C. K. (2021). Will the questions ever end? Person-level increases in careless responding during questionnaire completion. *Organizational Research Methods, 24*(4), 718-738.

Benson, M. J., & Campbell, J. P. (2007). To be, or not to be, linear: An expanded representation of personality and its relationship to leadership performance. *International Journal of Selection and Assessment, 15*(2), 232-249.

Carretta, T. R. (2010). Air Force Officer Qualifying Test validity for non-rated officer Specialties, *Military Psychology, 22,* 450-464.

Carretta, T. R., & Ree, M. J. (2003). Pilot selection methods. In B. H. Kantowitz (Series Ed.) & P. S. Tsang & M. A. Vidulich (Vol. Eds.). *Human factors in transportation: Principles and practices of aviation psychology* (pp. 357-396). Mahwah, NJ: Erlbaum.

De Corte, W., Lievens, F., & Sackett, P. R. (2007). Combining predictors to achieve optimal trade-offs between selection quality and adverse impact. *Journal of Applied Psychology, 92*(5), 1380.

De Corte, W., Lievens, F., & Sackett, P. R. (2022). A comprehensive examination of the cross-validity of pareto-optimal versus fixed-weight selection systems in the bi-objective selection context. *Journal of Applied Psychology, 107*(8), 1243.

DeSimone, J. A., DeSimone, A. J., Harms, P. D., & Wood, D. (2018). The differential impacts of two forms of insufficient effort responding. *Applied Psychology, 67*(2), 309-338.

Drasgow, F., Nye, C. D., Carretta, T. R., & Ree, M. J. (2010). Factor structure of the Air Force Officer Qualifying Test Form S: Analysis and comparison with previous forms. *Military Psychology, 22*(1), 68-85.

Huang, J. L., Curran, P. G., Keeney, J., Poposki, E. M., & DeShon, R. P. (2012). Detecting and deterring insufficient effort responding to surveys. *Journal of Business and Psychology*, 27, 99–114.

Lawley, D. N. (1943). A note on Karl Pearson's selection formulae. *Proceedings of the Royal Society of Edinburgh, Section A, 62(*1), 28–30.

Manley, G. G., & Weissmuller, J. (2017, April). *Development of the United States Air Force's Self-Description Inventory* [Poster]. Society for Industrial Organizational Psychology Annual Conference, Orlando, FL.

Rupp, D. E., Song, C., & Strah, N. (2020). Addressing the so-called validity-diversity trade-off: Exploring the practicalities and legal defensibility of Pareto-optimization for reducing adverse impact within personnel selection. *Industrial and Organizational Psychology: Perspectives on Science and Practice, 13,* 246-271.

Woolley, M. R., Walsh, J. L., Mann, K. J., Wilson, R. T., Carretta, T. R., Mouton, A. N., & Deregla, A. R. (2022). *Self-Description Inventory – Officer (SDI-O): Item-, facet-, and domain-level analyses,* AFRL-RH-WP-TR-2022-0091. Wright-Patterson AFB, OH: 711 Human Performance Wing, Airman Biosciences Division, Performance Optimization Branch.

Case number AFRL-2023-1540 was cleared for public release on 3 April 2023.

AUTHOR INDEX

Wright State University – Conferences & Events
3640 Colonel Glenn Highway
Dayton, Ohio 45435
USA

ISBN 979-8-3313-0346-4

CURRAN ASSOCIATES INC.
proceedings.com